AF773214

Springer Proceedings in Mathematics & Statistics

Volume 162

Springer Proceedings in Mathematics & Statistics

This book series features volumes composed of selected contributions from workshops and conferences in all areas of current research in mathematics and statistics, including operation research and optimization. In addition to an overall evaluation of the interest, scientific quality, and timeliness of each proposal at the hands of the publisher, individual contributions are all refereed to the high quality standards of leading journals in the field. Thus, this series provides the research community with well-edited, authoritative reports on developments in the most exciting areas of mathematical and statistical research today.

More information about this series at http://www.springer.com/series/10533

Patrícia Gonçalves · Ana Jacinta Soares
Editors

From Particle Systems to Partial Differential Equations III

Particle Systems and PDEs III, Braga, Portugal, December 2014

Springer

Editors
Patrícia Gonçalves
Centre of Mathematics
University of Minho
Braga
Portugal

Ana Jacinta Soares
Centre of Mathematics
University of Minho
Braga
Portugal

ISSN 2194-1009 ISSN 2194-1017 (electronic)
Springer Proceedings in Mathematics & Statistics
ISBN 978-3-319-32142-4 ISBN 978-3-319-32144-8 (eBook)
DOI 10.1007/978-3-319-32144-8

Library of Congress Control Number: 2014938099

Mathematics Subject Classification (2010): 35L67, 35R60, 35Q20, 35Q30, 35Q35, 60K35, 60G60, 60F17, 82C40

© Springer International Publishing Switzerland 2016
This work is subject to copyright. All rights are reserved by the Publisher, whether the whole or part of the material is concerned, specifically the rights of translation, reprinting, reuse of illustrations, recitation, broadcasting, reproduction on microfilms or in any other physical way, and transmission or information storage and retrieval, electronic adaptation, computer software, or by similar or dissimilar methodology now known or hereafter developed.
The use of general descriptive names, registered names, trademarks, service marks, etc. in this publication does not imply, even in the absence of a specific statement, that such names are exempt from the relevant protective laws and regulations and therefore free for general use.
The publisher, the authors and the editors are safe to assume that the advice and information in this book are believed to be true and accurate at the date of publication. Neither the publisher nor the authors or the editors give a warranty, express or implied, with respect to the material contained herein or for any errors or omissions that may have been made.

Printed on acid-free paper

This Springer imprint is published by Springer Nature
The registered company is Springer International Publishing AG Switzerland

Preface

This volume presents the proceedings of the third international conference on particle systems and partial differential equations, "PS-PDEs III", which was held at the Centre of Mathematics of the University of Minho in Braga, Portugal, during 17–19 December 2014.

The meeting was intended to bring together prominent active researchers working in the fields of probability and partial differential equations, so that they could present their latest scientific findings in both areas, and to promote discussion on some of their areas of expertise. Further, it was intended to introduce a vast and varied public, including young researchers, to the subject of interacting particle systems, its underlying motivation and its relation to partial differential equations.

This volume includes 16 contributed papers written by conference participants on essential and intriguing topics in the fields of probability theory, partial differential equations and kinetic theory.

We believe that this volume will be of great interest to probabilists, analysts and also to those mathematicians with a general interest in mathematical physics, stochastic processes and differential equations, as well as those physicists whose work intersects with statistical mechanics, statistical physics and kinetic theory.

We would like to take this opportunity to extend our thanks to all the speakers, and to the participants, for contributing to the success of this meeting.

Lastly, we wish to gratefully acknowledge the financial support provided by Fundação para a Ciência e a Tecnologia through the FCT-FACC funds, to the Centre of Mathematics of the University of Minho, to the Centre of Mathematics, Fundamental Applications and Operations Research of the University of Lisbon and to the Co-Lab initiative UT Austin-Portugal.

We really hope that you enjoy reading this book!

Braga, Portugal
February 2016

Patrícia Gonçalves
Ana Jacinta Soares

Contents

On Linear Hypocoercive BGK Models

Franz Achleitner, Anton Arnold and Eric A. Carlen

Abstract We study hypocoercivity for a class of linear and linearized BGK models for discrete and continuous phase spaces. We develop methods for constructing entropy functionals that prove exponential rates of relaxation to equilibrium. Our strategies are based on the entropy and spectral methods, adapting Lyapunov's direct method (even for "infinite matrices" appearing for continuous phase spaces) to construct appropriate entropy functionals. Finally, we also prove local asymptotic stability of a nonlinear BGK model.

Keywords Kinetic equations · BGK models · Hypocoercivity · Entropy method

1 Introduction

This paper is concerned with the large time behavior of linear BGK models (named after the physicists Bhatnagar-Gross-Krook [5]) for a phase space density $f(x, v, t)$; $x,\ v \in \mathbb{R}^d$, satisfying the kinetic evolution equation

$$f_t + v \cdot \nabla_x f - \nabla_x V \cdot \nabla_v f = \mathbf{Q}f := M_{T(t)}(v) \int_{\mathbb{R}^d} f(x, v, t)\ \mathrm{d}v - f(x, v, t)\,, \quad t \geq 0, \tag{1.1}$$

with some given confinement potential $V(x)$ and where M_T denotes the normalized Maxwellian at some temperature T:

F. Achleitner · A. Arnold (✉)
Vienna University of Technology, Institute of Analysis and Scientific Computing, Wiedner Hauptstr. 8-10, 1040 Wien, Austria
e-mail: anton.arnold@tuwien.ac.at

F. Achleitner
e-mail: franz.achleitner@tuwien.ac.at

E.A. Carlen
Department of Mathematics, Rutgers University, 110 Frelinghuysen Rd., Piscataway, NJ 08854, USA
e-mail: carlen@math.rutgers.edu

© Springer International Publishing Switzerland 2016

P. Gonçalves and A.J. Soares (eds.), *From Particle Systems to Partial Differential Equations III*, Springer Proceedings in Mathematics & Statistics 162, DOI 10.1007/978-3-319-32144-8_1

$$M_T(v) = (2\pi T)^{-d/2} e^{-|v|^2/2T}.$$

We assume that the initial condition is normalized as

$$\int_{\mathbb{R}^d \times \mathbb{R}^d} f(x, v, 0)\, \mathrm{d}x\mathrm{d}v = 1,$$

and this normalization persists under the flow of (1.1). The function $T(t)$ is defined so that the energy is conserved:

$$\int_{\mathbb{R}^d \times \mathbb{R}^d} \left[\frac{|v|^2}{2} + V(x)\right] f(x, v, t)\, \mathrm{d}x\mathrm{d}v = \int_{\mathbb{R}^d \times \mathbb{R}^d} \left[\frac{|v|^2}{2} + V(x)\right] f(x, v, 0)\, \mathrm{d}x\mathrm{d}v =: E_0.$$

This is achieved in case

$$T(t) := \frac{2}{d}\left[E_0 - \int_{\mathbb{R}^d} V(x)\rho(x, t)\, \mathrm{d}x\right], \tag{1.2}$$

where $\rho(x, t) := \int_{\mathbb{R}^d} f(x, v, t)\, \mathrm{d}v$, which completes the specification of the equation.

This model differs form the usual BGK model in that the Maxwellian M_T has a spatially constant temperature and zero momentum. This is already a simplification of the standard BGK model in which M_T would be replaced by the local Maxwellian corresponding to f; i.e., the local Maxwellian with the same hydrodynamic moments as f. However, (1.1)–(1.2) is still non-linear since $T(t)$ depends linearly on f, but then M_T depends nonlinearly on T. This simplified equation arises in certain models of thermostated systems [4]. Under sufficient growth assumptions on V as $|x| \to \infty$, the unique normalized steady state of (1.1) is

$$f^{\infty}(x, v) = \exp\left(-\frac{1}{T_\infty}\left[V(x) + \frac{|v|^2}{2}\right]\right),$$

where the normalization constant shall be included in V and T_∞ such that the energy associated to f^∞ is E_0.

In fact, we simplify the model further: We take $d = 1$, replace the spatial domain $\mathbb{R}^d$ by the unit circle $\mathbf{T}^1$, and then dispense with the confining potential. Thus we shall first investigate the linear BGK model

$$f_t + v\, f_x = \mathbf{Q}f := M_T(v) \int_{\mathbb{R}} f(x, v, t)\, \mathrm{d}v - f(x, v, t), \quad t \geq 0. \tag{1.3}$$

Let $\mathrm{d}\tilde{x}$ denote the normalized Lebesgue measure on $\mathbf{T}^1$, and consider normalized initial data $f(x, v, 0)$ such that $\int_{\mathbf{T}^1 \times \mathbb{R}} f(x, v, 0)\, \mathrm{d}\tilde{x}\mathrm{d}v = 1$ (a normalization which is conserved under the flow). In this case, Eq. (1.2) for the temperature reduces to $T(t) = 2E_0$, independent of t, with

$$E_0 := \int_{\mathbf{T}^1 \times \mathbb{R}} \frac{v^2}{2} f(x, v, 0)\, \mathrm{d}\tilde{x}\mathrm{d}v.$$

For the simplified linear equation (1.3), the unique steady state is $f^\infty = M_T$, uniform on the circle. We shall study the rate at which normalized solutions of (1.3) approach the steady state $f^\infty = M_T$ as $t \to \infty$. This problem is interesting since the collision mechanism drives the local velocity distribution towards M_T, but a more complicated mechanism involving the interaction of the streaming term $v\partial_x$ and the collision operator $\mathbf{Q}$ is responsible for the emergence of spatial uniformity.

To elucidate this key point, let us define the operator $\mathbf{L}$ by

$$\mathbf{L}f(x, v) := -v\, \partial_x f(x, v) + \mathbf{Q}f(x, v).$$

Then the evolution equation (1.3) can be written $f_t = \mathbf{L}f$. Let $\mathcal{H}$ denote the weighted space $L^2(\mathbf{T}^1 \times \mathbb{R}; M_T^{-1}(v)\, \mathrm{d}v)$. Then $\mathbf{Q}$ is self-adjoint on $\mathcal{H}$, $\mathbf{L}f^\infty = 0$, and a simple computation shows that if $f(t)$ is a solution of (1.3),

$$\frac{\mathrm{d}}{\mathrm{d}t}\|f(t) - f^\infty\|_{\mathcal{H}}^2 = 2\langle f(t), \mathbf{L}f(t)\rangle_{\mathcal{H}} = 2\langle f(t), \mathbf{Q}f(t)\rangle_{\mathcal{H}} = -2\|f - M_T\rho\|_{\mathcal{H}}^2,$$

where, as before, $\rho(x, t) := \int_{\mathbb{R}} f(x, v, t)\, \mathrm{d}v$. Thus, while the norm $\|f(t) - f^\infty\|_{\mathcal{H}}$ is monotone decreasing, the derivative is zero whenever $f(t)$ has the form $f(t) = M_T\rho$ for *any* smooth density ρ. In particular, the inequality

$$\langle f - f^\infty, \mathbf{L}(f - f^\infty)\rangle_{\mathcal{H}} \le -\lambda\|f - f^\infty\|_{\mathcal{H}}^2 \tag{1.4}$$

is valid in general for $\lambda = 0$, but for no positive value of λ. If (1.4) were valid for some $\lambda > 0$, we would have had $\|f(t) - f^\infty\|_{\mathcal{H}}^2 \le e^{-t\lambda}\|f(0) - f^\infty\|_{\mathcal{H}}^2$ for all solutions of our equation, and we would say that the evolution equation is *coercive*. However, while this is not the case, it does turn out that one still has constants $1 < C < \infty$ and $\lambda > 0$ such that

$$\|f(t) - f^\infty\|_{\mathcal{H}}^2 \le Ce^{-t\lambda}\|f(0) - f^\infty\|_{\mathcal{H}}^2\,. \tag{1.5}$$

(The fact that there exist initial data $f(0) \ne f^\infty$ for which the derivative of the norm is zero shows that necessarily $C > 1$.) In Villani's terminology (see Sect. 3.2 of [21]), this means that our evolution equation is *hypocoercive*.

Many hypocoercive equations have been studied in recent years [2, 10–12, 21], including BGK models in Sects. 1.4 and 3.1 of [10] (see also Sect. 4.1 below), but sharp decay rates were rarely an issue there. The fact that normalized solutions of (1.3) converge exponentially fast at *some* rate to f^∞ is a consequence of a probabilistic analysis of such equations in [4]: In fact, Eq. (1.3) is the Kolmogorov forward equation for a certain Markov process, and as shown in [4] an argument based on a Doeblin condition yields exponential convergence. However, this approach relies on compactness arguments and does not yield explicit values for C or λ. We shall discuss another approach to the problem of establishing hypocoercivity for such models that does yield explicit—and quite reasonable—values for C and λ. To this end, our main tool will be variants of the *entropy–entropy production method.* Our first main result will be a decay estimate for (1.3):

Theorem 1 (decay estimate for (1.3)) *Fix unit temperature $T = 1$. There exists an entropy functional $e(f)$ satisfying*

$$\frac{1}{2}e(f) \le \|f - M_1\|^2_{\mathcal{H}} \le 4e(f)$$

such that for all (normalized) solutions $f(t)$ of (1.3) with $e(f^I) < \infty$,

$$e(f(t)) \le e^{-t \cdot 0.547592\ldots} e(f^I), \qquad t \ge 0.$$

Finally, we shall study the linearization of a one dimensional BGK equation around a Maxwellian with some constant-in-x temperature. In one dimension, if collisions conserve both energy and momentum, they are trivial: The only kinematic possibilities are an exchange of velocities which has no effect at all at the kinetic level. Therefore, in one dimension the natural BGK equation, which would correspond for example to the Kac equation [13], uses Maxwellians determined by the density and temperature alone. The method will be applied to the three dimensional equation in a follow-up paper.

For a probability density $f(x, v)$ on $\mathbf{T}^1 \times \mathbb{R}$ we thus consider the nonlinear BGK equation

$$f_t(x, v, t) + v\, f_x(x, v, t) = M_f(x, v, t) - f(x, v, t), \qquad t \ge 0, \tag{1.6}$$

where M_f is the local Maxwellian having the same local density and "temperature" as f: The density is defined as $\rho(x, t) := \int_{\mathbb{R}} f(x, v, t)\,\mathrm{d}v$ and the pressure as $P(x, t) := \int_{\mathbb{R}} v^2 f(x, v, t)\,\mathrm{d}v$. In analogy to the situation with zero velocity we shall refer to the conditional second moment, $\tilde{T}(x, t) := P(x, t)/\rho(x, t)$ as temperature (with the gas constant scaled as $R = 1$). Then, for fixed t, the local Maxwellian M_f is defined as

$$M_f(x, v) = \frac{\rho(x)}{\sqrt{2\pi \tilde{T}(x)}} e^{-v^2/2\tilde{T}(x)} = \frac{\rho^{3/2}(x)}{\sqrt{2\pi P(x)}} e^{-v^2 \rho(x)/2P(x)}, \tag{1.7}$$

and we shall mostly use the second version of it in the sequel. The existence of global solutions for the Cauchy problem of similar nonlinear BGK models has been proven in [7, 16, 18].

We assume $\int_{\mathbf{T}^1} \rho(x)\,\mathrm{d}\tilde{x} = 1$ and define $T := \int_{\mathbf{T}^1} P(x)\,\mathrm{d}\tilde{x}$, which are both conserved by the flow of (1.6). Now we consider f close to the global equilibrium $M_T(v)$, with h defined by $f = M_T + h$. Then

$$\begin{aligned} \rho(x,t) &= 1+\sigma(x,t) \quad \text{with} \quad \sigma(x,t) := \int_{\mathbb{R}} h(x,v,t)\,\mathrm{d}v, \\ P(x,t) &= T+\tau(x,t) \quad \text{with} \quad \tau(x,t) := \int_{\mathbb{R}} v^2 h(x,v,t)\,\mathrm{d}v, \end{aligned} \tag{1.8}$$

which implies

$$\int_{\mathbf{T}^1} \sigma(x,t)\,\mathrm{d}\tilde{x} = 0 \quad \text{and} \quad \int_{\mathbf{T}^1} \tau(x,t)\,\mathrm{d}\tilde{x} = 0. \tag{1.9}$$

The perturbation h then satisfies

$$h_t(x,v,t) + v\,h_x(x,v,t) = [M_f(x,v,t) - M_T(v)] - h(x,v,t), \quad t \geq 0.$$

For σ and τ small we have

$$M_f(x,v) - M_T(v) = \frac{(1+\sigma)^{3/2}(x)}{\sqrt{2\pi(T+\tau(x))}} e^{-v^2(1+\sigma(x))/2(T+\tau(x))} - \frac{1}{\sqrt{2\pi T}} e^{-v^2/2T} \tag{1.10}$$

$$\approx \left(\frac{3}{2} - \frac{v^2}{2T}\right) M_T(v)\sigma(x) + \left(-\frac{1}{2T} + \frac{v^2}{2T^2}\right) M_T(v)\tau(x), \tag{1.11}$$

which yields the linearized BGK model that we shall analyze in this paper:

$$\begin{aligned} &h_t(x,v,t) + v\,h_x(x,v,t) \\ &= M_T(v)\left[\left(\frac{3}{2} - \frac{v^2}{2T}\right)\sigma(x,t) + \left(-\frac{1}{2T} + \frac{v^2}{2T^2}\right)\tau(x,t)\right] - h(x,v,t), \quad t \geq 0. \end{aligned} \tag{1.12}$$

Following the same approach as for Theorem 1 we shall obtain a decay estimate for (1.12), and then local asymptotic stability for the nonlinear BGK equation (1.6). For the latter purpose, we need to introduce another set of norms.

For $\gamma \geq 0$, let $H^\gamma(\mathbf{T}^1)$ be the Sobolev space consisting of the completion of smooth functions φ on $\mathbf{T}^1$ in the Hilbertian norm

$$\|\varphi\|_{H^\gamma}^2 := \sum_{k\in\mathbb{Z}} (1+k^2)^\gamma |\varphi_k|^2,$$

where φ_k is the kth Fourier coefficient of φ. Let $\mathscr{H}_\gamma$ denote the Hilbert space $H^\gamma(\mathbf{T}^1) \otimes L^2(\mathbb{R}; M_T^{-1})$. Then the inner product in $\mathscr{H}_\gamma$ is given by

$$\langle f, g\rangle_{\mathscr{H}_\gamma} = \int_{\mathbf{T}^1}\int_{\mathbb{R}} \overline{f}(x,v)\left[\left(1-\partial_x^2\right)^\gamma g(x,v)\right] M_T^{-1}(v)\mathrm{d}v\mathrm{d}\tilde{x}.$$

Theorem 2 (decay estimates for (1.12), (1.6)) *Fix unit temperature* $T = 1$.

(a) For all $\gamma \geq 0$ *there is an entropy functional* $e_\gamma(f)$ *satisfying*

$$\frac{2}{3} e_\gamma(f) \leq \|f - M_1\|^2_{\mathcal{H}_\gamma} \leq \frac{4}{3} e_\gamma(f) \tag{1.13}$$

such that if $h = f - M_1$ *is a solution of the linearized BGK equation* (1.12) *with initial data* $h^I = f^I - M_1$ *such that* $\int_{\mathbb{T}^1} \int_{\mathbb{R}} (1, v^2) f^I \, dv \, d\tilde{x} = (1, 1)$, *and* $e_\gamma(f^I) < \infty$, *then*

$$e_\gamma(f(t)) \leq e^{-t/25} e_\gamma(f^I), \qquad t \geq 0. \tag{1.14}$$

(b) Moreover, for all $\gamma > 1/2$, *there is an explicitly computable* $\delta_\gamma > 0$ *such that if* f *is a solution of the nonlinear BGK equation* (1.6) *with initial data* f^I *such that* $\int_{\mathbb{T}^1} \int_{\mathbb{R}} (1, v^2) f^I \, dv \, d\tilde{x} = (1, 1)$, *and* $\|f^I - M_1\|_{\mathcal{H}_\gamma} < \delta_\gamma$, *then for the same entropy function* e_γ, (1.14) *is again valid.*

Before turning to our main investigation, i.e. exponential decay in the BGK equations (1.3), (1.12), (1.6), we shall study some still simpler models with a finite number of positions and velocities: In Sect. 2 we analyze coercive BGK models with first two and then finitely many velocities using relative entropies. Since this approach fails for discrete hypocoercive BGK models (considered in Sect. 3), their analysis will be based on spectral methods and Lyapunov's direct method. Section 4 is concerned with space-inhomogeneous BGK models. We shall start with its discrete velocity analogs in Sects. 4.1–4.2, where the velocity modes will be expanded in Krawtchouk polynomials – a discrete analog of the Hermite polynomials. In Sect. 4.3 we shall finally analyze the exponential convergence of the linear BGK equation (1.3), using a Hermite expansion of the velocity modes and an adaption of Lyapunov's direct method, used here for "infinite matrices". This will yield the proof of Theorem 1. This strategy is modified in Sect. 4.4 for the linearized BGK equation (1.12), proving Theorem 2(a). Finally, in Sect. 4.5 we analyze the local asymptotic stability of the nonlinear BGK equation (1.6), as stated in Theorem 2(b).

2 Discrete Coercive BGK Models

In this section we consider space-homogeneous BGK models with a finite number of velocities. Our main tool in the investigation is the relative entropy, which is defined as follows (see Sect. 2.2 of [3] for more details):

Definition 1 (a) Let J be either $\mathbb{R}^+$ or $\mathbb{R}$. A scalar function $\psi \in C(\bar{J}) \cap C^2(J)$ satisfying the conditions

$$\psi(1) = 0, \quad \psi \geq 0, \quad \psi'' \geq 0, \text{ on } J \tag{2.1}$$

(and hence also $\psi'(1) = 0$) is called *entropy generator*.

(b) Let $f_1 \in L^1(\mathbb{R}^{2d})$, $f_2 \in L^1_+(\mathbb{R}^{2d})$ with $\iint f_1 \,\mathrm{d}x\,\mathrm{d}v = \iint f_2 \,\mathrm{d}x\,\mathrm{d}v = 1$ and $\frac{f_1}{f_2}(x,v) \in \bar{J}$ a.e. (w.r.t. the measure $f_2(\mathrm{d}x\,\mathrm{d}v)$). Then

$$e_\psi(f_1|f_2) := \iint_{\mathbb{R}^{2d}} \psi\left(\frac{f_1}{f_2}\right) f_2 \,\mathrm{d}x\,\mathrm{d}v \geq 0 \tag{2.2}$$

is called a *relative entropy* of f_1 with respect to f_2 with generating function ψ.

In applications, the most important examples are the logarithmic entropy $e_1(f_1|f_2)$, generated by

$$\psi_1(\sigma) := \sigma \ln \sigma - \sigma + 1,$$

and the power law entropies $e_p(f_1|f_2)$, generated by

$$\psi_p(\sigma) := \sigma^p - 1 - p(\sigma - 1), \quad p > 1. \tag{2.3}$$

Except for the quadratic entropy e_2 we shall always use $J = \mathbb{R}^+$. Below we shall use also a second family of power law entropies $\hat{e}_p(f_1|f_2)$ generated by

$$\hat{\psi}_p(\sigma) := |\sigma - 1|^p, \quad p > 1. \tag{2.4}$$

The above definition clearly shows that $e_\psi(f_1|f_2) = 0$ if and only if $f_1 = f_2$. In the next section we shall hence try to prove that solutions $f(t)$ to BGK models satisfy $e_\psi(f(t)|f^\infty) \to 0$ as $t \to \infty$. For the entropies e_p, $p \geq 1$ such a convergence in relative entropy then also implies L^1–convergence, due to the *Csiszár-Kullback inequality*:

$$\|f_1 - f_2\|^2_{L^1(\mathbb{R}^{2d})} \leq 2\, e_1(f_1|f_2) \leq \frac{2}{p(p-1)}\, e_p(f_1|f_2),$$

where we used $\psi_1(\sigma) \leq \psi_p(\sigma)/\psi_p''(1)$, $\sigma \geq 0$ in the second inequality. For the entropies defined in (2.4) one has a substitute for the Csiszár-Kullback inequality, namely the identity

$$\hat{e}_p(f_1|f_2) = \|f_1 - f_2\|^p_{L^p(f_2^{1-p})}.$$

To illustrate the standard entropy method on a very simple example, we first revisit the ODE (1.10) from [3] for the vector $f(t) = (f_1(t),\ f_2(t))^\top \in \mathbb{R}^2$:

$$\frac{\mathrm{d}}{\mathrm{d}t} f = \lambda \mathbf{A} f, \quad t \geq 0, \qquad f(0) = f^I \in \mathbb{R}^2, \tag{2.5}$$

with the parameter $\lambda > 0$, and the matrix $\mathbf{A}$ has BGK form:

$$\mathbf{A} := \begin{pmatrix} -1 & 1 \\ 1 & -1 \end{pmatrix} = 2\left[\begin{pmatrix} \frac{1}{2} \\ \frac{1}{2} \end{pmatrix} \otimes (1,\ 1) - \begin{pmatrix} 1 & 0 \\ 0 & 1 \end{pmatrix}\right]. \tag{2.6}$$

This ODE can be seen as an x–homogeneous variant of (1.3) with just two discrete velocities. In fact, on the right hand side of (2.6), the column vector $(\frac{1}{2},\ \frac{1}{2})^\top$ corresponds to the Maxwellian $M(v)$ in the BGK equation (1.3), and the row vector $(1,\ 1)$ corresponds to the velocity integral. The symmetric matrix $\mathbf{A}$ has an eigenvalue 0 with corresponding eigenvector $f^\infty := (\frac{1}{2},\ \frac{1}{2})^\top$ and an eigenvalue -2. Hence $\mathbf{A}$ is coercive on $\{f^\infty\}^\perp$. Since each column of $\mathbf{A}$ sums up to 0, the "total mass" of the system, i.e. $f_1(t) + f_2(t)$, stays constant in time. Hence, we shall assume w.l.o.g. that f^I is normalized, i.e. $f_1^I + f_2^I = 1$. Thus, as $t \to \infty$, $f(t) = f^\infty + (f^I - f^\infty)\, e^{-2\lambda t}$ converges to f^∞ exponentially with rate 2λ. For $f_{1,2}^I \geq 0$ we have $f_{1,2}(t) \geq 0$.

In analogy to Definition 1 we introduce for (2.5) (with $n = 2$) the relative entropy generated by ψ:

$$e_\psi(f(t)|f^\infty) := \sum_{j=1}^{n} \psi\left(\frac{f_j(t)}{f_j^\infty}\right) f_j^\infty\,. \tag{2.7}$$

Its time derivative under the flow of (2.5) reads

$$\begin{aligned}\frac{\mathrm{d}}{\mathrm{d}t} e_\psi(f(t)|f^\infty) &= -\lambda(f_1 - f_2)\left[\psi'\left(\frac{f_1(t)}{f_1^\infty}\right) - \psi'\left(\frac{f_2(t)}{f_2^\infty}\right)\right] \\ &=: -I_\psi(f(t)|f^\infty) = -2\lambda\psi''(\zeta)(f_1 - f_2)^2 \leq 0\,,\end{aligned} \tag{2.8}$$

where $\zeta = \zeta(t)$ is an intermediate value between $2f_1(t)$ and $2f_2(t)$. $I_\psi(f(t)|f^\infty)$ denotes the *Fisher information* (of $f(t)$ w.r.t. f^∞).

As pointed out in [3], it is not obvious to bound this Fisher information from below directly by a multiple of the relative entropy (except for quadratic entropies). The goal of such an estimate would be to establish the exponential decay of the relative entropy. Hence, it is the essence of the entropy method to consider the entropy dissipation rate: Differentiating (2.8) once more in time gives

$$\begin{aligned}R_\psi(f(t)|f^\infty) &:= -\frac{\mathrm{d}}{\mathrm{d}t} I_\psi(f(t)|f^\infty) \\ &= 2\lambda I_\psi(f(t)|f^\infty) + \lambda^2\big(f_1(t) - f_2(t)\big)^2\left[\psi''\left(\frac{f_1(t)}{f_1^\infty}\right)\frac{1}{f_1^\infty} + \psi''\left(\frac{f_2(t)}{f_2^\infty}\right)\frac{1}{f_2^\infty}\right].\end{aligned} \tag{2.9}$$

Due to $\psi'' \geq 0$ the second term is nonnegative. Hence,

$$-\frac{\mathrm{d}}{\mathrm{d}t} I_\psi(f(t)|f^\infty) \geq 2\lambda I_\psi(f(t)|f^\infty)\,.$$

And this yielded in [3] the exponential decay of $I_\psi(f(t)|f^\infty)$ and of $e_\psi(f(t)|f^\infty)$ at the *sub-optimal* rate 2λ. But this procedure can be improved easily to give the following sharp result:

Theorem 3 *Let the convex entropy generator ψ satisfy either: ψ'' is convex on J; or ψ' is concave on* $(0, 1)$ *along with ψ' is convex on* $(1, \infty)$. *Then the solution to* (2.5) *satisfies*

$$I_\psi(f(t)|f^\infty) \le e^{-4\lambda t}\, I_\psi(f^I|f^\infty), \quad t \ge 0, \tag{2.10}$$

$$e_\psi(f(t)|f^\infty) \le e^{-4\lambda t}\, e_\psi(f^I|f^\infty), \quad t \ge 0. \tag{2.11}$$

Proof Case 1 : ψ'' convex on J
We have for $0 \le s \le 1$:

$$s\psi''(\sigma_2) + (1-s)\psi''(\sigma_1) \ge \psi''\big(s\sigma_2 + (1-s)\sigma_1\big).$$

Integrating this inequality over $s \in [0, 1]$ yields $\forall\, \sigma_1 \ne \sigma_2 \in J$:

$$\frac{\psi''(\sigma_1) + \psi''(\sigma_2)}{2} \ge \kappa \frac{\int_{\sigma_1}^{\sigma_2} \psi''(\sigma)\,\mathrm{d}\sigma}{\sigma_2 - \sigma_1} = \kappa \frac{\psi'(\sigma_2) - \psi'(\sigma_1)}{\sigma_2 - \sigma_1}, \tag{2.12}$$

where κ is introduced only for later reference. Here we set $\kappa = 1$.

We now recall that $f_1^\infty = f_2^\infty$. Hence, (2.9) and (2.12) give

$$\frac{\mathrm{d}}{\mathrm{d}t} I_\psi(f(t)|f^\infty) \le -4\lambda I_\psi(f(t)|f^\infty), \tag{2.13}$$

and (2.10) follows. As usual in the entropy method, one next integrates (2.13) in time (from t to ∞) to obtain

$$\frac{\mathrm{d}}{\mathrm{d}t} e_\psi(f(t)|f^\infty) \le -4\lambda e_\psi(f(t)|f^\infty),$$

and this finishes the proof for the case ψ'' convex.

Case 2 : ψ' concave on $(0, 1)$ along with ψ' convex on $(1, \infty)$
We may assume without loss of generality that $f_1 > f_2$. Then $f_1/f_1^\infty > 1 > f_2/f_2^\infty$, and by the tangent line inequality for the concave function $\psi'\big|_{(0,1)}$,

$$0 = \psi'(1) \le \psi'\left(\frac{f_2}{f_2^\infty}\right) + \psi''\left(\frac{f_2}{f_2^\infty}\right)\left(\frac{f_2^\infty - f_2}{f_2^\infty}\right).$$

Likewise, using the tangent line inequality for the convex function $\psi'\big|_{(1,\infty)}$,

$$\psi'\left(\frac{f_1}{f_1^\infty}\right) \le \psi'(1) + \psi''\left(\frac{f_1}{f_1^\infty}\right)\left(\frac{f_1 - f_1^\infty}{f_1^\infty}\right) = \psi''\left(\frac{f_1}{f_1^\infty}\right)\left(\frac{f_1 - f_1^\infty}{f_1^\infty}\right).$$

Altogether we have

$$\psi''\left(\frac{f_1}{f_1^\infty}\right)\left(\frac{f_1 - f_1^\infty}{f_1^\infty}\right) \ge \psi'\left(\frac{f_1}{f_1^\infty}\right) \quad \text{and} \quad \psi''\left(\frac{f_2}{f_2^\infty}\right)\left(\frac{f_2^\infty - f_2}{f_2^\infty}\right) \ge -\psi'\left(\frac{f_2}{f_2^\infty}\right). \tag{2.14}$$

Now continuing to assume that $f_1 > f_2$, and using the fact that $f_1^\infty = f_2^\infty$ so that $f_1 - f_2 = 2(f_1 - f_1^\infty) = 2(f_2^\infty - f_2)$,

$$\begin{aligned}
&(f_1 - f_2)\left[\psi''\left(\frac{f_1}{f_1^\infty}\right)\frac{1}{f_1^\infty} + \psi''\left(\frac{f_2}{f_2^\infty}\right)\frac{1}{f_2^\infty}\right] \\
&\quad = 2(f_1 - f_1^\infty)\psi''\left(\frac{f_1}{f_1^\infty}\right)\frac{1}{f_1^\infty} + 2(f_2^\infty - f_2)\psi''\left(\frac{f_2}{f_2^\infty}\right)\frac{1}{f_2^\infty} \\
&\quad \ge 2\left[\psi'\left(\frac{f_1}{f_1^\infty}\right) - \psi'\left(\frac{f_2}{f_2^\infty}\right)\right].
\end{aligned}$$

Therefore,

$$\lambda^2(f_1 - f_2)^2\left[\psi''\left(\frac{f_1}{f_1^\infty}\right)\frac{1}{f_1^\infty} + \psi''\left(\frac{f_2}{f_2^\infty}\right)\frac{1}{f_2^\infty}\right] \ge 2\lambda I_\psi(f|f^\infty).$$

Again from (2.9) we obtain (2.13). □

Remark 1. Concerning the logarithmic and power law entropies from (2.3) one easily verifies: ψ_p satisfies the condition $\psi^{IV} \ge 0$ on J (or the inequality (2.12)) exactly for $p \in [1, 2] \cup [3, \infty)$.

2. For ψ_p with $p \in (2, 3)$, inequality (2.12) holds with $\kappa = \frac{p-1}{2}$ (but not for any larger constant κ). This follows from $g_p(z) := z^{p-2} + 1 - \frac{z^{p-1}-1}{z-1} > 0$ on $\mathbb{R}^+$ and $g_p(0) = 0$, which can be verified by elementary computations. Hence, for $p \in (2, 3)$, the entropy method yields exponential decay of $e_p(f(t)|f^\infty)$ with the reduced rate $2(\kappa + 1)\lambda = (p + 1)\lambda$:

$$e_p(f(t)|f^\infty) \le e^{-(p+1)\lambda t} e_p(f^I|f^\infty), \quad t \ge 0.$$

But the decay estimates (2.11), (2.10) are in general false for $p \in (2, 3)$.

In an alternative approach, one can verify for $2 < p < 3$ the estimates

$$\psi_p(\sigma) \le \psi_3(\sigma), \quad \forall \sigma \ge 0; \qquad \psi_3(\sigma) \le C_p\psi_p(\sigma), \quad \forall 0 \le \sigma \le 2,$$

where $[0, 2]$ is the maximum range of values for $\frac{f_1}{f_1^\infty}$ and $\frac{f_2}{f_2^\infty}$. Here the constant is $C_p = \frac{\psi_3(2)}{\psi_p(2)} = \frac{4}{2^p-1-p}$. With (2.11) this implies

$$e_p(f(t)|f^\infty) \le e^{-4\lambda t} e_3(f^I|f^\infty) \le C_p e^{-4\lambda t} e_p(f^I|f^\infty), \quad t \ge 0.$$

Hence, the entropies e_p, $p \in (2, 3)$ still decay with the optimal rate 4λ, but at the price of the multiplicative constant $C_p > 1$.

3. The relative entropies $\hat{e}_p$, $p \ge 2$ from (2.4) satisfy the second set of assumptions in Theorem 3. Note that ψ''' does not have to be continuous at $\sigma = 1$.

2.1 Multi-velocity BGK Models

Now, we consider discrete space-homogeneous BGK models in $\mathbb{R}^n$: The evolution of a vector $f(t) = (f_1(t), f_2(t), \dots, f_n(t))^\top \in \mathbb{R}^n$ is governed by

$$\begin{cases} \frac{\mathrm{d}}{\mathrm{d}t} f = 2\lambda \mathbf{A} f, \quad t \ge 0, \\ f(0) = f^I \in \mathbb{R}^n, \end{cases} \tag{2.15}$$

for some $\lambda > 0$ and a matrix $\mathbf{A} \in \mathbb{R}^{n\times n}$ in BGK form

$$\mathbf{A} = \begin{pmatrix} \rho_1 \\ \vdots \\ \rho_n \end{pmatrix} \otimes (1, \dots, 1) - \mathbf{I} \tag{2.16}$$

with $\rho = (\rho_1, \dots, \rho_n)^\top \in (0, 1)^n$ such that $\sum_{j=1}^n \rho_j = 1$.

Such a matrix $\mathbf{A}$ has a simple eigenvalue 0 with left eigenvector $l_1 = (1, \dots, 1)$ and right eigenvector $r_1 = \rho$, and an eigenvalue -1 with geometric multiplicity $n-1$. Since each column of $\mathbf{A}$ sums up to 0, the "total mass" of system (2.15) stays constant in time, i.e. $\sum_{j=1}^n f_j(t) = \sum_{j=1}^n f_j^I$.

Matrix $\mathbf{A} = (a_{jk})_{j,k=1,\dots,n}$ has only non-negative off-diagonal coefficients a_{jk} $(j \ne k)$; such matrices are called *essentially non-negative* or Metzler matrices [19]. An essentially non-negative matrix $\mathbf{A}$ induces via (2.15) a semi-flow which preserves non-negativity of its initial datum f^I, i.e. $f_j^I \ge 0$ for all $j = 1, \dots, n$, implies $f_j(t) \ge 0$ for all $t \ge 0$.

Remark An essentially non-negative matrix is called *Q-matrix* (or W-matrix in [20]) if it has an eigenvalue 0 with right eigenvector $(1, \dots, 1)^\top$. Q-matrices are the infinitesimal generators of continuous-time Markov processes with finite state space [15].

In the following, we consider normalized positive initial data f^I, i.e. $\sum_{j=1}^n f_j^I = 1$, such that the solution f of (2.15) is positive and normalized for all $t \ge 0$. Thus, as $t \to \infty$, $f(t) = f^\infty + (f^I - f^\infty)\, e^{-2\lambda t}$ converges to the normalized steady state $f^\infty := \rho$ exponentially with rate 2λ.

The study of the long-time behavior of solutions f to (2.15) is a classical topic, an approach via entropy methods can be found in [17, 20]. Note that Perthame [17, Sect. 6.3] considers essentially positive matrices (i.e. off-diagonal elements are positive) to simplify the presentation. However, the results generalize to irreducible Q-matrices, since only the non-negativity of off-diagonal elements is used, see also [17, Remark 6.2]. While [17, Proposition 6.5] establishes only exponential decay in entropy, we aim at the optimal decay rate in the entropy approach.

We consider the time derivative of the relative entropy (2.7) under the flow of (2.15)

$$\frac{\mathrm{d}}{\mathrm{d}t} e_\psi(f(t)|f^\infty) = \sum_{j=1}^{n} \psi'\Big(\tfrac{f_j(t)}{f_j^\infty}\Big) 2\lambda\, (f_j^\infty - f_j(t)) =: -I_\psi(f(t)|f^\infty) \le 0 \quad (2.17)$$

which is non-positive due to the properties (2.1) of an entropy generator (ψ' is an increasing function with $\psi'(1) = 0$). Next, we compute the second order derivative of $e_\psi(f(t)|f^\infty)$ w.r.t. time:

$$\begin{aligned} R_\psi(f(t)|f^\infty) &:= -\frac{\mathrm{d}}{\mathrm{d}t} I_\psi(f(t)|f^\infty) = \frac{\mathrm{d}}{\mathrm{d}t} \sum_{j=1}^{n} \psi'\Big(\tfrac{f_j(t)}{f_j^\infty}\Big) \frac{\mathrm{d}}{\mathrm{d}t} f_j \\ &= \sum_{j=1}^{n} \psi'\Big(\tfrac{f_j(t)}{f_j^\infty}\Big) \frac{\mathrm{d}^2}{\mathrm{d}t^2} f_j + \sum_{j=1}^{n} \psi''\Big(\tfrac{f_j(t)}{f_j^\infty}\Big) \tfrac{1}{f_j^\infty} \Big(\frac{\mathrm{d}}{\mathrm{d}t} f_j\Big)^2 \\ &= 2\lambda I_\psi(f(t)|f^\infty) + \sum_{j=1}^{n} \psi''\Big(\tfrac{f_j(t)}{f_j^\infty}\Big) \tfrac{1}{f_j^\infty} \Big(\frac{\mathrm{d}}{\mathrm{d}t} f_j\Big)^2 \ge 2\lambda I_\psi(f(t)|f^\infty), \end{aligned}$$

since $\mathbf{A}^2 = -\mathbf{A}$ and $\psi'' \ge 0$. This yields the non-optimal entropy dissipation rate 2λ. To obtain a better entropy dissipation rate, we want to estimate the neglected term via

$$\sum_{j=1}^{n} \psi''\Big(\tfrac{f_j(t)}{f_j^\infty}\Big) \tfrac{1}{f_j^\infty} \Big(\frac{\mathrm{d}}{\mathrm{d}t} f_j\Big)^2 \ge \mu I_\psi(f(t)|f^\infty) \ge 0 \quad (2.18)$$

for some $\mu > 0$.

Theorem 4 *Let $\rho = (\rho_1, \dots, \rho_n)^\top \in (0,1)^n$ such that $\sum_{j=1}^n \rho_j = 1$ and let the convex entropy generator $\psi \in C^2(J)$ satisfy for some $\mu > 0$ and all $u = (u_1, \dots, u_n)^\top \in [0,1]^n$ with $\sum_{j=1}^n u_j = 1$:*

$$\sum_{j=1}^{n} \psi''\Big(\tfrac{u_j}{\rho_j}\Big) \tfrac{1}{\rho_j} (\rho_j - u_j)^2 \ge \frac{\mu}{2\lambda} \sum_{j=1}^{n} \psi'\Big(\tfrac{u_j}{\rho_j}\Big)(u_j - \rho_j). \quad (2.19)$$

Then, for all non-negative normalized initial data f^I, the solution f to (2.15) *satisfies*

$$I_\psi(f(t)|f^\infty) \le e^{-(2\lambda+\mu)t}\, I_\psi(f^I|f^\infty), \quad t \ge 0, \tag{2.20}$$

$$e_\psi(f(t)|f^\infty) \le e^{-(2\lambda+\mu)t}\, e_\psi(f^I|f^\infty), \quad t \ge 0. \tag{2.21}$$

Proof The solution f to (2.15) is positive and normalized for all $t > 0$. Under Assumption (2.19) on ψ, we obtain the estimates (2.18), and

$$\frac{\mathrm{d}}{\mathrm{d}t} I_\psi(f(t)|f^\infty) \le -(2\lambda+\mu)\, I_\psi(f(t)|f^\infty), \tag{2.22}$$

hence (2.20) follows. Next, one integrates (2.22) in time (from t to ∞) to obtain

$$\frac{\mathrm{d}}{\mathrm{d}t} e_\psi(f(t)|f^\infty) \le -(2\lambda+\mu)\, e_\psi(f(t)|f^\infty),$$

and this finishes the proof. □

For the quadratic entropy generator ψ_2 inequality (2.19) holds with $\mu = 2\lambda$. Thus we recover the optimal decay rate 4λ in (2.20)–(2.21). For the logarithmic entropy generator ψ_1 an estimate for μ in (2.19) has been given in [6, 9] as

$$\frac{\mu}{2\lambda} \ge \sqrt{\rho_{min}(1-\rho_{min})} \qquad \text{with } \rho_{min} = \min_{j=1,\dots,n} \rho_j .$$

Next, we consider entropy generators ψ in the sense of Definition 1, such that ψ' is concave on $(0,1)$ along with ψ' convex on $(1,\infty)$. Thus, for $f_1 \ge f_1^\infty > 0$ and $f_2^\infty \ge f_2 > 0$, the inequalities (2.14) continue to hold. Distinguishing the cases $u_j < \rho_j$, $u_j > \rho_j$ and the trivial case $u_j = \rho_j$, we deduce for all $j = 1, \dots, n$,

$$\psi''\Big(\tfrac{u_j}{\rho_j}\Big)\tfrac{1}{\rho_j}(\rho_j - u_j)^2 \ge \psi'\Big(\tfrac{u_j}{\rho_j}\Big)(u_j - \rho_j),$$

hence (2.19) holds with $\mu = 2\lambda$. However, for the entropy generators $\hat\psi_p$ in (2.4) with $p \ge 2$ the optimal value is $\mu = (p-1)2\lambda$.

In the following, we restrict ourselves to $n = 2$ and determine the best constant for some polynomial entropy generators:

Lemma 1 *Let $\rho_1, \rho_2 \in (0,1)$ with $\rho_1 + \rho_2 = 1$. The entropy generator $\psi(\sigma)$ satisfies condition* (2.19) *with*

$$1 \ge \frac{\mu}{2\lambda} = \begin{cases} 1 & \text{for } \psi(\sigma) = \psi_2(\sigma), \\ 2\min\{\rho_1, \rho_2\} & \text{for } \psi(\sigma) = \psi_3(\sigma), \\ 2 - 2\sqrt{1 - 3\rho_2\,(1-\rho_2)} > 0 & \text{for } \psi(\sigma) = \psi_4(\sigma). \end{cases}$$

Proof For $n = 2$, the assumptions on ρ and u in (2.19) imply

$$-(\rho_2 - u_2) = \rho_1 - u_1 = \rho_1(u_1 + u_2) - u_1 = \rho_1 u_2 - \rho_2 u_1 = \rho_1\rho_2(\tfrac{u_2}{\rho_2} - \tfrac{u_1}{\rho_1}).$$

Thus condition (2.19) is equivalent to

$$\sum_{j=1}^{2} \psi''\Big(\tfrac{u_j}{\rho_j}\Big)\tfrac{1}{\rho_j}\rho_1^2\rho_2^2(\tfrac{u_2}{\rho_2} - \tfrac{u_1}{\rho_1})^2 \geq \frac{\mu}{2\lambda}\sum_{j=1}^{2} \psi'\Big(\tfrac{u_j}{\rho_j}\Big)(-1)^j \rho_1\rho_2(\tfrac{u_2}{\rho_2} - \tfrac{u_1}{\rho_1}) \geq 0.$$

Setting $v_1 := u_1/\rho_1$ and $v_2 := u_2/\rho_2$, we deduce for $\psi_p(\sigma)$, $p > 1$,

$$(p-1)\big[v_1^{p-2}\rho_2 + v_2^{p-2}\rho_1\big](v_1 - v_2)^2 \geq \frac{\mu}{2\lambda}\big[v_1^{p-1} - v_2^{p-1}\big](v_1 - v_2) \quad \forall v_1, v_2 \geq 0.$$

Moreover, for $v_2 > 0$, dividing by v_2^p and defining $z := v_1/v_2$, we obtain

$$(p-1)\big[z^{p-2}\rho_2 + \rho_1\big](z-1)^2 \geq \frac{\mu}{2\lambda}\big[z^{p-1} - 1\big](z-1) \qquad \forall z \geq 0.$$

We show the statement for the quartic entropy generator $\psi_4(\sigma)$, the (simpler) proof for quadratic and cubic entropy generators is omitted. For ψ_4, condition (2.19) is equivalent to

$$g(z) := z^2(3\rho_2 - \tilde{\mu}) - \tilde{\mu}z + 3\rho_1 - \tilde{\mu} \geq 0 \qquad \forall z \geq 0$$

with $\tilde{\mu} := \mu/(2\lambda)$. Evaluating $g(z)$ at $z = 0$ and taking the limit $z \to \infty$, we deduce the necessary conditions $3\rho_1 \geq \tilde{\mu}$ and $3\rho_2 > \tilde{\mu}$, respectively. The minimum of $g(z)$ on $z \in (0, \infty)$ is zero, iff $\tilde{\mu}$ solves $\tilde{\mu}^2 - 4(3\rho_1 - \tilde{\mu})\,(3\rho_2 - \tilde{\mu}) = 0$. This quadratic polynomial has a simple positive zero given by $\tilde{\mu}_0 = 2 - 2\sqrt{1 - 3\rho_2\,(1-\rho_2)} > 0$, since $\rho_1 + \rho_2 = 1$.

The expression $\tilde{\mu}_0 = 2 - 2\sqrt{1 - 3\rho_2\,(1-\rho_2)} > 0$ attains its maximum 1 for $\rho_2 \in (0, 1)$ at $\rho_2 = 1/2$. □

Remark The quadratic entropy $\psi_2(\sigma)$ satisfies Assumption (2.19) with $\mu = 2\lambda$ for all $f_1^\infty, f_2^\infty \in (0, 1)$. The cubic entropy $\psi_3(\sigma)$ and the quartic entropy $\psi_4(\sigma)$ satisfy (2.19) with $\mu = 2\lambda$ only for $f_1^\infty = f_2^\infty = \frac{1}{2}$.

3 A Discrete Hypocoercive BGK Model

In this section we consider an example for a discrete version (both in x and v) of (1.1). More precisely, we consider the evolution of a vector $f(t) = \big(f_j(t);\ j = 1, \ldots, 4\big)^\top \in \mathbb{R}^4$, where its four components may correspond to the following points in the $x - v$–phase space: $(1, 1)$, $(1, -1)$, $(-1, -1)$, $(-1, 1)$, in this order. Its evolution is given by

$$\frac{\mathrm{d}}{\mathrm{d}t} f = (\mathbf{A}+\mathbf{B}) f, \quad t \geq 0, \tag{3.1}$$
$$f(0) = f^I \in \mathbb{R}^4 .$$

Similarly to (2.6), the matrix $\mathbf{A}$ has BGK form:

$$\mathbf{A} := \frac{1}{2}\begin{pmatrix} -1 & 1 & 0 & 0 \\ 1 & -1 & 0 & 0 \\ 0 & 0 & -1 & 1 \\ 0 & 0 & 1 & -1 \end{pmatrix} = \frac{1}{2}\begin{pmatrix} 1 & 1 \\ 1 & 1 \end{pmatrix} \otimes \begin{pmatrix} 1 & 0 \\ 0 & 1 \end{pmatrix} - \mathbf{I}, \tag{3.2}$$

where the first summand on the r.h.s. is the projection onto the kernel of $\mathbf{A}$,

$$\ker \mathbf{A} = \operatorname{span}[(1\,1\,0\,0)^\top, (0\,0\,1\,1)^\top].$$

In (3.1), the matrix $\mathbf{B}$ is skew-symmetric and reads

$$\mathbf{B} := \begin{pmatrix} 0 & -1 & 0 & 1 \\ 1 & 0 & -1 & 0 \\ 0 & 1 & 0 & -1 \\ -1 & 0 & 1 & 0 \end{pmatrix} = \begin{pmatrix} 0 & -1 \\ 1 & 0 \end{pmatrix} \otimes \begin{pmatrix} 1 & -1 \\ -1 & 1 \end{pmatrix}. \tag{3.3}$$

$\mathbf{B}$ corresponds to a discretization of the transport operator in (1.1) by symmetric finite differences. We remark that (3.1) does not preserve positivity but, as we shall show, the hypocoercivity of (1.1). Motivated by the theory of hyperbolic systems, one may also replace the transport operator by an upwind discretization with a then non-symmetric matrix $\tilde{\mathbf{B}}$. Then, the resulting system would preserve positivity. But it would be coercive rather than hypocoercive. Here we opt to discuss the situation with $\mathbf{B}$ given in (3.3).

The spectrum of $\mathbf{A}+\mathbf{B}$ is given by $0, -\frac{1}{2} \pm \frac{\sqrt{15}}{2}i, -1$. The unique, (in the 1-norm) normalized steady state of (3.1) is given by $f^\infty = w_1 = \frac{1}{4}(1111)^\top$, which spans the kernel of $\mathbf{A}+\mathbf{B}$. Eigenvectors of the non-trivial eigenvalues are given by $w_{2,3} := (\sqrt{5}, \pm\sqrt{3}i, -\sqrt{5}, \mp\sqrt{3}i)^\top$ and $w_4 := (1, -1, 1, -1)^\top$, and all three of them have mass 0. This shows that $\frac{1}{2}$ is the sharp decay rate of any (normalized) $f(t)$ towards f^∞. But this "spectral gap" of size $\frac{1}{2}$ disappears in the symmetric part of the matrix: $\sigma\big(\mathbf{A} + \frac{\mathbf{B}+\mathbf{B}^\top}{2}\big) = \{0, 0, -1, -1\}$. Hence, the matrix $\mathbf{A}+\mathbf{B}$ is only *hypocoercive* on $\{f^\infty\}^\perp$ (as defined by Villani, see Sect. 3.2 of [21]). But using an appropriate similarity transformation of $\mathbf{A}+\mathbf{B}$ one can again recover the sharp decay rate of the hypocoercive BGK-model (3.1) via energy or entropy methods.

In particular, we shall use Lyapunov's direct method—see Lemma 3 in the following subsection—to prove decay to equilibrium for normalized solutions: If f^I is normalized, then the solution to (3.1) satisfies (for any norm on $\mathbb{R}^4$)

$$\|f(t) - f^\infty\| \leq c\, e^{-t/2} \|f^I - f^\infty\|, \quad t \geq 0,$$

with some generic constant $c \geq 1$.

3.1 Lyapunov's Direct Method

We consider an ODE for a vector $f(t) = (f_1(t), f_2(t), \dots, f_n(t))^\top \in \mathbb{R}^n$:

$$\begin{cases} \frac{\mathrm{d}}{\mathrm{d}t} f = \mathbf{A} f, & t \geq 0, \\ f(0) = f^I \in \mathbb{R}^n, \end{cases} \tag{3.4}$$

for some real (typically non-symmetric) matrix $\mathbf{A} \in \mathbb{R}^{n\times n}$. The origin 0 is a steady state of (3.4). The stability of the trivial solution $f^0(t) \equiv 0$ is determined by the eigenvalues of matrix $\mathbf{A}$:

Theorem 5 *Let $\mathbf{A} \in \mathbb{R}^{n\times n}$ and let λ_j $(j = 1, \dots, n)$ denote the eigenvalues of $\mathbf{A}$ (counted with their multiplicity).*

(S1) The equilibrium f^0 of (3.4) *is stable if and only (i) $\Re\lambda_j \leq 0$ for all $j = 1, \dots, n$; and (ii) all eigenvalues with $\Re\lambda_j = 0$ are non-defective.*[1]
(S2) The equilibrium f^0 of (3.4) *is asymptotically stable if and only if $\Re\lambda_j < 0$ for all $j = 1, \dots, n$.*
(S3) The equilibrium f^0 of (3.4) *is unstable in all other cases.*

To study the stability for f^0 via Lyapunov's direct method, a first guess for a Lyapunov function $V(f)$ is the (squared) Euclidean norm $V(f) = \|f\|_2^2$. The derivative of $V(f)$ along solutions $f(t)$ of (3.4) satisfies

$$\frac{\mathrm{d}}{\mathrm{d}t} V(f(t)) = \langle f(t), (\mathbf{A}^\top + \mathbf{A}) f(t) \rangle .$$

Thus the derivative depends only on the symmetric part $\frac{1}{2}(\mathbf{A}^\top + \mathbf{A})$ of a matrix $\mathbf{A}$. Hence the choice $V(f) = \|f\|_2^2$ is only suitable for symmetric matrices $\mathbf{A}$.

To study the stability of $f^0(t) \equiv 0$ w.r.t. (3.4) for a general $\mathbf{A}$, it is standard to consider the generalized (squared) norm

$$V(f) := \langle f, \mathbf{P} f \rangle \quad \text{for some symmetric, positive definite matrix } \mathbf{P} \in \mathbb{R}^{n\times n}.$$

The derivative of $V(f)$ along solutions $f(t)$ of (3.4) satisfies

$$\frac{\mathrm{d}}{\mathrm{d}t} V(f(t)) = \langle \mathbf{A} f(t), \mathbf{P} f(t) \rangle + \langle f(t), \mathbf{P}\mathbf{A} f(t) \rangle = \langle f(t), \mathbf{R} f(t) \rangle , \tag{3.5}$$

with matrix $\mathbf{R} := \mathbf{A}^\top\mathbf{P} + \mathbf{P}\mathbf{A}$. Conclusions on the stability of f^0 are possible, depending on the (negative) definiteness of $\mathbf{R}$, see e.g. [14, Proposition 7.6.1].

To determine the decay rate of an asymptotically stable steady state, we shall use the following algebraic result.

[1] An eigenvalue is defective if its geometric multiplicity is strictly less than its algebraic multiplicity.

Lemma 2 *For any fixed matrix* $\mathbf{C} \in \mathbb{C}^{n\times n}$, *let* $\mu := \min\{\Re\{\lambda\}|\lambda$ *is an eigenvalue of* $\mathbf{C}\}$. *Let* $\{\lambda_j|1 \le j \le j_0\}$ *be all the eigenvalues of* $\mathbf{C}$ *with* $\Re\{\lambda_j\} = \mu$, *only counting their geometric multiplicity.*

If all λ_j $(j = 1, \dots, j_0)$ *are non-defective, then there exists a Hermitian, positive definite matrix* $\mathbf{P} \in \mathbb{C}^{n\times n}$ *with*

$$\mathbf{C}^*\mathbf{P} + \mathbf{P}\mathbf{C} \ge 2\mu\mathbf{P}\,, \tag{3.6}$$

where $\mathbf{C}^*$ *denotes the Hermitian transpose of* $\mathbf{C}$. *Moreover, (non-unique) matrices* $\mathbf{P}$ *satisfying* (3.6) *are given by*

$$\mathbf{P} := \sum_{j=1}^{n} b_j\, w_j \otimes \overline{w_j}^\top\,, \tag{3.7}$$

where w_j $(j = 1, \dots, n)$ *denote the eigenvectors of* $\mathbf{C}^*$, *and* $b_j \in \mathbb{R}^+$ $(j = 1, \dots, n)$ *are arbitrary weights.*

Remark Lemma 2 is the complex analog of [2, Lemma 4.3] or [1, Lemma 2.6]. In particular, if $\mathbf{C} \in \mathbb{R}^{n\times n}$ is a real matrix, then the inequality (3.6) of Lemma 2 holds true for real, symmetric, positive definite matrices $\mathbf{P} \in \mathbb{R}^{n\times n}$. Moreover, the case of defective eigenvalues is also treated in [1, 2].

If $\mathbf{A} \in \mathbb{R}^{n\times n}$ has only eigenvalues with negative real parts, then the origin is the unique and asymptotically stable steady state $f^0 = 0$ of (3.4). Due to Lemma 2, there exists a symmetric, positive definite matrix $\mathbf{P} \in \mathbb{R}^{n\times n}$ such that $\mathbf{A}^\top\mathbf{P} + \mathbf{P}\mathbf{A} \le -2\mu\mathbf{P}$ where $\mu = \min|\Re\lambda_j|$. Thus, the derivative of $V(f) := \langle f\,,\,\mathbf{P}f\rangle$ along solutions of (3.4) satisfies

$$\frac{\mathrm{d}}{\mathrm{d}t}V(f(t)) \le -2\mu V(f(t)) \qquad \text{with} \quad \mu = \min|\Re\lambda_j|, \tag{3.8}$$

which implies $V(f(t)) \le e^{-2\mu t}V(f^I)$ and $\|f(t)\|^2 \le ce^{-2\mu t}\|f^I\|^2$ for some $c \ge 1$ by equivalence of norms on $\mathbb{R}^n$.

In contrast, we consider next matrices $\mathbf{A} \in \mathbb{R}^{n\times n}$ having only eigenvalues with non-positive real part. More precisely, let $\mathbf{A}$ satisfy

(A1) $\mathbf{A}$ has a simple eigenvalue $\lambda_1 = 0$ with left eigenvector $w_1^\top \in \mathbb{R}^n$ and right eigenvector $v_1 \in \mathbb{R}^n$;
(A2) the other eigenvalues λ_j $(j = 2, \dots, n)$ of $\mathbf{A}$ have negative real part.

Then, the space of steady states of (3.4) consists of $\operatorname{span}\{v_1\}$, and solutions to (3.4) will typically not decay to 0. More precisely, if f is a solution of ODE (3.4) with initial datum f^I satisfying $\langle w_1\,,\,f^I\rangle = c$ for some $c \in \mathbb{R}$, then $\langle w_1\,,\,f(t)\rangle = c$ for all $t \ge 0$. Therefore we aim to prove the convergence of solutions $f(t)$ of (3.4) for an initial datum f^I (normalized in the sense of $\langle w_1\,,\,f^I\rangle = 1$) to the unique steady state $f^\infty \in \operatorname{span}\{v_1\}$ (again normalized as $\langle w_1\,,\,f^\infty\rangle = 1$).

Lemma 3 *Let $\mathbf{A} \in \mathbb{R}^{n\times n}$ satisfy (A1)–(A2) with non-defective eigenvalues λ_j for $j = 1, \ldots, n$. If f is a solution of* (3.4) *for some normalized initial datum f^I (i.e. $\langle w_1\,,\ f^I\rangle = 1$), then*

$$\|f(t) - f^\infty\| \le c\,\|f^I - f^\infty\|e^{-\lambda_* t}\,, \quad t \ge 0\,, \tag{3.9}$$

where $\lambda_* := \min_{\lambda_j \neq 0} |\Re\lambda_j|$ and some constant $c \ge 1$.

Proof To present a unified approach for symmetric and non-symmetric matrices $\mathbf{A}$ satisfying (A1)–(A2), we consider again the "distorted" vector norm $\|f\|_{\mathbf{P}} := \sqrt{\langle f\,,\ \mathbf{P}f\rangle}$, and the relative entropy-type functional

$$E_{\psi_2}(f(t)|f^\infty) := \|f(t) - f^\infty\|_{\mathbf{P}}^2$$

with some real, symmetric and positive definite matrix $\mathbf{P}$ to be determined. Its derivative satisfies

$$\frac{\mathrm{d}}{\mathrm{d}t}E_{\psi_2}(f(t)|f^\infty) = \Big\langle (f - f^\infty)\,,\ (\mathbf{A}^\top\mathbf{P} + \mathbf{P}\mathbf{A})(f - f^\infty)\Big\rangle.$$

Every matrix $\mathbf{A} \in \mathbb{R}^{n\times n}$ induces an orthogonal decomposition of $\mathbb{R}^n$ via

$$\mathbb{R}^n = \ker(\mathbf{A}) \,\oplus\, \mathrm{ran}(\mathbf{A}^\top) = \ker(\mathbf{A}^\top) \,\oplus\, \mathrm{ran}(\mathbf{A}).$$

Thus, there exists an orthogonal projection from $\mathbb{R}^n$ onto $\mathrm{ran}(\mathbf{A})$, which is represented by a matrix $\mathbf{P}_1 \in \mathbb{R}^{n\times n}$ with $\mathbf{P}_1^2 = \mathbf{P}_1$. Due to assumption (A1), matrix $\mathbf{A}^\top$ has a one-dimensional kernel which is spanned by w_1, hence $\mathbf{P}_1 w_1 = 0$. Since $w_1^\top$ is a left eigenvector of $\mathbf{A}$ for the eigenvalue 0, a solution f of (3.4) for a normalized initial datum f^I (i.e. $\langle w_1\,,\ f^I\rangle = 1$) is again normalized, i.e. $\langle w_1\,,\ f(t)\rangle = 1$ for all $t \ge 0$. Thus, $\langle w_1\,,\ f(t) - f^\infty\rangle \equiv 0$ iff $\langle w_1\,,\ f^I - f^\infty\rangle = 0$, which implies $f(t) - f^\infty \in \mathrm{ran}(\mathbf{A})$ for all $t \ge 0$. Moreover,

$$\frac{\mathrm{d}}{\mathrm{d}t}E_{\psi_2}(f(t)|f^\infty) = \Big\langle \mathbf{P}_1(f - f^\infty)\,,\ \mathbf{P}_1^\top(\mathbf{A}^\top\mathbf{P} + \mathbf{P}\mathbf{A})\mathbf{P}_1\,\mathbf{P}_1(f - f^\infty)\Big\rangle.$$

In order to prove

$$\mathbf{P}_1^\top(\mathbf{A}^\top\mathbf{P} + \mathbf{P}\mathbf{A})\mathbf{P}_1 \le -2\lambda_*\mathbf{P}_1^\top\mathbf{P}\mathbf{P}_1 \tag{3.10}$$

we consider the modified matrix $\tilde{\mathbf{A}} := \mathbf{A} - \lambda_* v_1 \otimes w_1^\top \in \mathbb{R}^{n\times n}$. Due to (A1)–(A2) and the assumptions in Lemma 3, $\tilde{\mathbf{A}}$ has only non-defective eigenvalues with negative real part. Due to Lemma 2, there exists a real, symmetric, positive-definite matrix $\mathbf{P}$ such that $\tilde{\mathbf{A}}^\top\mathbf{P} + \mathbf{P}\tilde{\mathbf{A}} \le -2\lambda_*\mathbf{P}$. This implies (3.10) since $\mathbf{P}_1^\top\big((v_1 \otimes w_1^\top)^\top\mathbf{P} + \mathbf{P}(v_1 \otimes w_1^\top)\big)\mathbf{P}_1 = 0$. Therefore we conclude

$$\frac{\mathrm{d}}{\mathrm{d}t}E_{\psi_2}(f(t)|f^\infty) \le -2\lambda_* E_{\psi_2}(f(t)|f^\infty)\,, \tag{3.11}$$

and $E_{\psi_2}(f(t)|f^\infty) \le E_{\psi_2}(f^I|f^\infty)e^{-2\lambda_* t}$ follows. Moreover, $0 \le \lambda_{P,min}\mathbf{I} \le \mathbf{P} \le \lambda_{P,max}\mathbf{I}$, where $\lambda_{P,min} > 0$ is the smallest eigenvalue and $\lambda_{P,max} > 0$ is the biggest eigenvalue of $\mathbf{P}$. Therefore, $\lambda_{P,min}\|f\|_2^2 \le \|f\|_{\mathbf{P}}^2 \le \lambda_{P,max}\|f\|_2^2$ and (3.9) follows. □

Remark For a symmetric matrix $\mathbf{A}$, the choice $\mathbf{P} = \mathbf{I}$ is admissible and one recovers the optimal decay rate and constant $c = 1$ in estimate (3.9).

Remark Assume now that the matrix $\mathbf{A}$ from Lemma 3 satisfies also $\ker(\mathbf{A}) = \ker(\mathbf{A}^\top)$, which corresponds to *detailed balance* for the steady state. Then, Lemma 3 allows for a simpler proof: Let $w_1 = f^\infty \in \mathbb{R}^n$ be a normalized steady state. Then the orthogonal projector $w_1 \otimes \overline{w_1}^\top$ commutes with both $\mathbf{A}$ and $\mathbf{A}^\top$. Let $\mathbf{P}_1$ denote its complementary projection. Then $\mathrm{ran}(\mathbf{P}_1)$ is invariant under $e^{\mathbf{A}t}$, and (3.10) with $\mathbf{P}$ from (3.7) follows from Lemma 2 applied to $\mathbf{A}$ restricted to $\mathrm{ran}(\mathbf{P}_1)$.

4 Space-Inhomogeneous BGK Models

In this section we study the large-time behavior of the BGK equation (1.3) on $L^2(\mathbf{T}^1 \times \mathbb{R}; M_T^{-1}(v)\,dv)$ with periodic boundary conditions in x. We start with the x–Fourier series of f:

$$f(x, v, t) = \sum_{k\in\mathbb{Z}} f_k(v, t)\, e^{ikx}\,, \tag{4.1}$$

and obtain the following evolution equation for the spatial modes f_k, $k \in \mathbb{Z}$:

$$\partial_t f_k + ikv f_k = \mathbf{Q} f_k = M_T(v) \int_{\mathbb{R}} f_k(v, t)\,dv - f_k(v, t)\,, \quad k \in \mathbb{Z};\ t \ge 0\,. \tag{4.2}$$

Since the BGK operator $\mathbf{Q}$ projects onto the centered Maxwellian at temperature T, it is natural to consider (4.2) in the basis spanned by the Hermite functions (in v). This is natural for the following reason:

The Hermite polynomials (for temperature T) are the system of orthonormal polynomials that one obtains by applying the Gram-Schmidt orthonormalization procedure to the sequence of monomials $\{v^\ell\}$ in $L^2(M_T)$; let $P_\ell(v)$ denote the ℓth Hermite polynomial. The *Hermite functions* themselves are the functions of the form $\tilde{g}_\ell(v) = P_\ell(v)M_T(v)$, and evidently these are orthonormal in $L^2(M_T^{-1})$. This is the space in which we work.

The key fact concerning the Hermite functions is that multiplication by v acts on them in a very simple way, and this is relevant since the action of our streaming operator on the kth mode is multiplication by ikv. In fact, the reason for the simple nature of its action is very general and thus applies to generalizations of the Hermite functions. Since we use this below, we explain the simple action from a general point of view, using only the fact that M_T is even.

Note that multiplication by v is evidently self adjoint on $L^2(M_T^{-1})$. Also, for each ℓ, $v\tilde{g}_\ell(v)$ is in the span of $\{\tilde{g}_0, \dots, \tilde{g}_{\ell+1}\}$. Hence, for $m > \ell + 1$

$$0 = \langle \tilde{g}_m, v\tilde{g}_\ell \rangle_{L^2(M_T^{-1})} = \langle \tilde{g}_\ell, v\tilde{g}_m \rangle_{L^2(M_T^{-1})}$$

from which we conclude that the ℓ, m matrix elements of multiplication by v are zero for $|\ell - m| > 2$. Finally, by the symmetry of M_T, the diagonal matrix elements are all zero. Hence, *in the Hermite basis, multiplication by v is represented by a tridiagonal symmetric matrix that is zero on the main diagonal.* The operator $\mathbf{Q}$ is evidently diagonal in the Hermite basis. Hence the operator $\mathbf{L}_k := -ikv + \mathbf{Q}$ has a simple tridiagonal structure. We shall see that the matrix representing ikv is

$$ik\sqrt{T}\begin{pmatrix} 0 & \sqrt{1} & 0 & \cdots \\ \sqrt{1} & 0 & \sqrt{2} & 0 \\ 0 & \sqrt{2} & 0 & \sqrt{3} \\ \vdots & 0 & \sqrt{3} & \ddots \end{pmatrix}$$

while $\mathbf{Q} = \mathrm{diag}(0, -1, -1, \cdots)$.

The infinite tridiagonal matrix representing $\mathbf{L}_k = -ikv + \mathbf{Q}$ in the Hermite basis is still not easy to analyze directly. We cannot compute its eigenfunctions in closed form, and hence cannot apply formula (3.7) to implement Lyapunov's method.

However, we can do this for a related family of discrete velocity models, since then we are dealing with finite matrices. The discrete models, using the binomial approximation to the Gaussian distribution, are sufficiently close in structure to the continuous velocity BGK model that they suggest an ansatz for the $\mathbf{P}$ operator that specifies the entropy function norm. In fact, a complete solution of a 2-velocity model provides the essential hint for proving hypocoercivity of the continuous velocity BGK model.

We shall present the details of this analysis in Sect. 4.3 below. Here, the above remark only serves as a motivation for our analysis of discrete velocity models, which are velocity discretizations of the BGK equation (1.3). We shall start with the two velocity case, and then discuss its generalization to n velocities.

4.1 A Two Velocity BGK Model

In this section we revisit the following hyperbolic system, which can be considered as a kinetic equation with the two velocities $v = \pm\sigma$, and some parameter $\sigma > 0$:

$$\partial_t f_\pm \pm \sigma \partial_x f_\pm = \pm\frac{1}{2}(f_- - f_+), \quad t \geq 0, \tag{4.3}$$

for the distributions $f_\pm(x,t)$ of right- and left-moving particles, 2π–periodic in x. The matrix of the interaction term on the r.h.s. has the form

$$\frac{1}{2}\begin{pmatrix} -1 & 1 \\ 1 & -1 \end{pmatrix},$$

and hence (4.3) is also of BGK-form. Due to the conservation of the total mass $\int_0^{2\pi} \big(f_+(x,t) + f_-(x,t)\big)\, \mathrm{d}x$ of (4.3), its unique normalized steady state is $f_+^\infty = f_-^\infty = const = \frac{1}{4\pi}\int_0^{2\pi} \big(f_+^I(x) + f_-^I(x)\big)\, \mathrm{d}x$.

This toy model (with the choice $\sigma = 1$) was analyzed in Sect. 1.4 of [10] to illustrate the hypocoercivity method presented there. As for (4.2), we Fourier transform (4.3) in x and expand it in the discrete velocity basis $\{\binom{1}{1}, \binom{1}{-1}\}$. This yields for each mode $k \in \mathbb{Z}$ the following decoupled ODE-system:

$$\frac{\mathrm{d}}{\mathrm{d}t} u_k = -\mathbf{C}_k\, u_k, \quad \mathbf{C}_k = \begin{pmatrix} 0 & ik\sigma \\ ik\sigma & 1 \end{pmatrix}, \tag{4.4}$$

with $u_k(t) \in \mathbb{C}^2$, $k \in \mathbb{Z}$. The matrices $-\mathbf{C}_k$ have the eigenvalues $-\frac{1}{2} \pm \sqrt{\frac{1}{4} - k^2\sigma^2}$ in the case $|k| \le \frac{1}{2\sigma}$ and $-\frac{1}{2} \pm i\sqrt{k^2\sigma^2 - \frac{1}{4}}$ in the case $|k| > \frac{1}{2\sigma}$. Hence, as $t \to \infty$, $u_0(t)$ converges to an eigenvector of the 0-eigenvalue, i.e. $u_0^\infty = (f_+^\infty + f_-^\infty,\, 0)^\top$, with the exponential rate $\lambda_0 := 1$. All modes $u_k(t)$ with $k \neq 0$ converge to $u_k^\infty = 0$ with an exponential rate determined by the spectral gap of the matrix $\mathbf{C}_k$. For simplicity we shall assume here that $\frac{1}{2\sigma} \notin \mathbb{N}$. This avoids defective eigenvalues of the matrices $\mathbf{C}_k$, but they could be included as discussed in Lemma 4.3 of [2]. The spectral gap of the low modes (i.e. for $0 < |k| < \frac{1}{2\sigma}$) is $\lambda_k := \frac{1}{2} - \sqrt{\frac{1}{4} - k^2\sigma^2}$, and it is $\lambda_k := \frac{1}{2}$ for the high modes. Hence, the exponential decay rate of the sequence of modes $\{u_k(t)\}_{k\in\mathbb{Z}}$ is given by the decay of the modes $k = \pm 1$: $\lambda := \min_{k\in\mathbb{Z}}\{\lambda_k\} = \Re\big(\frac{1}{2} - \sqrt{\frac{1}{4} - \sigma^2}\big)$. By Plancherel's theorem this is then also the convergence rate of $f(t) = (f_+(t),\, f_-(t))^\top$ towards the steady state $f^\infty = (f_+^\infty,\, f_-^\infty)^\top$.

The goal of entropy methods is to prove this exponential decay towards equilibrium, possibly with the sharp rate, by constructing an appropriate Lyapunov functional. In the hypocoercive method developed in [10] the authors obtained, for the case $\sigma = 1$ and the quadratic entropy, a decay rate bounded above by $\frac{1}{5}$. But the sharp rate for this case is $\lambda = \frac{1}{2}$. We shall now construct a refined Lyapunov functional that captures the sharp decay rate.

Following Lemma 2(i) we introduce the positive definite transformation matrices $\mathbf{P}_0 := \mathbf{I}$,

$$\mathbf{P}_k := \begin{pmatrix} 4k^2\sigma^2 & -2ik\sigma \\ 2ik\sigma & 2 - 4k^2\sigma^2 \end{pmatrix}, \quad \text{for } 0 < |k| < \frac{1}{2\sigma},$$

and

$$\mathbf{P}_k := \begin{pmatrix} 1 & \frac{-i}{2k\sigma} \\ \frac{i}{2k\sigma} & 1 \end{pmatrix}, \quad \text{for } |k| > \frac{1}{2\sigma}. \tag{4.5}$$

In the latter case, $\mathbf{P}_k$ is unique only up to a multiplicative constant, which is chosen here such that $\operatorname{Tr} \mathbf{P}_k = n = 2$. We define the "distorted" vector norms for each mode u_k:

$$\|u_k\|_{\mathbf{P}_k} := \sqrt{\langle u_k, \mathbf{P}_k u_k \rangle}.$$

Due to the ODE (4.4) and the matrix inequality (3.6) it satisfies

$$\frac{\mathrm{d}}{\mathrm{d}t} \|u_k\|^2_{\mathbf{P}_k} = -\langle u_k, (\mathbf{C}_k^* \mathbf{P}_k + \mathbf{P}_k \mathbf{C}_k) u_k \rangle \le -2\lambda_k \|u_k\|^2_{\mathbf{P}_k}, \quad k \in \mathbb{Z} \setminus \{0\}, \tag{4.6}$$

and hence

$$\|u_k(t) - u_k^\infty\|_{\mathbf{P}_k} \le e^{-\lambda_k t} \|u_k(0) - u_k^\infty\|_{\mathbf{P}_k}, \quad t \ge 0, \ k \in \mathbb{Z}. \tag{4.7}$$

With this motivation we define the following norm as a Lyapunov functional for the sequence of modes:

$$E\big(\{u_k\}_{k\in\mathbb{Z}}\big) := \sqrt{\sum_{k\in\mathbb{Z}} \|u_k\|^2_{\mathbf{P}_k}}. \tag{4.8}$$

From (4.7) we obtain

$$E\big(\{u_k(t) - u_k^\infty\}\big) \le e^{-\lambda t} E\big(\{u_k(0) - u_k^\infty\}\big), \quad t \ge 0,$$

with $\lambda = \min_{k\in\mathbb{Z}}\{\lambda_k\}$. Due to Plancherel's theorem, this is also a norm for the corresponding distributions $f = (f_+, f_-)^\top$:

$$E\big(\{u_k\}\big) = \|Bf\|_{L^2(0,2\pi;\mathbb{R}^2)},$$

where B is a (nonlocal) bounded operator on $L^2(0, 2\pi; \mathbb{R}^2)$ with bounded inverse. More precisely, $B = I + K$, where K is a compact operator with $\|K\| < 1$, since $\mathbf{P}_k \overset{|k|\to\infty}{\longrightarrow} \mathbf{I}$ (cf. (4.5)). This implies the sought-for exponential decay of $f(t)$ with sharp rate:

Theorem 6 *Let* $\frac{1}{2\sigma} \notin \mathbb{N}$. *Then the solution to* (4.3) *satisfies*

$$\|f(t) - f^\infty\|_{L^2(0,2\pi;\mathbb{R}^2)} \le c\, e^{-\lambda t} \|f^I - f^\infty\|_{L^2(0,2\pi;\mathbb{R}^2)}, \quad t \ge 0,$$

with $\lambda = \Re\big(\frac{1}{2} - \sqrt{\frac{1}{4} - \sigma^2}\big)$ *and some generic constant* $c > 0$.

4.2 A Multi-velocity BGK Model

We now turn to a discrete velocity model analog of the linear BGK equation (1.3), and we shall establish its hypocoercivity. Fixing unit temperature T, recall that as a consequence of the Central Limit Theorem, the measure $M_1(v)\mathrm{d}v$ is the (weak) limit of a sequence of discrete probability measures $\{\mu_n\}$ where

$$\mu_n := \sum_{j=0}^{n} 2^{-n}\binom{n}{j}\delta_{(2j-n)/\sqrt{n}} ,$$

where δ_y denotes the unit mass at $y \in \mathbb{R}$. Each of the probability measures $\mu_n, n \in \mathbb{N}$, has zero mean and unit variance.

The Hermite polynomials have a natural discrete analog, namely the *Krawtchouk polynomials*. A good reference containing proofs of all of the facts we use below is the survey [8]. (We are only concerned with a special family of the more general Krawtchouk polynomials discussed in [8], namely the $s = 2$ case in the terminology used there.) The standard Krawtchouk polynomials of order m are a set of $n + 1$ polynomials $K_{n,m}$; $m = 0, ..., n$ that are orthogonal with respect to the probability measure

$$\omega_n = \sum_{j=0}^{n} 2^{-n}\binom{n}{j}\delta_j ,$$

and are given by the following generating function:

$$(1+t)^{n-v}(1-t)^v = \sum_{m=0}^{n} t^m K_{n,m}(v) . \tag{4.9}$$

The leading coefficient of $K_{n,m}$ has the sign $(-1)^m$. One has the orthogonality relations

$$\int_{\mathbb{R}} K_{n,m}K_{n,\ell}\,\mathrm{d}\omega_n = \begin{cases} \binom{n}{m} & m = \ell , \\ 0 & m \neq \ell . \end{cases} \tag{4.10}$$

Then the *discrete Hermite polynomials* $H_{n,m}$ are defined by

$$H_{n,m}(v) := (-1)^m \binom{n}{m}^{-1/2} K_{n,m}\left(\frac{n}{2} + \frac{\sqrt{n}}{2}v\right) \quad \text{for} \quad m = 0, 1, \ldots, n;\ v \in \mathbb{R} . \tag{4.11}$$

Then $\{H_{n,0}, \ldots, H_{n,n}\}$ is the set of $n + 1$ polynomials that are orthogonal with respect to μ_n, and hence are an orthonormal basis for $L^2(\mathbb{R}; \mu_n)$, and for each m and v, $\lim_{n\to\infty} H_{n,m}(v) = \frac{1}{\sqrt{m!}}H_m(v)$. The analog of the crucial Hermite–recurrence relation (4.16) for the Krawtchouk polynomials is

$$(m+1)K_{n,m+1} = (n-2v)K_{n,m} - (n-m+1)K_{n,m-1} \, .$$

Rewriting this in terms of the discrete Hermite polynomials, one obtains

$$vH_{n,m}(v) = \sqrt{m+1}\left(\frac{n-m}{n}\right)^{1/2} H_{n,m+1}(v) + \sqrt{m}\left(\frac{n-m+1}{n}\right)^{1/2} H_{n,m-1}(v) \, . \tag{4.12}$$

Notice that this reduces to (4.16) in the limit $n \to \infty$ (up to the multiplication by the standard Gaussian).

We are now ready to produce a discrete velocity analog of (1.3) in continuous x-space. The phase space is $\mathbf{T}^1 \times [v_0, \dots, v_n]$ where the discrete velocity $v_j = (2j-n)/\sqrt{n}$. Our phase space density at time t is a vector $\mathbf{f}(x,t)$ with $n+1$ non-negative entries $f_0(x,t), \dots, f_n(x,t)$, such that

$$\sum_{j=0}^{n}\left(\int_{\mathbf{T}^1} f_j(x,t)\mathrm{d}x\right) = 1 \, .$$

We associate to $\mathbf{f}(x,t)$ the probability measure on the phase space given by

$$\sum_{j=0}^{n} f_j(x,t)\delta_{(2j-n)/\sqrt{n}} \, .$$

The discrete unit Maxwellian (of order n) is the vector $\mathbf{m} = 2^{-n}\left(\binom{n}{0}, \binom{n}{1}, \dots, \binom{n}{n}\right)^{\top}$. Then the order n discrete analog of (1.3) is the equation

$$\partial_t \mathbf{f}(x,t) + \mathbf{V}\partial_x \mathbf{f}(x,t) = \mathbf{m}\left(\sum_{j=0}^{n} f_j(x,t)\right) - \mathbf{f}(x,t) \, , \quad t \ge 0;\ x \in \mathbf{T}^1 \, , \tag{4.13}$$

with the $(n+1) \times (n+1)$ matrix $\mathbf{V} = \mathrm{diag}(v_0, \dots, v_n)$. Proceeding as for (4.2) yields the evolution equation for the spatial modes $\mathbf{f}_k(t)$, $k \in \mathbb{Z}$. Expanding $\mathbf{f}_k$ in the discrete Hermite basis $\{H_{n,m}(v_j);\ j = 0, \dots, n\}_{m=0,\dots,n}$, we obtain for each k the equation

$$\partial_t \hat{\mathbf{f}}_k + ik\mathbf{L}_1\hat{\mathbf{f}}_k = \mathbf{L}_2\hat{\mathbf{f}}_k \, , \quad t \ge 0;\ k \in \mathbb{Z} \, ,$$

where the vector $\hat{\mathbf{f}}_k(t) \in \mathbb{C}^{n+1}$ represents the basis coefficients of $\mathbf{f}_k(t)$. As before $\mathbf{L}_2$ is the $(n+1) \times (n+1)$ matrix $\mathbf{L}_2 = \mathrm{diag}(0, -1, -1, \dots)$, and $\mathbf{L}_1$ is the symmetric tridiagonal matrix whose diagonal entries are all zero, and whose superdiagonal sequence is given by

$$[\mathbf{L}_1]_{m,m+1} = \sqrt{m+1}\left(\frac{n-m}{n}\right)^{1/2} ; \qquad m = 0, 1, \dots, n-1 \, .$$

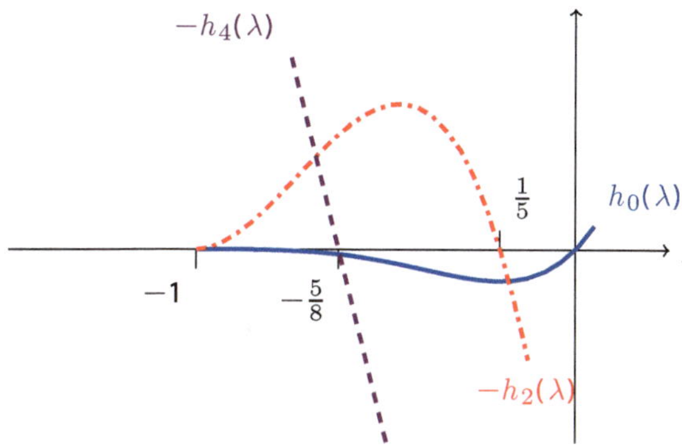

Fig. 1 Functions appearing in the eigenvalue equation of $-ik\mathbf{L}_1 + \mathbf{L}_2$; *solid blue curve*: $h_0(\lambda)$; *red dash-dotted curve*: $-h_2(\lambda)$; *purple dashed line*: $-h_4(\lambda)$

For example, with $n = 4$,

$$\mathbf{L}_1 = \begin{pmatrix} 0 & 1 & 0 & 0 & 0 \\ 1 & 0 & \sqrt{3/2} & 0 & 0 \\ 0 & \sqrt{3/2} & 0 & \sqrt{3/2} & 0 \\ 0 & 0 & \sqrt{3/2} & 0 & 1 \\ 0 & 0 & 0 & 1 & 0 \end{pmatrix}.$$

Next we discuss the time decay of the solution to (4.13) towards $\mathbf{f}^\infty = \mathbf{m}$. We shall focus on the example with order $n = 4$, but the other cases behave similarly. Computing for the modes $k = \pm 1$ the eigenvalues of $\mp i\mathbf{L}_1 + \mathbf{L}_2$ we find two complex pairs and one real eigenvalue $\lambda_0 = -0.526948302245121...$ which has the least negative real part, and hence determines the exponential decay rate of $\mathbf{f}_{\pm 1}(t)$. This situation for higher $|k|$ is similar, but even better, with faster decay. To see this we write the eigenvalue equation for the matrices $-ik\mathbf{L}_1 + \mathbf{L}_2,\ k \in \mathbb{Z}$ as

$$h_0(\lambda) := \lambda(\lambda+1)^4 = -k^2(\lambda+1)^2(5\lambda+1) - k^4(4\lambda + \frac{5}{2}) =: -k^2 h_2(\lambda) - k^4 h_4(\lambda).$$

The function h_0 is negative on $(-1, 0)$, $-h_2$ on $(-\frac{1}{5}, 0]$ and $-h_4$ on $(-\frac{5}{8}, 0]$ (cf. Fig. 1). For $k \neq 0$, the function $k^2 h_2(\lambda) + k^4 h_4(\lambda)$ has exactly one real zero, $\tilde{\lambda}(k)$, and it is nonnegative on $[\tilde{\lambda}(k), 0]$. For each fixed $k \in \mathbb{Z}$, the function $k^2 h_2 + k^4 h_4$ is strictly increasing w.r.t. λ. Hence, the above eigenvalue equation has exactly one real zero $\lambda_0(k)$, and it lies in $(-\frac{5}{8}, 0]$. For each fixed $\lambda \in [\tilde{\lambda}(k), 0]$, the function $k^2 h_2 + k^4 h_4$ is strictly increasing w.r.t. increasing $|k|$. Hence, $\lambda_0(k)$ decreases monotonically (w.r.t. $|k|$) towards $-\frac{5}{8}$.

This proves that the 5 velocity model is hypocoercive, at least in the norm E defined in (4.8) (with the transformation matrices $\mathbf{P}_k$ now corresponding to $-\mathbf{C}_k := \mp ik\mathbf{L}_1 + \mathbf{L}_2$). The sharp decay rate is given by $\lambda_0 = -0.526948302245121...$.

To establish a uniform-in-k spectral gap was already cumbersome for the case $n = 4$, and it becomes even more involved for larger n. In the following section we present a much simpler strategy, at the price of giving up sharpness of the decay rate. But more importantly, that strategy will also be applicable for the continuous velocity case, which is represented by a tridiagonal "infinite matrix".

4.3 A Continuous Velocity BGK Model

In this subsection we continue our discussion of the space-inhomogeneous BGK equation (1.3) or, equivalently, (4.2). This will yield the proof of Theorem 1.

Using the probabilists' Hermite polynomials,

$$H_m(v) := (-1)^m e^{\frac{v^2}{2}} \frac{\mathrm{d}^m}{\mathrm{d}v^m} e^{-\frac{v^2}{2}}, \quad m \in \mathbb{N}_0, \tag{4.14}$$

we define the normalized Hermite functions

$$g_m(v) := (2\pi m!)^{-1/2} H_m(v)\, e^{-\frac{v^2}{2}}, \quad \text{and} \quad \tilde{g}_m(v) := \frac{1}{\sqrt{T}} g_m\Big(\frac{v}{\sqrt{T}}\Big). \tag{4.15}$$

They satisfy

$$\int_{\mathbb{R}} \tilde{g}_m(v) \tilde{g}_n(v) M_T^{-1}(v)\, \mathrm{d}v = \delta_{mn}$$

and the recurrence relation

$$v\, \tilde{g}_m(v) = \sqrt{T}\big[\sqrt{m+1}\, \tilde{g}_{m+1}(v) + \sqrt{m}\, \tilde{g}_{m-1}(v)\big]. \tag{4.16}$$

In the basis $\{\tilde{g}_m\}_{m\in\mathbb{N}_0}$ Eq. (4.2) becomes

$$\partial_t \hat{\mathbf{f}}_k + ik\sqrt{T}\, \mathbf{L}_1 \hat{\mathbf{f}}_k = \mathbf{L}_2 \hat{\mathbf{f}}_k, \quad t \geq 0;\ k \in \mathbb{Z}. \tag{4.17}$$

Here, the "infinite vector" $\hat{\mathbf{f}}_k(t) \in l^2(\mathbb{N}_0)$ is the representation of the function $f_k(v,t) \in L^2(\mathbb{R}; M_T^{-1})$ in the Hermite function basis, and the operators $\mathbf{L}_1$, $\mathbf{L}_2$ are represented by "infinite matrices" as

$$\mathbf{L}_1 = \begin{pmatrix} 0 & \sqrt{1} & 0 & \cdots \\ \sqrt{1} & 0 & \sqrt{2} & 0 \\ 0 & \sqrt{2} & 0 & \sqrt{3} \\ \vdots & 0 & \sqrt{3} & \ddots \end{pmatrix}, \quad \mathbf{L}_2 = \operatorname{diag}(0, -1, -1, \ldots). \tag{4.18}$$

Next we shall prove the exponential decay of (4.17), using a modified strategy compared to Sect. 4.2. For the 5 velocity model there, it was possible (with some effort) to determine the sharp spectral gap of the matrices $-ik\mathbf{L}_1 + \mathbf{L}_2$, uniform in all modes k. But since this seems not (easily) possible for the infinite dimensional case in (4.17), we shall construct now approximate transformation matrices $\mathbf{P}_k$ that yield at least a (reasonable) lower bound on the spectral gap, and hence on the decay rate. For simplicity we set now $T = 1$, as the temperature could be "absorbed" into the parameter k by scaling.

Let $\mathbf{A}$ be an $(n+1) \times (n+1)$ tridiagonal matrix that is zero on the main diagonal. That is $A_{i,j} = 0$ unless $j = i+1$ or $i = j-1$. We further suppose that $\mathbf{A}$ is real and symmetric, so that $\mathbf{A}$ is characterized by the numbers $a_1, \ldots, a_n$ where $a_j = A_{j-1,j}$. Let $\mathbf{B} = \text{diag}(0, -1, \ldots, -1)$. Finally, for $k \in \mathbb{Z}$, consider the matrix $-\mathbf{C}_k := -ik\mathbf{A} + \mathbf{B}$.

In the simplest case $n = 1$ with $a = 1$, we obtain

$$\mathbf{A} = \begin{pmatrix} 0 & 1 \\ 1 & 0 \end{pmatrix}, \quad \mathbf{B} = \begin{pmatrix} 0 & 0 \\ 0 & -1 \end{pmatrix} \quad \text{and} \quad -\mathbf{C}_k = \begin{pmatrix} 0 & -ik \\ -ik & -1 \end{pmatrix}.$$

For this matrix $\mathbf{C}_k$, the transformation matrix $\mathbf{P}_k$ was already computed in (4.5) (with $\sigma = 1$). For $k \neq 0$ a simple computation yields

$$\mathbf{C}_k^*\mathbf{P}_k + \mathbf{P}_k\mathbf{C}_k = \mathbf{P}_k$$

so that with this choice of $\mathbf{P}_k$, Lyapunov's method yields exponential decay of the ODE-sequence $\frac{\mathrm{d}}{\mathrm{d}t}u_k = -\mathbf{C}_k u_k,\ k \in \mathbb{Z}$ at the optimal rate $e^{-t/2}$ (cf. Sect. 4.1).

We now turn to $n > 1$. For $k \neq 0$ define $\mathbf{P}_k$ to be the $(n+1) \times (n+1)$ matrix whose upper left 2×2 block is $\begin{pmatrix} 1 & -i\alpha/k \\ i\alpha/k & 1 \end{pmatrix}$, where $0 < \alpha < k$ is a parameter to be chosen below, and with the remaining diagonal entries being 1, and all other entries being 0. Then the eigenvalues of $\mathbf{P}_k$ are $(k+\alpha)/k$, 1 and $(k-\alpha)/k$, so that $\mathbf{P}_k$ is positive definite, and close to the identity for large k.

We take $-\mathbf{C}_k := -ik\mathbf{A} + \mathbf{B}$ as above. Then

$$\mathbf{C}_k^*\mathbf{P}_k + \mathbf{P}_k\mathbf{C}_k = -ik(\mathbf{A}\mathbf{P}_k - \mathbf{P}_k\mathbf{A}) - (\mathbf{B}\mathbf{P}_k + \mathbf{P}_k\mathbf{B}),$$

and its upper left 3×3 block reads

$$\begin{pmatrix} 2a_1\alpha & -i\alpha/k & a_2\alpha \\ i\alpha/k & 2 - 2a_1\alpha & 0 \\ a_2\alpha & 0 & 2 \end{pmatrix}. \tag{4.19}$$

The lower right $(n-2) \times (n-2)$ block is 2 times the identity, the off diagonal blocks are zero. In all of our finite dimensional approximations to (4.18) we have $a_1 = 1$. The value of a_2 is different for the different discrete velocity models, but to simplify matters, we only present calculations for $a_2 = \sqrt{2}$, which is the value for the limiting continuous velocity model.

The determinants of the upper left 2×2 and 3×3 blocks read, respectively,

$$\delta_2(\alpha, k) = \alpha\left(4 - \left(4 + \frac{1}{k^2}\right)\alpha\right) \quad \text{and} \quad \delta_3(\alpha, k) = 4\alpha\left((\alpha - 2)(\alpha - 1) - \frac{\alpha}{2k^2}\right).$$

For each k, $\delta_3(\alpha, k)/\alpha$ has two positive roots, and is negative between them. Hence our matrix is positive definite when α lies between zero and the smaller positive root of $\delta_3(\alpha, k)/\alpha$. This root is least when $k = 1$, when it has the value $\frac{7-\sqrt{17}}{4} \approx 0.719$. Hence, by Sylvester's criterion, $\mathbf{C}_k^*\mathbf{P}_k + \mathbf{P}_k\mathbf{C}_k$ is positive definite for all $k \neq 0$ if and only if $\alpha \in (0, \frac{7-\sqrt{17}}{4}) \approx (0, 0.719)$. Note that also $\delta_2(\alpha, k) > 0$ for these α, k.

When α is in this range, our 3×3 matrix (4.19) has three positive eigenvalues $\lambda_1, \lambda_2, \lambda_3$ which we may take to be arranged in increasing order. Then

$$\sqrt{\lambda_1} = \frac{\sqrt{\delta_3(\alpha, k)}}{\sqrt{\lambda_2\lambda_3}} \geq \frac{2\sqrt{\delta_3(\alpha, k)}}{\lambda_2 + \lambda_3} > \frac{\sqrt{\delta_3(\alpha, k)}}{2}$$

since the trace of our matrix is 4. Hence, the least eigenvalue λ_1 of our 3×3 matrix satisfies

$$\lambda_1 = \lambda_1(\alpha, k) \geq \frac{1}{4}\delta_3(\alpha, k).$$

Hence we choose $\alpha = \alpha_k$ to maximize $\delta_3(\alpha, k)$ between its first two roots. Its maximal value, $\delta_3(\alpha_k, k)$, depends on k, but it is easily seen to be least for $k = 1$ with $\alpha_1 = \frac{1}{3}$. Simple computations and estimates then yield $\lambda_1(\alpha_k, k) \geq \frac{1}{4}\delta_3(\alpha_1, 1) = 17/54$ for all k.

Since we always take $\alpha < 1$, the largest eigenvalue of the matrix $\mathbf{P}_k$ (defined with $\alpha = \alpha_k$) is no more than 2, uniformly in k. Hence

$$\mathbf{C}_k^*\mathbf{P}_k + \mathbf{P}_k\mathbf{C}_k \geq \frac{17}{54}\mathbf{I} \geq \frac{17}{108}\mathbf{P}_k \tag{4.20}$$

uniformly in k. Thus in each Fourier mode, we at least have exponential decay (of a quadratic type entropy) at the rate $17/108$ (by proceeding as in (4.6)).

Since this is also uniform in n, we obtain a bound for the continuous velocity model. Let the infinite matrix $\mathbf{P}_k$ be the positive matrix using the optimal value of α in the kth mode, and regarded as a bounded operator on $L^2(M_T^{-1})$ through its action on Hermite modes. Define the entropy function by

$$e(f) := \sum_{k\in\mathbb{Z}} \langle (f_k(v) - M_1(v)), \mathbf{P}_k(f_k(v) - M_1(v))\rangle_{L^2(M_1^{-1})}. \tag{4.21}$$

We obtain that, for solutions $f(t)$ of our BGK equation (1.3) or, equivalently, (4.2),

$$\frac{\mathrm{d}}{\mathrm{d}t}e(f(t)) \leq -\frac{17}{108}e(f(t)),$$

giving exponential relaxation.

The least eigenvalue of $\mathbf{P}_k$, $1 - \alpha/k$, is at least $1 - \frac{7-\sqrt{17}}{4} > \frac{1}{4}$ uniformly in k, and hence we have the inequality

$$e(f) \geq \frac{1}{4} \|f - M_1\|^2_{\mathscr{H}},$$

with $\mathscr{H} = L^2(\mathbf{T}^1 \times \mathbb{R}; M_T^{-1}(v)\,dv)$.

The above method to establish exponential decay is simple to apply but does not give the sharp decay rate (it is off by a factor of about 9, as indicated by numerical results). Hence we shall now sketch how to improve on it. The essence of the above method is to use an ansatz for the transformation matrix $\mathbf{P}_k$, namely to use for its upper left 2×2 block the matrix from the 2 velocity case. Using instead larger blocks, will most likely improve the decay rate.

As a second alternative we shall now present an improvement of the crucial matrix inequality (4.20), but we shall keep the same ansatz for the matrix $\mathbf{P}_k$: In the inequality

$$\mathbf{C}_k^* \mathbf{P}_k + \mathbf{P}_k \mathbf{C}_k - 2\mu \mathbf{P}_k \geq \mathbf{0} \tag{4.22}$$

we shall choose $\mu \in [0, 1]$ as large as possible (related to the matrix inequality (3.6)). The upper left 3×3 block of this matrix on the l.h.s. reads

$$\mathbf{D} := \begin{pmatrix} 2\alpha - 2\mu & -i\alpha(1-2\mu)/k & \sqrt{2}\alpha \\ i\alpha(1-2\mu)/k & 2 - 2\alpha - 2\mu & 0 \\ \sqrt{2}\alpha & 0 & 2 - 2\mu \end{pmatrix}.$$

We shall first derive strict inequalities on μ to obtain the positive definiteness of this matrix, using Sylvester's criterion. From $\mathbf{D}_{0,0}$ we deduce the first condition $0 \leq \mu < \alpha$. The determinant of the upper left 2×2 block reads

$$\delta_2(\mu; \alpha, k) = 4(\alpha - \mu)(1 - \alpha - \mu) - \frac{\alpha^2}{k^2}(1 - 2\mu)^2.$$

Since the last term increases with $|k|$, it suffices to consider δ_2 for $k = 1$. Next we want to establish the positivity of

$$\frac{\delta_2(\mu; \alpha, 1)}{4(1 - \alpha^2)} = \mu^2 - \mu + \alpha \frac{1 - 5\alpha/4}{1 - \alpha^2}.$$

The zero order term of this quadratic polynomial is positive on the relevant α–interval $(0, \frac{7-\sqrt{17}}{4}) \subset (0, \frac{4}{5})$, taking its maximum value $\frac{1}{4}$ at $\alpha = \frac{1}{2}$. For that limiting case, the r.h.s. reads $(\mu - \frac{1}{2})^2$, and for $0 < \alpha < \frac{1}{2}$, $\delta_2(\mu; \alpha, 1)$ always has a zero in the interval $(0, \frac{1}{2})$. This discussion yields the second condition $0 \leq \mu < \frac{1}{2}$, related to $\alpha < \frac{1}{2}$.

Next we consider the positivity of the determinant of the upper left 3×3 block, which reads

$$\delta_3(\mu; \alpha, k) = 8(1-\mu)(\alpha-\mu)(1-\alpha-\mu) - 4\alpha^2(1-\alpha-\mu) - 2\frac{\alpha^2}{k^2}(1-\mu)(1-2\mu)^2 .$$

For the same reason as before, we only have to consider the case $k = 1$. For the resulting cubic polynomial in μ we want to find its largest zero in the interval $[0, \frac{1}{2}]$ w.r.t. the parameter $\alpha \in [0, \frac{1}{2}]$. By numerical inspection we find that $\alpha_0 \approx 0.4684$ yields $\delta_3(\mu; \alpha_0, 1) \geq 0$ for $\mu \in [0, 0.273796...]$. This yields the third condition on μ and shows that the matrix inequality (4.22) holds with $\mu_0 := 0.273796...$, uniformly in $k \neq 0$. This somewhat more involved discussion shows that the decay rate can be improved to $2\mu_0 \approx 0.547592$. This finishes the proof of Theorem 1.

Remark To appreciate the above decay rate μ_0 (since $e(f)$ is a quadratic functional), we compare it to a numerical computation of the spectral gap of the "infinite matrices" $-ik\mathbf{L}_1 + \mathbf{L}_2$, $k \in \mathbb{Z}$ from (4.18). To this end we cut out the upper left $n \times n$ submatrix for large values of n. For increasing n the spectral gap approaches 0.6973. Hence our decay rate is off by only a factor of about 2.5. If one desired a closer bound, one could work with a $\mathbf{P}$ matrix with a larger block, say 3×3, in the upper left.

4.4 Linearized BGK Equation

Next we shall analyze here the linearized BGK equation (1.12) for the perturbation $h(x, v, t) = f(x, v, t) - M_T(v)$. We recall the definition of the normalized Hermite functions $\tilde{g}_m(v)$, $m \in \mathbb{N}_0$ from (4.15) and give explicit expressions for

$$\tilde{g}_0(v) = M_T(v) \quad \text{and} \quad \tilde{g}_2(v) = \frac{v^2 - T}{\sqrt{2}T} M_T(v) .$$

With this notation, (1.12) reads

$$h_t(x, v, t) + v h_x(x, v, t) = \left(\tilde{g}_0(v) - \frac{1}{\sqrt{2}}\tilde{g}_2(v)\right)\sigma(x, t) + \frac{1}{\sqrt{2}T}\tilde{g}_2(v)\tau(x, t) - h(x, v, t).$$

Fourier transforming in x, as in (4.1), each spatial mode $h_k(v, t)$ evolves as

$$\partial_t h_k + ikvh_k = \tilde{g}_0(v)\sigma_k(t) + \tilde{g}_2(v)\frac{1}{\sqrt{2}}\left(\frac{\tau_k(t)}{T} - \sigma_k(t)\right) - h_k , \quad k \in \mathbb{Z};\ t \geq 0 . \tag{4.23}$$

Here, σ_k and τ_k denote the spatial modes of the v–moments σ and τ defined in (1.8).

Next we expand $h_k(\cdot,t) \in L^2(\mathbb{R}; M_T^{-1})$ in the orthonormal basis $\{\tilde{g}_m(v)\}_{m\in\mathbb{N}_0}$:

$$h_k(v,t) = \sum_{m=0}^{\infty} \hat{h}_{k,m}(t)\, \tilde{g}_m(v)\,, \quad \text{with} \quad \hat{h}_{k,m} = \langle h_k(v), \tilde{g}_m(v)\rangle_{L^2(M_T^{-1})}\,,$$

and the "infinite vector" $\hat{\mathbf{h}}_k(t) = (\hat{h}_{k,0}(t),\, \hat{h}_{k,1}(t),\, \ldots)^\top \in \ell^2(\mathbb{N}_0)$ contains all Hermite coefficients of $h_k(\cdot,t)$, for each $k \in \mathbb{Z}$. In particular we have

$$\hat{h}_{k,0} = \int_{\mathbb{R}} h_k(v)\tilde{g}_0(v) M_T^{-1}(v)\,\mathrm{d}v = \sigma_k$$

and

$$\hat{h}_{k,2} = \int_{\mathbb{R}} h_k(v)\tilde{g}_2(v) M_T^{-1}(v)\,\mathrm{d}v = \frac{1}{\sqrt{2}}\left(\frac{\tau_k}{T} - \sigma_k\right)\,.$$

Hence, (4.23) can be written equivalently as

$$\partial_t h_k(v,t) + ikvh_k(v,t) = \tilde{g}_0(v)\hat{h}_{k,0}(t) + \tilde{g}_2(v)\hat{h}_{k,2}(t) - h_k(v,t)\,, \quad k \in \mathbb{Z};\ t \geq 0\,.$$

In analogy to (4.17), its Hermite coefficients satisfy

$$\partial_t \hat{\mathbf{h}}_k(t) + ik\sqrt{T}\,\mathbf{L}_1 \hat{\mathbf{h}}_k(t) = \mathbf{L}_3 \hat{\mathbf{h}}_k(t)\,, \quad k \in \mathbb{Z};\ t \geq 0\,, \tag{4.24}$$

where the operators $\mathbf{L}_1$, $\mathbf{L}_3$ are represented by "infinite matrices" on $\ell^2(\mathbb{N}_0)$ by

$$\mathbf{L}_1 = \begin{pmatrix} 0 & \sqrt{1} & 0 & \cdots \\ \sqrt{1} & 0 & \sqrt{2} & 0 \\ 0 & \sqrt{2} & 0 & \sqrt{3} \\ \vdots & 0 & \sqrt{3} & \ddots \end{pmatrix}, \quad \mathbf{L}_3 = \operatorname{diag}(0,\ -1,\ 0,\ -1,\ -1, \ldots)\,.$$

We remark that (4.24) simplifies for the spatial mode $k = 0$. One easily verifies that the flow of (1.12) preserves (1.9), i.e. $\sigma_0(t) = 0$, $\tau_0(t) = 0\ \forall t \geq 0$. Hence, (4.23) yields

$$\partial_t h_0(v,t) = -h_0(v,t)\,, \quad t \geq 0\,.$$

For $k \neq 0$, we note that the linearized BGK equation is very similar to the equation specified in (4.17) and (4.18): The only difference is that $\mathbf{L}_2$ is replaced by $\mathbf{L}_3$, which has one more zero on the diagonal. Our treatment of $ik\sqrt{T}\mathbf{L}_1 - \mathbf{L}_2$ in the previous section suggests the form of the positive matrix $\mathbf{P}_k$ that will provide our Lyapunov functional in this case. We obtained the matrix $\mathbf{P}_k$ in that case by replacing four entries around the location of the zero in $\mathbf{L}_2$ with the entries of $\begin{pmatrix} 1 & -i\alpha/k \\ i\alpha/k & 1 \end{pmatrix}$, the matrix that provides the optimal $\mathbf{P}_k$ for the two-velocity model. In the present case, we use two such matrices, one for each zero.

For parameters α and β to be chosen below, we define $\mathbf{P}_k$ to be the matrix that has

$$\begin{pmatrix} 1 & -i\alpha/k & 0 & 0 \\ i\alpha/k & 1 & 0 & 0 \\ 0 & 0 & 1 & -i\beta/2k \\ 0 & 0 & i\beta/2k & 1 \end{pmatrix} \tag{4.25}$$

as its upper-left 4×4 block, with all other entries being those of the identity. We define $-\mathbf{C}_k = -ik\mathbf{L}_1 + \mathbf{L}_3$, where, for the rest of this subsection, we use units in which $T = 1$.

Lemma 4 *Choosing $\alpha = \beta = 1/3$ in $\mathbf{P}_k$ uniformly in $|k| \in \mathbb{N}$, we have*

$$\mathbf{C}_k^*\mathbf{P}_k + \mathbf{P}_k\mathbf{C}_k \geq 2\mu\mathbf{P}_k \tag{4.26}$$

where

$$\mu = 0.0206\,. \tag{4.27}$$

Proof We compute that $\mathbf{C}_k^*\mathbf{P}_k + \mathbf{P}_k\mathbf{C}_k$ is twice the identity matrix whose upper left 5×5 block is replaced by

$$\mathbf{D}_{k,\alpha,\beta} = \begin{pmatrix} 2\alpha & -i\alpha/k & \sqrt{2}\alpha & 0 & 0 \\ i\alpha/k & 2-2\alpha & 0 & -\beta/\sqrt{2} & 0 \\ \sqrt{2}\alpha & 0 & \sqrt{3}\beta & -i\beta/2k & \beta \\ 0 & -\beta/\sqrt{2} & i\beta/2k & 2-\sqrt{3}\beta & 0 \\ 0 & 0 & \beta & 0 & 2 \end{pmatrix}.$$

We seek to choose α and β to make this matrix positive definite.

For $1 \leq j \leq 5$, let $\delta_j(k, \alpha, \beta)$ denote the determinant of the upper left $j \times j$ submatrix of $\mathbf{D}_{k,\alpha,\beta}$. For $\alpha = \beta$, the first and third column of $\mathbf{D}_{k,\alpha,\beta}$ have the common factor α. We then compute that

$$\delta_5(k, \alpha, \alpha) = \alpha^2 p_5(\alpha, k)\,,$$

where $p_5(\alpha, k)$ is a cubic polynomial in α with coefficients depending on k:

$$p_5(\alpha, k) = 16(\sqrt{3}-1) - \left[8\sqrt{3} + 16 + \frac{2+4\sqrt{3}}{k^2}\right]\alpha + \left[34 - 6\sqrt{3} + \frac{24k^2+1}{2k^4}\right]\alpha^2 - \left[4\sqrt{3} - 1 + \frac{\sqrt{3}}{k^2}\right]\alpha^3\,.$$

Next, we establish the bound

$$\begin{aligned} p_5(\alpha, k) &\geq p_5(\alpha, 1) \qquad (4.28)\\ &= 16(\sqrt{3}-1) - (12\sqrt{3}+18)\alpha + (46.5 - 6\sqrt{3})\alpha^2 - (5\sqrt{3}-1)\alpha^3 > 0 \end{aligned}$$

for $\alpha \in [0, \alpha_1]$ with $\alpha_1 \approx 0.555$ and $|k| \in \mathbb{N}$. To see the first inequality we consider

$$p_5(\alpha, k) - p_5(\alpha, 1) = \alpha(1 - \tfrac{1}{k^2})\varphi(\alpha, k)$$

with

$$\varphi(\alpha, k) := \sqrt{3}\alpha^2 - \left(\tfrac{1}{2}(1 + \tfrac{1}{k^2}) + 12\right)\alpha + 2 + 4\sqrt{3}.$$

It satisfies $\varphi(\alpha, 1) > 0$ for $\alpha \in [0, \alpha_2]$ with $\alpha_2 \approx 0.765$ and $\partial_k\varphi = \alpha/k^3$ for $\alpha \geq 0$ and $k \in \mathbb{N}$. The r.h.s. of (4.28) is easily seen to be monotone decreasing and evaluating it at $\alpha = 1/3$ and simplifying, we obtain $p_5(\alpha, k) \geq 2.5$ for $\alpha \in [0, 1/3]$. Finally, we then have

$$\delta_5(k, \alpha, \alpha) \geq 2.5\alpha^2$$

for $\alpha \in [0, 1/3]$ and all $k \neq 0$. A similar but simpler analysis shows that for $j = 1, 2, 3, 4$, $\delta_j(k, \alpha, \alpha) > 0$ for $\alpha \in [0, 1/3]$ and all $k \neq 0$.

Thus, we choose $\alpha = \beta = 1/3$ uniformly in k and this makes $\mathbf{D}_{k,\alpha,\beta}$ positive definite. Let $\{\lambda_1, \lambda_2, \lambda_3, \lambda_4, \lambda_5\}$ be the eigenvalues of $\mathbf{D}_{k,1/3,1/3}$ arranged in increasing order. We seek a lower bound on λ_1. Note that by the arithmetic-geometric mean inequality,

$$\begin{aligned} \lambda_1 = \frac{\delta_5(k, 1/3, 1/3)}{\lambda_2\lambda_3\lambda_4\lambda_5} &\geq \delta_5(k, 1/3, 1/3)\left(\frac{\lambda_2 + \lambda_3 + \lambda_4 + \lambda_5}{4}\right)^{-4}\\ &\geq 256\,\frac{\delta_5(k, 1/3, 1/3)}{(\mathrm{Tr}[\mathbf{D}_{k,1/3,1/3}])^4}\,. \end{aligned}$$

Since $\mathrm{Tr}[\mathbf{D}_{k,\alpha,\beta}] = 6$ independent of k, α and β, we finally obtain the bound $\lambda_1 \geq 0.0549$, and this means that, uniformly in $k \neq 0$,

$$\mathbf{C}_k^*\mathbf{P}_k + \mathbf{P}_k\mathbf{C}_k \geq 0.0549\,\mathbf{I}\,. \qquad (4.29)$$

A simple computation shows that the eigenvalues of $\mathbf{P}_k$ are $1, 1 \pm 1/6k$, and $1 \pm 1/3k$. Hence uniformly in k,

$$\frac{2}{3}\mathbf{I} \leq \mathbf{P}_k \leq \frac{4}{3}\mathbf{I}\,. \qquad (4.30)$$

Combining (4.30) with (4.29) yields the result. □

To deduce the first statement of Theorem 2 we consider a solution h of (1.12), and for $\gamma \geq 0$ the entropy functional $e_\gamma(f)$ defined by

$$e_\gamma(f) := \sum_{k\in\mathbb{Z}} (1+k^2)^\gamma \langle h_k(v), \mathbf{P}_k h_k(v)\rangle_{L^2(M_T^{-1})}, \tag{4.31}$$

with $f = M_1 + h$. Here the matrices $\mathbf{P}_0 = \mathbf{I}$ and $\mathbf{P}_k$ defined in (4.25) for $k \neq 0$ are regarded as bounded operators on $L^2(M_T^{-1})$. Then

$$\frac{\mathrm{d}}{\mathrm{d}t} e_\gamma(f) = -\sum_{k\in\mathbb{Z}} (1+k^2)^\gamma \langle h_k(v), (\mathbf{C}_k^*\mathbf{P}_k + \mathbf{P}_k\mathbf{C}_k) h_k(v)\rangle_{L^2(M_T^{-1})} \leq -0.0412\, e_\gamma(f), \tag{4.32}$$

which implies (1.14) and this finishes the proof of Theorem 2(a).

We note that the constant in (4.32) is within a factor of 18 of what numerical calculation shows is best possible. With more work, in particular not making the simplifying assumption $\alpha = \beta$ in the definition of $\mathbf{P}$, and also employing some of the ideas in the final part of Sect. 4.3, one can still better within this framework.

4.5 Local Asymptotic Stability for the BGK Equation

For $\gamma \geq 0$, let $H^\gamma(\mathbf{T}^1)$ be the Sobolev space consisting of the completion of smooth functions φ on $\mathbf{T}^1$ in the Hilbertian norm

$$\|\varphi\|_{H^\gamma}^2 := \sum_{k\in\mathbb{Z}} (1+k^2)^\gamma |\varphi_k|^2,$$

where φ_k is the kth Fourier coefficient of φ. Let $\mathscr{H}_\gamma$ denote the Hilbert space $H^\gamma(\mathbf{T}^1) \otimes L^2(\mathbb{R}; M_T^{-1}(v)\,\mathrm{d}v)$, where the inner product in $\mathscr{H}_\gamma$ is given by

$$\langle f, g\rangle_{\mathscr{H}_\gamma} = \int_{\mathbf{T}^1}\int_{\mathbb{R}} \overline{f}(x,v)\left[\left(1-\partial_x^2\right)^\gamma g(x,v)\right] M_T^{-1}(v)\mathrm{d}v\mathrm{d}\tilde{x},$$

where $\mathrm{d}\tilde{x}$ denotes the normalized Lebesgue measure on $\mathbf{T}^1$.

Then $\mathscr{H}_0$ is simply the weighted space $L^2(\mathbf{T}^1 \times \mathbb{R}; M_T^{-1}(v)\,\mathrm{d}v)$ and, for all $\gamma \geq 0$, $\mathbf{Q}$ is self-adjoint on $\mathscr{H}_\gamma$.

Let ρ, P, σ and τ be defined in terms of a density f as in (1.8). For all γ, $\|\sigma\|_{H^\gamma}^2 = \langle \sigma M_T, f - f^\infty\rangle_{\mathscr{H}_\gamma}$. Then by the Cauchy-Schwarz inequality,

$$\|\sigma\|_{H^\gamma}^2 \leq \|\sigma M_T\|_{\mathscr{H}_\gamma} \|f - f^\infty\|_{\mathscr{H}_\gamma} = \|\sigma\|_{H^\gamma} \|f - f^\infty\|_{\mathscr{H}_\gamma}. \tag{4.33}$$

Likewise, $\|\tau\|_{H^\gamma}^2 = \langle \tau v^2 M_T, f - f^\infty\rangle_{\mathscr{H}_\gamma}$, and by the Cauchy-Schwarz inequality,

$$\|\tau\|_{H^\gamma}^2 \le \|\tau v^2 M_T\|_{\mathcal{H}_\gamma}\|f - f^\infty\|_{\mathcal{H}_\gamma} = \sqrt{3}T\|\tau\|_{H^\gamma}\|f - f^\infty\|_{\mathcal{H}_\gamma} \,. \tag{4.34}$$

For $\gamma > 1/2$, functions in H^γ are Hölder continuous, and the H^γ norm controls their supremum norm. Combining this with the estimates proved above, we see that for all $\gamma > 1/2$, there is a finite constant C_γ such that the pressure and density satisfy

$$\|\sigma\|_\infty = \|\rho - 1\|_\infty \le C_\gamma\|f - f^\infty\|_{\mathcal{H}_\gamma} \quad \text{and} \quad \|\tau\|_\infty = \|P - T\|_\infty \le C_\gamma\|f - f^\infty\|_{\mathcal{H}_\gamma} \,. \tag{4.35}$$

Using these estimates it is a simple matter to control the approximation in (1.10). For $s \in [0, 1]$ and $(x, v) \in \mathbf{T}^1 \times \mathbb{R}$, define

$$F(s, x, v) := \frac{(1 + s\sigma(x))^{3/2}}{\sqrt{2\pi(T + s\tau(x))}} e^{-v^2(1+s\sigma(x))/2(T+s\tau(x))} \,,$$

so that the gain term in the linearized BGK equation (1.12) is $\partial_s F(0, x, v)$. In this notation,

$$\begin{aligned} &R_f(x, v) := \\ &M_f(x, v) - M_T(v) - \left[\left(\frac{3}{2} - \frac{v^2}{2T}\right) M_T(v)\sigma(x) + \left(-\frac{1}{2T} + \frac{v^2}{2T^2}\right) M_T(v)\tau(x)\right] \\ &= \int_0^1 [\partial_s F(s, x, v) - \partial_s F(0, x, v)]\, \mathrm{d}s = \int_0^1 \int_0^s \left[\partial_s^2 F(r, x, v)\right] \mathrm{d}r\mathrm{d}s \,. \end{aligned}$$

We compute

$$\begin{aligned} &\partial_s^2 F(s, x, v) \\ &= \frac{\tau - T\sigma}{(1 + s\sigma)^2}\left[-\frac{3\sigma}{4\theta_s} + \left(\frac{3}{2}v^2\sigma + \frac{3}{4}\tau\right)\frac{1}{\theta_s^2} - \left(\frac{1}{4}v^4\sigma + \frac{3}{2}v^2\tau\right)\frac{1}{\theta_s^3} + \frac{v^4\tau}{4\theta_s^4}\right] M_{\theta_s}(v) \end{aligned} \tag{4.36}$$

with the notations $\theta_s := \frac{T+s\tau}{1+s\sigma}$ and $M_{\theta_s}(v) := \frac{1}{\sqrt{2\pi\theta_s}} e^{-v^2/(2\theta_s)}$. Note that the r.h.s. of (4.36) is of the order $\mathcal{O}(\sigma^2 + \tau^2)$, which will be related to $\mathcal{O}((f - f^\infty)^2)$ due to the estimates (4.33)–(4.34).

Simple but cumbersome calculations now show that if $\gamma > 1/2$ and $\|f - f^\infty\|_{\mathcal{H}_\gamma}$ is sufficiently small, then there exists a finite constant $\tilde{C}_{\gamma,T}$ depending only on γ and T such that for all $s \in [0, 1]$,

$$\left\|\partial_s^2 F(s, x, v)\right\|_{\mathcal{H}_\gamma} \le \tilde{C}_{\gamma,T}\|f - f^\infty\|_{\mathcal{H}_\gamma}^2 \,, \tag{4.37}$$

and hence

$$\|R_f\|_{\mathcal{H}_\gamma} \le \tilde{C}_{\gamma,T}\|f - f^\infty\|_{\mathcal{H}_\gamma}^2 \,. \tag{4.38}$$

[The calculations are simplest for non-negative integer γ, in which case the Sobolev norms can be calculated by differentiation. For $\gamma > 1/2$ and sufficiently small $\|f - f^\infty\|_{\mathcal{H}_\gamma}$, the estimates (4.35) ensure for all $s \in [0, 1]$ the boundedness of $0 < \epsilon < \|1 + s\sigma\|_\infty$, $\|T + s\tau\|_\infty < \infty$ for some fixed $\epsilon > 0$ and the $L^2(\mathbb{R}; M_T^{-1}(v)\,dv)$-integrability of

$$e^{-v^2(1+s\sigma(x))/2(T+s\tau(x))} \le e^{-v^2/3T} \quad \text{for all } x\,.$$

In (4.37), higher powers of $\|f - f^\infty\|_{\mathcal{H}_\gamma}$ (arising due to derivatives of σ and τ) can be absorbed into the constant of the quadratic term.]

Now let f be a solution of the BGK equation (1.6) with constant temperature $T = 1$ and define $h(x, v, t) := f(x, v, t) - M_T(v)$ as in the introduction. Now define the linearized BGK operator

$$\mathbf{Q}_2 h(x, v, t) := \left(\frac{3}{2} - \frac{v^2}{2T}\right) M_T(v)\sigma(x) + \left(-\frac{1}{2T} + \frac{v^2}{2T^2}\right) M_T(v)\tau(x) - h(x, v, t) \tag{4.39}$$

where of course σ and τ are determined by f, and hence h. Then the nonlinear BGK equation (1.6) becomes

$$h_t(x, v, t) + v\, h_x(x, v, t) = \mathbf{Q}_2 h(x, v, t) + R_f(x, v, t)\,, \qquad t \ge 0\,, \tag{4.40}$$

which deviates from the linearized BGK equation (1.12) only by the additional term R_f.

It is now a simple matter to prove local asymptotic stability. We shall use here exactly the entropy functional $e_\gamma(f)$ defined in (4.31) with $f = M_1 + h$. Now assume that h solves (4.40). To compute $\frac{d}{dt} e_\gamma(f)$ we use the inequality (4.32) for the drift term and for $\mathbf{Q}_2 h$ in (4.40), as well as $\|\mathbf{P}_k\| \le \frac{4}{3}$ and (4.38) for the term R_f. This yields

$$\frac{d}{dt} e_\gamma(f) \le -0.0412\, e_\gamma(f) + \frac{8}{3}\tilde{C}_{\gamma,T} \|h\|^3_{\mathcal{H}_\gamma}\,, \tag{4.41}$$

(if $\|h\|_{\mathcal{H}_\gamma}$ is small enough) where we have used the fact that $h = f - f^\infty$. Then since

$$\frac{2}{3} e_\gamma(f) \le \|h\|^2_{\mathcal{H}_\gamma} \le \frac{4}{3} e_\gamma(f)\,,$$

which is simply a restatement of (4.30), it is now simple to complete the proof of Theorem 2(b): Estimate (4.41) shows that there is a $\delta_\gamma > 0$ so that if the initial data $f^I(x, v)$ satisfies $\|f^I - f^\infty\|_{\mathcal{H}_\gamma} < \delta_\gamma$, then the solution $f(t)$ satisfies

$$e_\gamma(f(t)) \le e^{-t/25} e_\gamma(f^I)\,.$$

Here we used that the linear decay rate in (4.41) is slightly better than $\frac{1}{25}$, to compensate the nonlinear term.

We expect that the strategy from this section can be adapted also to nonlinear kinetic Fokker-Planck equations; this will be the topic of a subsequent work.

Acknowledgments The second author (AA) was supported by the FWF-doctoral school "Dissipation and dispersion in non-linear partial differential equations". The third author (EC) was partially supported by U.S. N.S.F. grant DMS 1501007.

References

1. Achleitner, F., Arnold, A., Stürzer, D.: Large-time behavior in non-symmetric Fokker-Planck equations. Rivista di Matematica della Università di Parma **6**, 1–68 (2015)
2. Arnold, A., Erb, J.: Sharp entropy decay for hypocoercive and non-symmetric Fokker-Planck equations with linear drift (2014). arXiv:1409.5425
3. Arnold, A., Markowich, P., Toscani, G., Unterreiter, A.: On convex Sobolev inequalities and the rate of convergence to equilibrium for Fokker-Planck type equations. Commun. Partial. Differ. Equ. **26**(1–2), 43–100 (2001)
4. Bergmann, P., Lebowitz, J.L.: A new approach to nonequilibrium processes. Phys. Rev. **99**(2), 578–587 (1955)
5. Bhatnagar, P.L., Gross, E.P., Krook, M.: A model for collision processes in gases. I. Small amplitude processes in charged and neutral one-component systems. Phys. Rev. **94**(3), 511–525 (1954)
6. Bobkov, S.G., Tetali, P.: Modified logarithmic Sobolev inequalities in discrete settings. J. Theor. Probab. **19**(2), 289–336 (2006)
7. Bosi, R., Cáceres, M.J.: The BGK model with external confining potential: existence, long-time behaviour and time-periodic Maxwellian equilibria. J. Stat. Phys. **136**(2), 297–330 (2009)
8. Coleman, R.: On Krawtchouk polynomials (2011). arXiv:1101.1798
9. Diaconis, P., Saloff-Coste, L.: Logarithmic Sobolev inequalities for finite Markov chains. Ann. Appl. Probab. **6**(3), 695–750 (1996)
10. Dolbeault, J., Mouhot, C., Schmeiser, C.: Hypocoercivity for linear kinetic equations conserving mass. Trans. Am. Math. Soc. **367**(6), 3807–3828 (2015)
11. Dolbeault, J., Mouhot, C., Schmeiser, C.: Hypocoercivity for kinetic equations with linear relaxation terms. C. R. Math. Acad. Sci. Paris **347**(9–10), 511–516 (2009)
12. Hérau, F.: Hypocoercivity and exponential time decay for the linear inhomogeneous relaxation Boltzmann equation. Asymptot. Anal. **46**(3–4), 349–359 (2006)
13. Kac, M.: Foundations of kinetic theory. In: Neyman, J. (ed.) Proceedings of 3rd Berkeley Symposium on Mathematical Statistics and Probability, University of California, vol 3, pp. 171–197 (1956)
14. Michel, A.N., Hou, L., Liu, D.: Stability of Dynamical Systems, 2nd edn. Birkhäuser/Springer (2015)
15. Norris, J.R.: Markov Chains. Cambridge University Press, Cambridge (1997)
16. Perthame, B.: Global existence to the BGK model of Boltzmann equation. J. Diff. Equ. **82**(1), 191–205 (1989)
17. Perthame, B.: Transport Equations in Biology. Birkhäuser, Basel (2007)
18. Perthame, B., Pulvirenti, M.: Weighted L^∞ bounds and uniqueness for the Boltzmann BGK model. Arch. Rational Mech. Anal. **125**(3), 289–295 (1993)
19. Seneta, E.: Nonnegative matrices and Markov chains. Springer Series in Statistics, 2nd edn. Springer-Verlag, New York (1981)
20. van Kampen, N.G.: Stochastic Processes in Physics and Chemistry, 3rd edn. Elsevier (2007)
21. Villani, C.: Hypocoercivity. Mem. Amer. Math. Soc. **202**(950) (2009)

Hydrodynamic Limit of Quantum Random Walks

Alexandre Baraviera, Tertuliano Franco and Adriana Neumann

Abstract We discuss here the hydrodynamic limit of independent quantum random walks evolving on $\mathbb{Z}$. As main result, we obtain that the time evolution of the local equilibrium is governed by the convolution of the chosen initial profile with a rescaled version of the limiting probability density obtained in the law of large numbers for a single quantum random walk.

Keywords Quantum random walk · Hydrodynamic limit · Local equilibrium

1 Introduction

An important subject in Statistical Physics is the comprehension of the hydrodynamic behavior of interacting particle systems. Roughly speaking, given a discrete system that evolves in time, its hydrodynamic limit consists in the limit for the time trajectory of the spatial density of particles (as some parameters are rescaled, in general, space and time). Proving rigorously such scaling limit is often a mathematical problem of deep technical difficulty. As a guide book on the subject we cite [1] and references therein.

Since the seventies, the hydrodynamic limit has been developed and successfully proved for many interacting particle systems, for instance the symmetric (and the asymmetric) simple exclusion process, the zero range process, independent random walks, and many others. In particular, the hydrodynamic limit of independent copies of a stochastic process is quite well understood, as one can see for instance in [2, 3].

A. Baraviera · A. Neumann
Universidade Federal do Rio Grande do Sul, Porto Alegre, Brazil
e-mail: baravi@mat.ufrgs.br

A. Neumann
e-mail: aneumann@impa.br

T. Franco (✉)
Universidade Federal da Bahia, Salvador, Brazil
e-mail: tertu@ufba.br

© Springer International Publishing Switzerland 2016

P. Gonçalves and A.J. Soares (eds.), *From Particle Systems to Partial Differential Equations III*, Springer Proceedings in Mathematics & Statistics 162,
DOI 10.1007/978-3-319-32144-8_2

We devote this paper to the study of a particular case of the hydrodynamic limit of independent copies of a stochastic process. The stochastic process we are concerned with is the *Quantum Random Walk* (QRW), as proposed in [4]. Such paper gave origin to a vast literature, inspiring several connections with quantum optics and quantum computation, see for instance the excellent survey [5] on quantum random walks.

In the recent paper [6], it was proved a law of large numbers for the QRW. Differently from the classical random walk (see [7] for a definition), the limit for the QRW, in the ballistic scale, is not a deterministic number, but a probability distribution. This is in some sense a consequence of the fact that the quantum random walk evolves faster than its classical version. In average, after the same number of steps the distance from the starting point of a quantum walk is larger than its classical counterpart.

Here, we present a proof of the hydrodynamic limit for a system of independent copies of the QRW. The hydrodynamic limit of independent copies of a stochastic process is not at all a novelty in the literature, see [3]. Nevertheless, we present these notes with the aim of introducing the QRW subject in a simple way and to make some observations on the peculiar hydrodynamic behavior for independent QRWs.

It is supposed that each QRW starts from a localized state, and the number of independent copies of the QRW starting at each state is determined by independent Poisson random variables. The parameter of each Poisson random variable is a function of space and is called the *slowly varying parameter* driven by a smooth profile γ of compact support. Under these assumptions, we prove that the limiting profile is driven by a convolution of the initial profile γ with the probability density obtained in [6] from the law of large numbers for a single quantum random walk.

It is worth to mention that, if the initial profile has compact support, then the limiting profile at any positive time will also have compact support. This contrasts with the hydrodynamic limit for classical symmetric random walks, where the limiting profile evolves according to the heat equation. For the heat equation, it is well known that the diffusion has infinite speed of propagation. That is, even for initial profiles with compact support, for any positive time, the solution will be non-zero everywhere. Hence, we roughly deduce that: while the QRW is faster than its classical counterpart (in the scaling aspect), a system of independent QRW's is slower than a system of independent classical random walks (in the macroscopic diffusion aspect).

The outline of the paper is the following: in Sect. 2, we define the state space of a single QRW. In Sect. 3, we explain the dynamics of a QRW. In Sect. 4 we state the hydrodynamic limit. In Sect. 5 we state and prove the *local equilibrium*, which in its hand implies the hydrodynamic limit.

2 The State Space of the QRW

We define in this section the state space of a single QRW, which, in agreement with the postulates of the Quantum Mechanics, is a Hilbert space. Its meaning is discussed below in detail.

Definition 1 If $f_1 : H_1 \to \mathbb{C}$ and $f_2 : H_2 \to \mathbb{C}$ are two linear functionals over some vector spaces H_1 and H_2, the tensor product of f_1 and f_2 is the bilinear functional $f_1 \otimes f_2 : H_1 \times H_2 \to \mathbb{C}$ defined by

$$(f_1 \otimes f_2)(y_1, y_2) := f_1(y_1) \cdot f_2(y_2).$$

Notice that the tensor product is bilinear, whilst its Cartesian product is linear. By the Riesz Representation Theorem, a Hilbert space can be understood as a space of linear functionals. Then, it makes sense to speak on the tensor product of two Hilbert spaces.

Definition 2 We denote the QRW state space by Ω, which is defined as the tensor product of the Hilbert spaces $\mathcal{H}_P$ and $\mathcal{H}_C$:

$$\Omega := \mathcal{H}_P \otimes \mathcal{H}_C,$$

where $\mathcal{H}_P$ is taken as the Hilbert space of square summable complex double-sided sequences:

$$\mathcal{H}_P = \ell^2(\mathbb{Z}) := \left\{ (\ldots, x_{-2}, x_{-1}, x_0, x_1, x_2 \ldots) \, ; \, \sum_{k \in \mathbb{Z}} |x_k|^2 < \infty \right\},$$

being, $\forall k \in \mathbb{Z}$, $x_k \in \mathbb{C}$, and $\mathcal{H}_C = \mathbb{C}^2$.

The nomenclature $\mathcal{H}_P$ and $\mathcal{H}_C$, somewhat common in the literature, comes from the idea that $\mathcal{H}_P$ is the Hilbert space associated with the *position* of the quantum object and $\mathcal{H}_C$ is the Hilbert space associated with the state of a certain *coin*. In the case presented here, the simplest one, $\mathcal{H}_P$ is $\ell^2(\mathbb{Z})$ and $\mathcal{H}_C$ is $\mathbb{C}^2$.

From now on, elements of $\ell^2(\mathbb{Z})$ will be denoted by $|x\rangle$. To facilitate calculations, let $\{|e_k\rangle\}_{k \in \mathbb{Z}}$ be the canonical basis of $\ell^2(\mathbb{Z})$. Thus, if $x \in \ell^2(\mathbb{Z})$, then

$$|x\rangle = \sum_{k \in \mathbb{Z}} x_k \, |e_k\rangle.$$

According to the common notation in Quantum Mechanics, the canonical basis of $\mathbb{C}^2$ is denoted by $\{|+1\rangle, |-1\rangle\}$. Any element of the Hilbert space $\mathbb{C}^2$ is usually called a *qubit*. The qubit can be interpreted as the state of a coin (or spin).

Now, let us discuss the physical interpretation of the state space Ω. Suppose that the state of the quantum object at some time is

$$\psi := \sum_{k\in\mathbb{Z}} x_k\,|e_k\rangle \otimes |{+1}\rangle + \sum_{k\in\mathbb{Z}} y_k\,|e_k\rangle \otimes |{-1}\rangle.$$

It is a common sense in Quantum Mechanics that the particle is not at any particular place before an observation. Only after the observation, and thus after the consequently random result, one can say that the particle is at some place (more sophisticate physical interpretations are available but here we state only this pragmatic point of view).

If we perform a measurement to observe position/coin's value of the object, the outcome will be random, moreover localized, i.e., of the form $|e_k\rangle \otimes |\pm 1\rangle$, with probability proportional to the modulus' square of the respective component.

For instance, considering the state ψ above, the probability of observing the state $|e_k\rangle \otimes |+1\rangle$ as outcome (respectively $|e_k\rangle \otimes |-1\rangle$) will have probability proportional to $|x_k|^2$ (respectively $|y_k|^2$). If we perform an experiment to observe only the position, with probability proportional to $|x_k|^2 + |y_k|^2$, the outcome will be $|e_k\rangle$.

Analogously, if we perform a measurement to observe only the coin's value, the outcome will be $|{+1}\rangle$ with probability proportional to $\sum_{k\in\mathbb{Z}} |x_k|^2$ and the outcome will be $|{-1}\rangle$ with probability proportional to $\sum_{k\in\mathbb{Z}} |y_k|^2$.

3 The Dynamics of a Single QRW

The dynamics of the QRW is a function $U : \Omega \to \Omega$ which will be defined ahead, composed of two parts. Informally, the first part consists on an unitary operator[1] that acts on the coin. The second part is a translation to the right or to the left on the element of $\ell^2(\mathbb{Z})$, depending if the respective coin qubit is $|{+1}\rangle$ or $|{-1}\rangle$.

We recall the notation $U_2(\mathbb{C})$ for the set of unitary operators, that is, the set of linear operators on $\mathbb{C}^2$ preserving the canonical L^2-norm. In this work, we treat only the particular operator $H \in U_2(\mathbb{C})$, the *Hadamard* operator, whose matrix is given in the canonical basis by

$$H := \frac{1}{\sqrt{2}} \begin{bmatrix} 1 & 1 \\ 1 & -1 \end{bmatrix}, \tag{1}$$

whose effect emulates the evolution of an unbiased coin. For example, if the initial coin state is $|{-1}\rangle$, after the action of H we get $\frac{1}{\sqrt{2}}(|{+1}\rangle - |{-1}\rangle)$. In this final state we have one half of probability for finding one of the two possible coin states after a measurement.

[1]Unitary matrix: its columns (or lines) compound an orthonormal basis for the space.

We define now the part of the dynamics acting on the space. Let $\tau_m : \ell^2(\mathbb{Z}) \to \ell^2(\mathbb{Z})$ be the shift to the right of size $m \in \mathbb{Z}$, i.e., if $|x\rangle = \sum_{k\in\mathbb{Z}} x_k|e_k\rangle$, then

$$\tau_m|x\rangle := \sum_{k\in\mathbb{Z}} x_k|e_{k+m}\rangle .$$

The linear operator $S : \ell^2(\mathbb{Z}) \otimes \mathbb{C}^2 \to \ell^2(\mathbb{Z}) \otimes \mathbb{C}^2$ is defined by

$$S\Big(|x\rangle \otimes |{+1}\rangle\Big) := \tau_1|x\rangle \otimes |{+1}\rangle , \quad \forall x \in \ell^2(\mathbb{Z}) ,$$

and

$$S\Big(|x\rangle \otimes |{-1}\rangle\Big) := \tau_{-1}|x\rangle \otimes |{-1}\rangle , \quad \forall x \in \ell^2(\mathbb{Z}) .$$

Finally, denote by Id the identity operator over $\ell^2(\mathbb{Z})$ and define $U : \ell^2(\mathbb{Z}) \otimes \mathbb{C}^2 \to \ell^2(\mathbb{Z}) \otimes \mathbb{C}^2$ by the composition

$$U := S \circ (\,\mathrm{Id} \otimes H) .$$

The dynamics is defined as follows: if at time zero the state is some $\psi \in \ell^2(\mathbb{Z}) \otimes \mathbb{C}^2$, then the state at time $n = 1$ is given by $U\psi$, and at an arbitrary time $n \in \mathbb{N}$ is given by $U^n\psi$. As an example, if $\psi = |x\rangle \otimes |{+1}\rangle + |y\rangle \otimes |{-1}\rangle$, then

$$\begin{aligned} U\psi &= S\Big(|x\rangle \otimes H(|{+1}\rangle) + |y\rangle \otimes H(|{-1}\rangle)\Big) \\ &= S\Big(|x\rangle \otimes \Big(\tfrac{|+1\rangle+|-1\rangle}{\sqrt{2}}\Big) + |y\rangle \otimes \Big(\tfrac{|+1\rangle-|-1\rangle}{\sqrt{2}}\Big)\Big) \\ &= \frac{1}{\sqrt{2}}\Big[\Big(\tau_1|x\rangle + \tau_1|y\rangle\Big) \otimes |{+1}\rangle + \Big(\tau_{-1}|x\rangle - \tau_{-1}|y\rangle\Big) \otimes |{-1}\rangle\Big)\Big] . \end{aligned} \tag{2}$$

In general, for given $\psi \in \Omega = \ell^2(\mathbb{Z}) \otimes \mathbb{C}^2$ we will denote

$$U^n\psi = \Big(\sum_{k\in\mathbb{Z}} x_{nk}\,|e_k\rangle\Big) \otimes |{+1}\rangle + \Big(\sum_{k\in\mathbb{Z}} y_{nk}\,|e_k\rangle\Big) \otimes |{-1}\rangle , \tag{3}$$

always keeping in mind that the complex numbers x_{nk}, y_{nk} depend on ψ. Notice that the dynamics $(\psi, U\psi, U^2\psi, U^3\psi, \ldots)$ is deterministic.

Definition 3 Given $\psi \in \Omega$ with unitary norm, denote by X_n^ψ a random variable (on some probability space) assuming integers values with distribution given by

$$\mathbb{P}\Big(X_n^\psi = k\Big) = |x_{nk}|^2 + |y_{nk}|^2 , \qquad \forall k \in \mathbb{Z} ,$$

where $x_{nk}, y_{nk} \in \mathbb{C}$ are defined in (3).

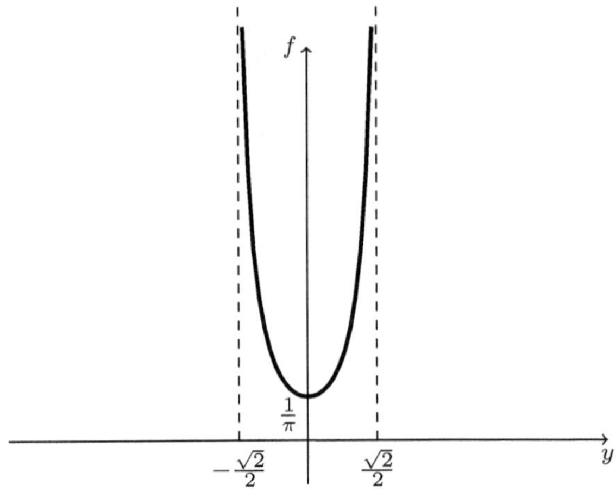

Fig. 1 Illustration of the density f obtained in the central limit theorem for the QRW, as stated in the Theorem 1

Given an initial state $\psi \in \Omega$, the state $U^n\psi$ obtained after n iterations of U gives all the information about the distribution of the position/coin of the particle at time n. Moreover, after an observation at time n of the position of the particle, the outcome of position is a random variable with the distribution presented in Definition 3 (Fig. 1).

Now we point out two remarks. First, although the Hilbert spaces are complex, if we multiply the initial state by a complex number, there is no change in the particle space distribution at final time. That is, for any $\zeta \in \mathbb{C}$ of unitary modulus, the distributions of position at time $n \in \mathbb{N}$, obtained from $U^n\psi$ and from $U^n(\zeta\psi)$ are the same. The role of complex numbers is noted in Quantum Mechanics when considering sums of states (giving rise to the phenomena known as *interference*). Second, the signs appearing in (2) generate cancellations and a very peculiar behavior[2] of the QRW, as one can see in the result below extracted from [6]:

Theorem 1 (Grimmett/Janson/Scudo'03) *For any $\psi \in \Omega$ which is a finite sum of localized states,*

$$\frac{X_n^\psi}{n} \overset{n\to\infty}{\longrightarrow} Y, \quad \textit{in distribution,}$$

where Y is a real random variable of density

$$f(y) = \begin{cases} \dfrac{1}{\pi(1-y^2)\sqrt{1-2y^2}}, & \textit{if } y \in [-\frac{\sqrt{2}}{2}, \frac{\sqrt{2}}{2}], \\ 0, & \textit{otherwise.} \end{cases} \tag{4}$$

[2]In comparison with the classical random walk.

4 Hydrodynamic Limit for a System of Independent Quantum Random Walks

We turn now our attention to a system of independent QRW's. Fix once and for all a continuous non-negative function $\gamma : \mathbb{R} \to \mathbb{R}_+$ with compact support.

Definition 4 Let μ_n be a product measure on $\mathbb{N}^{\mathbb{Z}}$ whose marginal at site $k \in \mathbb{Z}$ is a Poisson $(\gamma(\frac{k}{n}))$, i.e.,

$$\mu_n\{\eta \in \mathbb{N}^{\mathbb{Z}}\ ;\ \eta(k) = j\} \;=\; \frac{e^{-\lambda}\lambda^j}{j!}, \tag{5}$$

being $\lambda = \gamma(\frac{k}{n})$.

In hydrodynamic limit, this is usually called a product measure with *slowly varying parameter*, see [1]. Let $X_n^k(1), X_n^k(2), X_n^k(3), \ldots$ be independent copies of the random variable X_n^{ψ} given in Definition 3 choosing

$$\psi \;=\; |e_k\rangle \otimes |+1\rangle. \tag{6}$$

Remark 1 We consider initial states of the form (6) only for sake of clarity. For finite sums of localized states, all results remains in force (properly redefining the Poisson product measures above).

Denote by $\mathbb{P}$ and $\mathbb{E}$ the probability and the expectation, respectively, induced by μ_n and the random variables defined above. When considering a single random variable $X_n^k(j)$, we will write only P and E. Let $\mathbb{1}_A(\omega)$ be the function which is to one if $\omega \in A$ and zero otherwise.

Definition 5 For each $x \in \mathbb{Z}$, define the random variable

$$\xi_n(\ell) \;=\; \sum_{k\in\mathbb{Z}} \sum_{j=1}^{\eta(k)} \mathbb{1}_{[X_n^k(j)=\ell]}\,. \tag{7}$$

where $\eta \in \mathbb{N}^{\mathbb{Z}}$ is chosen according to μ_n, independently of all the random variables $X_n^k(j)$.

Intuitively, $\xi_n(x)$ is obtained by the following procedure: first we choose how many QRW's start at each localized state $|e_k\rangle \otimes |+1\rangle$ via the measure μ_n. Then we evolve each QRW n steps. After that, we observe where each QRW is. As explained before, the outcome is random, given by some $X_n^k(j)$. Looking at (7) we notice that the random variable $\xi_n(\ell)$ counts how many of those random variables gave as result the site $\ell \in \mathbb{Z}$.

Now, we are in position to the state our main result. We denote by $C_c(\mathbb{R})$ the set of continuous functions $H : \mathbb{R} \to \mathbb{R}$ with compact support. To not overload the notation, we will write $\lfloor tn \rfloor$, the integer part of tn, only by tn.

Theorem 2 (Hydrodynamic limit of QWR's) *For all $t > 0$ and for all $H \in C_c(\mathbb{R})$,*

$$\lim_{n\to\infty} \frac{1}{n} \sum_{\ell\in\mathbb{Z}} H(\tfrac{\ell}{n})\, \xi_{tn}(\ell) \;=\; \int_{\mathbb{R}} H(x)\rho(t,x)dx$$

in probability, where the function $\rho : \mathbb{R}_+ \times \mathbb{R} \to \mathbb{R}_+$ is given by

$$\rho(t,x) \;=\; (\gamma * f_t)(x) \;:=\; \int_{\mathbb{R}} \gamma(y)\, \tfrac{1}{t} f(\tfrac{x-y}{t})\, dy\,, \tag{8}$$

where f is the function defined in (4) *and $f_t(x) := \frac{1}{t} f(\frac{x}{t})$.*

We notice that the time $tn = \lfloor tn \rfloor$ appearing in the previous statement corresponds to the so-called *ballistic* scaling. For models where the time scaling is $\lfloor tn^2 \rfloor$, it is called the *diffusive* scaling. The limit for a system of independent quantum random walks occurs in the ballistic scaling, while for unbiased classical random walks it occurs in the diffusive scale. This is an intrinsic characteristic of quantum random walks: because of the aforementioned cancellations, they move faster than the classical random walks.

On the other hand, the time evolution of the initial profile γ according to $\gamma * f_t$ is somewhat slower than the equivalent time evolution obtained in the case of classical independent random walks, where the initial profile γ evolves through the heat equation's semigroup. For any positive time, the solution of heat equation starting from γ is positive everywhere,[3] but this does not happen with $\gamma * f_t$. Since f and γ have compact support, for any time $t > 0$, the function $\gamma * f_t$ has compact support as well, hence it is not positive everywhere. This means that the diffusion of mass through $\gamma * f_t$ has finite speed of propagation,[4] differently from the diffusion given by the heat equation.

5 Local Equilibrium

In this section, we prove a result usually called in the literature as the *conservation of local equilibrium*, which is its hand implies the hydrodynamic limit as stated in Theorem 2.

We begin with some topological considerations. In the space $\mathbb{N}^{\mathbb{Z}}$ endowed with the distance

$$d(\eta_1,\eta_2) \;=\; \sum_{k\in\mathbb{Z}} \frac{1}{2^{|k|}} \frac{|\eta_1(k)-\eta_2(k)|}{1+|\eta_1(k)-\eta_2(k)|}\,,$$

[3] The so-called infinite propagation speed in PDE's, see [8, p. 49].

[4] Physically, to speak about finite propagation speed for QRW's it is necessary to go further into the Lieb-Robinson bound, see [9]. We did not investigate such subject in this paper.

denote by $\{\tau_k \,;\, k \in \mathbb{Z}\}$ the group of translations. In other words, $\tau_k\eta$ is the configuration given by

$$(\tau_k\eta)(j) \;=\; \eta(j+k)\,.$$

The action of the translation group is naturally extended to the space of probability measures on $\mathbb{N}^{\mathbb{Z}}$. For $k \in \mathbb{Z}$ and a probability measure μ on $\mathbb{N}^{\mathbb{Z}}$, we denote by $\tau_k\mu$ the unique probability measure such that

$$\int f(\eta)(\tau_k\mu)(d\eta) \;=\; \int f(\tau_k\eta)\mu(d\eta)$$

for all integrable continuous functions f in the topology induced by the aforementioned distance.

For $c > 0$, define ν_c as the product probability measure on $\mathbb{N}^{\mathbb{Z}}$ whose marginals are Poisson probability measures with the same parameter $c > 0$, i.e.,

$$\nu_c\{\eta \,;\, \eta(k) = \ell\} \;=\; e^{-c}\frac{c^\ell}{\ell!}\,,$$

for any $k \in \mathbb{Z}$. Informally, the conservation of local equilibrium says that under suitable hypothesis on the initial distribution of particles, the distribution of the observed particles at time $\lfloor tn \rfloor$ is, in an asymptotic sense, locally a Poisson product measure whose parameter is a function of time and space. Its precise statement is

Theorem 3 (Conservation of local equilibrium) *Let α_{tn} be the probability measure on $\mathbb{N}^{\mathbb{Z}}$ induced by the random element*

$$\xi_{tn} \;:=\; (\ldots, \xi_{tn}(-2), \xi_{tn}(-1), \xi_{tn}(0), \xi_{tn}(1), \ldots),$$

see the Definition 5. For $x \in \mathbb{R}$ and $n \in \mathbb{N}$, denote $xn := \lfloor xn \rfloor$. Then, for any $x \in \mathbb{R}$ and any $t > 0$,

$$\lim_{n\to\infty} \tau_{xn}\alpha_{tn} \;=\; \nu_{\rho(t,x)}$$

in the sense of weak convergence of probability measures,[5] *where $\rho(t,x)$ is the function defined in* (8).

Proof In order to not overload notation, we start by considering the case $t = 1$. The general statement is postponed to the end of the proof.

The weak convergence of probability measures on $\mathbb{N}^{\mathbb{Z}}$ is equivalent to the convergence of its finite dimensional distributions, see [10]. Moreover, the convergence of the finite dimensional distributions is characterized by the convergence of their Laplace transforms. Hence, we concern our attention to the convergence of the Laplace transform

[5] See Ref. [10].

$$\mathbb{E}\Big[\exp\Big\{-\sum_{\ell\in\mathbb{Z}}\lambda(\ell)\,\xi_n(\ell)\Big\}\Big],$$

where $\lambda:\mathbb{Z}\to\mathbb{R}_+$ is a function that is non zero only on a finite subset of $\mathbb{Z}$. By (7),

$$\sum_{\ell\in\mathbb{Z}}\lambda(\ell)\xi_n(\ell) \;=\; \sum_{\ell\in\mathbb{Z}}\lambda(\ell)\sum_{k\in\mathbb{Z}}\sum_{j=1}^{\eta(k)}\mathbb{1}_{[X_n^k(j)=\ell]} \;=\; \sum_{k\in\mathbb{Z}}\sum_{j=1}^{\eta(k)}\lambda(X_n^k(j))\,.$$

Recalling the independence of the random variables and the equality above, we obtain

$$\begin{aligned}\mathbb{E}\Big[\exp\Big\{-\sum_{\ell\in\mathbb{Z}}\lambda(\ell)\,\xi_n(\ell)\Big\}\Big] &= \prod_{k\in\mathbb{Z}}\mathbb{E}\Big[\exp\Big\{\sum_{j=1}^{\eta(k)}\lambda(X_n^k(j))\Big\}\Big]\\ &= \prod_{k\in\mathbb{Z}}\int E\Big[\exp\Big\{-\lambda(X_n^k(1))\Big\}\Big]^{\eta(k)}\,d\mu_n\,,\end{aligned} \tag{9}$$

where E is the expectation over a single random variable $X_n^k(1)$. Let

$$\beta_k := E\Big[\exp\Big\{-\lambda(X_n^k(1))\Big\}\Big].$$

Under μ_n, the random variable $\eta(k)$ has Poisson distribution given by (5). Thus,

$$\int \beta_k^{\eta(k)}\,d\mu_n \;=\; \exp\Big\{\gamma(\tfrac{k}{n})(\beta_k-1)\Big\}.$$

Applying this to (9) we are lead to

$$\begin{aligned}\mathbb{E}\Big[\exp\Big\{-\sum_{\ell\in\mathbb{Z}}\lambda(\ell)\,\xi_n(\ell)\Big\}\Big] &= \prod_{k\in\mathbb{Z}}\exp\Big\{\gamma(\tfrac{k}{n})(\beta_k-1)\Big\}\\ &= \exp\Big\{\sum_{k\in\mathbb{Z}}\gamma(\tfrac{k}{n})(\beta_k-1)\Big\}.\end{aligned} \tag{10}$$

Denote by $p_n(k,\ell)$ the probability of the quantum random walk, starting at $\psi = e_k\otimes|+1\rangle$, after a time n, have been observed at the position $\ell\in\mathbb{Z}$. That is,

$$p_n(k,\ell) \;:=\; P\big[X_n^k(1)=\ell\big].$$

Therefore,

$$\beta_k := E\Big[\exp\Big\{-\lambda(X_n^k(1))\Big\}\Big] \;=\; \sum_{\ell\in\mathbb{Z}} e^{-\lambda(\ell)}\,p_n(k,\ell)\,.$$

Replacing previous the formula in (10) and interchanging summations, we get to

$$\mathbb{E}\Big[\exp\Big\{-\sum_{\ell\in\mathbb{Z}}\lambda(\ell)\,\xi_n(\ell)\Big\}\Big] = \exp\Big\{\sum_{\ell\in\mathbb{Z}}(e^{-\lambda(\ell)}-1)\sum_{k\in\mathbb{Z}}\gamma(\tfrac{k}{n})p_n(k,\ell)\Big\}.$$

The formula above characterizes the measure α_n on $\mathbb{N}^{\mathbb{Z}}$ (induced by the random element ξ_n) as a product measure whose marginal at the site $\ell \in \mathbb{Z}$ is a Poisson probability measure of parameter

$$B(\ell,n) := \sum_{k\in\mathbb{Z}}\gamma(\tfrac{k}{n})\,p_n(k,\ell).$$

As a consequence of symmetry of the Hadamard operator, it is easy to verify that $p_n(k,\ell) = p_n(\ell,k)$. This implies

$$B(\ell,n) = \sum_{k\in\mathbb{Z}}\gamma(\tfrac{k}{n})\,p_n(\ell,k) = E\Big[\gamma\Big(\tfrac{X_n^\ell(1)}{n}\Big)\Big].$$

Given $x \in \mathbb{R}$ and $n \in \mathbb{N}$, recall the notation $xn = \lfloor xn \rfloor$. By Theorem 1, since γ is smooth, and since f is an even function,

$$\begin{aligned}\lim_{n\to\infty}B(xn,n) &= \lim_{n\to\infty}E\Big[\gamma\Big(\tfrac{X_n^{xn}}{n}\Big)\Big]\\ &= \lim_{n\to\infty}E\Big[\gamma\Big(\tfrac{X_n^0+xn}{n}\Big)\Big]\\ &= \int_{\mathbb{R}}\gamma(y+x)\,f(y)\,dy\\ &= \int_{\mathbb{R}}\gamma(y)\,f(x-y)\,dy.\end{aligned}$$

Since α_n is a product measure, the limit above implies that

$$\lim_{n\to\infty}\tau_{xn}\alpha_n = \nu_{\rho(1,x)},$$

proving the statement for $t = 1$. For general $t > 0$, one has to replace $X_n^k(j)$ by $X_{tn}^k(j)$, keeping μ_n unchanged. Denote by α_{tn} the measure on $\mathbb{N}^{\mathbb{Z}}$ induced by ξ_{tn}. In this situation, it is straightforward to check that α_{tn} is also a product measure whose marginal has Poisson distribution and the limit for its parameter is given by

$$\begin{aligned}\lim_{n\to\infty}B(xn,tn) &= \lim_{n\to\infty}E\Big[\gamma\Big(\tfrac{X_{tn}^{xn}}{n}\Big)\Big]\\ &= \lim_{n\to\infty}E\Big[\gamma\Big(\tfrac{t\,X_{tn}^0}{tn}+\tfrac{xn}{n}\Big)\Big]\end{aligned}$$

$$= \int_{\mathbb{R}} \gamma(ty + x)\, f(y)\, dy$$
$$= \int_{\mathbb{R}} \gamma(y)\, \tfrac{1}{t} f(\tfrac{x-y}{t})\, dy\,.$$

Since α_{tn} is a product measure, the limit above implies that

$$\lim_{n\to\infty} \tau_{xn}\alpha_{tn} = \nu_{\rho(t,x)}\,,$$

concluding the proof. □

We shall prove Theorem 2 now.

Proof (*Proof of Theorem* 2) It is a known result that the conservation of local equilibrium, proved in Theorem 3, implies the hydrodynamic limit stated in Theorem 2, see for instance [1, Chap. 3]. □

References

1. Kipnis, C., Landim, C.: Scaling limits of interacting particle systems. Grundlehren der Mathematischen Wissenschaften. Fundamental Principles of Mathematical Sciences, vol. 320. Springer, Berlin (1999)
2. DeMasi, A., Presutti, E.: Mathematical methods for hydrodynamic limits. Lecture Notes in Mathematics, vol. 1501. Springer (1991)
3. Dobrushin, R., Siegmund-Schultze, R.: The hydrodynamic limit for systems of particles with independent evolution. Mathematische Nachrichten **105**(1), 199–224 (1982)
4. Aharonov, Y., Davidovich, L., Zagury, N.: Quantum random walks. Phys. Rev. A **48**(2), 1687–1690 (1993)
5. Kempe, J.: Quantum random walks—an introductory overview. Contemp. Phys. **44**(4), 307–327 (2003)
6. Grimmett, G., Janson, S., Scudo, P.: Weak limits of quantum random walks. Phys. Rev. E **69** (2004)
7. Durrett, R.: Probability: Theory and Examples, 4th edn. Cambridge University Press (2010)
8. Evans, L.: Partial Differential Equations. Graduate Studies in Mathematics V.19, 2nd edn. American Mathematical Society (1998)
9. Lieb, E., Robinson, D.: The finite group velocity of quantum spin systems. Commun. Math. Phys. **28**, 251–257 (1972)
10. Billingsley, P.: Convergence of Probability Measures. Wiley Series in Probability and Statistics: Probability and Statistics, 2nd edn. Wiley, New York (1999)

Sub-shock Formation in Reacting Gas Mixtures

Marzia Bisi, Fiammetta Conforto and Giorgio Martalò

Abstract The shock-wave structure problem is investigated for a gas mixture of four species, undergoing a reversible bimolecular reaction, modelled by a 10 moment Grad closure of reactive Boltzmann equations. The presence of jump discontinuities within the shock structure solution is discussed, the supersonic regime is characterized, and the critical values of Mach number allowing the formation of sub-shocks in the field variables of one or more components of the mixture are pointed out.

Keywords Shock-wave problem · Sub-shocks · Kinetic theory for gas mixtures · Grad 10 moment equations · Reactive mixtures

1 Introduction

Many recent papers, as among others [2, 5, 6, 12, 17, 18], show that the behaviour of multi-temperature models for gas mixtures exhibits very interesting features in the context of the shock wave structure problems.

M. Bisi
Department of Mathematics and Computer Science,
University of Parma, Parco Area delle Scienze, 53/A, Parma, Italy
e-mail: marzia.bisi@unipr.it

F. Conforto (✉)
Dipartimento di Scienze Matematiche e Informatiche,
Scienze Fisiche e Scienze della Terra, University of Messina,
v.le F. Stagno d'Alcontres 31, Messina, Italy
e-mail: fiammetta.conforto@unime.it; fiamma@unime.it

G. Martalò
Institut de Mathématiques de Bordeaux, CNRS and Université de Bordeaux,
351, cours de la Libération, 33405 Talence, France
e-mail: giorgio.martalo@u-bordeaux.fr

G. Martalò
INRIA Bordeaux Sud-Ouest, 200 Avenue de la Vieille Tour, 33405 Talence, France,
e-mail: giorgio.martalo@inria.fr

© Springer International Publishing Switzerland 2016

P. Gonçalves and A.J. Soares (eds.), *From Particle Systems to Partial Differential Equations III*, Springer Proceedings in Mathematics & Statistics 162,
DOI 10.1007/978-3-319-32144-8_3

For hyperbolic systems of both conservation and balance laws, compatible with an entropy principle, it is well-stated [7, 8, 21] that a continuous solution of the shock wave structure problem, between two equilibrium configurations, can not propagate with a speed, s, greater than the maximum characteristic velocity of the system, evaluated in the unperturbed state. Despite many efforts in this direction, the continuity of wave fronts propagating below this threshold is still an open problem, in the sense that, to authors' knowledge, no rigorous results are available. Many references on shock waves propagation in a single gas may be found in the literature, among which we mention [1, 9, 22].

For what concerns gaseous mixtures, shock structure solutions have been recently investigated for some specific problems, both from the numerical and the analytical points of view. In [17, 18] numerical examples, compared with experimental data, are shown for inert binary mixtures described in the framework of Extended Thermodynamics [19] at Euler level. On the other hand, in [15, 16, 20], the analysis is developed starting from suitable kinetic models of Boltzmann or BGK type.

For what concerns hydrodynamic closures derived from kinetic Boltzmann equations, a multi-temperature Euler system for an inert binary mixture of monoatomic gases has been rigorously dealt with in [6], in which the role in the continuity of travelling waves played by each singularity manifold related to each component of the mixture is pointed out, and numerical tests show the presence of one or two sub-shocks within the wave front for suitable values of the Mach number. In [5], the steady shock problem is faced for a reactive gas of four monoatomic components undergoing a reversible bimolecular reaction, described by a multi-temperature (and one-velocity) Euler closure of the reactive Boltzmann equations. The existence of discontinuous solutions is shown by numerical simulations in which the variables describing each component may suffer a jump. For the same reacting mixture, but modelled by a 13-moment Grad closure of the reactive Boltzmann equations [3, 4, 14], the discontinuous shock structure solutions are investigated in [12], and it is pointed out that, analogously to the simpler systems studied in [5, 6], each component may generate a sub-shock, on its own variables. Each component, in fact, defines its own singularity sub-manifold in the phase space across which only its own variables suffer a jump discontinuity, if the two (upstream and downstream) equilibria lie on opposite sides with respect to such sub-manifold. The role of the maximum characteristic velocities relevant to each species in the sub-shock formation is also discussed.

In [11], for an inert mixture of multi-velocity, multi-temperature Euler gases in the context of Extended Thermodynamics, the long-time behaviour of the solution of the Riemann problem is studied in order to characterize the corresponding shock structure, and the Rankine–Hugoniot conditions are used to identify each sub-shock within the wave front, confirming the results in [6].

Very recently, in [2], a 10-moment Grad closure of the reactive Boltzmann equations is used to describe a binary mixture of inert monoatomic gases, and the shock wave structure problem is faced within the whole hierarchy of hyperbolic subsystems of balance laws: Grad 10-moments equations, the multi-temperature Euler description recovered assuming vanishing viscous stress tensors (restoring the studies in [6]), and the Euler (single velocity and single temperature) equilibrium sub-system of the hierarchy. The complete description of possible solutions, exhibiting up to two sub-shocks, is given and the explicit expressions of all critical Mach numbers, as well as their admissibility conditions, are deduced in terms of particle masses of the species and of the unperturbed equilibrium state.

The present paper faces the problem of sub-shock formation in 10-moment Grad equations for the more interesting and realistic case of reactive gases, again for the same four-species mixture with the same chemistry considered in [5] and [12], in order to improve the preliminary available results. From the physical point of view, the 10-moment closure suffers the heavy drawback of neglecting the heat flux effects occurring in the evolution. However, with respect to the more realistic 13-moment approximation, it has the advantage of providing explicit expressions for all the involved quantities in terms of a field vector of lower dimension, as well as a wider hyperbolicity region. Therefore, since this paper represents a first attempt of handling chemical processes, the aim of highlighting their effects on the shock wave structure solutions, minimizing at the same time the both analytical and numerical difficulties, may justify the choice of confining again our investigation in the framework of the 10-moment level description adopted in [2], despite the loss of physical consistency. The extensions to the classical 13-moment Grad closure, including the evolution of heat flux vectors, are considered in forthcoming papers. Moreover, from the mathematical point of view, the presence of the reaction (involving exchanges of mass and internal energy in the transient, and imposing the mass action law of chemistry at equilibrium) does not allow to prove all the huge variety of features deduced in [2] for the inert mixture, which is here recovered in the limit of vanishing chemical bond energy. In spite of the drawback of handling the implicit function introduced by the law of mass action defining the chemical equilibrium, we are able to show the effects of the chemistry on the solutions. In order to stress the role of the chemistry, the comparison with the corresponding inert case is carried out. The presence of the reaction modifies the characteristic speeds of the system and of its equilibrium subsystem, which differs from the classical Euler equations establishing the equilibrium subsystem of the inert gas. It will be shown that this situation translates into a change in the supersonic regime, that for the inert gas is characterized by Mach numbers greater than 1, while in the present reactive frame the lower bound of supersonic flows is decreased below 1.

The paper is organized as follows. In Sect. 2 we briefly present the model, its equilibrium subsystem, and all the characteristic speeds involved in our analysis. In Sect. 3 the shock wave structure problem is studied for both inert and reactive gases, and the possible analytical investigation (allowed by the involved implicit function) on the admissible critical Mach numbers is presented. Some results on the existence and uniqueness of a perturbed equilibrium state defined by the unperturbed state are also given. Finally, in Sect. 4, a particular reaction is chosen in order to characterize the different ranges of all possible admitted shock structures, and some concluding remarks are discussed.

2 Governing Equations

Let us consider a 1-D 10-moment approximation of Grad type for a reactive mixture of gases undergoing a reversible bimolecular reaction

$$1 + 2 \rightleftharpoons 3 + 4. \tag{1}$$

Under the hypothesis of negligible heat fluxes, imposed on the 13-moment Grad description proposed in [3, 4], the sixteen macroscopic field variables are species densities n_i, mass velocities u_i, species temperatures T_i and stress deviators σ_i, for $i = 1, 2, 3, 4$, and they define the global quantities through the relationships

$$n = \sum_{i=1}^{4} n_i, \qquad \rho = \sum_{i=1}^{4} \rho_i = \sum_{i=1}^{4} m_i n_i, \qquad \rho u = \sum_{i=1}^{4} \rho_i u_i, \tag{2}$$

$$n k_B T = \sum_{i=1}^{4} n_i k_B T_i + \frac{1}{3} \sum_{i=1}^{4} \rho_i \left(u_i^2 - u^2 \right), \qquad \sigma = \sum_{i=1}^{4} \sigma_i + \frac{2}{3} \sum_{i=1}^{4} \rho_i \left(u_i^2 - u^2 \right), \tag{3}$$

where m_i is the particle mass for ith species and k_B is the Boltzmann constant.

In the vanishing heat flux limit of 13-moments Grad equations [3, 4], the hyperbolic system of partial differential equations, in conservative form, reads as

$$\partial_t n_i + \partial_x (n_i u_i) = Q_i, \tag{4}$$

$$\partial_t (m_i n_i u_i) + \partial_x \left(m_i n_i u_i^2 + n_i k_B T_i \right) = R_i + m_i u_i Q_i, \tag{5}$$

$$\partial_t \left(\frac{1}{2} m_i n_i u_i^2 + \frac{3}{2} n_i k_B T_i \right) + \partial_x \left[\left(\frac{1}{2} m_i n_i u_i^2 + \frac{5}{2} n_i k_B T_i + \sigma_i \right) u_i \right] = S_i + u_i R_i + \left(\frac{1}{2} m_i u_i^2 + \frac{3}{2} k_B T_i \right) Q_i, \tag{6}$$

$$\partial_t \left(\frac{2}{3} m_i n_i u_i^2 + \sigma_i \right) + \partial_x \left[\left(\frac{2}{3} m_i n_i u_i^2 + \frac{4}{3} n_i k_B T_i + \frac{7}{3} \sigma_i \right) u_i \right] = V_i + \frac{4}{3} u_i R_i + \frac{2}{3} m_i u_i^2 Q_i, \tag{7}$$

where

$$Q_i = \frac{2\nu_{12}^{34} l_i}{\sqrt{\pi}} \left[n_3 n_4 \left(\frac{m_{12}}{m_{34}} \right)^{\frac{3}{2}} \exp\left(\frac{\Delta E}{k_B T} \right) - n_1 n_2 \right] \Gamma\left(\frac{3}{2}, \frac{\Delta E}{k_B T} \right) \tag{8}$$

$$R_i = n_i \sum_{j=1}^{4} \nu_{1ij} m_{ij} n_j \left(u_j - u_i \right) - m_i \left(u_i - u \right) Q_i \tag{9}$$

$$\begin{aligned} S_i = {} & 2n_i \sum_{j=1}^{4} \frac{\nu_{1ij} m_{ij}}{m_i + m_j} n_j \left[\frac{3}{2} k_B \left(T_j - T_i \right) + \frac{1}{2} m_j \left(u_j - u_i \right)^2 \right] \\ & + \left[\frac{3}{2} \frac{m_i}{m} k_B T - \frac{3}{2} \left(1 - l_i \right) \Delta E \frac{m - m_i}{m} - \frac{3}{2} k_B T_i + \frac{1}{2} m_i \left(u_i - u \right)^2 \right] Q_i \\ & + \frac{2\nu_{12}^{34} l_i k_B T}{\sqrt{\pi}} \left[n_3 n_4 \left(\frac{m_{12}}{m_{34}} \right)^{\frac{3}{2}} \exp\left(\frac{\Delta E}{k_B T} \right) - n_1 n_2 \right] \frac{m - m_i}{m} \Gamma\left(\frac{5}{2}, \frac{\Delta E}{k_B T} \right) \end{aligned} \tag{10}$$

$$\begin{aligned} V_i = {} & \frac{2}{3} m_i \left(u_i - u \right)^2 Q_i + \sum_{j=1}^{4} \frac{2\nu_{1ij} m_{ij}}{m_i + m_j} \left[\frac{2}{3} m_j n_j n_i \left(u_j - u_i \right)^2 + n_i \sigma_j - n_j \sigma_i \right] \\ & - \sum_{j=1}^{4} \frac{3\nu_{2ij} m_j}{2 \left(m_i + m_j \right)^2} \left[\frac{2}{3} m_i m_j n_i n_j \left(u_j - u_i \right)^2 + m_i n_i \sigma_j + m_j n_j \sigma_i \right] \end{aligned} \tag{11}$$

and $l_1 = l_2 = -l_3 = -l_4 = 1$, $m = m_1 + m_2 = m_3 + m_4$, $m_{ij} = m_i m_j / (m_i + m_j)$, $\Delta E > 0$ is the activation energy for the endothermic reaction $1 + 2 \rightarrow 3 + 4$, ν_{1ij} and ν_{2ij} are suitably weighted elastic collision frequencies and ν_{12}^{34} is the chemical interaction frequency.

An equilibrium state of the system (4)–(7) is characterized by the so-called mass action law, and by a unique mean velocity and a unique temperature and by vanishing stress deviators

$$n_1 n_2 = n_3 n_4 \left(\frac{m_{12}}{m_{34}} \right)^{\frac{3}{2}} \exp\left(\frac{\Delta E}{k_B T} \right), \tag{12}$$

$$u_i = u, \qquad T_i = T, \qquad \sigma_i = \sigma = 0, \qquad i = 1, 2, 3, 4. \tag{13}$$

The conservation equations of total mass, momentum and energy, together with two independent partial conservation equations for number densities through the chemical reaction, represent the equilibrium sub-system, that is

$$\partial_t \left(n_1 + n_\ell \right) + \partial_x \left[\left(n_1 + n_\ell \right) u \right] = 0, \qquad \ell = 3, 4, \tag{14}$$

$$\partial_t \rho + \partial_x \left(\rho u \right) = 0, \tag{15}$$

$$\partial_t \left(\rho u \right) + \partial_x \left(\rho u^2 + n k_B T \right) = 0, \tag{16}$$

$$\partial_t \left(\frac{1}{2} \rho u^2 + \frac{3}{2} n k_B T - \Delta E n_1 \right) + \partial_x \left[\left(\frac{1}{2} \rho u^2 + \frac{5}{2} n k_B T - \Delta E n_1 \right) u \right] = 0, \tag{17}$$

coupled with the additional constraint due to the chemical equilibrium condition imposed by the mass action law (12).

To conclude this section, let us point out that the characteristic speeds of system (4)–(7) are

$$u_i - \sqrt{\frac{3\,(n_i k_B T_i + \sigma_i)}{m_i n_i}}, \qquad u_i, \qquad u_i + \sqrt{\frac{3\,(n_i k_B T_i + \sigma_i)}{m_i n_i}}. \tag{18}$$

Only the greater ones can be involved in the problem of sub-shock formation, denoted by

$$\lambda_i = u_i + \sqrt{\frac{3\,(n_i k_B T_i + \sigma_i)}{m_i n_i}}, \tag{19}$$

together with the highest characteristic velocity of the equilibrium sub-system [10]

$$\mu = u + \sqrt{\frac{5 n k_B T}{3\rho}\left(1 - \frac{2}{5\mathscr{C}}\right)}, \tag{20}$$

where

$$\mathscr{C} = 1 + \frac{3 n k_B^2 T^2}{2 \Delta E^2}\left(\frac{1}{n_1} + \frac{1}{n_2} + \frac{1}{n_3} + \frac{1}{n_4}\right), \tag{21}$$

which, suitably evaluated, represents the lower bound of the wave front speeds in such kind of problem. Moreover, notice that

$$\mu < \mu^{Euler} = u + \sqrt{\frac{5 n k_B T}{3\rho}},$$

meaning that the chemical reaction slows down the maximum characteristic speed of the equilibrium sub-system, which is now smaller than the one for a single Euler gas, μ^{Euler}, and this means an enlargement of the supersonic regime. Let us finally stress that in the limit as $\Delta E \to 0$, then $\mathscr{C} \to \infty$, and $\mu \to \mu^{Euler}$.

3 Shock-Wave Structure Solutions

As well-known, a shock-wave structure solution is a function $\mathbf{u}\,(\varphi)$ of the variable $\varphi = x - st$, $s > 0$ constant, connecting two equilibrium states, $\mathbf{u}_+$ and $\mathbf{u}_-$, of the system

$$\lim_{\varphi \to \pm\infty} \mathbf{u}\,(\varphi) = \mathbf{u}_\pm, \qquad \lim_{\varphi \to \pm\infty} \frac{d\mathbf{u}}{d\varphi}\,(\varphi) = 0 \tag{22}$$

so that the wave propagates, at constant velocity s, into the fixed unperturbed downstream equilibrium $\mathbf{u}_+$. As usual in such kind of problems, let us introduce the relative velocities v_i and Mach number M_+ as follows

$$v_i = s - u_i, \qquad M_+ = \frac{s - u_+}{c_+}, \qquad c_+^2 = \frac{5 n_+ k_B T_+}{3 \rho_+}, \tag{23}$$

where c_+ is the classical Euler unperturbed sound speed and

$$v_+ = s - u_+ = c_+ M_+ = v_{i+}, \qquad \forall i = 1, 2, 3, 4. \tag{24}$$

In the reference frame co-moving with the shock front, the set of balance equations (4)–(7) yields the following system of ODEs

$$\frac{d}{d\varphi}(n_i v_i) = -Q_i \tag{25}$$

$$\frac{d}{d\varphi}\left(m_i n_i v_i^2 + n_i k_B T_i + \sigma_i\right) = \Sigma_{v_i} \tag{26}$$

$$\frac{d}{d\varphi}\left[\left(\frac{1}{2} m_i n_i v_i^2 + \frac{5}{2} n_i k_B T_i + \sigma_i\right) v_i\right] = -\Sigma_{T_i} \tag{27}$$

$$\frac{d}{d\varphi}\left[\left(\frac{2}{3} m_i n_i v_i^2 + \frac{4}{3} n_i k_B T_i + \frac{7}{3}\sigma_i\right) v_i\right] = -\Sigma_{\sigma_i} \tag{28}$$

where

$$\Sigma_{v_i} = R_i - m_i v_i Q_i, \tag{29}$$

$$\Sigma_{T_i} = S_i - v_i R_i + \left(\frac{1}{2} m_i v_i^2 + \frac{3}{2} k_B T_i\right) Q_i, \tag{30}$$

$$\Sigma_{\sigma_i} = V_i - \frac{4}{3} v_i R_i + \frac{2}{3} m_i v_i^2 Q_i, \tag{31}$$

and Q_i, R_i, S_i and V_i have been properly rewritten in terms of the relative velocities v_i.

The equilibrium state $\mathbf{u}_-$ is related to the unperturbed state $\mathbf{u}_+$ through the Rankine–Hugoniot conditions applied to the equilibrium sub-system (14)–(17), yielding, besides the trivial case $\mathbf{u}_- = \mathbf{u}_+$,

$$n_{i-} = n_{i+} \frac{c_+ M_+}{v_-} + l_i N\left(\mathbf{u}_+, M_+; v_-\right), \qquad i = 1, 2, 3, 4 \tag{32}$$

$$n_- = n_+ \frac{c_+ M_+}{v_-}, \qquad \rho_- = \rho_+ \frac{c_+ M_+}{v_-}, \tag{33}$$

$$T_- = T_+ \frac{v_-}{c_+ M_+}\left[1 + \frac{5 M_+}{3 c_+}\left(c_+ M_+ - v_-\right)\right], \tag{34}$$

where

$$N\left(\mathbf{u}_+, M_+; v_-\right) = \frac{\rho_+ c_+}{2\Delta E v_-}\left(v_- - c_+ M_+\right)\left[c_+\left(M_+^2 + 3\right) - 4M_+ v_-\right], \tag{35}$$

and v_- is implicitly defined by the mass action law (12) imposed on n_{i-} and T_- provided in (32) and (34).

As pointed out in [12], the presence of a chemical reaction influences the upstream equilibrium, whose existence and uniqueness have now to be verified, since it has to be univocally determined for fixed unperturbed equilibrium and Mach number. It now differs from the one recovered in the inert case [2], coinciding with the classical one of the Euler equations, given by

$$n_{i-}^{inert} = n_{i+}\frac{4M_+^2}{M_+^2 + 3}, \qquad n_-^{inert} = n_+\frac{4M_+^2}{M_+^2 + 3}, \qquad \rho_-^{inert} = \rho_+\frac{4M_+^2}{M_+^2 + 3}, \tag{36}$$

$$v_-^{inert} = c_+\frac{M_+^2 + 3}{4M_+}, \qquad T_-^{inert} = T_+\frac{\left(M_+^2 + 3\right)\left(5M_+^2 - 1\right)}{16M_+^2}. \tag{37}$$

Let us observe that relations (32)–(35) coincide with the classical ones (36)–(37), if, and only if, $N\left(\mathbf{u}_+, M_+; v_-\right) = 0$, which, besides the trivial case $v_- = c_+ M_+$ corresponding to $\mathbf{u}_- = \mathbf{u}_+$, means that v_- has to be given by (37).

The physical constraints due to positivity of species number densities and mixture temperature of the upstream equilibrium given by (32)–(35) give some ranges of admissible values for the upstream velocity (different from $v_+ = c_+ M_+$), that from now on will be denoted as v_-^r (and the corresponding upstream state will be set $\mathbf{u}_-^r$). To be more precise, the positivity of temperature T_- yields the condition $v_-^r \in I_T$, with

$$I_T := \left(0, c_+\frac{5M_+^2 + 3}{5M_+}\right),$$

whereas, denoting

$$\xi_k^\pm = \frac{\rho_+ c_+\left(5M_+^2 + 3\right) \pm \sqrt{9\rho_+^2 c_+^2\left(M_+^2 - 1\right)^2 + 32\Delta E\, n_{k+}\rho_+ M_+^2}}{8\rho_+ M_+}, \qquad k = 1, 2,$$

$$\eta_\ell^\pm = \frac{\rho_+ c_+\left(5M_+^2 + 3\right) \pm \sqrt{9\rho_+^2 c_+^2\left(M_+^2 - 1\right)^2 - 32\Delta E\, n_{\ell+}\rho_+ M_+^2}}{8\rho_+ M_+}, \qquad \ell = 3, 4,$$

the positivity of species number densities gives the constraint $v_-^r \in I_n$, where, if $\eta_\ell^\pm \in \mathbb{R}$,

$$I_n := \left(\max\left\{\xi_1^-, \xi_2^-\right\}, \min\left\{\eta_3^-, \eta_4^-\right\}\right) \cup \left(\max\left\{\eta_3^+, \eta_4^+\right\}, \min\left\{\xi_1^+, \xi_2^+\right\}\right),$$

which, if $\eta_\ell^\pm \in \mathbb{C}$, reduces to

$$I_n = \left(\max\left\{\xi_1^-, \xi_2^-\right\}, \min\left\{\xi_1^+, \xi_2^+\right\}\right).$$

This feature has already been investigated numerically in [12] in the context of an analogous reactive gas mixture described at 13 moment Grad closure, in which the same equilibrium subsystem is obviously admitted.

3.1 Existence and Uniqueness of Upstream Equilibrium State

In order to prove the existence and uniqueness of a physical solution of the mass action law (12) as equation in v_-, let us denote

$$\mathscr{F}(v_-) := \bar{m}\,(n_{3-}v_-)\,(n_{4-}v_-)\exp\left(\frac{\Delta E}{k_B T_-}\right) - (n_{1-}v_-)\,(n_{2-}v_-) = 0 \tag{38}$$

where $\bar{m} = (m_{12}/m_{34})^{3/2}$ and n_{i-} and T_- are given by (32)–(35). Its derivative can be written as

$$\begin{aligned}\mathscr{F}'(v_-) = &-\mathscr{M}(v_-)\left[\bar{m}\exp\left(\frac{\Delta E}{k_B T_-}\right)(n_{3-}v_- + n_{4-}v_-) + n_{1-}v_- + n_{2-}v_-\right]\\ &- \bar{m}\,(n_{3-}v_-)\,(n_{4-}v_-)\exp\left(\frac{\Delta E}{k_B T_-}\right)\frac{\Delta E}{k_B T_-^2}\,\mathscr{T}(v_-)\end{aligned} \tag{39}$$

where

$$\mathscr{M}(v_-) := l_i\frac{d}{dv_-}(n_{i-}v_-) = \frac{\rho_+ c_+}{2\Delta E}\left[c_+\left(5M_+^2+3\right) - 8M_+v_-\right], \tag{40}$$

$$\mathscr{T}(v_-) := \frac{dT_-}{dv_-} = \frac{T_+}{3c_+^2 M_+}\left[c_+\left(5M_+^2+3\right) - 10M_+v_-\right]. \tag{41}$$

It can be immediately noticed that $\mathscr{F}(v_-)$ is decreasing for

$$0 < v_- < c_+\frac{5M_+^2+3}{10M_+}$$

and $\mathscr{F}(v_-)$ is increasing for

$$c_+\frac{5M_+^2+3}{8M_+} < v_- < c_+\frac{5M_+^2+3}{5M_+}.$$

Since $v_+ = c_+ M_+$, root of (38), satisfies

$$c_+ \frac{5M_+^2+3}{8M_+} < v_+ < c_+ \frac{5M_+^2+3}{5M_+}, \tag{42}$$

for all $M_+ > 1$, and in such interval $\mathscr{F}$ is increasing, an admissible root v_-^r of (38), different from v_+, must be

$$v_-^r < c_+ \frac{5M_+^2+3}{8M_+}.$$

Let us also point out that for all $M_+ > 1$

$$\xi_k^- < v_-^{inert} < \eta_\ell^-, \qquad \eta_\ell^+ < v_+ < \xi_k^+,$$

and

$$v_-^{inert} < c_+ \frac{5M_+^2+3}{8M_+}.$$

By noticing that

$$\mathscr{F}\left(\xi_k^\pm\right) = \bar{m}\left(n_{3-}\xi_k^\pm\right)\left(n_{4-}\xi_k^\pm\right)\exp\left(\frac{\Delta E}{k_B T_-}\right) > 0, \qquad k = 1, 2,$$

$$\mathscr{F}\left(v_-^{inert}\right) < \bar{m}\left(n_{3+}v_+\right)\left(n_{4+}v_+\right)\exp\left(\frac{\Delta E}{k_B T_+}\right) - \left(n_{1+}v_+\right)\left(n_{2+}v_+\right) = 0,$$

$$\mathscr{F}\left(\frac{3c_+}{5M_+}\right) > \bar{m}\left(n_{3+}v_+\right)\left(n_{4+}v_+\right)\exp\left(\frac{\Delta E}{k_B T_+}\right) - \left(n_{1+}v_+\right)\left(n_{2+}v_+\right) = 0$$

and by setting $v_1 = \max\left(\xi_1^-, \xi_2^-, 3c_+/(5M_+)\right)$, one can use the Bolzano's theorem to conclude about the existence of at least one physical solution v_-^r of (38) in $\left(v_1, v_-^{inert}\right)$ for $M_+ > 1$.

It can be numerically checked that the function $\mathscr{F}(v_-)$ is convex, therefore even uniqueness of a physical solution v_-^r is guaranteed. For $M_+ > \sqrt{9/5}$, uniqueness may be also proved analytically, by using the monotonicity of $\mathscr{F}(v_-)$ in the interval $\left(0, c_+ \frac{5M_+^2+3}{10M_+}\right)$ and by verifying that

$$\mathscr{F}(v_-) < 0 \qquad \forall v_- \in \left(c_+ \frac{5M_+^2+3}{10M_+}, c_+ \frac{5M_+^2+3}{8M_+}\right) \cap I_n. \tag{43}$$

In fact, under the hypothesis $M_+ > \sqrt{9/5}$, one can easily prove that $N(\mathbf{u}_+, M_+; v_-) > 0$ and, as consequence,

$$n_{k-}v_- > n_{k+}v_+, \qquad n_{\ell-}v_- < n_{\ell+}v_+ \qquad k = 1, 2, \quad \ell = 3, 4$$

and also

$$T_- > T_+$$

for every $v_- \in \left(c_+ \frac{5M_+^2+3}{10M_+}, c_+ \frac{5M_+^2+3}{8M_+}\right) \cap I_n$; therefore

$$\begin{aligned}\mathscr{F}(v_-) &= \bar{m}\,(n_{3-}v_-)\,(n_{4-}v_-)\exp\left(\frac{\Delta E}{k_B T_-}\right) - (n_{1-}v_-)\,(n_{2-}v_-) \\ &< \bar{m}\,(n_{3+}v_+)\,(n_{4+}v_+)\exp\left(\frac{\Delta E}{k_B T_+}\right) - (n_{1+}v_+)\,(n_{2+}v_+) = \mathscr{F}(v_+) = 0.\end{aligned}$$

Thanks to the previous analysis, it can also be easily shown that

$$v_-^r < v_-^{inert} \leq v_+, \tag{44}$$

for all $M_+ > 1$, meaning also that the mean velocity of the upstream equilibrium in presence of the chemical reaction is greater than in the inert case, $u_-^r > u_-^{inert}$, so that the intensity of the shock structure is increased by the chemical reaction, as confirmed by Fig. 1.

Moreover, it has to be stressed that in the inert case from (36) and (37), observing that $\lim_{M_+ \to 1} v_-^{inert}(\mathbf{u}_+, M_+) = v_+ = c_+ M_+$, it is easily obtained that $\lim_{M_+ \to 1} \mathbf{u}_-^{inert}(\mathbf{u}_+, M_+) = \mathbf{u}_+$, meaning that in the limit $M_+ = 1$ the null-shock solution, i.e. the constant solution $\mathbf{u} = \mathbf{u}_- = \mathbf{u}_+$, is admitted.

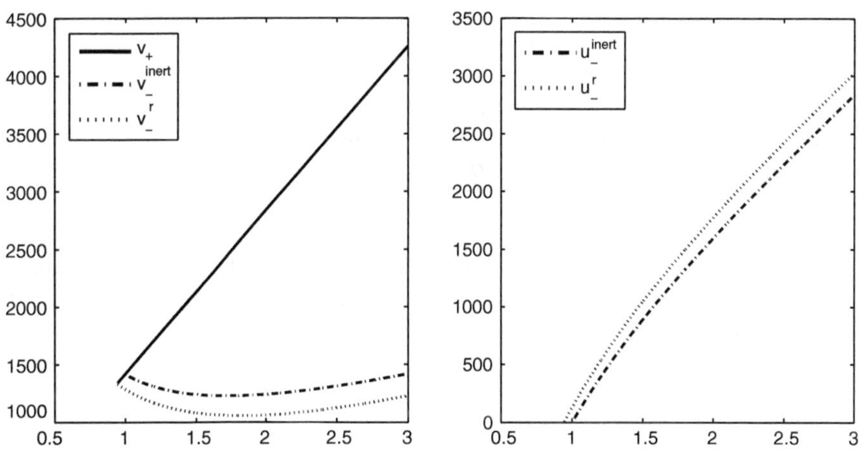

Fig. 1 Plots of v_+ (*continuous*), v_-^r (*dotted*) and v_-^{inert} (*dot-dashed*) (*left*), and of u_-^r (*dotted*) and u_-^{inert} (*dot-dashed*), with $u_+ = 0$ (*right*), versus Mach number M_+, the parameters are relevant to the second test for reaction (59) in Sect. 4

Conversely, in presence of the chemical reaction, by examining the limit as M_+ tends to 1 throughout the proof it is easy to prove that

$$\lim_{M_+\to 1} \mathbf{u}_-^r(\mathbf{u}_+, M_+) = \mathbf{u}_+ \qquad \Leftrightarrow \qquad v_-^r = v_+ = c_+,$$

meaning that the trivial constant solution $\mathbf{u} = \mathbf{u}_- = \mathbf{u}_+$ is still admitted, and obviously coincides with the corresponding one in the inert case. But, as far as non trivial solutions are concerned, in the limit as M_+ tends to 1, it is easily shown that

$$\lim_{M_+\to 1} v_-^r(\mathbf{u}_+, M_+) < v_-^{inert} = v_+ = c_+ M_+,$$

as clearly shown in Fig. 1, where the trend, versus M_+, of the perturbed equilibrium relative velocity v_-^r is plotted, together with the one in the inert case v_-^{inert} corresponding to the same unperturbed value v_+. The values of v_-^r keep lower than the corresponding ones of v_-^{inert}, and of v_+, for all $M_+ > 1$. Moreover, it can be checked that both the slopes of v_-^r and v_-^{inert} have the same asymptotic trend as M_+ tends to infinity. Unfortunately, for $M_+^{sonic} < M_+ < 1$ this analysis can be mainly carried out on the numerical ground.

From Rankine–Hugoniot conditions (32)–(35), it is then easily obtained that

$$\lim_{M_+\to 1} \mathbf{u}_-^r(\mathbf{u}_+, M_+) \neq \mathbf{u}_+,$$

meaning that in the limit $M_+ = 1$ the null-shock (constant) solution is not reached. The constraint of the mass action law limits the configurations of the two equilibria in such a way that the shock front speed is increased with respect to the corresponding either inert, or equilibrium cases.

In the Euler or Grad equations for both a single gas or an inert mixture, the supersonic regime is characterized by the condition

$$M_+ > 1 \qquad \Leftrightarrow \qquad s > \mu_+^{Euler} = u_+ + c_+$$

(which for an inert mixture is the maximum eigenvalue of the equilibrium subsystem), whereas in presence of the reversible bimolecular chemical reaction taken into account it modifies into

$$M_+ > M_+^{sonic} = \frac{C_+}{c_+} = \sqrt{1 - \frac{2}{5\mathscr{C}_+}} \qquad \Leftrightarrow \qquad s > \mu_+ = u_+ + C_+, \tag{45}$$

where, recalling (20) and (23),

$$C_+^2 = c_+^2 \left(1 - \frac{2}{5\mathscr{C}_+}\right), \qquad \sqrt{\frac{3}{5}} < \frac{C_+}{c_+} < 1, \tag{46}$$

(for the details [10] is referred to the reader) and then

$$M_{+}^{sonic} < 1, \qquad \mu_{+} < \mu_{+}^{Euler}.$$

Therefore the presence of a chemical reaction with a non-vanishing chemical energy gap ΔE enlarges the supersonic regime. It is worth to be stressed that, as already outlined at the end of Sect. 2, in the limit $\Delta E \to 0$, the behaviour is analogous to that of the inert mixture, since $M_{+}^{sonic} \to 1$, and $\mu_{+} \to \mu_{+}^{Euler}$, even if the mixture is still reacting: in fact, even in the particular case $\Delta E = 0$, equilibrium mass action law (12) is still valid, and provides a relation involving the four number densities, that remain non-hydrodynamic variables (not governed by conservation equations).

We can conclude that shock wave structure solutions for this reactive mixture exist for all $M_{+} > M_{+}^{sonic}$ and the limit of null-shock solution is reached at the boundary of the sonic regime, i.e.

$$\lim_{M_{+} \to M_{+}^{sonic}} \mathbf{u}_{-}^{r} \left(\mathbf{u}_{+}, M_{+}\right) = \mathbf{u}_{+}.$$

This allows to expect, as shown in Sect. 4, that jump discontinuities may arise even for $M_{+} < 1$, within the supersonic regime, $M_{+} > M_{+}^{sonic}$.

3.2 Singular Barriers and Critical Mach Numbers

We now turn our attention to the so-called singular barrier [2, 12, 13]. The ODEs system (25)–(28) can be written in block diagonal matrix form [2], each block being

$$\mathbf{B}_{i}\left(\mathbf{u}_{i}\right) \cdot \frac{d\mathbf{u}_{i}}{dx} = \mathbf{g}_{i}\left(\mathbf{u}\right), \qquad (47)$$

where $\mathbf{u}_{i} = \left(n_{i}, v_{i}, T_{i}, \sigma_{i}\right)^{T}$, and the singular barrier, i.e. the locus of the singularities of the system, is the manifold characterized by the vanishing of the determinant of $\mathbf{B}_{i}$, whose equation is

$$\frac{3}{2} k_{B} \prod_{i=1}^{4} n_{i} v_{i}^{2} B_{i}\left(\mathbf{u}_{i}\right) = 0, \qquad (48)$$

where

$$B_{i}\left(\mathbf{u}_{i}\right) := m_{i} n_{i} v_{i}^{2} - 3\left(n_{i} k_{B} T_{i} + \sigma_{i}\right) = 0, \qquad i = 1, 2, 3, 4, \qquad (49)$$

are the singularities sub-manifolds, each one related to a component within the mixture.

As shown in [12] and [2], the presence of sub-shocks within a shock structure solution is due to the singularities sub-manifolds lying between the equilibrium states $\mathbf{u}_{+}$ and $\mathbf{u}_{-}^{r}$.

It is also worth to point out the validity of Lax stability conditions across any sub-shock. In fact, for a fixed M_+, if the singular barrier $B_{i^*}(\mathbf{u}) = 0$, relevant to whatever species i^*, lies between $\mathbf{u}_+$ and $\mathbf{u}^r_-$, and if the signs of B_{i^*} evaluated at $\mathbf{u}_+$ and $\mathbf{u}^r_-$ (surely opposite) satisfy the additional requirement $B_{i^*}\left(\mathbf{u}^r_-\right) < 0 < B_{i^*}(\mathbf{u}_+)$, then the shock front speed s and the characteristic speed λ_{i^*} fulfil the relation $\lambda_{i^*}(\mathbf{u}_+) < s < \lambda_{i^*}\left(\mathbf{u}^r_-\right)$. Let us also observe that, denoting by φ_{i^*} the value of φ at which the relevant sub-shock occurs, and by $\mathbf{u}_R$ and $\mathbf{u}_L$ the two generic limiting states ahead and behind the sub-shock front, $\lim_{\varphi\to\varphi_{i^*+}} \mathbf{u} = \mathbf{u}_R$, $\lim_{\varphi\to\varphi_{i^*-}} \mathbf{u} = \mathbf{u}_L$, then it must be $\mathbf{u}_{i^*R} \neq \mathbf{u}_{i^*L}$ and $\mathbf{u}_{jR} = \mathbf{u}_{jL}$, for all $j \neq i^*$, since only the variables relevant to the component i^* may suffer a jump discontinuity, all the others being continuous. Therefore, smoothness arguments on $B_i(\mathbf{u})$ and $\lambda(\mathbf{u})$, together with the information on the sign of $B_i(\mathbf{u})$ on each of the two half-spaces separated by $B_{i^*}(\mathbf{u}) = 0$, assure that $\lambda_{i^*}(\mathbf{u}_R) < s < \lambda_{i^*}(\mathbf{u}_L)$, and hence guaranteeing the stability in the sense of Lax of the sub-shock generated by the species i^*. We can conclude that the shock wave structure solution can be either continuous, or discontinuous with at most four (as many as the components in the mixture) different and stable sub-shocks, located in different positions, φ_i, within the shock front.

As far as the position of $\mathbf{u}_+$ with respect to the barriers $B_i = 0$ is concerned, the same results found in the inert case continue to hold [2]. In fact, it is easy to show that

$$B_i(\mathbf{u}_+) = m_i n_{i+} c_+^2 M_+^2 - 3 n_{i+} k_B T_+ \geq 0 \qquad \Leftrightarrow \qquad M_+ \geq M_{i+}^* := \sqrt{\frac{9}{5}\xi_{i+}}, \tag{50}$$

where

$$\xi_{i+} = \frac{\rho_+}{m_i n_+} = \frac{1}{m_i}\sum_{j=1}^{4} m_j \chi_{j+}, \qquad \chi_{j+} = \frac{n_{j+}}{n_+}.$$

In the inert gas, the quantities M_{i+}^* can be possible critical values for Mach number if they are greater than one; otherwise $B_i(\mathbf{u}_+) > 0$ for all $M_+ > 1$; moreover,

$$M_{i+}^* > 1 \qquad \Leftrightarrow \qquad \xi_{i+} > \frac{5}{9}.$$

For the reactive mixture, the admissibility condition becomes $M_{i+}^* > M_+^{sonic}$; otherwise, $B_i(\mathbf{u}_+) > 0$ for any Mach number; moreover,

$$M_{i+}^* > M_+^{sonic} \qquad \Leftrightarrow \qquad \xi_{i+} > \frac{5}{9} - \frac{2}{9\mathscr{C}_+}.$$

Because of the implicit form of v^r_- due to the presence of the chemical reaction, it is not possible to carry out an explicit analysis concerning the position of $\mathbf{u}^r_-$ with respect to the barriers $B_i = 0$, as done in the inert case [2]. In fact, $B_i\left(\mathbf{u}_-^r\right)$ can be rewritten in terms of Mach number as follows

$$B_i\left(\mathbf{u}_-^r\right) = n_{i-}v_-^r\left[\left(m_i + \frac{3\rho_+}{n_+}\right)v_-^r - \frac{3\rho_+ c_+}{5n_+M_+}\left(5M_+^2+3\right)\right] \tag{51}$$

so, the vanishing of $B_i\left(\mathbf{u}_-^r\right)$ can only be explicitly expressed in terms of the relative velocity

$$B_i\left(\mathbf{u}_-^r\right) \leq 0 \qquad \Leftrightarrow \qquad 0 < v_-^r \leq v_{i-}^* := c_+\frac{5M_+^2+3}{5M_+}\frac{3\xi_{i+}}{3\xi_{i+}+1}, \tag{52}$$

and the corresponding critical value of the Mach number, say M_{i-}^{*r}, can only be evaluated numerically.

However, one can observe that

$$M_{i+}^2 = \frac{9}{5}\xi_{i+} > 1 \qquad \Leftrightarrow \qquad \frac{3\xi_{i+}}{3\xi_{i+}+1} > \frac{5}{8} \tag{53}$$

and that this last inequality implies that

$$v_{i-}^* = c_+\frac{5M_+^2+3}{5M_+}\frac{3\xi_{i+}}{3\xi_{i+}+1} > c_+\frac{5M_+^2+3}{8M_+} > v_-^{inert} > v_-^r \tag{54}$$

for any Mach number $M_+ > 1$; this allows to conclude that

$$M_{i+}^* > 1 \qquad \Rightarrow \qquad B_i\left(\mathbf{u}_-^r\right) < 0, \qquad \forall M_+ > 1. \tag{55}$$

Moreover, let i^* be the index such that $m_{i^*} = \min_{j=1,\dots,4} m_j$; then, since

$$m_{i^*}n_+ < \rho_+ \qquad \Rightarrow \qquad m_{i^*} + 3\frac{\rho_+}{n_+} < 4\frac{\rho_+}{n_+}, \tag{56}$$

hence

$$\frac{3\xi_{i^*+}}{3\xi_{i^*+}+1} > \frac{3}{4} > \frac{5}{8}, \tag{57}$$

and then, like in the inert case [2], for the lightest species it is always $M_{i^*+}^* > 1$, and $B_{i^*}\left(\mathbf{u}_-^r\right) < 0$. Therefore, the lightest component can always generate a stable sub-shock for $M_+ > M_{i^*+}^*$, since

$$B_{i^*}\left(\mathbf{u}_-^r\right) < 0 < B_{i^*}\left(\mathbf{u}_+\right), \qquad \forall M_+ > M_{i^*+}^*.$$

In order to clarify the role of the chemical reaction, it can be interesting to carry out a comparison with the ideal inert mixture case, the unperturbed equilibrium being an arbitrary constant state. As already discussed in [2], in this case the perturbed equilibrium is related to the unperturbed one through the classical Euler Rankine–Hugoniot conditions given by (36) and (37), yielding that

$$B_i \left(\mathbf{u}_{-}^{inert}\right) \leq 0 \quad \Leftrightarrow \quad M_{+} \geq M_{i-}^{*inert} := \sqrt{\frac{3\,(3\xi_{i+}+5)}{5\,(9\xi_{i+}-1)}}, \tag{58}$$

the critical Mach numbers being real and greater than one for $1/9 < \xi_{i+} < 5/9$.

Therefore, even if in presence of the chemical reaction it is not possible to obtain explicit expressions of the critical Mach numbers related to the upstream equilibrium state, the situation is somehow analogous to that holding for an inert mixture [2].

In fact, as it will be shown by some examples in the next section, it is still true that if the critical Mach number generated by the ith component in one of the two, unperturbed or perturbed, equilibria is admissible, in the sense that it is greater than M_{+}^{sonic}, then the critical Mach number related to the same component in the other equilibrium is certainly not admissible, in the sense that it will be less than M_{+}^{sonic}.

The presence of the chemical reaction modifies, with respect to the inert case, the upstream equilibrium which is now constrained to belong to the sub-manifold of the phase space defined by the additional restriction imposed through the mass action law. In the inert case, any pair of constant states related by the classical Rankine–Hugoniot conditions (36), (37) for an Euler gas, with the additional relations for the species number densities, can represent the boundary values in the problem; in the reactive case, both the equilibria are constrained to lie on the chemical equilibrium manifold, and this obviously influences the signs of the barriers, as well as the critical Mach numbers, and therefore the admitted shock structure solutions.

In order to appreciate the effect of chemistry on the solution, we choose a particular chemical reaction and we analyze, through the sign of the different barriers, the behaviour of the critical Mach numbers in various examples (varying the equilibrium configuration) in both inert and reactive cases.

4 Chemical Reactions and Concluding Remarks

As already said, in presence of the chemical reaction, due to the implicit definition of the upstream equilibrium, it is not possible to carry on the analysis in order to characterize all the possible solutions in all the ranges of Mach number, as in the inert case [2]. Moreover, the presence of more than two components in the mixture does not even allow to deduce general informations on behaviour of solutions with respect to equilibrium concentrations and mass ratios, as it was possible in the case of binary mixtures [2, 6].

Therefore, in order to go deeper into the problem of sub-shock formation, we have to fix particle masses and the internal energy gap, that will be chosen to be

$$m_1 = 0.018, \quad m_2 = 0.001, \quad m_3 = 0.017, \quad m_4 = 0.002, \quad \Delta E = 63300,$$

where $m_2 < m_4 < m_3 < m_1$. These parameters are consistent with the chemical reaction

$$H_2O + H \rightleftharpoons OH + H_2 \tag{59}$$

that however involves even polyatomic species and should be suitably modelled by kinetic (and hydrodynamic) equations with a suitable internal energy variable, which is here neglected.

As a first test, the unperturbed equilibrium configuration $\mathbf{u}_+$ is set to be

$$n_{1+} = 0.4, \quad n_{2+} = 0.3, \quad n_{3+} = 0.2, \quad n_{4+} = 0.1, \quad u_+ = 0, \quad T_+ = 2772.7,$$

the temperature being determined through the mass action law (12). With this choice, the supersonic regime is bounded by $M_+^{sonic} = 0.96$, and the admissible critical Mach numbers are all related to the downstream equilibrium

$$M_{1+}^* = 1.05, \qquad M_{2+}^* = 4.47, \qquad M_{3+}^* = 1.08, \qquad M_{4+}^* = 3.16,$$

with $M_+^{sonic} < 1 < M_{1+}^* < M_{3+}^* < M_{4+}^* < M_{2+}^*$, whose order is reverse with respect to that of the corresponding masses. Then, the critical Mach numbers M_{i-}^{*inert}, if real, are surely less than M_+^{sonic}; in particular,

$$M_{1-}^{*inert} = 0.95, \qquad M_{2-}^{*inert} = 0.48, \qquad M_{3-}^{*inert} = 0.92, \qquad M_{4-}^{*inert} = 0.52,$$

and $M_{2-}^{*inert} < M_{4-}^{*inert} < M_{3-}^{*inert} < M_{1-}^{*inert} < M_+^{sonic} < 1$.

Figure 2 clearly shows that only the barriers $B_i(\mathbf{u}_+)$ change their signs for $M_+ > M_{i+}^*$, respectively. More precisely, for any $M_+ > M_{i+}^*$, it is $B_i(\mathbf{u}_-) < 0 < B_i(\mathbf{u}_+)$.

This allows to characterize the five different ranges of possible solutions as follows

- if $M_+^{sonic} < M_+ < M_{1+}^*$, only continuous solutions exist, since $B_i(\mathbf{u}_+)\, B_i\left(\mathbf{u}_-^r\right) > 0$, for all $i = 1, 2, 3, 4$;
- if $M_{1+}^* < M_+ < M_{3+}^*$, then a sub-shock may appear in the variables relevant to species 1, since $B_1(\mathbf{u}_+)\, B_1\left(\mathbf{u}_-^r\right) < 0$, and $B_i(\mathbf{u}_+)\, B_i\left(\mathbf{u}_-^r\right) > 0$, for all $i = 2, 3, 4$;
- if $M_{3+}^* < M_+ < M_{4+}^*$, then two different sub-shocks may appear in the variables relevant to species 1 and 3, since $B_i(\mathbf{u}_+)\, B_i\left(\mathbf{u}_-^r\right) < 0$, for all $i = 1, 3$, and $B_i(\mathbf{u}_+)\, B_i\left(\mathbf{u}_-^r\right) > 0$, for all $i = 2, 4$;
- if $M_{4+}^* < M_+ < M_{2+}^*$, then three different sub-shocks may appear in the variables relevant to species 1, 3 and 4, since in this case $B_i(\mathbf{u}_+)\, B_i\left(\mathbf{u}_-^r\right) < 0$, for all $i = 1, 3, 4$, and $B_2(\mathbf{u}_+)\, B_2\left(\mathbf{u}_-^r\right) > 0$;
- if $M_+ > M_{2+}^*$, then four different sub-shocks, one for each component, may appear, since in this last case $B_i(\mathbf{u}_+)\, B_i\left(\mathbf{u}_-^r\right) < 0$, for all $i = 1, 2, 3, 4$.

In the inert case, the only difference in the ranges, with respect to the reactive case, is that $M_+^{sonic} \equiv 1$.

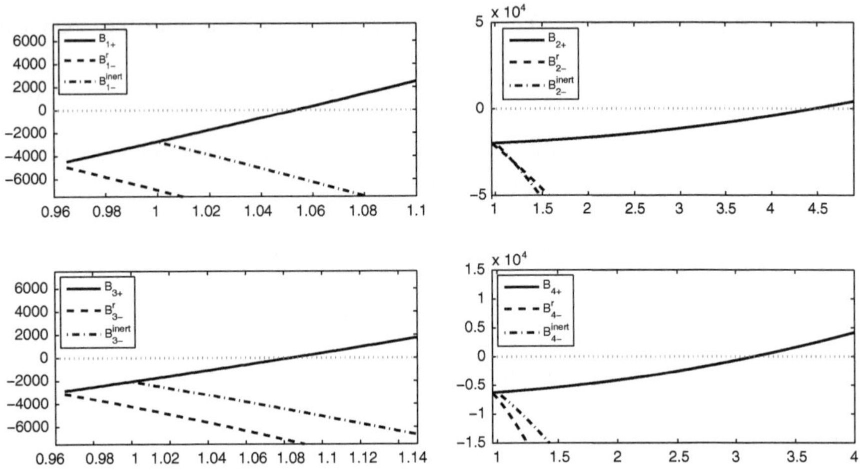

Fig. 2 Plots of $B_i(\mathbf{u}_+)$ (*continuous*), $B_i\left(\mathbf{u}_-^r\right)$ (*dashed*) and $B_i\left(\mathbf{u}_-^{inert}\right)$ (*dot-dashed*) versus Mach number M_+, for reaction (59), with $n_{1+} = 0.4$, $n_{2+} = 0.3$, $n_{3+} = 0.2$, $n_{4+} = 0.1$, $u_+ = 0$, $T_+ = 2772.7$

The results become much more interesting in a second test performed by considering the following equilibrium configuration

$$n_{1+} = 0.8, \quad n_{2+} = 0.9, \quad n_{3+} = 0.05, \quad n_{4+} = 0.1, \quad u_+ = 0, \quad T_+ = 1285.2,$$

in which the concentrations of species 1 and 2, which are the products of the exothermic reaction, are both larger than those of both the other species. As it will be shown, the possible behaviours of the shock wave structure solutions are different in the two cases.

With this choice, in fact, the sonic Mach number is $M_+^{sonic} = 0.942$, and the critical Mach numbers due to the unperturbed equilibrium are

$$M_{1+}^* = 0.94, \qquad M_{2+}^* = 3.99, \qquad M_{3+}^* = 0.97, \qquad M_{4+}^* = 2.82,$$

with $M_{1+}^* < M_+^{sonic} < M_{3+}^* < 1 < M_{4+}^* < M_{2+}^*$, and

$$M_{1-}^{*inert} = 1.07, \qquad M_{2-}^{*inert} = 0.49, \qquad M_{3-}^{*inert} = 1.03, \qquad M_{4+}^{*inert} = 0.53,$$

with $M_{2-}^{*inert} < M_{4-}^{*inert} < M_+^{sonic} < 1 < M_{3-}^{*inert} < M_{1-}^{*inert}$.

In this case, as far as the inert case is concerned [2], only M_{2+}^* and M_{4+}^* are greater than one and, as shown in [2], this allows to deduce that $B_{2,4}\left(\mathbf{u}_-^{inert}\right) < 0$, for all $M_+ > 1$. This circumstance allows to expect that both species 2 and 4 generate a stable sub-shock, since it will be, analogously to the previous test, $B_i(\mathbf{u}_-) < 0 < B_i(\mathbf{u}_+)$, for all $M_+ > M_{i+}^*$, with $i = 2, 4$, both in the inert and reactive case. Figure 3 clearly

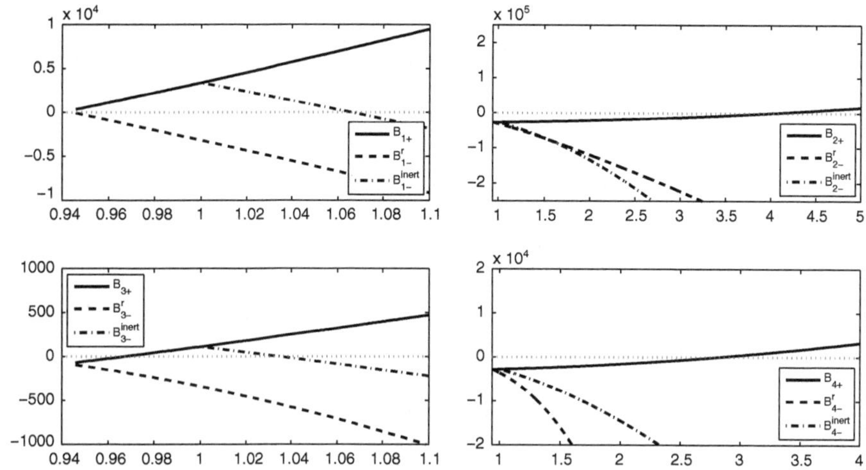

Fig. 3 Plots of $B_i(\mathbf{u}_+)$ *(continuous)*, $B_i\left(\mathbf{u}_-^r\right)$ *(dashed)* and $B_i\left(\mathbf{u}_-^{inert}\right)$ *(dot-dashed)* versus Mach number M_+, for reaction (59), with $n_{1+} = 0.8$, $n_{2+} = 0.9$, $n_{3+} = 0.05$, $n_{4+} = 0.1$, $u_+ = 0$, $T_+ = 1285.2$

shows the intersections of B_2 and B_4 with the horizontal axes at $M_+ = M_{2+}^*$ and $M_+ = M_{4+}^*$, respectively.

Moreover, the fact that M_{1+}^* and M_{3+}^* are both less than one means that $B_1(\mathbf{u}_+)$, $B_3(\mathbf{u}_+) > 0$, for all $M_+ > 1$, as shown in Fig. 3, guaranteeing that $s > \lambda_{1,3}(\mathbf{u}_+)$, for all $M_+ > 1$.

As shown in [2], it happens that $M_{1,3+}^* < 1$ implies $M_{1,3-}^{*inert} > 1$, and then, as in the previous test (except for the nature and the order of the critical Mach numbers), it is $B_{1,3}\left(\mathbf{u}_-^{inert}\right) < 0 < B_{1,3}(\mathbf{u}_+)$, or equivalently $s < \lambda_{1,3}\left(\mathbf{u}_-^{inert}\right)$, for all $M_+ > M_{1,3-}^{*inert}$, and each component, for suitable values of M_+, can generate a sub-shock. In details, since $1 < M_{3-}^{*inert} < M_{1-}^{*inert} < M_{4+}^* < M_{2+}^*$, the possible shock structure solutions are the following

- if $1 < M_+ < M_{3-}^{*inert}$, only continuous solutions exist, since $B_i(\mathbf{u}_+)\, B_i\left(\mathbf{u}_-^{inert}\right) > 0$, for all $i = 1, 2, 3, 4$;
- if $M_{3-}^{*inert} < M_+ < M_{1-}^{*inert}$, a sub-shock may appear in the variables relevant to species 3 since $B_3\left(\mathbf{u}_-^{inert}\right) < 0 < B_3(\mathbf{u}_+)$;
- if $M_{1-}^{*inert} < M_+ < M_{4+}^*$, then two different sub-shocks may appear in the variables relevant to species 1 and 3 since $B_{1,3}\left(\mathbf{u}_-^{inert}\right) < 0 < B_{1,3}(\mathbf{u}_+)$;
- if $M_{4+}^* < M_+ < M_{2+}^*$, then three different sub-shocks may arise since in this case $B_{1,3,4}\left(\mathbf{u}_-^{inert}\right) < 0 < B_{1,3,4}(\mathbf{u}_+)$;
- if $M_+ > M_{2+}^*$, then four different sub-shocks, one for each component, may appear since in this case $B_i\left(\mathbf{u}_-^{inert}\right) < 0 < B_i(\mathbf{u}_+)$, for all $i = 1, 2, 3, 4$.

In the reactive case, the situation is different.

Figure 3 clearly shows the intersections of B_2, B_3 and B_4 with the horizontal axes at $M_+ = M_{2+}^*$, $M_+ = M_{3+}^*$ and $M_+ = M_{4+}^*$, respectively. The admissible unperturbed critical Mach numbers are now M^*_{2+}, M^*_{3+}, and M^*_{4+}, since $M^{sonic}_+ < M^*_{3+} < 1 < M^*_{4+} < M^*_{2+}$, and this, through (55) and since species 2 is the lightest, allows to deduce that

$$B_i\left(\mathbf{u}^r_-\right) < 0, \qquad \forall M_+ > 1, \quad i = 2, 4.$$

Numerical simulations show that it is also

$$B_{2,3,4}\left(\mathbf{u}^r_-\right) < 0, \qquad \forall M_+ > M^{sonic}_+.$$

This circumstance allows to expect that species 2, 3 and 4 generate a stable sub-shock, since it will be, analogously to the previous test,

$$B_i\left(\mathbf{u}^r_-\right) < 0 < B_i\left(\mathbf{u}_+\right), \qquad \forall M_+ > M^*_{i+}, \qquad i = 2, 3, 4.$$

Moreover, the relation $M^*_{1+} < M^{sonic}_+$ again assures that

$$B_1\left(\mathbf{u}_+\right) > 0, \qquad \forall M_+ > M^{sonic}_+,$$

as in the inert case. Differently from the inert case, from the relation $M^*_{1+} < 1$, it is not possible to derive analytically any information either on the existence of the critical Mach number M^{*r}_{1-}, or on its position with respect to 1.

In Fig. 3 the numerical investigation carried out on $B_1\left(\mathbf{u}^r_-\right)$, varying M_+, shows that

$$B_1\left(\mathbf{u}^r_-\right) < 0, \qquad \forall M_+ > M^{*r}_{1-},$$

with $M^{*r}_{1-} > M^{sonic}_+$, and it should be appreciated that $B_1\left(\mathbf{u}^r_-\right)$ intersects the horizontal axis for a Mach which is very close to M^{sonic}_+.

So for all $M_+ > M^{*r}_{1-}$ the presence of the chemical reaction yields that

$$\lambda_1\left(\mathbf{u}_+\right) < s < \lambda_1\left(\mathbf{u}^r_-\right), \qquad \forall M_+ > M^{*r}_{1-},$$

and a stable shock may appear in the variables relevant to constituent 1, even for Mach numbers M_+ lower than 1.

The second test also shows how the unperturbed configuration affects the solution. With respect to the first, in the second test, in fact, the unperturbed concentration of one of the heavier (but not the heaviest), component 1, and of the lightest species 2 are increased, and the others are decreased. The effect is that, due to its higher concentration, even if it is not the heaviest, species 1 may generate a sub-shock for Mach numbers, $M_+ > M^{*r}_{1-}$, very close to M^{sonic}_+, and still lower than those, $M_+ > M^*_{3+}$, at which the heaviest species 3 can produce a jump discontinuity.

In presence of the chemical reaction, since $M^{sonic}_+ < M^{*r}_{1-} < M^*_{3+} < M^*_{4+} < M^*_{2+}$, the scenario modifies as follows

- if $M^{sonic}_+ < M_+ < M^{*r}_{1-}$, only continuous solutions exist, since $B_i\left(\mathbf{u}_+\right) B_i\left(\mathbf{u}^r_-\right) > 0$, for all $i = 1, 2, 3, 4$;

- if $M^{*r}_{1-} < M_+ < M^*_{3+}$, then a sub-shocks may appear in the variables relevant to species 1 since $B_1\left(\mathbf{u}^r_-\right) < 0 < B_1\left(\mathbf{u}_+\right)$;
- if $M^*_{3+} < M_+ < M^*_{4+}$, analogously to the inert case, two different sub-shocks may appear in the variables relevant to species 1 and 3, since $B_{1,3}\left(\mathbf{u}^r_-\right) < 0 < B_{1,3}\left(\mathbf{u}_+\right)$;
- if $M^*_{4+} < M_+ < M^*_{2+}$, the situation is analogous to the corresponding one in the inert case: three different sub-shocks may appear in the variables relevant to species 1, 3 and 4, since $B_{1,3,4}\left(\mathbf{u}^r_-\right) < 0 < B_{1,3,4}\left(\mathbf{u}_+\right)$;
- if $M_+ > M^*_{2+}$, again the situation is analogous to the corresponding one in the inert case: four different sub-shocks, one for each component, may appear since $B_i\left(\mathbf{u}^r_-\right) < 0 < B_i\left(\mathbf{u}_+\right)$, for all $i = 1, 2, 3, 4$.

As already pointed out, in the present reactive frame it is not possible to produce the same general overview of the different solutions, in terms of Mach numbers, mass ratios, and equilibrium concentrations shown in [2] for the inert binary mixture. However, even in presence of a chemical reaction we may deduce that the solution may exhibit up to four different sub-shocks within the shock front, as many as the components are, depending on how many critical Mach numbers are admitted, and on which are exceeded by the fixed value of the Mach number (or of the wave front speed), or, in other words, on which of the singularity sub-manifolds, $B_i = 0$, lie between the two equilibria $\mathbf{u}_+$ and $\mathbf{u}^r_-$.

The constant solution $\mathbf{u}^{inert}_- = \mathbf{u}_+$, which for the inert mixture represents the null shock limit as $M_+ \to 1$, in the reactive case is reached (providing $\mathbf{u}^r_- = \mathbf{u}_+$) at sonic threshold M^{sonic}_+, which is now lower than 1 due to the chemical processes. The range of Mach numbers allowing only continuous solutions is then a right neighbourhood of M^{sonic}_+, and sub-shocks may arise for Mach numbers lower than one, if the minimum of the admissible critical values of Mach number is below one. We can finally stress that the inert case can be recovered in the limit as $\Delta E \to 0$ of the reactive mixture.

Acknowledgments This work is performed in the frame of activities sponsored by INdAM–MIUR and by the Universities of Messina and Parma. The work of G. Martalò has been carried out with financial support from the French State, managed by the French National Research Agency (ANR) in the frame of the "Investments for the future" Programme IdEx Bordeaux—CPU (ANR-10-IDEX-03-02). F. Conforto and G. Martalò are grateful to GNFM for the financial support of the research project "Sub-shock formation in gas mixtures". M. Bisi and G. Martalò have been supported also by the French-Italian program Galileo, project G14-34, "Kinetic modelling and numerical simulations of reactive gaseous mixtures and plasmas for nuclear fusion". Fruitful discussions with A. Mentrelli and T. Ruggeri on the subject of the present work are gratefully acknowledged.

References

1. Bernhoff, N., Bobylev, A.: Weak shock waves for the general discrete velocity model of the Boltzmann equation. Commun. Math. Sci. **5**, 815–832 (2007)
2. Bisi, M., Conforto, F., Martalò, G.: Sub-shock formation in Grad 10 moment equations for a binary gas mixture. Continuum Mech. Thermodyn. (2015). doi:10.1007/s00161-015-0476-8
3. Bisi, M., Groppi, M., Spiga, G.: Grad's distribution functions in the kinetic equations for a chemical reaction. Continuum Mech. Thermodyn. **14**, 207–222 (2002)

4. Bisi, M., Groppi, M., Spiga, G.: Kinetic approach to chemically reacting gas mixtures. In: Pareschi, L., Russo, G., Toscani, G. (eds.) Modelling and Numerics of Kinetic Dissipative Systems, pp. 85–104. Nova Science, New York (2006)
5. Bisi, M., Martalò, G., Spiga, G.: Multi-temperature fluid-dynamic model equations from kinetic theory in a reactive gas: the steady shock problem. Comput. Math. Appl. **66**, 1403–1417 (2013)
6. Bisi, M., Martalò, G., Spiga, G.: Shock wave structure of multi-temperature Euler equations from kinetic theory for a binary mixture. Acta Appl. Math. **132**, 95–105 (2014)
7. Boillat, G., Ruggeri, T.: Hyperbolic principal subsystem: entropy convexity and subcharacteristic conditions. Arch. Rat. Mech. Anal. **137**, 305–320 (1997)
8. Boillat, G., Ruggeri, T.: On the shock structure problem for hyperbolic system of balance laws and convex entropy. Continuum Mech. Thermodyn. **10**(5), 285–292 (1998)
9. Cercignani, C., Frezzotti, A., Grosfils, P.: The structure of an infinitely strong shock wave. Phys. Fluids **11**, 2757–2764 (1999)
10. Conforto, F., Groppi, M., Monaco, R., Spiga, G.: Steady detonation problem for slow and fast chemical reactions. In: Pareschi, L., Russo, G., Toscani, G. (eds.) Modelling and Numerics of Kinetic Dissipative Systems, pp. 105–117. Nova Science, New York (2006)
11. Conforto, F., Mentrelli, A., Ruggeri, T.: Shock structure and multiple sub-shocks in hyperbolic systems of balance laws: the case of a multi-temperature mixture of Eulerian fluids. Preprint (2016)
12. Conforto, F., Monaco, R., Ricciardello, A.: Discontinuous shock structure in a reacting mixture modelled by Grad 13 moment approximation. Acta Appl. Math. **132**, 225–236 (2014)
13. Currò, C., Fusco, D.: Discontinuous travelling wave solutions for a class of dissipative hyperbolic models. Rend. Mat. Acc. Lincei series 9, **16**(1), 61–71 (2005)
14. Fernandes, A.S., Reinecke, S., Kremer, G.M.: A combined Chapman-Enskog and Grad method. *III* Polyatomic gases in magnetic fields. Continuum Mech. Thermodyn. **9**, 309–322 (1997)
15. Groppi, M., Rjasanow, S., Spiga, G.: A kinetic relaxation approach to fast reactive mixtures: shock wave structure. J. Stat. Mech.—Theory Exp. P10010 (2009)
16. Kosuge, S., Aoki, K., Takata, S.: Shock wave structure of a binary gas mixture: finite-difference analysis of the Boltzmann equation for hard-sphere molecules. Eur. J. Mech. B—Fluids **20**, 87–126 (2001)
17. Madjarević, D., Ruggeri, T., Simić, S.: Shock structure and temperature overshoot in macroscopic multi-temperature model of mixtures. Phys. Fluids **26**, 106102 (2014)
18. Madjarević, D., Simić, S.: Shock structure in Helium-Argon mixture—a comparison of hyperbolic multi-temperature model with experiment. Europhys. Lett. **102**, 44002 (2013)
19. Müller, I., Ruggeri, T.: Rational Extended Thermodynamics. Springer, New York (1988)
20. Raines, A.: Study of a shock wave structure in gas mixtures on the basis of the Boltzmann equation. Eur. J. Phys. B. **21**, 599–610 (2002)
21. Ruggeri, T.: Breakdown of shock-wave-structure solutions. Phys. Rev. E **47**(6), 4135–4140 (1993)
22. Tcheremissine, F.G., Kolobov, V.I., Arslanbekov, R.R.: Simulation of shock wave structure in nitrogen with realistic rotational spectrum and molecular interaction potential. In: Ivanov, M.S., Rebrov, A.K. (eds.) Proceedings of 25th Rarefied Gas Dynamics Symposium, pp. 203–208. Siberian Branch of the Russian Academy of Sciences, Novosibirsk (2007)

Compactness of Linearized Kinetic Operators

Laurent Boudin and Francesco Salvarani

Abstract This article reviews various results on the compactness of the linearized Boltzmann operator and of its generalization to mixtures of non-reactive monatomic gases.

Keywords Boltzmann equation · Boltzmann system for mixtures · Linearized Boltzmann operator · Compactness properties · Grad's procedure

1 Introduction

Investigating the compactness properties of linearized operators arising in kinetic theory is a crucial step to establish fluid dynamical approximations to solutions of the corresponding kinetic equations.

The starting point of this research line was given by Hilbert, in the same paper in which he introduced what we now call the Hilbert expansion [23]. Since then, a significant number of results allowed to clarify the main aspects of these compactness properties, both for the archetypal of all kinetic models, the classical Boltzmann

This work was partially funded by the French ANR-13-BS01-0004 project *Kibord* and by the French ANR-14-ACHN-0030-01 project *Kimega*.

L. Boudin
Laboratoire Jacques-Louis Lions (LJLL, CNRS UMR 7598),
REO Project-team, Sorbonne Universités, CNRS, UPMC Univ. Paris 06, INRIA,
4, Place Jussieu, 75005 Paris, France
e-mail: laurent.boudin@upmc.fr

F. Salvarani (✉)
Ceremade, UMR CNRS 7534, Université Paris-Dauphine, PSL Research University,
75775 Paris Cedex 16, France
e-mail: francesco.salvarani@unipv.it

F. Salvarani
Dipartimento di Matematica, Università Degli Studi di Pavia,
27100 Pavia, Italy

© Springer International Publishing Switzerland 2016

P. Gonçalves and A.J. Soares (eds.), *From Particle Systems to Partial Differential Equations III*, Springer Proceedings in Mathematics & Statistics 162,
DOI 10.1007/978-3-319-32144-8_4

equation, and, more recently, for the variant of the system of Boltzmann equations describing the behaviour of non-reactive mixtures constituted with monatomic gases.

In the mixture case, the interaction between the different species induces some peculiarities in the structure of the linearized kinetic operators, which can reflect some specific physical phenomena (such as uphill diffusion in the purely diffusive case, see [3, 5, 16, 24, 26, 29]). Consequently, it is not surprising that compactness properties in the mixture case cannot be deduced through a straightforward adaptation of the standard methods of proof from the mono-species case. In [4], the authors indeed observed that, when there are different involved molecular masses, the standard approach, which is mostly due to Grad [20], degenerates. A new method of proof is needed to recover the linearized operator compactness. Let us point out that this new argument does not hold either when the molecular masses become equal. Hence, both aforementioned strategies must be seen as complementary when dealing with mixtures.

The study of compactness properties for mixtures is only at its beginning, and there are still many unexplored situations. We quote for example the study of linearized kinetic operators for mixtures of polyatomic gases: the non-reactive case, for instance as defined in [7, 14], and the chemical reacting one as in [15]. Those models require a supplementary internal variables, such as the internal energy of the molecules. The presence of such a variable induces significant difficulties for the analysis of the compactness properties.

The article is divided into two parts. The next section is dedicated to the study of the compactness properties of the linearized standard Boltzmann operator for a monatomic perfect gas, including discussions related to Grad's angular cut-off assumption. Then, in Sect. 3, we consider the extension to a non-reactive mixture of ideal monatomic gases.

2 The Classical Boltzmann Equation Case

This section deals with the compactness properties of the classical linearized Boltzmann kernel. The subject has a long history, since the first study comes back to Hilbert [23], who applied his new theory of integral operators to this specific problem.

2.1 Boltzmann's Equation

As a starting point, we briefly introduce the classical Boltzmann equation. This equation is the first and most studied kinetic model since the nineteenth century, after the pioneering works of Boltzmann himself [1, 2] and Maxwell [26]. Since this equation has been widely studied, we only introduce its basic aspects and refer to the many reference texts on the Boltzmann equation, for instance [11, 12, 30], where the arguments below are more accurately discussed.

The Boltzmann equation describes the time evolution of a system composed by a large number of particles, described by a distribution function f defined on the phase space of the system. The particles are supposed to be identical and monatomic. They follow the classical mechanics laws, with only translational degrees of freedom. If the particles are contained in a domain $\Omega_x \subseteq \mathbb{R}^3$, the quantity $f(t, x, v)$ can be defined for any $(t, x, v) \in \mathbb{R}^+ \times \Omega_x \times \mathbb{R}^3$, and, for all t, the integral

$$\iint_{X \times V} f(t, x, v) \, \mathrm{d}v \, \mathrm{d}x$$

can be interpreted as the number of particles in the space volume $X \subseteq \Omega_x$ with velocity in $V \subseteq \mathbb{R}^3$. A reasonable assumption on f is

$$f(t, \cdot, \cdot) \in L^1_{\text{loc}}(\Omega_x; L^1(\mathbb{R}^3)), \qquad \forall\, t \in \mathbb{R}^+,$$

which ensures that there is always a finite number of particles in a bounded domain of the space. For the sake of simplicity, we also assume that the system is isolated, so that there is no external effect on the particles.

If, moreover, the particles do not interact with each other, the time evolution of f is driven by the so-called free transport equation

$$\frac{\partial f}{\partial t} + v \cdot \nabla_x f = 0. \tag{1}$$

When the interaction between particles cannot be neglected, Eq. (1) does not hold any more, and one has to add in (1) a right-hand-side term. When only binary and local collisions are allowed, the effect of the interactions is described by a quadratic (with respect to f) collision operator $Q(f, f)$.

If the pairwise interactions between particles of the system are elastic, then momentum and kinetic energy are conserved during the interaction process. Hence, if we denote by v' and v'_* the pre-collisional velocities, and by v and v_* the post-collisional ones, the following microscopic conservation laws hold:

$$v + v_* = v' + v'_*, \qquad v^2 + {v_*}^2 = v'^2 + {v'_*}^2.$$

These conservation laws allow to fix four of the six degrees of freedom of the interaction. The remaining degrees of freedom of the binary interaction can be described in several ways. For our purposes, we only consider two possible descriptions. The first one, the so-called σ-representation, is the representation of the pre-post velocities in the centre of mass of two particles: we introduce $\sigma \in S^2$, and write

$$v' = \frac{1}{2}(v + v_* + |v - v_*|\sigma), \qquad v'_* = \frac{1}{2}(v + v_* - |v - v_*|\sigma). \tag{2}$$

The second one is the ω-representation, defined as

$$v' = v - (\omega \cdot (v - v_*))\,\omega, \qquad v'_* = v_* + (\omega \cdot (v - v_*))\,\omega, \tag{3}$$

where $\omega \in S^2$. From a geometrical point of view, the unit vector σ represents the direction of the pre-collisional relative velocity, whereas the reflection with respect to the plane $\omega^\perp$ orthogonal to ω changes $v - v_*$ into $v' - v'_*$.

Hence, the time evolution of f is governed by the Boltzmann equation

$$\frac{\partial f}{\partial t} + v \cdot \nabla_x f = Q(f, f), \tag{4}$$

where the collision operator Q can be defined either in the σ-representation (2) by

$$Q(f,g) = \int_{\mathbb{R}^3} \int_{S^2} B(\sigma, v, v_*) \left[f(t,x,v')g(t,x,v'_*) - f(t,x,v)g(t,x,v_*) \right] \, d\sigma \, dv_* \tag{5}$$

or in the ω-representation

$$Q(f,g) = \int_{\mathbb{R}^3} \int_{S^2} B(\omega, v, v_*) \left[f(t,x,v')g(t,x,v'_*) - f(t,x,v)g(t,x,v_*) \right] \, d\omega \, dv_*. \tag{6}$$

Either way, in the study of the Boltzmann equation, particular care has to be given to the properties of the collision cross-section $B : S^2 \times \mathbb{R}^3 \times \mathbb{R}^3 \to \mathbb{R}^+$ of the system, which describes the details of the interactions between the particles.

In general, by symmetry arguments and thanks to the Galilean invariance, it is possible to prove that B, which is nonnegative, in fact depends on $|v - v_*|$ and $\cos\theta := \sigma \cdot (v - v_*)/|v - v_*|$, where $|\cdot|$ denotes the Euclidean norm in $\mathbb{R}^3$ and θ represents the deviation angle between the pre- and post-collisional relative velocities. For the sake of simplicity from now on, we write B as a function of σ and $V := v - v_*$. If we assume the collisions to be microreversible, we can state that

$$B(\sigma, v - v_*) = B(\sigma, v' - v'_*), \quad \forall \sigma \in S^2, \ \forall v, v_* \in \mathbb{R}^3. \tag{7}$$

The choice of collision kernel B has a deep influence on the properties of the Boltzmann equation. By limiting ourselves to the classical elastic case, it is possible to prove that

$$B(\sigma, v - v_*) = K\,|v - v_*|, \qquad K > 0, \tag{8}$$

for a gas of three-dimensional hard spheres. In the case of inverse s-power binary forces between particles (for example, $s = 2$ corresponding to Coulomb interactions and $s = 7$ to Van der Waals interactions, see [30] for more details), B can be factorized as

$$B(\sigma, v - v_*) = \Phi(|v - v_*|)b(\cos\theta), \tag{9}$$

where, in three space dimensions,

$$\Phi(|V|) = |V|^\gamma, \quad \gamma = (s-5)/(s-1),$$

and b is a locally smooth function with a non integrable singularity when θ tends to 0, i.e.

$$b(\cos\theta)\sin\theta \sim K\theta^{-(1+\eta)}, \quad \eta = 2/(s-1).$$

Factorized collision kernels like (9) are very popular in the study of the classical Boltzmann equation. By convention, Φ is named the *kinetic* collision kernel, and b the *angular* one. The class of kinetic collision kernels of the form $\Phi(|V|) = |V|^\gamma$ is usually split in three sub-classes, depending on the value of γ. When $\gamma > 0$, the kernel is said to derive from hard potentials, when $\gamma < 0$, the kernel is said to derive from soft potentials and, when $\gamma = 0$, the kinetic collision kernel does not play any role. In this latter situation, the corresponding Boltzmann equation describes the behaviour a gas of Maxwell molecules. Even if it is only a mathematical model, it is very popular in the literature, since it considerably simplifies the study of the Boltzmann equation. Let us point out that Maxwell and Boltzmann themselves used this model, because it allows to carry out many explicit computations.

In order to handle more easily the angular cross section, Grad [19] (see also [11]) suggested a working hypothesis, nowadays known as the *Grad angular cut-off assumption*. It consists in postulating that the collision kernel is integrable with respect to the angular variable. Note that the great majority of mathematical works about the Boltzmann equation is based on this Grad cut-off assumption, which could be considered, from the physical point of view, as a short-range assumption [30].

To conclude these considerations about the cross-section, let us emphasize that, whenever we use B in this article, we shall use a notation abuse for the sake of simplicity. We may as well write the variables of B, as (v, v_*, ω), (V, ω), (v, v_*, σ), (V, σ) or $(|V|, \cos\theta)$.

Now define the normalized centred Maxwellian

$$M(v) = \left(\frac{1}{2\pi}\right)^{3/2} e^{-v^2/2}, \quad v \in \mathbb{R}^3,$$

and a perturbation g to M as

$$f = M + M^{1/2}g.$$

The linearized collision operator $\mathscr{L}$ is studied for instance in [17] and can be defined by

$$\mathscr{L}g = \frac{1}{\sqrt{M}}\left[Q(\sqrt{M}g, M) + Q(M, \sqrt{M}g)\right]. \tag{10}$$

More precisely, $\mathscr{L}$ can be written as

$$\mathscr{L} = \mathscr{K} - \nu\,\mathrm{Id},$$

where $\mathscr{K}$ is given by

$$\mathscr{K} g(v) = \left(\frac{1}{2\pi}\right)^{3/2} \iint_{\mathbb{R}^3\times S^2} B(\sigma, v - v_*)\, e^{-v^2/4}\, e^{-v_*^2/2} \left[e^{v_*'^2/4} g(v_*') - e^{v_*^2/4} g(v_*) + e^{v'^2/4} g(v')\right] \mathrm{d}\sigma\, \mathrm{d}v_*, \quad (11)$$

and the collision frequency ν by

$$\nu(v) = \left(\frac{1}{2\pi}\right)^{3/2} \iint_{\mathbb{R}^3\times S^2} B(\sigma, v - v_*) e^{-v_*^2/2}\, \mathrm{d}\sigma\, \mathrm{d}v_*. \quad (12)$$

2.2 Earlier Compactness Results

The first result of compactness for $\mathscr{K}$ is given by Hilbert in [23] for the three-dimensional hard sphere case: Hilbert uses the now so-called Hilbert's theory of integral operators to write $\mathscr{K}$ as a kernel operator, and then obtains a compactness property.

Then, in [22], and in the same cross-section setting, Hecke presents a variant of the previous result: he proves that the linearized Boltzmann kernel is roughly of Hilbert-Schmidt type.

Carleman [10] provides an improvement to the latter result and significantly simplifies the proof. We must emphasize that those various compactness results are established in different L^2 settings, which may involve ν in the weights.

2.3 Grad's Procedure

In [20], Grad presents an extension of Hilbert's result in both Maxwell and hard potential cases, by supposing $\gamma \in [0, 1]$, and by using his angular cut-off assumption [19]. He requires that the form of the cross-section B is either (8) or (9), with a uniformly bounded angular cross-section b. More precisely, Grad imposes the following general condition on B:

$$B(\omega, V) \le a\,|\sin\theta|\,|\cos\theta| \left(|V| + \frac{1}{|V|^{1-\delta}}\right), \quad \forall\, \omega \in S^2,\ \forall\, V \in \mathbb{R}^3, \quad (13)$$

where $a > 0$, $0 < \delta < 1$. This allows to adapt Hecke's argument and prove that the kernel of $\mathcal{K}$ is Hilbert-Schmidt in $L^2(M\mathrm{d}v)$.

Let us provide more details about Grad's procedure. In order to prove its compactness in L^2, the operator $\mathcal{K}$ is written as the sum of two operators, $\mathcal{K}_1$ and $\mathcal{K}_2$, where, for any $v \in \mathbb{R}^3$,

$$\mathcal{K}_1 g(v) = -\left(\frac{1}{2\pi}\right)^{3/2} \iint_{\mathbb{R}^3\times S^2} B(\omega, v - v_*)\, e^{-\frac{1}{4}v^2}\, e^{-\frac{1}{4}{v_*}^2}\, g(v_*)\, \mathrm{d}\omega\, \mathrm{d}v_*,$$

$$\mathcal{K}_2 g(v) = \left(\frac{1}{2\pi}\right)^{3/2} \iint_{\mathbb{R}^3\times S^2} B(\omega, v - v_*)\, e^{-\frac{1}{4}v^2}\, e^{-\frac{1}{2}{v_*}^2} \times \left[e^{\frac{1}{4}{v_*'}^2}\, g(v_*') + e^{\frac{1}{4}v'^2}\, g(v')\right] \mathrm{d}\omega\, \mathrm{d}v_*.$$

Both operators $\mathcal{K}_1$ and $\mathcal{K}_2$ have a kernel form. This is straightforward for $\mathcal{K}_1$: indeed, if we set

$$k_1(v, v_*) = -\left(\frac{1}{2\pi}\right)^{3/2} \int_{S^2} B(\omega, v - v_*)\, e^{-\frac{1}{4}v^2}\, e^{-\frac{1}{4}v_*^2}\, \mathrm{d}\omega \qquad \forall\, v, v_* \in \mathbb{R}^3,$$

we clearly have

$$\mathcal{K}_1 g(v) = \int_{\mathbb{R}^3} k_1(v, v_*)\, g(v_*)\, \mathrm{d}v_*, \qquad \forall\, v \in \mathbb{R}^3.$$

The analogous result for $\mathcal{K}_2$ is more intricate and is detailed in the next lines.

We begin by using the microscopic collision rules (3) to write $\mathcal{K}_2 g$ in terms of v, v_* and v' only (hence, any dependence on v_*' disappears). The following lemma holds:

Lemma 1 *There exists a nonnegative function $\tilde{B}$ satisfying (13), such that, for all $v \in \mathbb{R}^3$,*

$$\mathcal{K}_2 g(v) = \iint_{\mathbb{R}^3\times S^2} \tilde{B}(\omega, v - v_*)\, e^{-\frac{1}{4}v^2 - \frac{1}{2}v_*^2 + \frac{1}{4}v'^2}\, g(v')\, \mathrm{d}\omega\, \mathrm{d}v_*. \tag{14}$$

Proof The key point of the proof lies in some geometrical properties of symmetry in the microscopic collision process. By involving the relative velocity $V = v - v_*$, it is possible to choose a unit vector $\omega^\perp \in \mathrm{Span}(V, \omega)$, orthogonal to ω. Then we clearly have

$$V = \omega(\omega \cdot V) + \omega^\perp(\omega^\perp \cdot V).$$

The previous equality allows to write that

$$v - (\omega \cdot V)\omega = v_* + (\omega^\perp \cdot V)\omega^\perp, \qquad v_* + (\omega \cdot V)\omega = v - (\omega^\perp \cdot V)\omega^\perp. \tag{15}$$

Note that the pre-collision relative velocity for the same post-collisional one V, but with respect to $\omega^\perp$, are obtained by a simple exchange between v' and v'_*. This means that the transformation $\omega \mapsto \omega^\perp$ induces $v' \mapsto v'_*$ and $v'_* \mapsto v'$.

Hence, by replacing ω by $\omega^\perp$, we get

$$\iint_{\mathbb{R}^3\times S^2} B(\omega, V) e^{-\frac{1}{2}v_*^2} e^{\frac{1}{4}(v_*+(\omega\cdot V)\omega)^2} g(v_* + (\omega\cdot V)\omega)\,\mathrm{d}\omega\,\mathrm{d}v_* =$$

$$\iint_{\mathbb{R}^3\times S^2} B(\omega^\perp, V) e^{-\frac{1}{2}v_*^2} e^{\frac{1}{4}(v_*+(\omega^\perp\cdot V)\omega^\perp)^2} g(v_* + (\omega^\perp\cdot V)\omega^\perp)\,\mathrm{d}\omega^\perp\,\mathrm{d}v_*.$$

The change of variables from ω to $\omega^\perp$ is a rotation, so that its Jacobian equals 1. By (15), the previous integral becomes

$$\iint_{\mathbb{R}^3\times S^2} B(\omega^\perp, V) e^{-\frac{1}{2}v_*^2} e^{\frac{1}{4}v'^2} g(v')\,\mathrm{d}\omega^\perp\,\mathrm{d}v_*$$
$$= \iint_{\mathbb{R}^3\times S^2} B(\omega^\perp, V) e^{-\frac{1}{2}v_*^2} e^{\frac{1}{4}v'^2} g(v')\,\mathrm{d}\omega\,\mathrm{d}v_*.$$

Let us set

$$\tilde{B}(\omega, V) = \left(\frac{1}{2\pi}\right)^{3/2} \left[B(\omega, V) + B(\omega^\perp, V)\right].$$

The estimate (13) on B guarantees that

$$\tilde{B}(\omega, V) \le 2\,a \left(\frac{1}{2\pi}\right)^{3/2} |\sin\theta|\,|\cos\theta| \left(|V| + |V|^{\delta-1}\right). \tag{16}$$

The thesis of the lemma is hence proved. □

The previous lemma allows to prove the following result.

Proposition 1 *There exists* $C > 0$ *such that*

$$\mathscr{K}_2 g(v) \le C \int_{\mathbb{R}^3} g(\eta)\, k_2(\eta, v)\,\mathrm{d}\eta, \quad \forall\, v \in \mathbb{R}^3,$$

where

$$k_2(\eta, v) = e^{-\frac{1}{8}(\eta-v)^2 - \frac{1}{8}\frac{(\eta^2-v^2)^2}{(\eta-v)^2}} |\eta - v|^{-1}, \quad \forall\, \eta, v \in \mathbb{R}^3.$$

Proof Using the change of variables $v_* \mapsto V_* = v_* - v$, of Jacobian equal to 1, in (14), we can write

$$\mathscr{K}_2 g(v) = \iint_{\mathbb{R}^3\times S^2} e^{-\frac{1}{4}v^2} e^{-\frac{1}{2}(V_*+v)^2} e^{\frac{1}{4}v'^2} g(v')\,\tilde{B}(\omega, V_*)\,\mathrm{d}\omega\,\mathrm{d}V_*. \tag{17}$$

Then, let us denote $V_* = p + q$, with $p = \omega(\omega \cdot V_*)$ and $q = V_* - \omega(\omega \cdot V_*)$. Note that the component q belongs to the plane $\Pi = \{\omega\}^\perp = \{p\}^\perp$.

Consider now the change of variables

$$(V_*, \omega) \mapsto (p, q), \qquad \mathbb{R}^3 \times S^2 \to \mathbb{R}^3 \times \Pi. \tag{18}$$

We have to be very careful with the integration order in the change of variables, because q strongly depends on p. More precisely, we first integrate with respect to q since $\Pi = \{p\}^\perp$, then we combine the one-dimensional integration in the direction ω with the integral on $\omega \in S^2$ to obtain a three-dimensional integration over all rectangular components of $|p|\omega$. Moreover, the Jacobian of (18) is given by

$$\mathrm{d}V_* \, \mathrm{d}\omega = \frac{2}{p^2 \sin(p, p+q)} \, \mathrm{d}p \, \mathrm{d}q.$$

Since it is clear that $v' = v + p$, (17) becomes, abused by the abuse of notation $\tilde{B}(p, p+q) = \tilde{B}(\omega, V_*)$,

$$\mathscr{K}_2 g(v) = 2 \int_{\mathbb{R}^3} \int_\Pi \tilde{B}(p, p+q) e^{-\frac{1}{4}v^2 - \frac{1}{2}(p+q+v)^2 + \frac{1}{4}(v+p)^2} \, g(v+p) \, |p|^{-2} \, |\sin(p, p+q)|^{-1} \, \mathrm{d}q \, \mathrm{d}p. \tag{19}$$

Using the fact that $p \cdot q = 0$, we deduce

$$-\frac{1}{4} v^2 + \frac{1}{4}(v+p)^2 - \frac{1}{2}(p+q+v)^2 = -\frac{1}{8} p^2 - \frac{1}{2}\left[q + \frac{1}{2}(2v+p)\right]^2,$$

which allows to write

$$\mathscr{K}_2 g(v) = 2 \int_{\mathbb{R}^3} \int_\Pi \tilde{B}(p, p+q) e^{-\frac{1}{8}p^2 - \frac{1}{2}[q + \frac{1}{2}(2v+p)]^2} \, g(v+p) \, |p|^{-2} \, |\sin(p, p+q)|^{-1} \, \mathrm{d}q \, \mathrm{d}p.$$

Let $z = v + p/2$, and consider z_1 its component parallel to ω, and denote $z_2 = z - z_1 \in \Pi$. Then, using the straightforward equality $(q + v + p/2)^2 = {z_1}^2 + (q + z_2)^2$, $\mathscr{K}_2 g$ becomes

$$\mathscr{K}_2 g(v) = 2 \int_{\mathbb{R}^3} e^{-\frac{1}{8}p^2 - \frac{1}{2}{z_1}^2} \, g(v+p) \, |p|^{-2} \int_\Pi \tilde{B}(p, p+q) e^{-\frac{1}{2}(q+z_2)^2} \, |\sin(p, p+q)|^{-1} \, \mathrm{d}q \, \mathrm{d}p. \tag{20}$$

We are led to prove that the integral

$$\Delta := \frac{1}{|p|} \int_{\Pi} \tilde{B}(p, p+q) e^{-\frac{1}{2}(q+z_2)^2} \, |\sin(p, p+q)|^{-1} \, dq$$

is upper-bounded, uniformly with respect to $p \in \mathbb{R}^3$ and $z_2 \in \Pi$. From (16), we obtain, for some constant $C > 0$,

$$\frac{\tilde{B}(p, p+q)}{|\sin(p, p+q)|} \le C \, |\cos(p, p+q)| \left(|p+q| + |p+q|^{\delta-1} \right).$$

Using $|\tan(p, p+q)| = |q|/|p|$, we can write

$$\frac{\tilde{B}(p, p+q)}{|\sin(p, p+q)|} \le C \left(1 + \frac{q^2}{p^2} \right)^{-\frac{1}{2}} \left[(p^2+q^2)^{\frac{1}{2}} + (p^2+q^2)^{\frac{\delta-1}{2}} \right].$$

This implies that

$$\frac{\tilde{B}(p, p+q)}{|p| \, |\sin(p, p+q)|} \le C \left[1 + (p^2+q^2)^{\frac{\delta}{2}-1} \right] \le C \left[1 + |q|^{\delta-2} \right],$$

using the fact that $\delta < 1$. It is now convenient to split the range of integration, in the expression of Δ, into $|q| \le 1$ and $|q| \ge 1$, and get

$$\Delta \le C \left(\int_{|q| \le 1} (1 + |q|^{\delta-2}) dq + \int_{|q| \ge 1} e^{-\frac{1}{2}(q+z_2)^2} dq \right).$$

The right-hand-side of the estimate is clearly upper-bounded by a universal constant.

To conclude the proof, we perform the change of variable $p \mapsto \eta = p + v$ in (20) after using the uniform upper bound of Δ. Then the thesis of the proposition is a consequence of the following equalities:

$${z_1}^2 = \left(z \cdot \frac{\eta - v}{|\eta - v|} \right)^2 = \left(\frac{1}{2}(\eta + v) \cdot \frac{(\eta - v)}{|\eta - v|} \right)^2 = \frac{1}{4} \frac{(\eta^2 - v^2)^2}{|\eta - v|^2}.$$

□

The compactness of $\mathscr{K}$ then appears as a consequence of the following properties:

- uniform decay at infinity:

$$\|\mathscr{K} g\|_{L^2(B(0,R)^c)} \le \zeta(R) \, \|g\|_{L^2(\mathbb{R}^3)}, \qquad \forall R > 0,$$

where $B(0, R)$ is the open ball of $\mathbb{R}^3_v$ centred at 0 and of radius R, and $\zeta(R)$ goes to 0 when R goes to $+\infty$;

- equicontinuity: for any $\varepsilon > 0$, there exists $\rho > 0$ such that, for all $w \in B(0, \rho)$,

$$\|(\tau_w - \mathrm{Id})\mathscr{K} g\|_{L^2(\mathbb{R}^3)} \le \varepsilon \|g\|_{L^2(\mathbb{R}^3)},$$

where Id is the identity and τ_w the translation operator

$$\tau_w \mathscr{K} g\,(v) = \mathscr{K} g\,(v + w), \qquad \forall v, w \in \mathbb{R}^3.$$

In [20], Grad provided the required estimates on the kernels k_1 and k_2, which allows to prove the compactness of $\mathscr{K}$.

2.4 *Extensions in the Cut-off Case*

Caflisch [8, 9] extended Grad's result to the soft potential case by treating Grad cut-off kernels with $\gamma \in (-1, 1]$ in three space dimensions.

In [18], Golse and Poupaud are interested in studying the stationary solutions of the three-dimensional linearized Boltzmann equation in a half-space. An important step in their proof strategy consists in obtaining the compactness of the linearized Boltzmann operator in $L^2(M\mathrm{d}v)$ for Grad cut-off kernels with $\gamma \in (-2, 1]$. By defining

$$f^* := M^{1/2} f \text{ and } L^*(f*) = M^{-1/2}\mathscr{L}(f),$$

they introduce the operator

$$K^*(f^*) := f^* - L^*(f^*).$$

Subsequently, to prove the compactness of K^*, they use some growth estimates for the operator and an iteration technique, which allows to deduce that $(K^*)^4$ is a Hilbert-Schmidt operator on $L^2(\mathrm{d}v)$. Since K^* is self-adjoint on $L^2(\mathrm{d}v)$, the linearized Boltzmann operator itself is compact in $L^2(M\mathrm{d}v)$.

More recently, Guo [21] extended Caflisch's result for Grad cut-off kernels to the range $\gamma \in (-3, 1]$ in three space dimensions.

The last result that we quote in this subsection is due to Levermore and Sun [25]. They prove a L^p compactness result for the gain parts of the linearized Boltzmann collision operator (in any dimension D) associated with weakly cut-off collision kernels that derive from a power-law intermolecular potential. In their proof, they assume that the cross-section has the form

$$B(|v - v_*|, \cos\theta) = |v - v_*|^\gamma\, b(\cos\theta), \quad \gamma \in (-D, +\infty)$$

where $b \in L^1(S^{D-1})$ is an even function. We really need these assumptions on b in order for B to be locally integrable in all its variables, which allows to give sense to both the gain and loss parts of the collision operator. In fact, the linearized Boltzmann operator $\mathscr{L}$ is split in the following way:

$$\mathscr{L} = \nu \times (\mathrm{Id} + \mathscr{K}_1 - \mathscr{K}_2 - \mathscr{K}_3) f,$$

where the loss operator $\mathscr{K}_1$ and the gain operators $\mathscr{K}_2$ and $\mathscr{K}_3$ are respectively given by

$$\mathscr{K}_1 f(v) = \frac{1}{\nu(v)} \iint_{\mathbb{R}^D \times S^{D-1}} B(\sigma, v - v_*) f(v_*) M(v_*) \mathrm{d}\sigma \, \mathrm{d}v_*,$$
$$\mathscr{K}_2 f(v) = \frac{1}{\nu(v)} \iint_{\mathbb{R}^D \times S^{D-1}} B(\sigma, v - v_*) f(v') M(v_*) \mathrm{d}\sigma \, \mathrm{d}v_*,$$
$$\mathscr{K}_3 f(v) = \frac{1}{\nu(v)} \iint_{\mathbb{R}^D \times S^{D-1}} B(\sigma, v - v_*) f(v'_*) M(v_*) \mathrm{d}\sigma \, \mathrm{d}v_*,$$

and the collision frequency ν is the D-dimensional analogous of (12).

They first prove that, under the assumptions written above, the operators $\mathscr{K}_j : L^p(\nu M \mathrm{d}v) \to L^p(\nu M \mathrm{d}v)$ are compact, $1 \le j \le 3$. Once proved the compactness result for L^p with $p = 2$, the result for every $p \in (1, \infty)$ is deduced thanks to a straightforward interpolation argument and the following compactness criterion, which generalizes the classical Hilbert-Schmidt property.

Lemma 2 *Let K be an integral operator given by*

$$Kf(v) = \int_{\mathbb{R}^D} k(v, v') f(v') \, \mathrm{d}\mu(v'),$$

where $\mathrm{d}\mu$ is a σ-finite, positive measure over $\mathbb{R}^D$. Let the kernel $k(v, v')$ be symmetric in v and v' and, for some $r \in [1, 2]$, satisfy the bound

$$\|k\|_{L^s(L^r)} := \left(\int_{\mathbb{R}^D} \left(\int_{\mathbb{R}^D} |k(v, v')|^r \mathrm{d}\mu(v') \right)^{s/r} \mathrm{d}\mu(v) \right)^{1/s} < +\infty,$$

where $s \in [2, +\infty]$ is defined by $1/r + 1/s = 1$. Let $p, q \in [r, s]$ such that $1/p + 1/q = 1$. Then, for any $f \in L^p(\mathrm{d}\mu)$ and $g \in L^q(\mathrm{d}\mu)$, the following estimate holds:

$$\int_{\mathbb{R}^D} |g(v) K f(v)| \mathrm{d}\mu(v) \le \iint_{\mathbb{R}^D \times \mathbb{R}^D} |k(v, v') f(v') g(v)| \mathrm{d}\mu(v) \mathrm{d}\mu(v')$$
$$\le \|k\|_{L^s(L^r)} \|f\|_{L^p} \|g\|_{L^q}.$$

Consequently, $K : L^p(\mathrm{d}\mu) \to L^p(\mathrm{d}\mu)$ is bounded and satisfies $|||K|||_{L^p} \le \|k\|_{L^s(L^r)}$. Moreover, if $r \in (1, 2]$ then $K : L^p(\mathrm{d}\mu) \to L^p(\mathrm{d}\mu)$ is compact.

2.5 Extensions to Kernels Without Cut-off

In [27], among other topics, Mouhot and Strain investigate compactness properties of both linearized Boltzmann and Landau operators to obtain explicit spectral gap and coercivity estimates. In this section, we only discuss the Boltzmann case, since the study of the Landau operator is not the purpose of this review article. They improve an earlier result of Pao [28], by using a completely different approach, and establish the Fredholm alternative for a broad class of collision kernels without any small deflection cut-off assumption. First note that it is well-known that $\mathscr{L}$ is an unbounded symmetric operator on L^2, see, for example, [11].

The cross-sections considered in their article have the form

$$B(|v - v_*|, \cos\theta) = |v - v_*|^\gamma\, b(\cos\theta), \qquad \gamma \in (-3, +\infty), \tag{21}$$

where b behaves as follows:

$$b(\cos\theta) \underset{\theta\to 0}{\sim} b^*(\theta)\ (\sin\theta/2)^{-2-\alpha}, \qquad \alpha \in [0, 2), \tag{22}$$

where b^* is a nonnegative function, bounded and non-zero near $\theta = 0$. When $\alpha \geq 0$, the angular singularity is not integrable: hence, we indeed deal with the non cut-off case.

By using the change of variable $\sigma \mapsto -\sigma$, the angular cross-section b can be replaced by its symmetric form

$$\tilde{b}(\cos\theta) = \frac{\mathbf{1}_{[0,\pi/2]}(\theta)}{2}\ [b(\cos\theta) + b(\cos(\pi - \theta))]\,.$$

In what follows, we are mostly interested in establishing the compactness of the collisional operator, and we do not take into account the (of course interesting) consequences on the spectral gap estimate.

The first part of the proof consists in a technical estimate on the linearized collision operator by assuming that the cross-section B is of variable hard spheres type, i.e. it does not depend on the angular variable:

$$B_q(|v - v_*|) = |v - v_*|^q, \qquad q \in (-3, +\infty).$$

The linearized collision operator $\mathscr{L}$ corresponding to B_q is then written in the following form:

$$\mathscr{L}g = \nu\, g - Kg,$$

where the multiplicative local part ν can be seen as the convolution

$$\nu(v) := \iint_{\mathbb{R}^3\times S^2} M(v)\, B_q(|v - v_*|)\, \mathrm{d}v_*\, \mathrm{d}\sigma = \left|S^2\right| \left(|\cdot|^q * M\right)(v),$$

and the non-local part writes

$$Kg(\upsilon) := \iint_{\mathbb{R}^3\times S^2} B_q(|\upsilon-\upsilon_*|)\Big[g(\upsilon')\,M^{1/2}(\upsilon'_*) + g(\upsilon'_*)\,M^{1/2}(\upsilon') \\ - g(\upsilon_*)\,M^{1/2}(\upsilon)\Big]\,M^{1/2}(\upsilon_*)\,\mathrm{d}\upsilon_*\,\mathrm{d}\sigma.$$

In fact, K itself is composed by a pure convolution part

$$K^c g(\upsilon) := |S^2|\,\big[|\cdot|^q * (M^{1/2}\,g)\big](\upsilon)\,M^{1/2}(\upsilon)$$

and by the remainder

$$K^+g(\upsilon) := \iint_{\mathbb{R}^3\times S^2} B_q(|\upsilon-\upsilon_*|)\Big[g(\upsilon')\,M^{1/2}(\upsilon'_*) + g(\upsilon'_*)\,M^{1/2}(\upsilon')\Big] \\ M^{1/2}(\upsilon_*)\,\mathrm{d}\upsilon_*\,\mathrm{d}\sigma.$$

Using the change of variable $\sigma \mapsto -\sigma$ in a part of the integral, the previous expression becomes

$$K^+g(\upsilon) := 2\iint_{\mathbb{R}^3\times S^2} B_q(|\upsilon-\upsilon_*|)g(\upsilon')\,M^{1/2}(\upsilon'_*)\,M^{1/2}(\upsilon_*)\,\mathrm{d}\upsilon_*\,\mathrm{d}\sigma.$$

At the formal level, and rigorously only when B_q is locally integrable with respect to the angular variable, we can associate to B_q a kernel $k_q := k_q(\upsilon,\upsilon')$ such that

$$K^+g(\upsilon) = \int_{\mathbb{R}^3} g(\upsilon')\,k_B(\upsilon,\upsilon')\,\mathrm{d}\upsilon', \qquad \upsilon\in\mathbb{R}^3.$$

Hence, the authors can apply Grad's strategy, as in Sect. 2.3, by first studying again the locally integrable case. In that situation, they prove the following preliminary results. The first one provides an explicit expression to the kernel.

Lemma 1 *For $q > -1$, the explicit formula holds:*

$$k_q(\upsilon,\upsilon') = \frac{8}{|\upsilon'-\upsilon|\,(2\pi)^{3/2}}\exp\left[-\frac{|\upsilon'-\upsilon|^2}{8} - \frac{|\upsilon'-\upsilon+2(\upsilon\cdot\omega)\omega|^2}{8}\right] \\ \times\left(\int_{\{\omega\}^\perp} |\upsilon'-\upsilon+z|^{q-(3-2)}\exp\left[-\frac{|z+(\upsilon-(\upsilon\cdot\omega)\omega)\,|^2}{2}\right]\mathrm{d}z\right).$$

The second one gives an a priori estimate on the kernel.

Proposition 2 *The kernel k_q is symmetric with respect to υ and υ', and, for any $q > -1$ and $s\in\mathbb{R}$, satisfies the estimate*

$$\int_{\mathbb{R}^3} k_q(v, v')\,(1 + |v'|)^s\, dv' \le C_{q,s}\,(1 + |v|)^{q+s-2}, \qquad v \in \mathbb{R}^3,$$

where $C_{q,s}$ is a constant which only depends on q and s.

The previous results are then extended to the non locally integrable case. More precisely, consider a cross-section B satisfying a condition of the type

$$B(|v - v_*|, \cos\theta) \ge K\, B_{\gamma,\alpha}(|v - v_*|, \cos\theta)\, \mathbf{1}_{[0,\theta_0]}(\theta), \qquad v,\ v_* \in \mathbb{R}^3, \quad \theta \in [0, \pi],$$

where $K > 0$ and $\theta_0 \in (0, \pi]$ are constants, and $B_{\gamma,\alpha}$ is given, for any $\gamma \in (-3, +\infty)$, $\alpha \in [0, 2)$, by

$$B_{\gamma,\alpha}(|v - v_*|, \cos\theta) = |v - v_*|^\gamma\, \sin^{-2-\alpha}(\theta/2), \qquad v,\ v_* \in \mathbb{R}^3, \quad \theta \in [0, \pi].$$

In order to use their preliminary results, the authors focus on a fictitious self-adjoint operator on L^2 defined by

$$\hat{L}g(v) = \iint_{\mathbb{R}^3 \times S^2} |v - v_*|^{\gamma+\alpha+2}\, \mathbf{1}_{[0,1]}(|v - v'|)\, \mathbf{1}_{[0,\theta_0]}(\theta)\, M^{1/2}(v)\, M(v_*) \left[-\frac{g(v')}{M(v')^{1/2}} - \frac{g(v'_*)}{M(v'_*)^{1/2}} + \frac{g(v)}{M(v)^{1/2}} + \frac{g(v_*)}{M(v_*)^{1/2}} \right] dv_*\, d\sigma,$$

where both v' and θ are considered as functions of v, v_* and σ.

This operator is then written as the sum of several operators

$$\hat{L} = \hat{\nu}\ \mathrm{Id} - \hat{K}^+ + \hat{K}^c,$$

and the authors prove that the right-hand side is the sum of Hilbert-Schmidt operators, which implies the operator compactness in L^2.

In fact, this Hilbert-Schmidt property is straightforward for the multiplicative operator $\hat{\nu}$ Id. Indeed, it is clear that there exists a constant $C > 0$ such that

$$\hat{\nu}(v) = \iint_{\mathbb{R}^3 \times S^2} |v - v_*|^{\gamma+\alpha+2}\, \mathbf{1}_{[0,1]}(|v - v'|)\, \mathbf{1}_{[0,\theta_0]}(\theta)\, M(v_*)\, dv_*\, d\sigma \ge C\,(1 + |v|)^{\gamma+\alpha}.$$

Note that an analogous result is immediate for $\hat{K}^c$.

Unfortunately, the situation is more intricate with the operator $\hat{K}^+$. In the case when $\gamma + \alpha = 0$, $\hat{K}^+$ can be written as a limit of Hilbert-Schmidt operators. Indeed, the kernel of $\hat{K}^+$ is, by simple inspection, $\hat{k} := k_2(v, v')\, \mathbf{1}_{[0,1]}(|v - v'|)\, \mathbf{1}_{[0,\theta_0]}(\theta)$, and hence,

$$\hat{k} = \frac{8}{|v'-v|\,(2\pi)^{3/2}} \exp\left\{-\frac{|v'-v|^2}{8} - \frac{|v'-v+2(v\cdot\omega)\omega|^2}{8}\right\}$$
$$\times \left(\int_{\omega^\perp} |v'-v+z| \exp\left\{-\frac{|z+(v-(v\cdot\omega)\omega)|^2}{2}\right\} \mathrm{d}z\right)$$
$$\times\, \mathbf{1}_{[0,1]}(|v-v'|)\, \mathbf{1}_{[0,\theta_0]}(\theta).$$

This kernel is then approximated as follows: it is split into

$$\hat{k} = \hat{k}^c_\varepsilon + \hat{k}^r_\varepsilon,$$

with

$$\hat{k}^c_\varepsilon = \left[\mathbf{1}_{[\varepsilon,1]}(|v-v'|) \times \mathbf{1}_{[\varepsilon,1]}\left(\left|\frac{v}{|v|}\cdot\frac{(v-v')}{|v-v'|}\right|\right)\right]\hat{k},$$

and, obviously,

$$\hat{k}^r_\varepsilon = \hat{k} - \hat{k}^c_\varepsilon.$$

The authors prove that $\hat{k}^r_\varepsilon$ is symmetric in v, v' and that

$$\lim_{\varepsilon\to 0}\ \sup_{v\in\mathbb{R}^3} \int_{\mathbb{R}^3} |\hat{k}^r_\varepsilon|\,\mathrm{d}v' = 0.$$

Therefore, the sequence of operators $\hat{K}^{+,c}_\varepsilon$ associated to kernels $\hat{k}^c_\varepsilon$ converges to $\hat{K}^+$ in L^2 when ε goes to 0. Hence, we only have to prove that each $\hat{K}^{+,c}_\varepsilon$ is compact. Note, then, that the kernel $\hat{k}^c_\varepsilon$ satisfies

$$\iint_{\mathbb{R}^3\times\mathbb{R}^3} \left(\hat{k}^c_\varepsilon\right)^2 \mathrm{d}v\,\mathrm{d}v'$$
$$\le C \int_{\mathbb{R}^3}\int_\varepsilon^1 (1+r)^2\,(1+|v|\,\sin\theta)^2\, e^{-\frac{r^2}{4}}$$
$$\int_0^\pi e^{-\frac{(r+2|v|\cos\theta)^2}{4}} \sin\theta\, \mathbf{1}_{[\varepsilon,1]}(|\cos\theta|)\,\mathrm{d}\theta\,\mathrm{d}r\,\mathrm{d}v$$
$$\le C\int_{\mathbb{R}^3} (1+|v|)^2\, e^{-\varepsilon^2|v|^2}\,\mathrm{d}v,$$

which clearly is a finite quantity. Consequently, $\hat{K}^{+,c}_\varepsilon$ is a Hilbert-Schmidt operator.

When $\gamma+\alpha\neq 0$, one considers the following symmetric weighted modification of $\hat{L}$:

$$\tilde{L} = (1+|\cdot|)^{-(\gamma+\alpha)/2}\,\hat{L}\left((1+|\cdot|)^{-(\gamma+\alpha)/2}\,\cdot\right)$$

and the corresponding decomposition $\tilde{L} = \tilde{\nu} - \tilde{K}^+ + \tilde{K}^c$. Then $\tilde{\nu}$ is uniformly strictly positive and upper-bounded.

The authors conclude their argument by proving that $\tilde{K}^c$ is a Hilbert-Schmidt operator. They first focus on the term $\tilde{K}^+$. Its kernel is

$$(1 + |v|)^{-(\gamma+\alpha)/2}\, k_{\gamma+\alpha+2}(v, v')\, (1 + |v'|)^{-(\gamma+\alpha)/2}\, \mathbf{1}_{[0,1]}(|v - v'|)\, \mathbf{1}_{[0,\theta_0]}(\theta)$$

and similar computations as above allow to prove again that $\tilde{K}^+$ can be written as a limit of Hilbert-Schmidt operators.

3 The Compactness Properties for the Linearized Kinetic Operators for Mixtures

In this section, mainly following [13] for the definition of the linearized operator, and [4] for the compactness result, we investigate the case of an ideal gas mixture, with monatomic species.

The main difficulty in the mixture case lies in the fact that we have to deal with species with different masses. Indeed, in this situation, we loose the symmetry between pre- and post-collisional velocities, which was crucial in Grad's strategy. Consequently, we need a new argument to recover the compactness of the linearized Boltzmann operator. Note that this result, detailed in Proposition 2 below, appears as an equivalence of the Euclidean norms of the variables (v, v'_*) and (v', v_*), which degenerates when the masses become equal. This means that the mono-species and multi-species cases must really be treated in two different ways.

3.1 Building the Linearized Collision Operator for Mixtures

Each of the $I \geq 2$ species are described through a distribution function f_i, $1 \leq i \leq I$. As in the mono-species case, this function depends on time $t \in \mathbb{R}_+$, space position $x \in \mathbb{R}^3$ and velocity $v \in \mathbb{R}^3$. In the following, we also use the macroscopic density of species i, defined by

$$n_i(t, x) = \int_{\mathbb{R}^3} f_i(t, x, v)\, \mathrm{d}v.$$

The interactions between molecules are assumed to remain elastic, so that two colliding molecules of species i and j, $1 \leq i, j \leq I$, with respective molecular (or molar) masses m_i and m_j, see their velocities modified through the collision rules

$$v' = \frac{m_i v + m_j v_*}{m_i + m_j} + \frac{m_j}{m_i + m_j} T_\omega(v - v_*), \tag{23}$$

$$v'_* = \frac{m_i v + m_j v_*}{m_i + m_j} - \frac{m_i}{m_i + m_j} T_\omega(v - v_*), \tag{24}$$

where $\omega \in S^2$ and T_ω denotes the symmetry with respect to the plane $\{\omega\}^\perp$, i.e.

$$T_\omega z = z - 2(\omega \cdot z)\omega, \qquad \forall z \in \mathbb{R}^3.$$

Then the collision operator related to species i and j is given by

$$Q_{ij}(f,g)(v) = \iint_{\mathbb{R}^3 \times S^2} \left[f(v')g(v'_*) - f(v)g(v_*) \right] B_{ij}(\omega, v - v_*) \, \mathrm{d}\omega \, \mathrm{d}v_*, \tag{25}$$

where f and g are general functions depending on the velocity variable. The cross-sections B_{ij}, $1 \le i, j \le I$, satisfy an analogous property to (7) and a similar condition to the one in the classical case (13), namely

$$B_{ij}(\omega, V) \le a\,|\sin\theta|\,|\cos\theta| \left(|V| + \frac{1}{|V|^{1-\delta}} \right), \quad \forall \omega \in S^2,\ \forall V \in \mathbb{R}^3, \tag{26}$$

where $a > 0$ and $0 < \delta < 1$ are again given constants which do not depend on i, j, and θ denotes the oriented angle between ω and V.

The time evolution of each distribution function f_i, $1 \le i \le I$, is then given by

$$\frac{\partial f_i}{\partial t} + v \cdot \nabla_x f_i = \sum_{j=1}^{I} Q_{ij}(f_i, f_j). \tag{27}$$

One can write weak forms of the collision operators using the changes of variables $(v, v_*) \mapsto (v_*, v)$ and $(v, v_*) \mapsto (v', v'_*)$ with a fixed $\omega \in S^2$. It is worth noticing that cases $i = j$ and $i \neq j$ are intrinsically different, see [6, 15] for more details. Moreover, we can formally write, for any i and j, and any functions (of v) f and g

$$\int_{\mathbb{R}^3} Q_{ij}(f,g)(v) \, \mathrm{d}v = 0, \tag{28}$$

$$\int_{\mathbb{R}^3} Q_{ij}(f,g)(v) \begin{pmatrix} m_i v \\ m_i v^2/2 \end{pmatrix} \mathrm{d}v + \int_{\mathbb{R}^3} Q_{ji}(g,f)(v) \begin{pmatrix} m_j v \\ m_j v^2/2 \end{pmatrix} \mathrm{d}v = 0. \tag{29}$$

One can also write an H-theorem [6, 15], which allows to obtain Maxwell functions as equilibrium. From now on, let us denote M_i the normalized, centred Maxwell function related to species i

$$M_i(v) = \left(\frac{m_i}{2\pi} \right)^{3/2} e^{-\frac{m_i}{2} v^2}, \quad \forall v \in \mathbb{R}^3. \tag{30}$$

Now we are ready to write the linearized collision operator $\mathscr{L}$ for mixtures. We shall work in a L^2 setting again. More precisely, for any function $g \in L^2(\mathbb{R}^3)^I$ of v, we shall write the L^2 norm of g as

$$\|g\|_{L^2}^2 = \sum_{j=1}^{I} \|g_j\|_{L^2}^2 = \sum_{j=1}^{I} \int_{\mathbb{R}^3} g_j(v)^2 \, \mathrm{d}v.$$

Consider now macroscopic densities $(n_1, \ldots, n_I)$ as given and define the standard perturbation $g = (g_1, \ldots, g_I)$ to $M = (M_1, \ldots, M_I)$ by

$$f_i = n_i M_i + n_i M_i^{1/2} g_i, \qquad 1 \le i \le I.$$

By defining the ith component of $\mathscr{L} g$ as

$$[\mathscr{L} g]_i = M_i^{-1/2} \sum_{j=1}^{I} n_i n_j \left(Q_{ij}(M_i, M_j^{1/2} g_j) + Q_{ij}(M_i^{1/2} g_i, M_j) \right), \tag{31}$$

for any function $g = (g_1, \ldots, g_I)$ and $1 \le i \le I$, as well as the ith component of $\mathscr{Q}(g, g)$ by

$$[\mathscr{Q}(g, g)]_i = M_i^{-1/2} \sum_{j=1}^{I} n_i n_j Q_{ij}(M_i^{1/2} g_i, M_j^{1/2} g_j), \qquad 1 \le i \le I, \tag{32}$$

the Boltzmann equations on the components of the perturbation g write

$$\partial_t g_i + v \cdot \nabla_x g_i + [\mathscr{L} g]_i = [\mathscr{Q}(g, g)]_i, \qquad 1 \le i \le I. \tag{33}$$

If we introduce the operator $\mathscr{K}$ such that the ith component of $\mathscr{K} g$ is given by

$$[\mathscr{K} g]_i\,(v) = \sum_{j=1}^{I} \left(\frac{m_i m_j}{4\pi^2} \right)^{3/4} n_i n_j \iint_{\mathbb{R}^3 \times S^2} B_{ij}(\omega, v - v_*)\, e^{-\frac{1}{4} m_i v^2}\, e^{-\frac{1}{2} m_j v_*^2} \tag{34}$$

$$\left[\left(\frac{m_i}{2\pi} \right)^{3/4} \left(e^{\frac{1}{4} m_j v_*'^2}\, g_j(v_*') - e^{\frac{1}{4} m_j v_*^2} g_j(v_*) \right) + \left(\frac{m_j}{2\pi} \right)^{3/4} e^{\frac{1}{4} m_i v'^2}\, g_i(v') \right] \mathrm{d}\omega\, \mathrm{d}v_* \tag{35}$$

and the positive function $\nu = \nu(v)$, whose ith component writes

$$\nu_i\,(v) = \sum_{j=1}^{I} n_i n_j \left(\frac{m_i m_j}{4\pi^2} \right)^{3/2} \iint_{\mathbb{R}^3 \times S^2} e^{-\frac{1}{2} m_j v_*^2}\, B_{ij}(\omega, v - v_*)\, \mathrm{d}\omega\, \mathrm{d}v_*, \tag{36}$$

we can immediately state that

$$\mathscr{L} = \mathscr{K} - \nu \,\mathrm{Id}\,. \tag{37}$$

The following result holds.

Theorem 1 *The operator $\mathscr{K}$, defined by (34), is compact from $L^2(\mathbb{R}^3)^I$ to $L^2(\mathbb{R})^I$.*

The detailed proof can be found in [4]. We discuss below its main features, emphasizing on the major strategy differences with respect to Grad's proof in the mono-species case.

3.2 Elements of Proof for the Compactness

First, we write

$$\mathscr{K} = \mathscr{K}_1 + \mathscr{K}_2 + \mathscr{K}_3 + \mathscr{K}_4,$$

where the ith component of each $\mathscr{K}_\ell g$, $1 \le \ell \le 4$, is given by

$$[\mathscr{K}_1 g]_i\,(v) = -\sum_{j=1}^{I} n_i\, n_j \left(\frac{m_i m_j}{4\pi^2}\right)^{3/4} \times \iint_{\mathbb{R}^3\times S^2} e^{-\frac{1}{4}m_i v^2}\, e^{-\frac{1}{4}m_j v_*^2}\, g_j(v_*)\, B_{ij}(\omega, v - v_*)\,\mathrm{d}\omega\,\mathrm{d}v_*,$$

$$[\mathscr{K}_2 g]_i\,(v) = \sum_{j\neq i} n_i\, n_j \left(\frac{m_i m_j}{4\pi^2}\right)^{3/4} \times \iint_{\mathbb{R}^3\times S^2} e^{-\frac{1}{4}m_i v^2}\, e^{-\frac{1}{2}m_j v_*^2}\, e^{\frac{1}{4}m_j v_*'^2}\, g_j(v_*')\, B_{ij}(\omega, v - v_*)\,\mathrm{d}\omega\,\mathrm{d}v_*,$$

$$[\mathscr{K}_3 g]_i\,(v) = {n_i}^2 \left(\frac{m_i}{2\pi}\right)^{3/2} \times \iint_{\mathbb{R}^3\times S^2} e^{-\frac{1}{4}m_i v^2}\, e^{-\frac{1}{2}m_i v_*^2} \left[e^{\frac{1}{4}m_i v_*'^2}\, g_i(v_*') + e^{\frac{1}{4}m_i v'^2}\, g_i(v')\right] B_{ii}(\omega, v - v_*)\,\mathrm{d}\omega\,\mathrm{d}v_*,$$

$$[\mathscr{K}_4 g]_i\,(v) = \sum_{j\neq i} n_i\, n_j \left(\frac{m_j}{2\pi}\right)^{3/2} \times \iint_{\mathbb{R}^3\times S^2} e^{-\frac{1}{4}m_i v^2}\, e^{-\frac{1}{2}m_j v_*^2}\, e^{\frac{1}{4}m_i v'^2}\, g_i(v')\, B_{ij}(\omega, v - v_*)\,\mathrm{d}\omega\,\mathrm{d}v_*.$$

The compactness for $\mathscr{K}$ is obtained by successively proving the compactness property for each $\mathscr{K}_\ell$. It is crucial to dissociate the cases when $i = j$ or not, because the proofs are quite different.

3.2.1 Compactness of $\mathscr{K}_1$

Denote, for any i, j,

$$k_1^{ij}(v, v_*) = \int_{S^2} e^{-\frac{1}{4}m_i v^2} e^{-\frac{1}{4}m_j v_*^2} B_{ij}(\omega, v - v_*)\, \mathrm{d}\omega, \quad \forall\, v, v_* \in \mathbb{R}^3.$$

We immediately have, for any i,

$$[\mathscr{K}_1 g]_i(v) = -\sum_{j=1}^{I} n_i\, n_j \left(\frac{m_i m_j}{4\pi^2}\right)^{3/4} \int_{\mathbb{R}^3} g_j(v_*)\, k_1^{ij}(v, v_*)\, \mathrm{d}v_*, \quad \forall\, v \in \mathbb{R}^3.$$

Hence, $\mathscr{K}_1$ has a kernel structure. Its compactness can be deduced thanks to the integrability properties of the associated kernels k_1^{ij}.

3.2.2 Compactness of $\mathscr{K}_2$

The proof strategy here is very different from Grad's [20]. Once again, we aim to recover the Hilbert-Schmidt structure for $\mathscr{K}_2$. We first write $\mathscr{K}_2$ in another form. Thanks to the microscopic conservation of kinetic energy during a collision, we have

$$-\frac{1}{4}m_i v^2 - \frac{1}{2}m_j v_*^2 + \frac{1}{4}m_j v_*'^2 = -\frac{1}{4}m_j v_*^2 - \frac{1}{4}m_i v'^2.$$

Consequently, $[\mathscr{K}_2]_i$ can be rewritten as

$$\begin{aligned}[\mathscr{K}_2 g]_i\,(v) = &\sum_{j \neq i} n_i\, n_j \left(\frac{m_i m_j}{4\pi^2}\right)^{3/4} \\ &\times \iint_{\mathbb{R}^3 \times S^2} e^{-\frac{1}{4}m_j v_*^2} e^{-\frac{1}{4}m_i v'^2} g_j(v_*')\, B_{ij}(\omega, v - v_*)\, \mathrm{d}\omega\, \mathrm{d}v_*. \quad (38)\end{aligned}$$

To recover a kernel form in (38), it would be very convenient to replace v_* and v' by v and v_*' in the exponential terms, and then perform a change of variables $v_* \mapsto v_*'$, ω remaining unchanged.

It is indeed possible thanks to the following result, which only holds when $m_i \neq m_j$ (in the monatomic case, that is equivalent to $i \neq j$).

Proposition 2 *There exists $\rho > 0$ such that, for any i, j with $i \neq j$,*

$$m_i v'^2 + m_j v_*^2 \geq \rho \left(m_i v^2 + m_j v_*'^2 \right) \tag{39}$$

for any v, $v_* \in \mathbb{R}^3$ and v', v_*' given by (23).

Proof The proof of Proposition 2 is quite simple. Let us choose i and $j \neq i$. Collision rules (23) can be rewritten as

$$v' = \left(\mathrm{I}_3 - 2\frac{m_j}{m_i + m_j} \omega\, \omega^\mathsf{T} \right) v + 2\frac{m_j}{m_i + m_j} \omega\, \omega^\mathsf{T} v_*, \tag{40}$$

$$v_*' = \left(\mathrm{I}_3 - 2\frac{m_i}{m_i + m_j} \omega\, \omega^\mathsf{T} \right) v_* + 2\frac{m_i}{m_i + m_j} \omega\, \omega^\mathsf{T} v, \tag{41}$$

where I_3 is the identity matrix of $\mathbb{R}^3$. Now we set

$$A(\omega) := \mathrm{I}_3 - 2\frac{m_i}{m_i + m_j} \omega\, \omega^\mathsf{T} \in \mathbb{R}^{3\times 3}.$$

From (41), we easily get

$$A(\omega) v_* = v_*' - 2\frac{m_i}{m_i + m_j} \omega\, \omega^\mathsf{T} v.$$

Fortunately, $A(\omega)$ is an invertible matrix, since

$$\det A(\omega) = \frac{m_j - m_i}{m_i + m_j}$$

and $j \neq i$. Note that, in the mono-species case, the proof already fails at this stage, since the corresponding matrix $A(\omega)$ is not invertible.

Consequently, we can write v_* in terms of both v and v_*':

$$v_* = \left(\mathrm{I}_3 - A(\omega)^{-1} \right) v + A(\omega)^{-1} v_*', \tag{42}$$

where we used the equality

$$-2\frac{m_i}{m_i + m_j} A(\omega)^{-1} \omega\, \omega^\mathsf{T} = \mathrm{I}_3 - A(\omega)^{-1}.$$

Then we obtain an expression of v' with respect to v and v_*' by putting (42) in (40):

$$v' = \left(\frac{m_i + m_j}{m_i} \mathrm{I}_3 - \frac{m_j}{m_i} A(\omega)^{-1} \right) v - \frac{m_j}{m_i} \left(\mathrm{I}_3 - A(\omega)^{-1} \right) v_*'.$$

Consider now the following block matrix in $\mathbb{R}^{6\times 6}$

$$\mathbb{A}(\omega) = \begin{bmatrix} \dfrac{m_i+m_j}{m_i}\,\mathrm{I}_3 - \dfrac{m_j}{m_i} A(\omega)^{-1} & -\sqrt{\dfrac{m_j}{m_i}}\left(\mathrm{I}_3 - A(\omega)^{-1}\right) \\ \sqrt{\frac{m_j}{m_i}}\left(\mathrm{I}_3 - A(\omega)^{-1}\right) & A(\omega)^{-1} \end{bmatrix}. \tag{43}$$

The previous matrix is invertible (check that $\det \mathbb{A}(\omega) = -1$) and we have $\mathbb{A}(\omega)^{-1} = \mathbb{A}(\omega)$. Moreover, it is clear that

$$\begin{bmatrix} \sqrt{m_i}\, v' \\ \sqrt{m_j}\, v_* \end{bmatrix} = \mathbb{A}(\omega) \begin{bmatrix} \sqrt{m_i}\, v \\ \sqrt{m_j}\, v'_* \end{bmatrix}.$$

The best constant ρ satisfying (39) is obtained by computing

$$\inf_{v, v'_* \in \mathbb{R}^3} \frac{\left|\mathbb{A}(\omega)\left[\sqrt{m_i}\, v \quad \sqrt{m_j}\, v'_*\right]^{\mathsf{T}}\right|^2}{\left|\left[\sqrt{m_i}\, v \quad \sqrt{m_j}\, v'_*\right]^{\mathsf{T}}\right|^2} = \|\mathbb{A}(\omega)^{-1}\|_2{}^{-2} = \|\mathbb{A}(\omega)\|_2{}^{-2}.$$

Since $\omega \mapsto \|\mathbb{A}(\omega)\|_2{}^{-2}$ is clearly a continuous positive function of ω on the compact set S^2, it reaches its minimum. Hence, we are led to set

$$\rho = \min_{j\neq i} \min_{\omega \in S^2} \|\mathbb{A}(\omega)\|_2{}^{-2} > 0$$

to satisfy (39). □

Remark 1 Note that, in fact, we can compute the explicit value of ρ, i.e.

$$\rho = \min_{j \neq i} \frac{(\sqrt{m_i} + \sqrt{m_j})^2}{|m_i - m_j|}.$$

Using Proposition 2 and (26) for each B_{ij}, we obtain the existence of a constant $C > 0$, only depending on all the molecular masses, such that, for any i,

$$\begin{aligned} [\mathscr{K}_2 g]_i\,(v) &\le C \sum_{j\neq i} n_i\, n_j\, e^{-\frac{\rho}{4} m_i v^2} \\ &\quad \times \iint_{\mathbb{R}^3 \times S^2} e^{-\frac{\rho}{4} m_j v_*'^2}\, g_j(v'_*) \left(|v - v_*| + |v - v_*|^{\delta - 1}\right) \mathrm{d}v_*\, \mathrm{d}\omega. \end{aligned}$$

We then perform the change of variable $v_* \mapsto v'_*$, whose Jacobian is $1/\det A(\omega)$. Noticing that

$$v - v_* = A(\omega)^{-1}\left(v - v'_*\right) \quad \text{and} \quad \|A(\omega)\|_2{}^{-1} \le \frac{\left|A(\omega)^{-1}\left(v - v'_*\right)\right|}{\left|\left(v - v'_*\right)\right|} \le \|A(\omega)^{-1}\|_2,$$

we can state that

$$|v - v_*| + |v - v_*|^{\delta-1} \leq \|A(\omega)^{-1}\|_2 \left|v - v_*'\right| + \|A(\omega)\|_2{}^{1-\delta} \left|v - v_*'\right|^{\delta-1}.$$

Eventually, we write

$$[\mathscr{K}_2 g]_i\,(v) \leq C \sum_{j \neq i} n_i\, n_j \iint_{\mathbb{R}^3 \times S^2} e^{-\frac{\beta}{4} m_i v^2}\, e^{-\frac{\beta}{4} m_j v_*'^2}\, g_j(v_*') \times \left(|v - v_*'| + |v - v_*'|^{\delta-1}\right) \mathrm{d}\omega\, \mathrm{d}v_*'.$$

We thus recover a kernel form for $\mathscr{K}_2$, which allows to conclude on the compactness with the same kind of arguments (integrability of the kernels) as in Sect. 3.2.1.

3.2.3 Compactness of $\mathscr{K}_3$ and $\mathscr{K}_4$

Since $\mathscr{K}_3$ appears as a mono-species operator, it can be straightforwardly treated following Grad's strategy. The compactness of $\mathscr{K}_4$, though it deals with several species, requires to recover a kernel form of $\mathscr{K}_4$, again following Grad's procedure. The detailed computations, which are similar to the ones in Sect. 2.3, can be found in [4].

Acknowledgments The authors thank Bérénice Grec for her careful proof-reading and the fruitful discussions about the compactness of the linearized Boltzmann operator.

References

1. Boltzmann, L.: Weitere studien über das wärmegleichgewicht unter gasmolekülen. Sitzungsberichte Akad. Wiss. **66**, 275–370 (1873)
2. Boltzmann, L.: Lectures on Gas Theory. Translated by Stephen G. Brush. University of California Press, Berkeley (1964)
3. Bothe, D.: On the Maxwell-Stefan approach to multicomponent diffusion. In: Parabolic problems, Progress in Nonlinear Differential Equations Applications, vol. 80, pp. 81–93. Birkhäuser/Springer Basel AG, Basel (2011)
4. Boudin, L., Grec, B., Pavić, M., Salvarani, F.: Diffusion asymptotics of a kinetic model for gaseous mixtures. Kinet. Relat. Models **6**(1), 137–157 (2013)
5. Boudin, L., Grec, B., Salvarani, F.: A mathematical and numerical analysis of the Maxwell-Stefan diffusion equations. Discrete Contin. Dyn. Syst. Ser. B **17**(5), 1427–1440 (2012)
6. Boudin, L., Grec, B., Salvarani, F.: The Maxwell-Stefan diffusion limit for a kinetic model of mixtures. Acta Appl. Math. **136**, 79–90 (2015)
7. Bourgat, J.-F., Desvillettes, L., Le Tallec, P., Perthame, B.: Microreversible collisions for polyatomic gases and Boltzmann's theorem. European J. Mech. B Fluids **13**(2), 237–254 (1994)
8. Caflisch, R.E.: The Boltzmann equation with a soft potential. I. Linear, spatially-homogeneous. Comm. Math. Phys. **74**(1), 71–95 (1980)
9. Caflisch, R.E.: The Boltzmann equation with a soft potential. II. Nonlinear, spatially-periodic. Comm. Math. Phys. **74**(2), 97–109 (1980)

10. Carleman, T.: Problèmes mathématiques dans la théorie cinétique des gaz. Publlisher Science Institute Mittag-Leffler. 2. Almqvist & Wiksells Boktryckeri Ab, Uppsala (1957)
11. Cercignani, C.: The Boltzmann equation and its applications. Applied Mathematical Sciences, vol. 67. Springer, New York (1988)
12. Cercignani, C., Illner, R., Pulvirenti, M.: The mathematical theory of dilute gases. Applied Mathematical Sciences, vol. 106. Springer, New York (1994)
13. Daus, E.S., Jüngel, A., Mouhot, C., Zamponi, N.: Hypocoercivity for a linearized multi-species Boltzmann system. ArXiv e-prints, April 2015
14. Desvillettes, L.: Sur un modèle de type Borgnakke-Larsen conduisant à des lois d'énergie non linéaires en température pour les gaz parfaits polyatomiques. Ann. Fac. Sci. Toulouse Math. (6), **6**(2):257–262 (1997)
15. Desvillettes, L., Monaco, R., Salvarani, F.: A kinetic model allowing to obtain the energy law of polytropic gases in the presence of chemical reactions. Eur. J. Mech. B Fluids **24**(2), 219–236 (2005)
16. Duncan, J.B., Toor, H.L.: An experimental study of three component gas diffusion. AIChE J. **8**(1), 38–41 (1962)
17. Glassey, R.T.: The Cauchy Problem in Kinetic Theory. Society for Industrial and Applied Mathematics (SIAM), Philadelphia (1996)
18. Golse, F., Poupaud, F.: Stationary solutions of the linearized Boltzmann equation in a half-space. Math. Methods Appl. Sci. **11**(4), 483–502 (1989)
19. Grad, H.: Principles of the kinetic theory of gases. In: Handbuch der Physik (herausgegeben von S. Flügge), Bd. 12, Thermodynamik der Gase, pp. 205–294. Springer, Berlin (1958)
20. Grad, H.: Asymptotic theory of the Boltzmann equation. II. In: Rarefied Gas Dynamics (Proc. 3rd Internat. Sympos., Palais de l'UNESCO, Paris, 1962), vol. I, pp. 26–59. Academic Press, New York, (1963)
21. Guo, Y.: Classical solutions to the Boltzmann equation for molecules with an angular cutoff. Arch. Ration. Mech. Anal. **169**(4), 305–353 (2003)
22. Hecke, E.: Über die Integralgleichung der kinetischen Gastheorie. Math. Z. **12**(1), 274–286 (1922)
23. Hilbert, D.: Begründung der kinetischen Gastheorie. Math. Ann. **72**(4), 562–577 (1912)
24. Hutridurga, H., Salvarani, F.: On the Maxwell-Stefan diffusion limit for a mixture of monatomic gases. ArXiv e-prints (2015)
25. Levermore, C.D., Sun, W.: Compactness of the gain parts of the linearized Boltzmann operator with weakly cutoff kernels. Kinet. Relat. Models **3**(2), 335–351 (2010)
26. Maxwell, J.C.: On the dynamical theory of gases. Phil. Trans. R. Soc. **157**, 49–88 (1866)
27. Mouhot, C., Strain, R.M.: Spectral gap and coercivity estimates for linearized Boltzmann collision operators without angular cutoff. J. Math. Pures Appl. (9), **87**(5), 515–535 (2007)
28. Pao, Y.P.: Boltzmann collision operator with inverse-power intermolecular potentials. I, II. Comm. Pure Appl. Math. **27**, 407–428; ibid. 27, 559–581 (1974)
29. Stefan, J.: Ueber das Gleichgewicht und die Bewegung insbesondere die Diffusion von Gasgemengen. Akad. Wiss. Wien **63**, 63–124 (1871)
30. Villani, C.: A review of mathematical topics in collisional kinetic theory. In: Handbook of Mathematical Fluid Dynamics, vol. I, pp. 71–305. North-Holland, Amsterdam (2002)

Asymptotics for FBSDES with Jumps and Connections with Partial Integral Differential Equations

André de Oliveira Gomes

Abstract It is our intention to survey the asymptotic study of a certain class of coupled forward-backward stochastic differential equations (FBSDEs for short) when the noise terms in the forward diffusion have small intensities that converge to zero. The system of FBSDEs discussed can be used to give a probabilistic representation for the solution in the viscosity sense of an associated system of partial-integral differential equations (PIDEs) with a terminal condition. The asymptotic study of this PIDE is done probabilistically using the FBSDE system. Secondly we present a large deviations principle for the laws of the forward and backward processes of the stochastic system.

Keywords FBSDEs with jumps · Viscosity solutions · Partial-integral differential equations · Large deviations principles

1 Motivation and Some Words About FBSDEs

The complexity of the phenomena studied by modern applied sciences created the need to take into account the randomness factor in their description. Randomness can occur in the imprecision of measurements or in the presence of the evolution of certain microscopic scales processes, which may have not only one direction, but several (often infinitely) ones in which they evolve. As the result of modelling with randomness the description of the evolution of natural/social phenomena, stochastic differential equations (SDEs) are the natural mathematical tool to use. Areas in which their use is very popular are, besides theoretical and applied physics, biology, neurophysiology and finance. The stochastic models that started to be employed were differential equations perturbed by an external force of randomness called white noise. But those models perturbed by white noise are continuous and they seem

A. de Oliveira Gomes (✉)
Instititut für Mathematik, Humboldt-Universität zu Berlin,
Rudower Chaussee 25 - Johann von Neumann Haus, 12489 Berlin, Germany
e-mail: andredeoliveiragomes2@gmail.com

© Springer International Publishing Switzerland 2016

P. Gonçalves and A.J. Soares (eds.), *From Particle Systems to Partial Differential Equations III*, Springer Proceedings in Mathematics & Statistics 162,
DOI 10.1007/978-3-319-32144-8_5

inappropriate to capture, for example, phenomena where it is observed the occurrence of abrupt variations in very small time scales. Due to these limitations, the use of dynamical systems perturbed by Lévy processes (stochastic processes with independent and stationary increments that allow the occurrence of jumps/discontinuities) has been receiving increasingly attention in other sciences, such as economics, natural sciences and physics.

If we consider particle systems whose motion is governed by "Lévy flights", popular name in physics for random walks in which the step lengths U have a probability distribution that is heavy-tailed, i.e., $P(U > u) = O(u^{-\alpha})$, with $\alpha \in (1, 2)$, and perform the hydrodynamic limit, in the presence of some assumptions on the system, we end up with a fractal Burgers equation

$$\partial_t u = \nu(-\Delta)^{\frac{\alpha}{2}} u - \langle u, \nabla u \rangle + f,$$

where $\nu > 0$ is the viscosity parameter, $\langle .,. \rangle$ stands for the inner product and f is an external force acting on the fluid.

Those kind of non-local models are becoming very popular not only in mathematics but in physics too, with the study of the so called anomalous diffusions (see [28] and further references there) and with the growing demand of using differential equations that take into account nonlocal effects of interaction and non-isotropic propagation of energy.

$f = f(t, x, u, \nabla u, u(t, \beta(x,.)) - u(t, x))$ captures nonlocal sources of interaction on the particles of the fluid, being β a displacement function in space. $(-\Delta)^{\frac{\alpha}{2}}$ is the usual fractional laplacian, a integro-differential operator, defined as

$$(-\Delta)^{\frac{\alpha}{2}} f(x) = c_{d,\alpha} \lim_{\varepsilon \to 0} \int_{|y-x|>\varepsilon} \frac{|f(x) - f(y)|}{|x - y|^{d+\alpha}} dy,$$

where $c_{d,\alpha}$ is a renormalization constant. We observe that the definition above generalizes the usal laplacian ($\alpha = 2$).

The presence of $(-\Delta)^{\frac{\alpha}{2}}$ is not surprising, since, via Kolmogorov's functional limit theorem, the distance from the origin of the "Lévy flights" converges, after a large number of steps, to an α-stable distribution.

The fractal Burgers equation has been studied by Biler et al. [3] and by Aschterenberg et al. in [1]. In [32] Zhang studied probabilistically the fractal Navier Stokes equation, which turns as an example in favor of probabilistic approaches to the study of nonlocal hydrodynamic models, as was made before to the Navier Stokes systems. We refer the reader to [7, 8, 11] as examples of probabilistic studies of Navier-Stokes equations.

We will associate to a certain class of partial-integral differential equations, where the fractal Burgers equation is included, a certain system of stochastic differential equations and, via this probabilistic object, we will address the problem of the vanishing viscosity limit $\nu \to 0$.

More specifically, in Sect. 3, it is presented the asymptotic study when $\varepsilon, \delta \to 0$ of the following coupled system of forward-backward stochastic differential equations with jumps, whose conditions on the coefficients will be explicited in the next section,

$$\begin{cases} X_s^{\varepsilon,\delta} &= x + \int_t^s b(r, X_r^{\varepsilon,\delta}, Y_r^{\varepsilon,\delta})\mathrm{d}r + \sqrt{\varepsilon}\int_t^s \sigma(r, X_r^{\varepsilon,\delta}, Y_r^{\varepsilon,\delta})\mathrm{d}Wr \\ &\quad + \delta \int_t^s \int_{R^l-\{0\}} \beta(X_r^{\varepsilon,\delta}, e)\widetilde{\mu}^{\frac{1}{\delta}}(\mathrm{de}, \mathrm{dr}) \\ Y_s^{\varepsilon,\delta} &= g(X_T^{\varepsilon,\delta}) + \int_s^T f\Big(r, X_r^{\varepsilon,\delta}, Y_r^{\varepsilon,\delta}, Z_r^{\varepsilon,\delta}, \int_{R^l-\{0\}} V_r^{\varepsilon,\delta}(e)\gamma^\delta(e)\frac{1}{\delta}\nu(\mathrm{de})\Big)\mathrm{dr} \\ &\quad - \int_s^T Z_r^{\varepsilon,\delta}\mathrm{d}Wr - \int_s^T \int_{R^l-\{0\}} V_r^{\varepsilon,\delta}\widetilde{\mu}^{\frac{1}{\delta}}(\mathrm{de}, \mathrm{dr}). \end{cases}$$

Here $(W_t)_{t\geq 0}$ is a Brownian motion with values in a euclidean space R^d and $\tilde{\mu}^{\frac{1}{\delta}}$ is a compensated Poisson random measure defined on the same probability space, independent of $(B_t)_{t\geq 0}$, with values in $(R^l - \{0\}, \mathscr{B}(R^l - \{0\}))$ and with compensator $\frac{1}{\delta}ds \otimes \nu$. The random measure $\tilde{\mu}^{\frac{1}{\delta}}$ captures the occurrence of discontinuities in the model. Here ds stands for the Lebesgue measure on R^+ and ν is a Lévy measure on some R^l, i.e. a measure that does not charge the origin, $\nu(\{0\}) = 0$, and with the integrability condition $\int_{R^l-\{0\}} 1 \wedge |e|^2\nu(de) < \infty$. The asymptotic study of the FBSDE system above carries information about the convergence of the viscosity solutions of the associated terminal value problem

$$\begin{cases} (\partial_t + \mathscr{L}^{\varepsilon,\delta})u^{\varepsilon,\delta}(t, x) \\ +h^\delta\Big(t, x, u^{\varepsilon,\delta}(t, x), \varepsilon\sigma^T(t, x, u^{\varepsilon,\delta}(t, x))\nabla_x u^{\varepsilon,\delta}(t, x), \\ u^{\varepsilon,\delta}(t, x + \delta\beta(x)) - u^{\varepsilon,\delta}(t, x)\Big) = 0 \\ u^{\varepsilon,\delta}(T, x) = g(x) \end{cases} \tag{1}$$

where $\mathscr{L}^{\varepsilon,\delta} = \mathscr{K}_1^\varepsilon + \mathscr{K}_2^\delta$ is the second order differential-integral operator associated to the jump diffusion, with

$$\mathscr{K}_1^\varepsilon\varphi(t, x) = \langle b(t, x, \varphi(t, x)), \nabla_x\varphi(t, x)\rangle + \frac{\varepsilon}{2}tr(a(t, x, \varphi(t, x))\nabla^2\varphi(t, x))$$

$$\mathscr{K}_2^\delta\varphi(t, x) = \int_{R^l-\{0\}} \frac{\varphi(t, x + \delta\beta(x)) - \varphi(t, x) - \langle\delta\beta(x), \nabla_x\varphi(t, x)\rangle}{\delta}\nu(\mathrm{de}),$$

for every $\varphi \in C^{1,2}\big([0, T] \times R^d, R^n\big)$, continuously differentiable function in the time variable $t \in [0, T]$ and two times continuously differentiable in the space variable $x \in R^d$, with values in R^n.

The terminal value function g is assumed to be Lipschitz, the diffusion matrix, assumed to be non-degenerate, is defined as $a_{i,j} = [\sigma\sigma^T]_{i,j}$, where σ^T is the transpose matrix of σ, and the semilinear term, assumed to be Lipschitz, is given, for every

$\delta > 0$, by

$$h^{\delta}(s,x,y,z,k) = f\Big(s,x,y,z,\int_{R^l-\{0\}} k(e)\gamma^{\delta}(e)\frac{1}{\delta}\nu(\mathrm{d}e)\Big).$$

In Sect. 4 we present a large deviations principle for the laws of the forward and backward processes $(X^{\varepsilon}, Y^{\varepsilon})_{\varepsilon>0}$ when $\varepsilon = \delta \to 0$. We mention for the Brownian case the works [26], where the same problem was addressed for a FBSDE decoupled system (i.e. the forward equation does not depend on the backward process), [10], for the Brownian coupled case and [15], where the authors studied the vanishing viscosity limit of quadratic Burgers nonlinearities via FBSDES. For the jump case we mention [29] where the decoupled case is addressed in the case of non-lipschitz coefficients. In contrast with this work that deals with coupled FBSDE systems where the forward process is a jump diffusion, in [29] the forward diffusion is only driven by a Brownian Motion, not covering interesting classes of quasilinear PIDEs associated to the stochastic system such as the fractal Burgers equation.

It is a natural question to ask why it appears two stochastic integral terms of two processes $(Z_s^{\varepsilon,\delta})_{t\le s\le T}$ and $(V_s^{\varepsilon,\delta})_{t\le s\le T}$ in the backward equation. Their role is to make the backward equation well-posed in the Itô sense, adapting the process $(Y_s^{\varepsilon,\delta})_{t\le s\le T}$ to the filtration considered. We illustrate this statement with an example.

We fix a finite time horizon $T > 0$ and $(\Omega, \mathscr{F}, P, \{\mathscr{F}_t\}_{t\in[0,T]})$ a complete filtered probability space in which we construct a Brownian motion $(W_t)_{t\ge 0}$ with values in R^d. Let us consider the following terminal value problem in the Itô sense:

$$\begin{cases} dY_t = 0 \\ Y_T = \xi, \end{cases} \tag{2}$$

where $\xi \in L^2_{\mathscr{F}_T}(\Omega)$ is a square integrable random variable, measurable with respect to the terminal σ-algebra $\mathscr{F}_T$ of the filtration. The natural problem of adaptedness is a obstacle to claim $Y_t \equiv \xi$ is the solution of (2). The natural way to adapt ξ is to project the terminal value in the σ-algebra of the filtration of each time, $Y_t \equiv E[\xi\,|\mathscr{F}_t]$. Now the stochastic process built like this is a solution in the Itô sense of (2).

If the filtration is Brownian, i.e. generated by the Brownian Motion $(W_t)_{t\ge 0}$, due to the martingale representation theorem (see [18]), there exists $(Z_t)_{0\le t\le T} \in L^2(\Omega, P, \{\mathscr{F}_t\}_{0\le t\le T})$ such that

$$Y_t = Y_0 + \int_0^t Z_s dW_s, \quad P-\text{a.s.}, \textit{ for all } t \in [0,T].$$

Hence,

$$Y_0 = Y_T - \int_0^t Z_s \mathrm{d}W_s$$

and

$$Y_t = \xi - \int_t^T Z_s dW_s \quad P - \text{a.s.}, \quad \textit{for all } t \in [0, T]$$

turns to be the way to look at the terminal value problem above, which is the simplest example of a backward stochastic differential equation (BSDE for short). $(Z_t)_{0 \le t \le T}$ appears as part of the solution to the problem, in order to guarantee adaptedness of the process, if we concern to solve such problem with Itô's theory. In the FBSDE presented associated to (1), two processes $(Z_s^{\varepsilon,\delta})_{t \le s \le T}$, $(V_s^{\varepsilon,\delta})_{t \le s \le T}$ are required to make the backward process adapted, since we have two independent sources of noise that generate the filtration, one Brownian and the other Poissonian.

Let us discuss formally the natural link between FBSDEs and PDEs in the Brownian case, in order to understand the link between the FBSDE system and (1).

Fixed $T > 0$ and a complete filtered probability space $(\Omega, \mathscr{F}, P, \{\mathscr{F}_t\}_{t \in [0,T]})$, where it is defined a Brownian motion $(W_t)_{t \in [0,T]}$ with values in R^d, and the filtration $\{\mathscr{F}_t\}_{t \in [0,T]}$ is the natural filtration generated by $(W_t)_{t \in [0,T]}$ and enlarged with the sets of null measure. Let $(X_t)_{t \in [0,T]}$ be the strong solution of the SDE,

$$\begin{cases} dX_t &= b(X_t)dt + \sigma(X_t)dW_t \\ X_0 &= x \in R^d. \end{cases} \tag{3}$$

where the vector fields $b : R^d \longrightarrow R^d$ and $\sigma : R^d \longrightarrow R^{d \times d}$ are regular enough to allow existence and uniqueness of a strong solution $(X_t)_{t \in [0,T]}$. For example, it is enough to assume that the vector fields are Lipschitz and the matrix diffusion is not degenerate.

Associated to the SDE, we define the infinitesimal generator of the diffusion $(X_t)_{0 \le t \le T}$,

$$\mathscr{L}\varphi(x) = \sum_{i=1}^d b_i(x) \frac{\partial}{\partial_{x_i}} \varphi(x) + \frac{1}{2} \sum_{i,j=1}^d a_{i,j}(x) \frac{\partial^2}{\partial_{x_i} \partial_{x_j}} \varphi \quad \textit{for all } \varphi \in C^2(R^d),$$

where $a_{i,j} = [\sigma\sigma^T]_{i,j}$.

We consider now the following terminal value problem for a semilinear parabolic PDE of the type

$$\begin{cases} \partial_t u(t,x) + \mathscr{L}u(t,x) + f(t,x,u(t,x),\nabla u(t,x)) = 0 \\ u(T,x) = g(x). \end{cases} \tag{4}$$

We assume regularity on f and g to ensure existence and uniqueness of a continuous solution with regularity enough for our purposes. Let u be the unique strong solution of the terminal value problem above, i.e. a continuous function, regular enough that solves the PDE (4) and with terminal condition $u(T, x) = g(x)$, for all $x \in R^d$. Let us define

$$Y_t = u(t, X_t).$$

If follows, from Itô's formula,

$$\begin{aligned} dY_t &= (\partial_t u(t, X_t) + \mathscr{L}u(t, X_t))dt + \nabla_x u(t, X_t)\sigma(X_t)dW_t \\ &= -f(t, X_t, Y_t, \nabla_x u(t, X_t))dt + \nabla_x u(t, X_t)\sigma(X_t)dW_t, \end{aligned}$$

that suggests that we should consider PDEs of the type

$$\partial_t u(t, x) + \mathscr{L}u(t, x) + f(t, x, u(t, x), \nabla_x u(t, x)\sigma(t, x)) = 0.$$

If u solves the equation above,

$$\begin{aligned} Y_t &= u(t, X_t) \\ Z_t &= \nabla_x u(t, X_t)\sigma(X_t) \end{aligned}$$

solve the equation

$$dY_t = -f(t, X_t, Y_t, Z_t)dt + Z_t dW_t,$$

which should be interpreted as a BSDE

$$Y_t = g(X_T) + \int_t^T f(s, X_s, Y_s, Z_s)ds - \int_t^T Z_s dW_s.$$

Given $\varepsilon, \delta > 0$, if (1) has a unique solution in some suitable functional space, $u^{\varepsilon,\delta}$, with enough regularity to use Itô's formula (we refer to [27]), the stochastic processes defined as

$$\begin{aligned} Y_s^{\varepsilon,\delta} &= u^{\varepsilon,\delta}(s, X_s^{\varepsilon,\delta}) \\ Z_s^{\varepsilon,\delta} &= \sqrt{\varepsilon}\nabla_x u^{\varepsilon}(t, X_s^{\varepsilon,\delta})\sigma(s, X_s^{\varepsilon,\delta}, Y_s^{\varepsilon,\delta}) \\ V_s^{\varepsilon,\delta}(.) &= u^{\varepsilon,\delta}(s, X_{s-}^{\varepsilon,\delta} + \delta\beta(X_{s-}^{\varepsilon,\delta})) - u^{\varepsilon}(s, X_{s-}^{\varepsilon,\delta}) \end{aligned}$$

solve the BSDE

$$\begin{aligned} Y_s^{\varepsilon,\delta} &= g(X_T^{\varepsilon,\delta}) + \int_s^T f\Big(r, X_r^{\varepsilon,\delta}, Y_r^{\varepsilon,\delta}, Z_r^{\varepsilon,\delta}, \int_{R^l-\{0\}} V_{r^-}^{\varepsilon,\delta}(e)\gamma^{\delta}(e)\frac{1}{\delta}\nu(\mathrm{d}e)\Big)\mathrm{d}r \\ &- \int_s^T Z_r^{\varepsilon,\delta}\mathrm{d}Wr - \int_s^T \int_{R^l-\{0\}} V_{r^-}^{\varepsilon,\delta}\widetilde{\mu}^{\frac{1}{\delta}}(\mathrm{d}e, \mathrm{d}r). \end{aligned} \tag{5}$$

The FBSDE system satisfied by $(X_t^{\varepsilon,\delta}, Y_t^{\varepsilon,\delta})_{0\leq t\leq T}$ is a coupled system of stochastic differential equations, in opposition to the previous example, since the equation satisfied by the forward jump diffusion $(X_t^{\varepsilon,\delta})_{t\in[0,T]}$ depends on the backward process $(Y_t^{\varepsilon,\delta})_{t\in[0,T]}$.

Forward backward stochastic differential equations became very popular in the last 20 years due to the huge range of applications and interactions with other mathematical fields. Besides the link with PDEs that we illustrated here with our simple example, it is very well known the strong link with stochastic optimal control. Stochastic optimal control solution theory has two important methodologies: dynamic programming principle and Pontryagin maximum principle. The first one deals with the Hamilton-Jacobi-Bellman (HJB) equation associated. HJB is a deterministic PDE whose solution is the value function for the stochastic optimization problem. The Pontryagin maximum principle involves the maximization of an hamiltonian and solving the adjoint equation, which is a BSDE (see [23] for details). With a stochastic optimal control problem in mind, Bismut [5] introduced in 1973 a linear BSDE asociated to the Pontryagin maximum principle. General nonlinear BSDE theory in the Brownian case was introduced by Pardoux and Peng in [24]. There are three main methods to solve FBSDEs: the contraction mapping, which assures existence and uniqueness of solution in a small time interval; the four step scheme, developed by Ma et al. [22], which requires although more strict assumptions, such as deterministic coefficients and non-degeneracy of the matrix diffusion for the forward equation, but producing a existence and uniqueness result of solution in a arbitrarily large time interval, being the backward process a function of the forward process via the HJB equation associated; and the method of continuation, investigated by Hu, Peng and Yong in [16, 25, 31] and that allows the FBSDEs systems to have random coefficients. We refer the reader to the book [23] for the solution theories for FBSDEs in the Brownian case and [14] to the jump case. As a natural generalization FBSDEs driven by a jump-diffusion become incrinsigly a natural object of study. BSDEs with jumps were first discussed by Li and Tang [20] and its connections with viscosity solutions of the associated system of parabolic integral-differential equations were first discussed in [2].

In the next section we present the conditions on the FBSDE system in order to guarantee existence and uniqueness of solution and the rigorous definition of viscosity solution of (1).

2 Functional Setting and Preliminary Results for the FBSDE System

Let $T > 0$ be a finite time and $(\Omega, \mathscr{F}, P)$ a complete probability space on which we define two independent processes:

- a d-dimensional Brownian Motion $W_t = (W_t^1, \ldots, W_t^d)$,
- an integer-valued random measure $\mu = \mu(\mathrm{d}t, \mathrm{d}e)$ on $([0, T]) \times E, \mathscr{B}(E), \nu)$ where $E = R^l - \{0\}$ equipped with its Borel σ-field $\mathscr{B}(E)$ and ν is a σ-finite Lévy measure on $(E, \mathscr{B}(E))$.

Moreover we assume the random measure is Poissonian with compensator defined by $\widehat{\mu}(\mathrm{d}t, \mathrm{d}e) := \mathrm{d}t\, \nu(\mathrm{d}e)$. So $\{(\widetilde{\mu}(0,T] \times A) := (\mu - \widehat{\mu}((0,T] \times A)\}_{t\geq 0}$ is a martingale for all $A \in \mathscr{B}(E)$ such that $\nu(A) < \infty$. The measure $\widetilde{\mu}^{\frac{1}{\varepsilon}}$ is the compensated Poisson random measure with intensity $\frac{1}{\varepsilon}\nu(de) \otimes dr$. Let $\widetilde{\mathscr{P}} = \mathscr{P} \otimes \mathscr{B}(E)$, where $\mathscr{P}$ is the predictable σ-field on $\Omega \times [0,T]$. For details about definitions of predictable σ-fields and for the definition of integrals with respect to a Poisson Random Measure we refer to [17]. Let $(\mathscr{F})_{t\geq 0}$ be the filtration generated by the two independent processes above and augmented by the P-null sets of $\mathscr{F}$. We remark that $(\mathscr{F}_t)_{t\geq 0}$ satisfies the usual hypothesis of right continuity and completeness.

For any $n \in N$, given $x, y \in R^n$, $\langle x, y\rangle = \sum_{i=1}^{d} x_i y_i$ and $|x| = \sqrt{\langle x, x\rangle}$. Given a matrix A, A^T denotes the transpose matrix of A.

Let us define the following functional spaces that we will use in the sequel. Fix $t \in [0,T]$ and $\tau > 0$. We define

$$\mathscr{M}^2(t,T,R^k) := \Big\{\varphi : \Omega \times [t,T] \longrightarrow R^k \ \ \textit{is an } (\mathscr{F}_t)_{t\geq 0} \textit{ predictable process}$$
$$\textit{such that } E\Big[\int_t^T |\varphi_s|^2 \mathrm{d}s\Big] < \infty\Big\},$$

$$\mathscr{S}^2(t,T,R^k) := \Big\{\varphi : \Omega \times [t,T] \longrightarrow R^k \ \ \textit{is an adapted cadlag process such}$$
$$\textit{that } E\Big[\sup_{t\leq s\leq T} |\varphi_s|^2 \mathrm{d}s\Big] < \infty\Big\},$$

and

$$\mathscr{K}^2_{\nu,\tau}(t,T,R^k) := \Big\{K : \Omega \times [t,T] \longrightarrow R^k \ \widetilde{\mathscr{P}} \ -\textit{measurable such that}$$
$$E\Big[\int_t^T \int_E |K_s(e)|^2 \frac{1}{\tau}\nu(\mathrm{d}e)\mathrm{d}s\Big] < \infty\Big\}.$$

We write $\mathscr{K}^2_{\nu}(t,T,R^k) = \mathscr{K}^2_{\nu,1}(t,T,R^k)$. These spaces are naturally complete normed spaces.

We are interested in the asymptotic study when $\varepsilon \to 0$ of the following coupled FBSDE system wih jumps:

$$\begin{cases} X_s^{\varepsilon,x,t} &= x + \displaystyle\int_t^s b(r, X_r^{t,x,\varepsilon}, Y_r^{t,x,\varepsilon})\mathrm{d}r \\ & \quad + \sqrt{\varepsilon}\displaystyle\int_t^s \sigma(r, X_r^{t,x,\varepsilon}, Y_r^{t,x,\varepsilon})\mathrm{d}W_r + \varepsilon \int_t^s \int_{R^l-\{0\}} \beta(X_r^{t,x,\varepsilon}, e)\widetilde{\mu}^{\frac{1}{\varepsilon}}(\mathrm{d}e, \mathrm{d}r) \\ Y_r^{t,x,\varepsilon} &= g(X_T^{t,x,\varepsilon}) + \displaystyle\int_s^T f\Big(r, X_r^{t,x,\varepsilon}, Y_r^{t,x,\varepsilon} Z_r^{t,x,\varepsilon}, \int_{R^l-\{0\}} V_r^{t,x,\varepsilon}(z)\gamma^{\varepsilon}(e)\nu(de)\Big)dr \\ & \quad - \int_s^T Z_r^{t,x,\varepsilon}\mathrm{d}W_r - \int_s^T \int_{R^l-\{0\}} V_r^{t,x,\varepsilon}(e)\widetilde{\mu}^{\frac{1}{\varepsilon}}(\mathrm{d}e\mathrm{d}r) \end{cases} \tag{6}$$

The solution process $(U_r^{t,x,\varepsilon})_{t\le s\le T} = (X_s^{t,x,\varepsilon}, Y_s^{t,x,\varepsilon}, Z_s^{t,x,\varepsilon}, V_s^{t,x,\varepsilon})_{t\le s\le T}$ takes values in $R^d \times R^n \times R^{n\times d} \times R^n$, being the dependence on the time t, initial condition of the forward jump diffusion x and in the parameter ε explicit in the notation. For every $\tau > 0$, the coefficients of the system are defined as follows,

$$\begin{aligned}
&b : [0,T] \times R^d \times R^n \longrightarrow R^d \\
&\sigma : [0,T] \times R^d \times R^n \longrightarrow R^{d\times d} \\
&f : [0,T] \times R^d \times R^n \times R^{n\times d} \times R^n \longrightarrow R^n \\
&\beta : R^d \times R^l - \{0\} \longrightarrow R^d \\
&\gamma^\tau : R^l - \{0\} \longrightarrow R \\
&g : R^d \longrightarrow R^n.
\end{aligned}$$

We remark that the vector field f depends on $V_r^{t,x,\varepsilon}$ in a very special way. The main reason for this restriction is that it is employed the comparison theorem to link the backward process of the system (1) to the viscosity solutions of the correspondent PIDE (in [2] it is proved the comparison theorem for BSDEs with jumps).

We are given a full-rank matrix G, $n \times d$. If $n = d$, G is the identity matrix and we use the notation:

$$\pi = \begin{pmatrix} x \\ y \\ z \end{pmatrix}, \qquad G^\varepsilon(t,u) = \begin{pmatrix} -G^t h^\tau \\ Gb \\ \sqrt{\varepsilon} G\sigma \end{pmatrix} (t, \pi, k).$$

where, for every $\tau > 0$,

$$h^\tau(s,x,y,z,k) = f\Big(s,x,y,z, \int_{R^l-\{0\}} k(e)\gamma^\tau(e)\frac{1}{\varepsilon}\nu(\mathrm{d}e)\Big)$$

Definition 1—Solution of (6): $(X_s^{t,x,\varepsilon}, Y_s^{t,x,\varepsilon}, Z_s^{t,x,\varepsilon}, V_s^{t,x,\varepsilon})_{t\le s\le T}$ is a solution of (6) if it is an adapted process to the filtration $(\mathscr{F}_s)_{t\le s\le T}$ belonging to $\mathscr{N}_{t,\varepsilon} := \mathscr{S}^2(t,T,R^d) \times \mathscr{S}^2(t,T,R^n) \times \mathscr{M}^2(t,T,R^{n\times d}) \times \mathscr{K}^2_{\nu,\varepsilon}(t,T,R^n)$ and solving (6) P-a.s.

In what follows we explicit the first set of assumptions that we will consider in the sequel of this work.

A.1.

- The coefficients b, σ, f are uniformly Lipschitz with respect to (x,y,z,k), with some constant $L > 0$, and there exists a function $\rho : R^l - \{0\} \longrightarrow R^+$ in $L^2(R^l - \{0\}, \mathscr{B}(R^l - \{0\}), \nu)$ such that, for all $t \in [0,T]$, $x, \bar{x} \in R^d$ and $e \in R^l - \{0\}$

$$|\beta(x) - \beta(\bar{x})| \le \rho(e)|x - \bar{x}|.$$

- The function $k \longrightarrow f(t,x,y,z,k)$ is non decreasing for all $(t,x,y,z) \in [0,T] \times R^d \times R^n \times R^{n\times d}$.
- For every $\tau > 0$, $0 \le \gamma^\tau(e) \le \tau(1 \wedge |e|)$ for all $e \in R^l - \{0\}$.
- The function g is uniformly Lipschitz with respect to $x \in R^d$, with constant $L > 0$.
- It is valid the following integrability condition,

$$E\Big[\int_0^T |b(s,0,0)|^2 \mathrm{d}s\Big] + E\Big[\int_0^T |f(s,0,0,0,0)|^2 \mathrm{d}s\Big] + E\Big[\int_0^T |\sigma(s,0,0)|^2 \mathrm{d}s\Big] + E\Big[\int_0^T \int_{R^l-\{0\}} |\beta(0,e)|^2 \nu(\mathrm{d}e)\mathrm{d}s\Big] < \infty.$$

The next set of assumptions states natural conditions of monotonicity on the coefficients in the following way.
A.2. There exists $\eta > 0$ such that:

- for all $\pi = (x,y,z)$, $\bar{\pi} = (\bar{x},\bar{y},\bar{z})$, $\hat{x} = x - \bar{x}$, $\hat{y} = y - \bar{y}$, $\hat{z} = z - \bar{z}$ and $\hat{h} = h - \bar{h}$, where $h = h(t,\pi,k,e)$ and $\bar{h} = h(t,\bar{\pi},\bar{k},e)$

$$\langle G^\varepsilon(t,\pi,k) - G^\varepsilon(t,\bar{\pi},\bar{k}), \pi - \bar{\pi}\rangle + \int_{R^l-\{0\}} \langle G^\varepsilon(\beta(x) - \beta(\bar{x})), \hat{k}(e)\rangle \nu(\mathrm{d}e)$$
$$\le -(\eta + \sqrt{\varepsilon})[|(G^\varepsilon)\hat{x}|^2 + |(G^\varepsilon)^T\hat{y}|^2 + |(G^\varepsilon)^T\hat{z}|^2 + \int_{R^l-\{0\}} |(G^\varepsilon)^T\hat{h}(e)|^2|\nu(\mathrm{d}e)]$$

- and for all $\hat{x} = x - \bar{x}$,

$$\langle g(x) - g(\bar{x}), G^\varepsilon(x - \bar{x})\rangle \ge \eta|(G^\varepsilon)\hat{x}|^2.$$

Under (A.1) and (A.2), given $\varepsilon > 0$, for any $x \in R^d$, there exists a unique adapted solution $(X_s^{t,x,\varepsilon}, Y_s^{t,x,\varepsilon}, Z_s^{t,x,\varepsilon}, V_s^{t,x,\varepsilon})_{t\le s\le T} \in \mathscr{N}_{t,\varepsilon}$ of (6).

A proof can be found in [30].

Futhermore, under the assumptions (A.1) and (A.2) we have the following a-priori estimate for the energy of the solutions of (6), reflecting the dependence to different initial conditions $x, x' \in R^d$, for some constant $C > 0$, only dependent of L and η, and independent of ε,

$$E\Big[\sup_{t\le s\le T} |X_s^{t,x,\varepsilon} - X_s^{t,x',\varepsilon}|^2 + \sup_{t\le s\le T} |Y_s^{t,x,\varepsilon} - Y_s^{t,x',\varepsilon}|^2 + \int_t^T |Z_r^{t,x,\varepsilon} - Z_r^{t,x',\varepsilon}|^2 \mathrm{d}r + \int_t^T \int_{R^l-\{0\}} |V_r^{t,x,\varepsilon} - V_r^{t,x',\varepsilon}|^2 \frac{1}{\varepsilon}\nu(\mathrm{d}e)\mathrm{d}r\Big] \le C|x - x'|^2. \tag{7}$$

For the proof of that estimate we refer the reader to Proposition 3.1 in [21].

For each $\varepsilon > 0$, we define the function $u^\varepsilon(t,x) := Y_t^{t,x,\varepsilon}$.

We recall that u^ε is a deterministic function P-a.s. of (t, x), since it is $\mathscr{F}_t$-measurable and due to the deterministic nature of the coefficients given in the FBSDE (no dependence in $\omega \in \Omega$). This can be argued with *Blumenthal's 0–1 Law* in the same way to the FBSDEs driven by Brownian Motion (see Remark 1.2 in [13] and Corollary 2.1 in [2]).

From (7), it is obvious that, for each $\varepsilon > 0$, u^ε is uniformly Lipschitz in space; for some $C = C(L, \eta) > 0$ and independent of $\varepsilon > 0$, for all $x, y \in R^d$ we have

$$|u^\varepsilon(t,x) - u^\varepsilon(t,y)| \le C|x-y|$$
$$|u^\varepsilon(t,x)| \le C(1+|x|).$$

Using the argument exposed in [19], we have the markovian dependency of the backward process with respect to the forward one as follows:

$$u^\varepsilon(t, X_s^{s,X_s^{t,x,\varepsilon}}) = u^\varepsilon(s, X_s^{t,x,\varepsilon}) = Y_s^{t,x,\varepsilon}.$$

Next, it follows the last set of assumptions that will be needed, in order to have representation formulas for the processes $(Z_s^{\varepsilon,\delta})_{0\in[0,T]}$ and $(V_t^{\varepsilon,\delta})_{t\in[0,T]}$ in terms of the deterministic function $u^{\varepsilon,\delta}$.

A.3

- The coefficient β is uniformly continuous in $x \in R^d$ and there exists a function $\rho : R^l - \{0\} \longrightarrow R^+$ in $L^2(E, \mathscr{B}(R^l - \{0\}), \nu)$ such that for all $t \in [0, T]$, $x, \bar{x} \in R^d$ and $e \in R^l - \{0\}$

$$|\beta(x)| \le \rho(e)|x|.$$

- The diffusion matrix $\sigma(t, x, y)$ is a bounded continuously twice differentiable function with its derivatives in x and y Hölder continuous.

Under (A.1)–(A.2)–(A.3), similarly to the decoupled case, since the forward jump-diffusion only depends on the backward process $(Y_s^\varepsilon)_{t\le s\le T}$ and the property $Y_s^{t,x,\varepsilon} = u^\varepsilon(s, X_s^{t,x,\varepsilon})$ decouples the system, it follows that there exists a constant $\kappa > 0$, only depending on L, η, ρ, independent of ε (Theorem 4.3.1 in [14]) such that

$$\begin{aligned}
&| u^\varepsilon(t,x) | \le \kappa;\\
&u^\varepsilon \in C_b^{1,1}([0,T]\times R^d);\\
&\sup_{(t,x)\in[0,T]\times R^d} | \nabla_x u^\varepsilon(t,x) | \le \kappa;\\
&Z_s^{t,\varepsilon,x} = \sqrt{\varepsilon}\nabla_x u^\varepsilon(t, X_s^{t,\varepsilon,x})\sigma(s, X_s^{t,\varepsilon}, Y_s^{t,\varepsilon,x});\\
&V_s^{t,x,\varepsilon}(.) = u^\varepsilon(s, X_{s-}^{t,x,\varepsilon} + \varepsilon\beta(X_{s-}^{t,x,\varepsilon})) - u^\varepsilon(s, X_{s-}^{t,x,\varepsilon}).
\end{aligned}$$

In a similar way to the decoupled case (see [2]) since the property $Y_s^{t,x,\varepsilon} = u^\varepsilon(s, X_s^{t,x,\varepsilon})$ decouples the system, we can state the following result.

Theorem 1 *Under the set of assumptions (A.1), (A.2) and (A.3), u^ε is a viscosity solution of the following terminal value problem:*

$$\begin{cases} (\partial_t + \mathscr{L}^\varepsilon)u^\varepsilon(t,x) \\ +h^\varepsilon\Big(t,x,u^\varepsilon(t,x),\sigma^T(t,x,u^\varepsilon(t,x))\nabla_x u^\varepsilon(t,x), \\ u^\varepsilon(t,x+\varepsilon\beta(x)) - u^\varepsilon(t,x)\Big) = 0 \\ u^\varepsilon(T,x) = g(x), \end{cases} \tag{8}$$

where $\mathscr{L}^\varepsilon = \mathscr{K}_1^\varepsilon + \mathscr{K}_2^\varepsilon$ is the decomposition of the operator associated to the jump diffusion in the local and nonlocal components with

$$\mathscr{K}_1^\varepsilon\varphi(t,x = \langle b(t,x,\varphi(t,x)),\nabla_x\varphi(t,x)\rangle + \frac{\varepsilon}{2}tr(a(t,x,\varphi(t,x))\nabla^2\varphi(t,x))$$
$$\mathscr{K}_2^\varepsilon\varphi(t,x) = \int_{R^l-\{0\}} \frac{\varphi(t,x+\varepsilon\beta(x)) - \varphi(t,x) - \langle\varepsilon\beta(x),\nabla_x\varphi(t,x)\rangle}{\varepsilon}\nu(de) \tag{9}$$

for all $\varphi \in C^{1,2}\big([0,T]\times R^d, R^n\big)$.

For sake of completeness, we present the notion of viscosity solution for the terminal value problem (8). Let us introduce, for every $\tau > 0$, the following integral operator,

$$B^\tau\varphi(t,x) = \int_{R^l-\{0\}} (\varphi(t,x+\varepsilon\beta(x,e)) - \varphi(t,x))\gamma^\tau(e)\nu(de).$$

Definition 2—Viscosity Solution of (8)

(i) The function $u^\varepsilon \in C([0,T]\times R^d)$ is a viscosity subsolution of (8) if, for all $1 \le i \le n$, $u_i^\varepsilon(T,x) \le g_i(x)$, for all $x \in R^d$ and if, for any $1 \le i \le n$, for all $\varphi \in C^{1,2}([0,T]\times R^d)$, whenever (t,x) is a point of global maximum of $u_i^\varepsilon - \varphi$, it holds

$$-\frac{\partial\varphi}{\partial t}(t,x) - \mathscr{K}_1^\varepsilon\varphi(t,x)$$
$$-\mathscr{K}_{2,\sigma}^\varepsilon(u^\varepsilon,\varphi(t,x)) - h_i^\varepsilon(t,x,\varphi,\sigma^t(t,x,\varphi)\nabla_x\varphi, B_\sigma^\tau(u^{\varepsilon,\varphi})) \le 0,$$

for any $\sigma > 0$, where

$$\mathscr{K}_{2,\sigma}^\varepsilon(u^\varepsilon,\varphi) = \int_{E_\sigma} \frac{\varphi(t,x+\varepsilon\beta(x)) - \varphi(t,x) - \langle\beta(x),\nabla_x\varphi(t,x)\rangle}{\varepsilon}\nu(de)$$
$$+\int_{E_\sigma^c} \frac{u^\varepsilon(t,x+\varepsilon\beta(x)) - u^\varepsilon(t,x) - \langle\beta(x),\nabla_x\varphi(t,x)\rangle}{\varepsilon}\nu(de),$$

and

$$B^{\tau}_{\sigma}(u^{\varepsilon},\varphi)=\int_{E_{\sigma}}\frac{(\varphi(t,x+\varepsilon\beta(x,e))-\varphi(t,x))\gamma^{\tau}(e)}{\varepsilon}\nu(\mathrm{d}e)$$
$$+\int_{E^{c}_{\sigma}}\frac{(u^{\varepsilon}(t,x+\varepsilon\beta(x,e))-u^{\varepsilon}(t,x))\gamma^{\tau}(e)}{\varepsilon}\nu(\mathrm{d}e),$$

with $E_{\sigma}=\{e\in R^{l}\mid |e|<\sigma\}$.

(ii) The function u^{ε} is a viscosity supersolution of (8) if, for all $1\leq i\leq n$, $u^{\varepsilon}_{i}(T,x)\geq g_{i}(x)$, for all $x\in R^{d}$, and if, for any $1\leq i\leq n$, for all $\varphi\in C^{1,2}([0,T]\times R^{d})$, whenever (t,x) is a point of global minimum of $u^{\varepsilon}_{i}-\varphi$ we have

$$-\frac{\partial\varphi}{\partial t}(t,x)-\mathscr{K}^{\varepsilon}_{1}\varphi(t,x)$$
$$-\mathscr{K}^{\varepsilon}_{2,\sigma}(u^{\varepsilon},\varphi(t,x))-h^{\varepsilon}_{i}(t,x,\varphi,\sigma^{t}(t,x,\varphi)\nabla_{x}\varphi,B^{\tau}_{\sigma}(u^{\varepsilon,\varphi}))\geq 0,$$

for any $\delta>0$.

(iii) The function u^{ε} is a viscosity solution of (8) if it is both a viscosity subsolution and a viscosity supersolution of (8).

3 The Asymptotic Study

We present the asymptotic study of the FBSDE when the noise intensities vanish, $\varepsilon,\delta\to 0$. The process $(X^{\varepsilon,\delta},Y^{\varepsilon,\delta},Z^{\varepsilon,\delta},V^{\varepsilon,\delta})\in\mathscr{S}^{2}(t,T,R^{d})\times\mathscr{S}^{2}(t,T,R^{n})\times\mathscr{M}^{2}(t,T,R^{n\times d})\times\mathscr{K}^{2}_{\nu,\delta}(t,T,R^{n})_{t\leq s\leq T}$ solves the FBSDE system:

$$\begin{cases}X^{\varepsilon,\delta}_{s}&=x+\int_{t}^{s}b(r,X^{\varepsilon,\delta}_{r},Y^{\varepsilon,\delta}_{r})\mathrm{d}r\\&\quad+\sqrt{\varepsilon}\int_{t}^{s}\sigma(r,X^{\varepsilon,\delta}_{r},Y^{\varepsilon,\delta}_{r})\mathrm{d}W r+\delta\int_{t}^{s}\int_{R^{l}-\{0\}}\beta(X^{\varepsilon,\delta}_{r},e)\widetilde{\mu}^{\frac{1}{\delta}}(\mathrm{d}e,\mathrm{d}r)\\Y^{\varepsilon,\delta}_{s}&=g(X^{\varepsilon,\delta}_{T})+\int_{s}^{T}f\Big(r,X^{\varepsilon,\delta}_{r},Y^{\varepsilon,\delta}_{r},Z^{\varepsilon,\delta}_{r},\int_{R^{l}-\{0\}}V^{\varepsilon,\delta}_{r}(e)\gamma^{\delta}(e)\frac{1}{\delta}\nu(\mathrm{d}e)\Big)\mathrm{d}r\\&\quad-\int_{s}^{T}Z^{\varepsilon,\delta}_{r}\mathrm{d}W r-\int_{s}^{T}\int_{R^{l}-\{0\}}V^{\varepsilon,\delta}_{r}\widetilde{\mu}^{\frac{1}{\delta}}(\mathrm{d}e,\mathrm{d}r).\end{cases}\tag{10}$$

The corresponding PIDE problem is

$$\begin{cases}(\partial_{t}+\mathscr{L}^{\varepsilon,\delta})u^{\varepsilon,\delta}(t,x)\\+h^{\delta}\Big(t,x,u^{\varepsilon,\delta}(t,x),\varepsilon\sigma^{T}(t,x,u^{\varepsilon,\delta}(t,x))\nabla_{x}u^{\varepsilon,\delta}(t,x),\\u^{\varepsilon,\delta}(t,x+\delta\beta(x))-u^{\varepsilon,\delta}(t,x)\Big)=0\\u^{\varepsilon}(T,x)=g(x),\end{cases}\tag{11}$$

where $\mathscr{L}^{\varepsilon} = \mathscr{K}_1^{\varepsilon} + \mathscr{K}_2^{\delta}$ is

$$\mathscr{K}_1^{\varepsilon}\varphi(t,x) = \langle b(t,x,\varphi(t,x)), \nabla_x\varphi(t,x)\rangle + \frac{\varepsilon}{2} tr(a(t,x,\varphi(t,x))\nabla^2\varphi(t,x))$$
$$\mathscr{K}_2^{\delta}\varphi(t,x) = \int_{R^l-\{0\}} \frac{\varphi(t,x+\delta\beta(x)) - \varphi(t,x) - \langle\delta\beta(x), \nabla_x\varphi(t,x)\rangle}{\delta}\nu(\mathrm{d}e) \quad (12)$$

for every function $\varphi \in C^{1,2}\big([0,T]\times R^d, R^n\big)$, continuously differentiable in $t \in [0,T]$ and twice continuously differentiable in $x \in R^d$, with values in R^n, with

$$h^{\delta}(s,x,y,z,k) = f\left(s,x,y,z, \int_{R^l-\{0\}} k(e)\gamma^{\delta}(e)\frac{1}{\delta}\nu(\mathrm{d}e)\right).$$

The result of this section is the following.

Theorem 2

(i) Under (A1), (A.2), (A.3), taking $\varepsilon = \delta$, *the solution* $(X_s^{\varepsilon}, Y_s^{\varepsilon}, Z_s^{\varepsilon}, V_s^{\varepsilon})_{t\le s\le T}$ *of (10) converges in* $\mathscr{S}^2(t,T,R^d)\times\mathscr{S}^2(t,T,R^n)\times\mathscr{M}^2(t,T,R^{n\times d})\times\mathscr{K}^2_{\nu,\varepsilon}(t,T,R^n)_{t\le s\le T}$ *to* $(X_s^0, V_s^0, 0, 0)_{t\le s\le T}$, *where* $(X_s^0, Y_s^0)_{t\le s\le T}$ *solves the following system of ordinary differential equations, for all* $t \le s \le T$,

$$\begin{cases} \frac{\partial X_s^{t,x}}{\partial s} &= b(s, X_s^{t,x}, Y_s^{t,x},) \\ \frac{\partial Y_s^{t,x}}{\partial s} &= -f(s, X_s^{t,x}, Y_s^{t,x}, 0, 0), \\ X_t^{t,x} &= x, \\ Y_T^{t,x} &= g(X_T^{t,x}) \end{cases} \quad (13)$$

(ii) Fixed $\delta > 0$, *considering now the convergence when* $\varepsilon \to 0$, $(X_s^{\varepsilon,\delta}, Y_s^{\varepsilon,\delta}, Z_s^{\varepsilon,\delta}, V_s^{\varepsilon,\delta})_{t\le s\le T}$ *converges in* $\mathscr{S}^2(t,T,R^d)\times\mathscr{S}^2(t,T,R^n)\times\mathscr{M}^2(t,T,R^{n\times d})\times\mathscr{K}^2_{\nu,\delta}(t,T,R^n)_{t\le s\le T}$ *to* $(X_s^{0,\delta}, Y_s^{0,\delta}, 0, V_s^{0,\delta})_{t\le s\le T}$ *solution of (10) with* $\varepsilon = 0$. *Moreover, the function* $u^{\varepsilon,\delta}(t,x) := Y_t^{t,x,\varepsilon,\delta}$, *viscosity solution of (11), converges uniformly in compact sets of* $[0,T]\times R^d$ *to* $u^{0,\delta}$, *viscosity solution of (11) with* $\varepsilon = 0$.

(iii) Moreover, fixed $\varepsilon > 0$, *if* $\delta \to 0$, $(X_s^{\varepsilon,\delta}, Y_s^{\varepsilon,\delta}, Z_s^{\varepsilon,\delta}, V_s^{\varepsilon,\delta})_{t\le s\le T}$ *converges in* $\mathscr{S}^2(t,T,R^d)\times\mathscr{S}^2(t,T,R^n)\times\mathscr{M}^2(t,T,R^{n\times d})\times\mathscr{K}^2_{\nu,\delta}(t,T,R^n)_{t\le s\le T}$ *to* $(X_s^{\varepsilon,0}, Y_s^{\varepsilon,0}, Z_s^{\varepsilon,0}, 0)_{t\le s\le T}$, *solution of (10) when* $\delta = 0$ *and the corresponding viscosity solution* $u^{\varepsilon,\delta}$ *of (11) converges to* $u^{\varepsilon,0}$ *viscosity solution of (11) uniformly in compact sets of* $[0,T]\times R^d$, *when* $\delta = 0$.

(iv) When $\varepsilon, \delta \to 0$, *the limit function* $u(t,x) = Y_t^{t,x}$ *of* $u^{\varepsilon,\delta}$ *is a viscosity solution of the first order terminal value problem, for all* $x \in R^d$ *and for all* $t \in [0,T]$.

$$\begin{cases} \frac{\partial u}{\partial t} + \langle b, \nabla u\rangle + f(t,x,u(t,x),0,0) &= 0 \\ u^{\varepsilon}(T,x) &= g(x); \end{cases} \quad (14)$$

Furthermore, if $u \in C_b^{1,1}([0,T] \times R^d)$ is a continuous function with continuous bounded derivatives, then u is the unique classical strong solution of (14).

Proof Let $(X_s^\varepsilon, Y_s^\varepsilon, Z_s^\varepsilon, V_s^\varepsilon)_{t\le s\le T}$ be the solution in $\mathscr{S}^2(t,T,R^d) \times \mathscr{S}^2(t,T,R^n) \times \mathscr{M}^2(t,T,R^{n\times d}) \times \mathscr{K}^2_{\nu,\varepsilon}(t,T,R^n)_{t\le s\le T}$ of

$$\begin{cases} X_s^{t,x,\varepsilon} & = x + \int_t^s b(r, X_r^\varepsilon, Y_r^\varepsilon)\mathrm{d}r \\ & \quad + \sqrt{\varepsilon}\int_t^s \sigma(r, X_r^\varepsilon, Y_r^\varepsilon)\mathrm{d}Wr + \varepsilon \int_t^s \int_{R^l-\{0\}} \beta(X_r^\varepsilon, e)\widetilde{\mu}^{\frac{1}{\varepsilon}}(\mathrm{d}e, \mathrm{d}r) \\ Y_s^\varepsilon & = g(X_T^\varepsilon) + \int_s^T f\Big(r, X_r^\varepsilon, Y_r^\varepsilon, Z_r^\varepsilon, \int_{R^l-\{0\}} V_r^\varepsilon(e)\frac{1}{\varepsilon}\gamma^\varepsilon(e)\nu(\mathrm{d}e)\Big)\mathrm{d}r \\ & \quad - \int_s^T Z_r^\varepsilon \mathrm{d}Wr - \int_s^T \int_{R^l-\{0\}} V_r^\varepsilon \widetilde{\mu}^{\frac{1}{\varepsilon}}(\mathrm{d}e, \mathrm{d}r), \end{cases} \tag{15}$$

and respectively $(X_s^{\varepsilon'}, Y_s^{\varepsilon'}, Z_s^{\varepsilon'}, V_s^{\varepsilon'})_{t\le s\le T}$ be the solution in $\mathscr{S}^2(t,T,R^d) \times \mathscr{S}^2(t,T,R^n) \times \mathscr{M}^2(t,T,R^{n\times d}) \times \mathscr{K}^2_{\nu,\varepsilon'}(t,T,R^n)_{t\le s\le T}$ of

$$\begin{cases} X_s^{\varepsilon'} & = x + \int_t^s b(r, X_r^{\varepsilon'}, Y_r^{\varepsilon'})\mathrm{d}r \\ & \quad + \sqrt{\varepsilon'}\int_t^s \sigma(r, X_r^{\varepsilon'}, Y_r^{\varepsilon'})\mathrm{d}Wr + \varepsilon' \int_t^s \int_{R^l-\{0\}} \beta(X_r^{\varepsilon'}, e)\widetilde{\mu}^{\frac{1}{\varepsilon'}}(\mathrm{d}e, \mathrm{d}r) \\ Y_s^{\varepsilon'} & = g(X_T^{\varepsilon'}) + \int_s^T f\Big(r, X_r^{\varepsilon'}, Y_r^{\varepsilon'}, Z_r^{\varepsilon'}, \int_{R^l-\{0\}} V_r^{\varepsilon'}(e)\gamma^\varepsilon(e)\frac{1}{\varepsilon'}\nu(\mathrm{d}e)\mathrm{d}r\Big) \\ & \quad - \int_s^T Z_r^{\varepsilon'} \mathrm{d}Wr - \int_s^T \int_{R^l-\{0\}} V_r^{\varepsilon'} \widetilde{\mu}^{\frac{1}{\varepsilon'}}(\mathrm{d}e, \mathrm{d}r). \end{cases} \tag{16}$$

Due to condition (A1) and condition (A.2), after applying Itô's formula, rotine calculations yield to

$$\lim_{|\varepsilon-\varepsilon'|\to 0} E\Big[|Y_t^\varepsilon - Y_t^{\varepsilon'}|^2 + \int_t^s |Y_r^\varepsilon - Y_r^{\varepsilon'}|^2\mathrm{d}r \\ + |Z_r^\varepsilon - Z_r^{\varepsilon'}|^2\mathrm{d}r + \int_{R^l-\{0\}} |V_r^\varepsilon - V_r^{\varepsilon'}|^2\frac{1}{\varepsilon}\nu(\mathrm{d}e)\mathrm{d}r\Big] = 0.$$

Due to the lipschitz nature of the coefficients, we conclude

$$\lim_{|\varepsilon-\varepsilon'|\to 0} E\Big[\int_t^T |X_r^\varepsilon - X_r^{\varepsilon'}|^2\mathrm{d}r\Big] = 0.$$

Using the monotonicity asusmption (A.2) we derive that

$$\lim_{|\varepsilon-\varepsilon'|\to 0} E\Big[|X_T^\varepsilon - X_T^{\varepsilon'}|^2\Big] = 0.$$

Making use of the martingale inequalities for stochastic integrals applying once again Itô's formula, we conclude that

$$\lim_{|\varepsilon-\varepsilon'|\to 0} E\Big[\sup_{t\le s\le T}|X_r^{\varepsilon}-X_r^{\varepsilon'}|^2+\sup_{t\le s\le T}|Y_r^{\varepsilon}-Y_r^{\varepsilon'}|^2\Big]=0.$$

In conclusion, $(X_s^{\varepsilon}, Y_s^{\varepsilon}, Z_s^{\varepsilon}, V_s^{\varepsilon})_{t\le s\le T}$ converges in $\mathscr{S}^2(t,T,R^d)\times\mathscr{S}^2(t,T,R^n)\times\mathscr{M}^2(t,T,R^{n\times d})\times\mathscr{K}^2_{\nu,\varepsilon}(t,T,R^n)$ up to a subsequence. We call the sublimit

$(X_s^0, Y_s^0, Z_s^0, V_s^0)_{s\le t\le T}$. Under the assumptions (A.1), (A.2) and (A.3) we have the following representation formulas

$$Z_s^{t,\varepsilon,x}=\sqrt{\varepsilon}\nabla_x u^{\varepsilon}(t,X_s^{t,\varepsilon,x})\sigma(s,X_s^{t,\varepsilon},Y_s^{t,\varepsilon,x})$$
$$V_s^{t,x,\varepsilon}(.)=u^{\varepsilon}(s,X_{s-}^{t,x,\varepsilon}+\varepsilon\beta(X_{s-}^{t,x,\varepsilon},.))-u^{\varepsilon}(s,X_{s-}^{t,x,\varepsilon}).$$

It is assumed in (A.3) that σ is bounded by a constant and β is bounded by a square integrable function. Hence, we conclude that

$$E\Big[\int_t^T |Z_r^{\varepsilon}|^2 dr\Big]\to 0 \quad as \quad \varepsilon\to 0$$

and

$$E\Big[\int_t^T\int_{R^l-\{0\}}|V_r^{\varepsilon}|^2\frac{1}{\varepsilon}\nu(de)dr\Big]\to 0 \quad as \quad \varepsilon\to 0.$$

Due to the boundedness of σ and β, and using the continuity of b, we can take the pointwise limit of the forward equation of (15), and conclude that

$$X_s^0=x+\int_s^s b(r,X_r^0,Y_r^0)dr \quad P-a.s.$$

As we remarked, as $\varepsilon\to 0$, $(Z_s^{\varepsilon})_{t\le s\le T}\to 0$ in $\mathscr{M}^2(t,T,R^{n\times d})$ and $(\frac{V_s^{\varepsilon}}{\varepsilon})_{t\le s\le T}\to 0$ in $\mathscr{K}^2_{\nu}(t,T,R^n)$ and since

$$E\Big[\Big|\int_t^T Z_r^{\varepsilon}dW_r\Big|^2\Big]=E\int_t^T|Z_r^{\varepsilon}|^2dr=0,$$
$$E\Big[\Big|\int_t^S\int_{R^l-\{0\}}V_r^{\varepsilon}\widetilde{\mu}(de,dr)\Big|^2=E\int_t^T\int_{R^l-\{0\}}|V_r^{\varepsilon}|^2\frac{1}{\varepsilon}\nu(de)\Big]=0,$$

we can take similarly the limit pointwise in the backward equation of (15) and using the continuity of the functions f and g we conclude P-a.s

$$Y_s^0=g(X_T)+\int_s^T f(r,x_r^0,Y_r^0,0,0),$$

which proves the statement *(i)* of the theorem.

In a similar way of what was done before, when we proved that $(X_s^{\varepsilon}, Y_s^{\varepsilon}, Z_s^{\varepsilon}, \frac{1}{\varepsilon}V_s^{\varepsilon})_{t\leq s\leq T}$, solution of (15), is a Cauchy sequence in $\mathcal{N}_t$, one can easily prove that when $\varepsilon \to 0$, $(X^{\varepsilon,\delta}, Y^{\varepsilon,\delta}, Z^{\varepsilon,\delta}, V^{\varepsilon,\delta})_{t\leq s\leq T}$ converges in $\mathcal{N}_t$ to $(X^{0,\delta}, Y^{0,\delta}, 0, V^{0,\delta})_{t\leq s\leq T}$ which solves (10) with $\varepsilon = 0$. Using Theorem 1, we know that $u^{0,\delta}(t,x) := Y_t^{0,\delta}$ is a viscosity solution of (11) with $\varepsilon = 0$. Similar conclusion holds when $\delta \to 0$.

Let us assume again that $\varepsilon = \delta$ and study the regularity of the limit function $u(t,x) = Y_t^{t,x,0}$. Using the argument of Corollary 1.4 in [13], given $t, t' \in [0,T]$, $x, x' \in R^d$, consider $(X_s^{t,x,\varepsilon}, Y_s^{t,x,\varepsilon}, Z_s^{t,x,\varepsilon}, V_s^{t,x,\varepsilon})_{t\leq s\leq T}$ the unique solution of

$$\begin{cases} X_s^{t,x,\varepsilon} &= x + \int_t^s b(r, X_r^{t,x,\varepsilon}, Y_r^{t,x,\varepsilon})\mathrm{d}r \\ &\quad + \sqrt{\varepsilon}\int_t^s \sigma(r, X_r^{t,x,\varepsilon}, Y_r^{t,x,\varepsilon})\mathrm{d}Wr + \varepsilon\int_t^s\int_{R^l-\{0\}} \beta(X_r^{t,x,\varepsilon}, e)\widetilde{\mu}^{\frac{1}{\varepsilon}}(\mathrm{d}e, \mathrm{d}r) \\ Y_s^{t,x,\varepsilon,} &= g(X_T^{t,\varepsilon,x}) + \int_s^T f\Big(r, X_r^{t,x,\varepsilon}, Y_r^{t,x,\varepsilon}, Z_r^{t,x,\varepsilon}, \int_{R^l-\{0\}} V_r^{t,x,\varepsilon}(e)\gamma^{\varepsilon}(e)\frac{1}{\varepsilon}\nu(\mathrm{d}e)\Big)\mathrm{d}r \\ &\quad - \int_s^T Z_r^{t,x,\varepsilon}\mathrm{d}Wr - \int_s^T\int_{R^l-\{0\}} V_r^{t,x,\varepsilon}\widetilde{\mu}^{\frac{1}{\varepsilon}}(\mathrm{d}e, \mathrm{d}r) \end{cases}$$

extended to the whole interval $[0,T]$ by setting, for all $0 \leq s \leq t$, $X_s^{t,x,\varepsilon} = x$, $Y_s^{t,x,\varepsilon} = Y_t^{t,x,\varepsilon}$ and $Z_s^{t,x,\varepsilon} = V_s^{t,x,\varepsilon} = 0$. Respectively let $(X_s^{t',x',\varepsilon}, Y_s^{t',x',\varepsilon}, Z_s^{t',x',\varepsilon}, V_s^{t',x',\varepsilon})_{t\leq s\leq T}$ be the unique solution of

$$\begin{cases} X_s^{t',x',\varepsilon} &= x' + \int_{t'}^s b(r, X_r^{t',x',\varepsilon}, Y_r^{t',x',\varepsilon})\mathrm{d}r \\ &\quad + \sqrt{\varepsilon}\int_{t'}^s \sigma(r, X_r^{t',x',\varepsilon}, Y_r^{t',x',\varepsilon})\mathrm{d}Wr + \varepsilon\int_{t'}^s\int_{R^l-\{0\}} \beta(X_r^{t',x,\varepsilon}, e)\widetilde{\mu}^{\frac{1}{\varepsilon}}(\mathrm{d}e, \mathrm{d}r) \\ Y_s^{t',x',\varepsilon,} &= g(X_T^{t',x',\varepsilon}) + \int_s^T f\Big(r, X_r^{t',x',\varepsilon}, Y_r^{t',x',\varepsilon}, Z_r^{t',x',\varepsilon}, \\ &\quad \int_{R^l-\{0\}} V_r^{t',x',\varepsilon}(e)\gamma^{\varepsilon}(e)\frac{1}{\varepsilon}\nu(\mathrm{d}e)\Big)\mathrm{d}r - \int_s^T Z_r^{t',x',\varepsilon}\mathrm{d}Wr \\ &\quad - \int_s^T\int_{R^l-\{0\}} V_r^{t',x',\varepsilon}\widetilde{\mu}^{\frac{1}{\varepsilon}}(\mathrm{d}e, \mathrm{d}r), \end{cases}$$

extended to the whole interval $[0,T]$ by setting, for all $0 \leq s \leq t'$, $X_s^{t',x',\varepsilon} = x$, $Y_s^{t',x',\varepsilon} = Y_t^{t',x',\varepsilon}$ and $Z_s^{t',x',\varepsilon} = V_s^{t',x',\varepsilon} = 0$. Similarly as in Proposition 3.1 in [21], we can prove that there exists some constants $\alpha, \beta > 0$ only depending on L, η, T, ρ such that

$$\begin{aligned} &E\Big[\sup_{0\leq s\leq T} \mid X_s^{t,x,\varepsilon} - X_s^{t',x',\varepsilon} \mid^2\Big] + E\sup_{0\leq s\leq T}\Big[\mid Y_s^{t,x,\varepsilon} - Y_s^{t',x',\varepsilon} \mid^2\Big] \\ &+ E\Big[\int_0^T \mid Z_s^{t,x,\varepsilon} - Z_s^{t',x,\varepsilon'} \mid^2 \mathrm{d}s\Big] + E\Big[\int_0^T\int_{R^l-\{0\}} \mid V_s^{t,x,\varepsilon}(e) - V_s^{t',x,\varepsilon'}(e) \mid^2 \frac{1}{\varepsilon}\nu(de)ds\Big] \\ &\leq \alpha \mid x - x' \mid^2 + \beta(1 + \mid x \mid^2 \vee \mid x' \mid^2) \mid t - t' \mid^2 . \end{aligned}$$

Hence,

$$|u^{\varepsilon}(t,x)-u^{\varepsilon}(t',x')|^2 \le \alpha \mid x-x' \mid^2 + \beta(1+\mid x\mid^2 \vee \mid x'\mid^2)\mid t-t'\mid^2, \tag{17}$$

which proves that $(u^{\varepsilon})_{\varepsilon>0}$ is a family of equicontinuous functions on the compact sets of $[0,T]\times R^d$. Using *Arzelà-Ascoli theorem*, up to a subsequence, we conclude the uniform convergence of u^{ε} to u in the compact sets of $[0,T]\times R^d$. Taking the limit in (17) we have proven that u is Lipschitz continuous in x and uniformly continuous in t.

A similar proof to the one of Theorem 5.1 of [9] yields to the conclusion that u is a viscosity solution in $[0,T]\times R^d$ of (14).

Moreover, let $v:[0,T]\times R^d \to R^k$ be a $C_b^{1,1}([0,T],R^k)$ solution, continuous Lipschitz in x and uniformly continuous in t of (14). Fixing $(t,x)\in[0,T]\times R^d$, we take the following function:

$\psi:[t,T]\to R^k$

$\psi(s):=v(s,X_s^{t,x})$.

Computing its time derivative, it follows that

$$\begin{aligned}
\frac{d\psi}{ds}(s) &= \frac{\partial v}{\partial s}(s,X_s^{t,x}) + \sum_{i=1}^{d}\frac{\partial v}{\partial x_i}(s,X_s^{t,x})\frac{\partial(X_s^{t,x})}{\partial t}\\
&= \frac{\partial v}{\partial s}(s,X_s^{t,x}) + \sum_{i=1}^{d}\frac{\partial v}{\partial x_i}(s,X_s^{t,x})f(s,X_s^{t,x},Y_s^{t,x})\\
&= -g(s,X_s^{t,x},v(s,X_s^{t,x}),0)\\
\psi(T) &= v(T,X_T^{t,x}) = h(x).
\end{aligned}$$

As a consequence, $v(t,x)=v(t,X_t^{t,x})=u(t,x)$, since, under A.1 and A.2, (13) has a unique continuous solution, which yields to a uniqueness property of solution for (14) in the class of $C_b^{1,1}([0,T]\times R^d)$.

4 Statement of a Large Deviations Principle

Large deviations theory (LDT for short) deals with probabilities of rare events in a exponential scale, under the variation of a parameter. We consider again the FBSDE (6) and we will state a LDP for the laws of $(X^{\varepsilon},Y^{\varepsilon})_{\varepsilon>0}$. The main property employed to establish the LDP for $(X_s^{\varepsilon},Y_s^{\varepsilon})_{t\le s\le T}$ is

$$Y_s^{t,x,\varepsilon} = u^{\varepsilon}(s,X_s^{t,\varepsilon,x}) \quad P-a.s. \tag{18}$$

The identity (18) is the key to prove the large deviation principle, since it decouples the FBSDE. To state the LDP to $(X^{\varepsilon})_{\varepsilon>0}$, we need the following integrability condition on the measure ν. There exists $\delta_0 > 0$ such that for all $\delta > \delta_0$

$$\int_{R^l-\{0\}} e^{-\delta z}\nu(dz) < \infty \tag{19}$$

We use the following result of (LDT), which is a refinement of Theorem 4.3 in [6] and its proof follows from the same arguments.

Theorem 3 *Given $t \in [0,T]$ and $x \in R^d$, consider $b : [t,T] \times R^d \to R^d$, $\sigma : [t,T] \times R^d \to R^{d\times d}$ and $\beta : R^d \times R^l - \{0\} \longrightarrow R^d$ Lipschitz continuous functions, with sublinear growth. For each $\varepsilon > 0$, let $b^\varepsilon : [t,T] \times R^d \to R^d$, $\sigma^\varepsilon : [t,T] \times R^d \to R^{d\times d}$ and $\beta^\varepsilon : R^d \times R^l - \{0\} \longrightarrow R^d$ be Lipschitz continuous and with sublinear growth, with the aditional assumption that β satisfies (A.1) and such that*

$$\lim_{\varepsilon\to 0} \mid b^\varepsilon - b \mid = \lim_{\varepsilon\to 0} \mid \sigma^\varepsilon - \sigma \mid = \lim_{\varepsilon\to 0} \| \beta^\varepsilon - \beta \|_{L^2(R^l-\{0\},\mathscr{B}(R^l-\{0\}),\nu)} = 0$$

uniformly in the compact sets of $[t,T] \times R^d$. We assume the exponential integrability property of the measure ν (19).

Let us define the product space $S = L^2([t,T], R^d) \times \mathscr{V}$, where

$$\mathscr{V} := \{g : [0,T] \times R^l - \{0\} \longrightarrow [0,\infty) \;\; Borel-measurable\}.$$

Then the process $(X_s^\varepsilon)_{t\le s\le T}$, unique strong solution of the stochastic differential equation, that has position x at time t,

$$dX_s^\varepsilon = b^\varepsilon(s, X_s^\varepsilon)\mathrm{d}s + \sqrt{\varepsilon}\sigma^\varepsilon(s, X_s^\varepsilon)\mathrm{d}W_s + \varepsilon\int_{R^l-\{0\}} \beta^\varepsilon(X_s^\varepsilon, e)\widetilde{\mu}^{\frac{1}{\varepsilon}}(\mathrm{de}, \mathrm{d}s)$$

satisfies a large deviations principle in $D_x([t,T], R^d)$, the space of càdlàg functions endowed with the Skorokhod metric (see [4] for the study of the Skorokhod Space), $\varphi : [t,T] \to R^d$ starting in x, with the good rate function:

$$\begin{aligned} I(\Phi) = \inf\Big\{ &\frac{1}{2}\int_t^T |f(s)|^2 \mathrm{d}s + \int_t^T \int_{R^l-\{0\}} (g(s,e)\ln g(s,e) - g(s,e) + 1)\nu(\mathrm{de})\mathrm{d}s : \\ &(f,g) \in \mathscr{S} \text{ such that} : \Phi_s = x + \int_t^s b(r,\Phi_r)\mathrm{d}r + \int_t^s \sigma(r,\Phi_r)f_r \\ &+ \int_t^T \int_{R^l-\{0\}} \beta(\Phi(r), e)(g(r,e) - 1)\nu(\mathrm{de})\mathrm{d}r \quad t \le s \le T \Big\}. \end{aligned}$$

This means that the level sets of I are compact and

$$\limsup_{\varepsilon\to 0} \varepsilon \ln P(X^\varepsilon \in F) \le -\inf_{\psi\in F} I(\psi),$$

$$\liminf_{\varepsilon \to 0} \varepsilon \ln P(X^{\varepsilon} \in G) \geq - \inf_{\psi \in G} I(\psi),$$

for every closed set $F \in D_x([t, T], R^d)$ and for every open set $G \in D_x([t, T], R^d)$

From this theorem, it follows the large deviations principle for $(X^{\varepsilon})_{\varepsilon>0}$. To establish the large deviations principle for $(Y^{\varepsilon})_{\varepsilon>0}$, we will use a classical tool to transfer one large deviations principle from a topologic vector space to a complete separable metric space, called contraction principle and whose proof can be found in [12].

Theorem 4 (Contraction Principle) *Let $f : \mathscr{S} \to \mathscr{T}$ be a continuous mapping from a topological vector space $\mathscr{S}$ to a metric space $(\mathscr{T}, d)$. We suppose that the family of probability measures $\{\mu_{\varepsilon}\}_{\varepsilon>0}$ satisfies a large deviations principle with a good rate function $I : X \to [0, \infty]$ and that, for each $\varepsilon > 0, f_{\varepsilon} : \mathscr{S} \to \mathscr{T}$ is a continuous function, uniformly convergent to f in all compact sets of $\mathscr{S}$. Then it follows that $\{\mu_{\varepsilon} \circ f_{\varepsilon}^{-1}\}_{\varepsilon>0}$ satisfies a large deviation principle with good rate function $J(y) = \inf\{I(x) : x \in X$ and $f(x) = y\}$.*

We are ready to state the large deviations principle for the laws of the family $(X^{\varepsilon}, Y^{\varepsilon})_{\varepsilon>0}$, when $\varepsilon \to 0$.

Theorem 5 (A large deviations principle) *When $\varepsilon \to 0$, $(X^{\varepsilon})_{\varepsilon>0}$ satisfies a LDP in $D_x([t, T], R^d)$ with the good rate function*

$$I(\Phi) = \inf \left\{ \frac{1}{2} \int_t^T |f(s)|^2 \mathrm{d}s + \int_t^T \int_{R^l - \{0\}} (g(s, e) \ln g(s, e) - g(s, e) + 1) \nu(\mathrm{d}e) \mathrm{d}s \mid \right.$$
$$(f, g) \in \mathscr{S} \text{ such that } \Phi_s = x + \int_t^s b(r, \Phi_r) \mathrm{d}r + \int_t^s \sigma(r, \Phi_r) f_r$$
$$\left. + \int_t^T \int_{R^l - \{0\}} \beta(\Phi(r), e)(g(r, e) - 1) \nu(\mathrm{d}e) \mathrm{d}r \quad t \leq s \leq T \right\}.$$

and $(Y^{\varepsilon})_{\varepsilon>0}$ satisfies a LDP $D([t, T], R^n)$ with the good rate function

$$J(\psi) = \inf \Big\{ I(\varphi) : F(\varphi) = \psi \quad if \quad (f, g) \in \mathscr{V} \Big\},$$

where $F(\varphi)(s) = u(s, \varphi_s)$ for all $\varphi \in D([t, T], R^d)$.

Proof Since $Y_s^{t,\varepsilon,x} = u^{\varepsilon}(s, X_s^{t,\varepsilon,x})$, the first equation on the FBSDE (6) is in the differential form given by (we omit the dependence on the initial condition and initial time), for all $t \leq s \leq T$,

$$dX_s^{\varepsilon} = b^{\varepsilon}(s, X_s^{\varepsilon}) \mathrm{d}s + \sqrt{\varepsilon} \sigma^{\varepsilon}(s, X_s^{\varepsilon}) \mathrm{d}Ws + \varepsilon \int_{R^l - \{0\}} \beta(X_s^{\varepsilon}, e) \widetilde{\mu}^{\frac{1}{\varepsilon}}(\mathrm{d}e, \mathrm{d}s),$$

where we write $b^{\varepsilon}(s, X_s^{\varepsilon}) = b(s, X_s^{\varepsilon}, u^{\varepsilon}(s, X_s^{\varepsilon}))$ and $\sigma^{\varepsilon}(s, X_s^{\varepsilon}) = \sigma(s, X_s^{\varepsilon}, u^{\varepsilon}(s, X_s^{\varepsilon}))$. We are in conditions to apply Theorem 3 and state that the family $(X^{\varepsilon})_{\varepsilon>0}$ satisfies a large deviations principle, with the good rate function

$$I(\Phi) = \inf\left\{\frac{1}{2}\int_t^T |f(s)|^2 ds + \int_t^T \int_{R^l-\{0\}} (g(s,e)\ln g(s,e) - g(s,e) + 1)\nu(de)ds :\right.$$
$$(f,g) \in \mathcal{V} \textit{ such that } \Phi_s = x + \int_t^s b(r,\Phi_r)dr + \int_t^s \sigma(r,\Phi_r)f_r$$
$$\left. + \int_t^T \int_{R^l-\{0\}} \beta(\Phi(r),e)(g(r,e)-1)\nu(de)dr, \quad t \le s \le T.\right\},$$

for all $\Phi \in D_x([t,T], R^d)$.

In what follows, in order to prove a LDP to $(Y_s^\varepsilon)_{t\le s\le T}$, we consider the following family of operators, defined by $F^\varepsilon : D([t,T], R^d) \to D([t,T], R^n)$

$$F^\varepsilon(\varphi)(s) = u^\varepsilon(s, \varphi_s).$$

We observe that $Y_s^\varepsilon = F^\varepsilon(X_s^\varepsilon)$ for all $s \in [t,T]$. The continuity of F^ε and the uniform convergence in the compact sets of $C([t,T], R^d)$ of F^ε to F, when $\varepsilon \to 0$, where

$$F(\varphi)(s) = u(s, \varphi_s),$$

for all $\varphi \in C([t,T], R^d)$, follows from the regularity conditions of the solution of (8). Using the contraction principle in the form presented in Theorem 4 the result follows.

Acknowledgments The author would like to thank the editors and the anonymous reviewers for their comments, suggestions and several corrections. The author acknowledges the financial support by the project IRTG 1740/ TRP 2011/50151-0 funded by the DFG/FAPESP.

References

1. Aschterenberg, R., Kiselev, A., Nazarov, F.: Blow up and regularity for fractal burgers equation. Dyn. PDE **5**(3), 211–240 (2008)
2. Barles, G., Buckdahn, R., Pardoux, E.: Backward stochastic differential equations and integral-partial differential equations. Stoch. Stoch. Rep. **60**, 57–83 (1996)
3. Biler, P., Funaki, T., Woyczynki, A.: Fractal Burgers equations. J. Differ. Equ. **148**, 9–46 (1998)
4. Billingsley, P.: Convergence of Probability Measures, 2nd edn. Wiley, New York (1999)
5. Bismut, J.M.: Théorie Probabilistique du Controle des Diffusions. Memoirs of the American Mathematical Society, vol. 176. Providence, Rhode Island (1973)
6. Budhiraja, A., Dupuis, P., Maroulas, M.: Variational representation for continuous time processes. Annales de l'Institut Henri Poincaré (B), Probabilités et Statistiques **47(3)**, 725–747 (2011)
7. Busnello, B., Flandoli, F., Romito, M.: A probabilistic representation for the vorticity of a three dimensional viscuous fluid and for general systems of parabolic equations. Proc. Edinb. Math. Soc. (2). **48(2)**, 295–336 (2005)
8. Constantin, P., Iyer, G.: A stochastic Lagrangian representation for the 3D incompressible Navier Stokes equation. Commun. Pure Appl. Maths. **LXI**, 330–345 (2008)
9. Crandall, M., Ishii, H., Lions, P.L.: User's guide to viscosity solutions of second order partial differential equations. Bull. Am. Math. Soc. **27**(1), 1–67 (1992)

10. Cruzeiro, A., Gomes, A., Zhang, L.: Asymptotic properties of coupled forward backward stochastic differential equations. Stochast. Dyn. **14–3**, 1–42 (2014)
11. Cruzeiro, A., Shamarova, E.: Navier Stokes equation and forward backward SDEs in the group of diffeomorfisms of a torus. Stoch. Process. Appl. **119**, 4034–4060 (2009)
12. Dembo, A., Zeitouni, O.: Large Deviations Techniques and Applications, 2nd edn. Springer, New York (1998)
13. Delarue, F.: On the existence and uniqueness of solutions to FBSDEs in a non degenerate case. Stoch. Process. Appl. **99**, 209–286 (2002)
14. Delong, L.: Backward Stochastic Differential Equations with Jumps and Their Actuarial and Financial Applications. Springer, London (2013)
15. Frei, C., Reis, G.: Quadratic FBSDE with generalized burgers type nonlinearities, PDE perturbation and large deviations. Stoch. Dyn. **13(2)**, 42pp (2013)
16. Hu, Y., Peng, S.: Solutions of forward backward stochastic differential equations. Prob. Theory Relat. Fields **103**, 273–283 (1995)
17. Jacod, J., Shyriaev, A.N.: Limit Theorems for Stochastic Processes. Springer, Heidelberg (2003)
18. Karatzas, S.: Brownian Motion and Stochastic Calculus. Graduate Studies in Mathematics, vol. 113 (1991)
19. Li, J., Peng, G.: Stochastic optimization theory of BSDEs with jumps and viscosity theory of HJB equation. Nonlinear Anal. **70**, 1776–1796 (2009)
20. Li, J., Tang, S.: Necessary conditions for optimal control of stochastic with random jumps. SIAM J. Control Optim. **32**(5), 1447–1475 (1999)
21. Li, J., Wei, M.: L^p estimates for fully coupled FBSDEs with jumps. Stoch. Process. Appl. **124-4**, 1582–1611 (2014)
22. Ma, J., Protter, P., Yong, J.: Solving forward-backward stochastic differential equations explicitly—a four step scheme. Prob. Theory Relat. Fields **103**, 273–283 (1995)
23. Ma, J., Yong, J.: Forward Backward Stochastic Differential Equations and Their Applications. Springer, Berlin (1999)
24. Pardoux, E., Peng, S.: Adapted solution of a backward stochastic differential equation. Syst. Control Lett. **14**, 55–61 (1990)
25. Peng, S., Wu, Z.: Fully coupled forward backward stochastic differential equations and applications to optimal control. SIAM J. Control Optim. **37**, 825–843 (1999)
26. Rainero, S.: Un principe de grandes déviations pour une équation différentielle stochastique progressive rétrograde. C. R. Acad. Sci. Paris, Ser. I **343(2)**, 141–144 (2006)
27. Situ, R.: On solutions of backward stochastic differential equations with jumps and applications. Stoch. Process. Appl. **66**, 209–236 (1997)
28. Sokolov, I.: Models of anomalous diffusion in crowded environments. Soft Matter **8**, 9043 (2012)
29. Sow, A.: BSDE with jumps and non-lipschitz coefficients: application to large deviations. Braz. J. Prob. Stat. **28–1**, 96–108 (2014)
30. Wu, Z.: Fully coupled FBSDE with Brownian motion and Poisson process in stopping time duration. J. Aust. Math. Soc. **74**, 249–266 (2003)
31. Yong, J.: Finding adapted solutions of forward backward stochastic differential equations: method of continuation. Prob. Theory Relat. Fields **107**, 537–542 (1997)
32. Zhang, X.: Stochastic Lagrangian particle approach to fractal Navier-Stokes equations. Comm. Math. Phys. **311**, 133–155 (2012)

Entropy Dissipation Estimates for the Landau Equation: General Cross Sections

Laurent Desvillettes

Abstract We present here an extension to the case of general cross sections of the lower bound obtained in Desvillettes (J Funct Anal 269:1359–1403, 2015, [14]) on the entropy dissipation of Landau's collision kernel (in the case of soft potentials, including the Coulomb potential). We also simplify somewhat the proof of the lower bound proposed in Desvillettes (J Funct Anal 269:1359–1403, 2015, [14]).

Keywords Landau equation · Landau operator · Entropy dissipation

1 Introduction and Main Result

1.1 Description of the Landau Operator and Equation

We are concerned here with the Landau operator appearing in plasma theory (cf. [11, 21]), defined by

$$Q_\psi(f,f)(v) = \nabla_v \cdot \left\{ \int_{\mathbb{R}^3} a^\psi(v-w) \Big(f(w)\, \nabla f(v) - f(v)\, \nabla f(w) \Big)\, dw \right\}. \tag{1}$$

Here, $a^\psi := a^\psi(z) := (a^\psi_{ij}(z))_{ij}$ $(z \in \mathbb{R}^3)$ is a (nonnegative symmetric) matrix-valued function with only one degenerate direction, namely that of z. More precisely,

$$a^\psi_{ij}(z) = \Pi_{ij}(z)\, \psi(|z|), \tag{2}$$

L. Desvillettes (✉)
Institut de Mathématiques de Jussieu - Paris Rive Gauche,
UMR 7586, CNRS, Université Paris Diderot, Sorbonne Paris Cité,
F-75013 Paris, France
e-mail: desvillettes@math.univ-paris-diderot.fr

© Springer International Publishing Switzerland 2016

P. Gonçalves and A.J. Soares (eds.), *From Particle Systems to Partial Differential Equations III*, Springer Proceedings in Mathematics & Statistics 162,
DOI 10.1007/978-3-319-32144-8_6

where ψ is a (scalar valued) nonnegative function, and

$$\Pi_{ij}(z) = \delta_{ij} - \frac{z_i z_j}{|z|^2} \tag{3}$$

is the i, j-component of the orthogonal projection Π onto $z^\perp := \{y \,/\, y \cdot z = 0\}$.

We observe that at the formal level (that is, when both f and φ are smooth functions having a reasonable behavior at infinity), the (symmetric) weak version of the Landau operator can be defined by the following formula:

$$\int_{\mathbb{R}^3} Q_\psi(f, f)(v)\, \varphi(v)\, dv \tag{4}$$

$$= -\frac{1}{2} \iint_{\mathbb{R}^3\times\mathbb{R}^3} f(v)\, f(w)\, a^\psi(v-w) \left(\frac{\nabla f}{f}(v) - \frac{\nabla f}{f}(w)\right)$$

$$\left(\nabla\varphi(v) - \nabla\varphi(w)\right) dvdw,$$

where the symmetric matrix a^ψ acts as a bilinear form on two vectors.

Using the test functions $\varphi(v) = 1$, v_i (for $i = 1, \dots, 3$), $\frac{|v|^2}{2}$, we see that (still at the formal level), the Landau operator conserves mass, momentum and kinetic energy, that is:

$$\int_{\mathbb{R}^3} Q(f, f)(v) \begin{pmatrix} 1 \\ v_i \\ |v|^2/2 \end{pmatrix} dv = 0. \tag{5}$$

We also get (once again at the formal level) the formula for the entropy dissipation $D_\psi := D_\psi(f)$ (defined on functions f from $\mathbb{R}^3$ to $\mathbb{R}_+$) by considering $\varphi(v) = \ln f(v)$:

$$D_\psi(f) := -\int_{\mathbb{R}^3} Q_\psi(f, f)(v) \ln f(v)\, dv \tag{6}$$

$$= \frac{1}{2} \iint_{\mathbb{R}^3\times\mathbb{R}^3} f(v)\, f(w)\, \psi(|v-w|)\, \Pi(v-w) \left(\frac{\nabla f}{f}(v) - \frac{\nabla f}{f}(w)\right)$$

$$\left(\frac{\nabla f}{f}(v) - \frac{\nabla f}{f}(w)\right) dvdw \geq 0.$$

The most physically relevant function ψ appearing in operator (1), (2) is $\psi(z) = |z|^{-1}$. It corresponds to the case when f is the density of charged particles (moving according to Coulomb interaction) in a plasma, cf. [21]. It also naturally appears in the so-called weak coupling asymptotics of Boltzmann equation (cf. [6] and the older reference [7]).

It is however also interesting, at least from the mathematical viewpoint, to consider more general functions ψ. We refer for example to [13] to see how the Landau kernel with arbitrary ψ can be obtained from the Boltzmann kernel (with arbitrary cross section) through a scaling involving the concept of grazing collisions.

We shall use in this paper the terminology of [14], which is very close to that of [26]. When the dimension of the space is $N = 3$ (we shall always make that assumption in the sequel), if ψ is given by a power law, we say that

$$\psi(|z|) = |z|^{\gamma+2} \tag{7}$$

is coming out of hard potentials when $\gamma \in]0, 1]$, Maxwell molecules when $\gamma = 0$, moderately soft potentials when $\gamma \in [-2, 0]$, and very soft potentials when $\gamma \in]-4, -2[$. We also shall call general soft potentials the case $\gamma < 0$ (that is, γ can be smaller than -4), and general hard potentials the case $\gamma \in]0, 2[$ (that is, γ can be larger than 1). Note that the Coulomb case falls within the category of very soft potentials.

We now introduce the spatially homogeneous Landau equation

$$\partial_t f(t, v) = Q_\psi(f(t, \cdot), f(t, \cdot)). \tag{8}$$

with initial data

$$f(0, v) = f_{in}(v). \tag{9}$$

As a consequence of formula (5), the solutions of the Landau equation (8), (9) satisfy (at least formally) the conservation of mass, momentum and energy, that is

$$\int_{\mathbb{R}^3} f(t, v) \begin{pmatrix} 1 \\ v_i \\ |v|^2/2 \end{pmatrix} dv = \int_{\mathbb{R}^3} f_{in}(v) \begin{pmatrix} 1 \\ v_i \\ |v|^2/2 \end{pmatrix} dv. \tag{10}$$

They also satisfy (at the formal level) the entropy identity (first part of Boltzmann's H-theorem)

$$\frac{d}{dt} H(f(t, \cdot)) = -D_\psi(f(t, \cdot)) \le 0, \tag{11}$$

where $H := H(f)$ is the entropy functional (defined on functions from $\mathbb{R}^3$ to $\mathbb{R}_+$):

$$H(f) := \int_{\mathbb{R}^3} f(v) \ln f(v) \, dv, \tag{12}$$

and D_ψ is the entropy dissipation functional defined in (6).

As stated in detail in [14], identities (10) and (11) naturally furnish an a priori estimate (when the initial data have a finite mass, energy and entropy): indeed

$$\sup_{t\in[0,T]} \int_{v\in\mathbb{R}^3} f(t,v) \left(1 + \frac{|v|^2}{2} + |\ln f(t,v)|\right) dv \tag{13}$$

$$+ \int_0^T D_\psi(f(t,\cdot))\, dt \le C(T, \mathscr{M}_{in}),$$

where the constant $C(T, \mathscr{M}_{in})$ only depends on T and

$$\mathscr{M}_{in} = \int_{v\in\mathbb{R}^3} f_{in}(v) \left(1 + \frac{|v|^2}{2} + |\ln(f_{in}(v))|\right) dv.$$

As a consequence (remembering that $D_\psi(f)$ is a nonnegative quantity), any (nonnegative) lower bound of $D_\psi(f)$ will naturally yield an a priori estimate for the solutions of the Landau equation (when the initial data have a finite mass, energy and entropy).

One of the first such lower bounds appeared in [17] in the case when ψ is a so-called "overMaxwellian" cross section, that is $\psi(z) \ge c_0\,|z|^2$, for some c_0. The context there was the study of the large time behavior of the Landau equation, and the lower bound was the relative Fisher information of f.

This result was substantially improved in [14], in order to include soft potentials including the Coulomb potential, but with the (non relative) Fisher information as a lower bound. An estimate of the same type but involving the relative Fisher information will be provided in a paper in preparation, cf. [10], and is related to Cercignani's conjecture (cf. [15] and the references therein).

In the sequel, we denote by $L^1_p(\mathbb{R}^3)$ the set of functions which have a moment of order p, that is $f\,(1+|v|^p)$ in $L^1(\mathbb{R}^3)$, and by $L \ln L$ the set of functions such that $f\,\ln f$ is in $L^1(\mathbb{R}^3)$.

The main theorem of [14] writes (in dimension 3 here, cf. [14] for the same result in higher dimension).

Theorem 1 *Let $f := f(v) \ge 0$, belonging to $L^1_2 \cap L\,\ln L(\mathbb{R}^3)$, be such that $\int f\,|\ln f|\,dv \le \bar{H}$, for some $\bar{H} > 0$. Let ψ satisfy*

$$\forall z \in \mathbb{R}^3, \qquad \psi(z) \ge c_0\, \inf(1, |z|^{\gamma_1+2}),$$

for some $c_0 > 0$ and $\gamma_1 \le 0$.

Then, there exists a constant $C := C(\int f\,dv, \int f\,v\,dv, \int f\,|v|^2/2\,dv, \bar{H}, \gamma_1, c_0) > 0$ which (explicitly) depends only on the mass, momentum, energy, (an upper bound of the) entropy and the parameters of the lower bound on ψ (that is, γ_1 and c_0), such that

$$\int_{\mathbb{R}^3} |\nabla\sqrt{f(v)}|^2\, (1+|v|^2)^{\inf(\gamma_1/2,-1)}\, dv \le C\,(1 + D_\psi(f)),$$

where $D_\psi(f)$ is defined in (6).

The inequality in this theorem is an entropy dissipation estimate which enables to control a weighted $H^1(\mathbb{R}^3)$ norm of $\sqrt{f}$ (that is, a weighted Fisher information) by the Landau entropy dissipation D_ψ of the Landau operator.

It is related to some other results linking smoothness to the dissipation of a Lyapounov (entropy) functional. In the case of the Boltzmann equation without cutoff, such estimates were proven in [1–4, 18–20, 22, 25] (cf. also the older attempts, more related to the large time behavior of the equation than to the issue of smoothness) in [8, 9, 12], and used for example in [5] (formula (70) p. 30, Lemma 13, p. 35, and Remark p. 36).

This theorem can be seen as a mixture of the estimates proven in [2], where the Boltzmann equation with general cross sections is considered, but where the large velocities are not part of the estimate, and the estimates proven in [17], in which the large velocities are treated, but only the Landau operator with overMaxwellian molecules is considered. It is also related to the Remark p. 36 in [5].

An important feature of Theorem 1 is the fact that the constant C appearing in the estimate only depends on quantities which are known to be controlled in the evolution of the spatially homogeneous Landau equation (8), (9), provided that they are initially finite (namely, the mass, momentum and (an upper bound of the) entropy, cf. the a priori estimate (13)). This feature ensures that applying Theorem 1 to the solution at time t of this equation (and with such initial data), we end up with a new a priori estimate for its solutions. In this way, it was possible to improve in [14] the existence theory for the Landau equation with very soft potentials (including the Coulomb case) as well as to recover recent results obtained on moderately soft potentials by Wu (cf. [26]).

Our goal in this work is to establish an extension of Theorem 1 to the case of more general cross sections ψ (that is which are not moderately soft potentials or very soft potentials like in [14], and not of Maxwell molecules type like in [17]). We indeed would like to be able to treat general soft potentials, that is ψ which decay at infinity like a negative power law (more precisely, cross sections ψ which are such that $\psi(z) \sim |z|^{\gamma+2}$ when $z \to 0$, for $\gamma < 0$), or even ψ which decay very quickly at infinity (like a negative exponential of a power law, or even a negative exponential of an exponential of a power law). We would also like to be able to treat cross sections ψ which are such that $\psi(z) \sim |z|^{\gamma+2}$ when $z \to 0$, for $\gamma \in]0, 2[$, that is, general hard potentials.

Though those extensions have no direct applications to physics, they enable to understand the proof of Theorem 1 more deeply than in [14] (especially the treatment of the determinant appearing in the denominator of Cramer's formula, see below), and provide the occasion of computing explicitly bounds for the constants appearing in the estimates of $D_\psi(f)$ (something that was done in [14] only for radially symmetric functions f).

We propose first an abstract (functional) result, which holds for any cross section $\psi \geq 0$, and will then be used in order to provide estimates for specific cases of ψ.

Theorem 2 *Let $f := f(v) \geq 0$, $M := M(v) \geq 0$, and $\phi := \phi(|v|^2/2) \geq 0$ be functions such that the right-hand side of inequality (14) below is finite.*
Then

$$\int_{\mathbb{R}^3} |\nabla\sqrt{f(v)}|^2\, M(v)\, dv \leq 3\, \Delta_\phi(f)^{-2} \left(\int_{\mathbb{R}^3} \phi(|w|^2/2)\, f(w) \,< w >^2 dw \right)^4 \tag{14}$$

$$\times \Bigg\{ 12 \left[\int_{\mathbb{R}^3} f(v) \,< v >^2\, M(v)\, dv \right] \Bigg[3 \left(\int_{\mathbb{R}^3} f(w) \,< w >\, \phi(|w|^2/2)\, dw \right)^2$$

$$+ 8 \left(\int_{\mathbb{R}^3} f(w) \,< w >^2\, |\phi'(|w|^2/2)|\, dw \right)^2 \Bigg]$$

$$+ 24\, D_\psi(f) \sup_{v\in\mathbb{R}^3} M(v) \left(\int_{\mathbb{R}^3} f(w)\, \phi^2(|w|^2/2)\, \frac{|v-w|^2}{\psi(|v-w|)} \,< w >^2 dw \right) \Bigg\},$$

where (here and in all the rest of the paper) $< v >= (1 + |v|^2)^{1/2}$, and

$$\Delta_\phi(f) := Det \left(\int_{\mathbb{R}^3} f(w)\, \phi(|w|^2/2) \begin{bmatrix} 1 & w_i & w_j \\ w_i & w_i^2 & w_i\, w_j \\ w_j & w_i\, w_j & w_j^2 \end{bmatrix} dw \right).$$

This functional estimate leads to the following corollary, which still holds for any cross section $\psi \geq 0$, but can in practice be used only when $z \mapsto \psi(z)/|z|^2$ is bounded below by a strictly positive constant on each bounded set of $\mathbb{R}^3$ (typically, for general soft potentials, but not (general or not) hard potentials).

Corollary 2.1 *Let $f := f(v) \geq 0$, belonging to $L^1_2 \cap L \ln L(\mathbb{R}^3)$ and such that $D_\psi(f)$ is finite. We also suppose that $M := M(v) \geq 0$ is bounded, and that $\phi \geq 0$ is C^1, bounded, with ϕ' also bounded.*
Then

$$\int_{\mathbb{R}^3} |\nabla\sqrt{f(v)}|^2\, M(v)\, dv \leq 72\, \Delta_\phi(f)^{-2}\, ||\phi||_\infty^4\, \mathscr{E}_f^5 \tag{15}$$

$$\times \left[||M||_\infty \left(\frac{3}{2}\, ||\phi||_\infty^2 + 4\, ||\phi'||_\infty^2 \right) \mathscr{E}_f^2 + \beta\, D_\psi(f) \right],$$

where (here and in all the rest of the paper)

$$\mathscr{E}_f := \int_{\mathbb{R}^3} f(v)\, (1 + |v|^2)\, dv,$$

and $\beta > 0$ is any number such that

$$\forall v, w \in \mathbb{R}^3, \qquad M(v)\,\phi^2(|w|^2/2) \le \beta\,\frac{\psi(|v-w|)}{|v-w|^2}\,. \tag{16}$$

Note also that in this result and its corollaries below (Corollaries 2.2, 2.3 and 2.4), up to the quantity $\Delta_\phi(f)$ which will be discussed later, the only dependence of the constants w.r.t. f is that of $\mathscr{E}_f$, that is, a dependence through quantities which are constant in the evolution of the spatially homogeneous Landau equation (8), (9).

This estimate leads in turn to a family of corollaries (Corollaries 2.2, 2.3 and 2.4), which hold for functions ψ satisfying various lower bounds.

We start with the case when ψ satisfies a lower bound corresponding to a nonpositive power law including general soft potentials.

Corollary 2.2 *Let $f := f(v) \ge 0$, belonging to $L^1_2 \cap L \ln L(\mathbb{R}^3)$, and ψ satisfying the lower bound $\psi(z) \ge c_0\,|z|^{2+\gamma}$ for some $c_0 > 0$, $\gamma \le 0$. We assume that $D_\psi(f)$ is finite.*

Then

$$\int_{\mathbb{R}^3} |\nabla\sqrt{f(v)}|^2 < v >^{\gamma}\, dv \le 72\,\Delta_\phi(f)^{-2}\,\mathscr{E}_f^5 \tag{17}$$

$$\times \left[(\frac{3}{2} + |\gamma|^2)\,\mathscr{E}_f^2 + c_0^{-1}\, 2^{\sup(0,|\gamma|-1)+\sup(2-|\gamma|,0)}\, D_\psi(f) \right],$$

and

$$\phi(z) = (1+2z)^{\gamma/4}.$$

Note that Corollary 2.2 (together with Proposition 4 below) gives a completely explicit estimate in Theorem 1 of [14]. Our feeling is that the exponent γ in the weight appearing in estimate (17) is optimal. This result (like those of Corollaries 2.3 and 2.4 below), together with the bound appearing in Proposition 4 on $\Delta_\phi(f)$, enables the building of an existence theory of standard weak solutions (that is, the concept of H-solutions appearing in [24] is not needed here) for the related spatially homogeneous Landau equations, provided that ψ has no too strong singularities (for example singularities at point 0 strictly weaker than $\psi(z) \sim |z|^{-2}$ can be handled). We refer to [14] for that kind of applications to the spatially homogeneous Landau equation.

Next we turn to the case when ψ can decay much more rapidly at infinity, namely like an exponential of a power.

Corollary 2.3 *Let $f := f(v) \ge 0$, belonging to $L^1_2 \cap L \ln L(\mathbb{R}^3)$, and ψ satisfying the lower bound: $\frac{\psi(z)}{|z|^2} \ge c_0\, e^{-c_1\,|v|^\delta}$, for some $c_0, c_1, \delta > 0$. We assume that $D_\psi(f)$ is finite.*

Then

$$\int_{\mathbb{R}^3} |\nabla\sqrt{f(v)}|^2\, e^{-\tilde{c_1}\,|v|^\delta}\, dv \le 72\,\Delta_\phi(f)^{-2}\, e^{-2\tilde{c_1}} \tag{18}$$

$$\times\, \mathscr{E}_f^5 \left(\left[\frac{3}{2}\, e^{-\tilde{c_1}} + \sup\left(4, 4\, (\tilde{c_1}\, \delta)^{2-4/\delta}\, (2\,\delta - 2)^{4/\delta}\, e^{-2+2/\delta}\right) \right] \mathscr{E}_f^2 + c_0^{-1}\, D_\psi(f) \right),$$

and

$$\phi(z) = e^{-\frac{\tilde{c_1}}{2}\,(1+2z)^\delta}, \qquad \tilde{c_1} = c_1\, 2^{\sup(0,\delta-1)}.$$

In estimate (18), it is not clear whether the weight $e^{-c_1\, 2^{\sup(\delta-1,0)}\, |v|^\delta}$ is optimal. We believe however that the optimal weight, if it exists, should be of the same general shape (that is, an exponential of a power δ), or close to such a shape.

Finally we turn to the case when ψ can decay even more rapidly at infinity, namely like an exponential of an exponential of a power. For this extreme situation, rather than giving a result concerning all possible decays, we focus on a special case, namely when $\psi(z) \geq e^{-e^{|z|}}$, for which it is possible to write a quite simple estimate.

Corollary 2.4 *Let $f := f(v) \geq 0$, belonging to $L^1_2 \cap L \ln L(\mathbb{R}^3)$, and ψ satisfying the lower bound: $\frac{\psi(z)}{|z|^2} \geq \exp(-e^{|z|})$. We assume that $D_\psi(f)$ is finite.*

Then

$$\int_{\mathbb{R}^3} |\nabla\sqrt{f(v)}|^2\, e^{-3\,e^{3\,|v|}}\, dv \leq 72\, \Delta_\phi(f)^{-2}\, \mathscr{E}_f^5\, e^{-9} \left(e^{-3} \left(\frac{3}{2}\, e^{-3} + 81 \right) \mathscr{E}_f^2 + 3\, D_\psi(f) \right), \tag{19}$$

and

$$\phi(z) = \exp(-\frac{3}{2}\, e^{3\sqrt{2z}}).$$

As in the previous corollary, one can observe that the weight appearing in the Fisher information is different from the cross section $z \mapsto \frac{\Psi(|z|)}{|z|^2}$, it is also not optimal.

We then consider the case of cross sections which are not strictly positive at point 0, so that Corollary 2.1 cannot be used, and we have to come back to Theorem 2. We propose first the following result, which enables to treat as a special case hard potentials.

Corollary 2.5 *Let $f := f(v) \geq 0$, belonging to $L^1_2 \cap L \ln L(\mathbb{R}^3)$, and ψ satisfying the following lower bound: $\psi(z) \geq c_0\, \inf(|z|^2, |z|^{\gamma+2})$, with $c_0 > 0$ and $\gamma \in]0, 3[$. We assume that $D_\psi(f)$ is finite.*

Then

$$\int_{\mathbb{R}^3} |\nabla\sqrt{f(v)}|^2\, dv \leq 3\, \Delta_\phi(f)^{-2}\, \mathscr{E}_f^4 \Bigg\{ 132\, \mathscr{E}_f^3 \tag{20}$$

$$+ 24\, c_0^{-1}\, D_\psi(f) \left(\mathscr{E}_f + \left(\frac{4\pi\,(p-1)}{3\,(p-1) - \gamma\, p} \right)^{1-1/p} ||f||_{L^p} \right) \Bigg\},$$

for all $p > \frac{3}{3-\gamma}$, and

$$\phi(z) = (1 + 2z)^{-1/2}.$$

In Corollary 2.5, we made no effort to obtain a better weight (than 1) in the r.h.s. of estimate (20). We will discuss this issue in Corollary 2.7. Note also that an L^p (with $p > 1$) norm of f appears in the constants of the estimate. Such a quantity is not constant in the evolution of the Landau equation, but is sometimes propagated (or even created), cf. for example [16].

When γ is not too large (that is, for general hard potentials, i.-e. $\gamma < 2$), it is possible to use a Sobolev estimate and an interpolation inequality, in order to get rid of the L^p norm in the result above. The price to pay is the appearance of an exponent larger than 1 for the entropy dissipation $D(f)$ appearing in the estimate.

Corollary 2.6 *Let $f := f(v) \geq 0$, belong to $L^1_2 \cap L \ln L(\mathbb{R}^3)$, and ψ satisfying the following lower bound: $\psi(z) \geq c_0 \inf(|z|^2, |z|^{\gamma+2})$, with $c_0 > 0$ and $\gamma \in]0, 2[$. We assume that $D_\psi(f)$ is finite.*

Then for any $p \in]\frac{3}{3-\gamma}, 3[$,

$$\int_{\mathbb{R}^3} |\nabla \sqrt{f(v)}|^2 \, dv \leq 792 \, \Delta_\phi(f)^{-2} \, \mathscr{E}_f^7 + 144 \, \Delta_\phi(f)^{-2} \, \mathscr{E}_f^5 \, c_0^{-1} \, D_\psi(f) \tag{21}$$

$$+144^{1/(1-\theta)} \, (1-\theta) \, \theta^{\theta/(1-\theta)} \left(\frac{4\pi \, (p-1)}{3\,(p-1) - \gamma \, p} \right)^{\frac{1-1/p}{1-\theta}} c_0^{-1/(1-\theta)} \, C_s^{\theta/(1-\theta)} \, \mathscr{E}_f^{1+4/(1-\theta)}$$

$$\times \, \Delta_\phi(f)^{-2/(1-\theta)} \, D_\psi(f)^{1/(1-\theta)},$$

where $\theta := \frac{3\,(p-1)}{2p}$, and C_s is the constant appearing in a Sobolev estimate (cf. proof of Corollary 2.6).

This result, together with the bound appearing in Proposition 4 on $\Delta_\phi(f)$, enables to obtain a new a priori estimate for the solutions $f(t, v)$ of the Landau equation with (general) hard potentials, when the initial data have a finite mass, energy and entropy. It gives indeed a bound for $\left(\int |\nabla_v \sqrt{f(t, v)}|^2 dv \right)^{1-\theta}$ in $L^1([0, T])$ for all $T > 0$. This result, related to the regularization effect of the Landau equation, is to be compared with the results of the same kind obtained in [16]. There, much more informations on the smoothness of the solution are provided, but only under extra assumptions on the initial data (and on the cross section ψ).

The interpolation procedure used here is reminiscent of those used in [17], or, in the context of the Boltzmann equation, in [23].

If we now suppose that ψ is growing at infinity at least as fast as $| \cdot |^{\gamma+2}$ (like in general hard potentials), then we can get a slightly better estimate than (20), in which the weight $< \cdot >^\gamma$ appears. Namely we obtain the

Corollary 2.7 *Let $f := f(v) \geq 0$, belong to $L^1_2 \cap L \ln L(\mathbb{R}^3)$, and ψ satisfying the following lower bound: $\psi(z) \geq c_0 \, |z|^{\gamma+2}$, with $c_0 > 0$ and $\gamma \in]0, 3[$. We assume that $D_\psi(f)$ is finite.*

Then

$$\int_{\mathbb{R}^3} |\nabla \sqrt{f(v)}|^2 < v >^{\gamma} dv \le 36\, \Delta_\phi(f)^{-2}\, \mathscr{E}_f^4 \left\{ \left(3 + 8\left(\frac{\gamma}{2}+1\right)^2\right) \mathscr{E}_f^2 \int_{\mathbb{R}^3} f(w) < w >^{2+\gamma} dw \right. \tag{22}$$

$$\left. + 24\, c_0^{-1}\, D_\psi(f) \left([2^{\frac{3}{2}\gamma} + 2^{\gamma}]\, \mathscr{E}_f + 3^{\gamma/2} \left(\frac{4\pi\,(p-1)}{(3-\gamma)\,p-3)} \right)^{1-1/p} ||f||_{L^p} \right) \right\},$$

where $p > \frac{3}{3-\gamma}$, *and*

$$\phi(z) = (1+2z)^{-\frac{\gamma}{4}-\frac{1}{2}}.$$

Note that in this result, the moment $\int f(w) < w >^{2+\gamma} dw$ appears in the estimate (as well as $||f||_{L^p}$, which already appeared in estimate (20)). This moment is not constant in the evolution of the spatially homogeneous Landau equation (with general hard potentials). It is however sometimes propagated, or even created (cf. [16]).

We now complete the estimates appearing in Corollaries 2.1–2.7 by a lower bound on $\Delta_\phi(f)$. We start with a proposition showing that $\Delta_\phi(f)$ is somehow bounded below by a quantity which can be equal to 0 only when f is concentrated on a hyperplane (provided that $\phi > 0$ a.e.).

Proposition 3 *Let* $f := f(v) \ge 0$ *belong to* $L^1_2(\mathbb{R}^3)$.
Then, for all $i, j \in \{1,2,3\}$ *such that* $i \ne j$, *and* $\varepsilon > 0$, $R > 0$,

$$\Delta_\phi(f) := Det \left(\int_{\mathbb{R}^3} \phi(|v|^2/2)\, f(v) \begin{bmatrix} 1 & v_i & v_j \\ v_i & v_i^2 & v_i\, v_j \\ v_j & v_i\, v_j & v_j^2 \end{bmatrix} dv \right)$$

$$\ge \varepsilon^6 \inf_{B(0,R)} \phi\left(\frac{|\cdot|^2}{2}\right)^3 \left(\int_{B(0,R)} f(v)\, dv \right. \tag{23}$$

$$\left. - \sup_{\lambda^2+\mu^2+\nu^2=1} \int_{B(0,R)} f(v)\, 1_{\{|\lambda+\mu\, v_i + \nu\, v_j| \le \varepsilon\}}\, dv \right)^3.$$

With this result in mind, any estimate on f which prevents concentration on zero-measure sets or at infinity can now be used to bound $\Delta_\phi(f)$ from below. Concentration on large velocities will be prevented by using the energy of f (remember that this quantity is constant during the evolution of Landau's equation). Concentration on zero-measure sets can be achieved (with efficiency from the point of view of numerical constants) if one uses some L^p estimate for f (cf. [17]). Though L^p regularity is known to be propagated (or even created) in some cases for the spatially homogeneous Landau equation, we however prefer to use $L \ln L$ regularity, which is much less efficient (from the point of view of numerical constants), but which can be obtained for all solutions of the spatially homogeneous Landau equation, as soon as the initial mass, energy and entropy are finite.

We provide therefore the following estimate:

Proposition 4 *Let $f := f(v) \geq 0$ belong to $L^1_2(\mathbb{R}^3)$, and assume that $H(f) \leq \bar{H}$. Then, for all $i, j \in \{1, 2, 3\}$ such that $i \neq j$, we have the estimate*

$$\Delta_\phi(f) \geq \left(\frac{1}{4}\int f(v)\,dv\right)^3 \inf_{B\left(0,\sup\left(1,2\,(\int f(v)\,v^2\,dv/\int f(v)\,dv)^{1/2}\right)\right)} \phi\left(\frac{|\cdot|^2}{2}\right)^3 \tag{24}$$

$$\times \inf\left[2^{-6}, 2^{-42}\left(\int f(v)\,dv\right)^6 e^{-24\bar{H}\,(\int f(v)\,dv)^{-1}} \sup\left(1, 2^{-18}\left(\frac{\int f(v)\,dv}{\int f(v)\,v^2\,dv}\right)^9\right)\right].$$

As can be seen, this estimate can be used (together with Corollaries 2.1–2.6) in order to yield a priori estimates for the solutions of the spatially homogeneous Landau equation, since it involves (when ϕ is strictly positive a.e.) only the mass, energy and (an upper bound of the) entropy of f, all quantities which are known to be controlled for those solutions.

We now briefly explain what are the possible extensions of the results presented in this section.

We first observe that for cross sections ψ which are bounded below (by a strictly positive constant) on all bounded subsets, it is most probably possible to extend the results stated in Corollaries 2.2, 2.3 and 2.4 to some ψ which are decaying at infinity even more rapidly than an exponential of exponential. Our feeling is that the more ψ rapidly decays at infinity, the less optimal the final weight appearing in the estimate of the entropy dissipation will be, if one uses Corollary 2.1. It becomes indeed more and more difficult to bound from below a function of $v - w$ by a tensor product (that is, a function of v multiplied by a function of w) when this function tends quickly towards 0 at infinity.

One can also deal with functions ψ which have more than one point of cancellation, at least if those points constitute a finite set, and if the cancellation at each point is not stronger than $|z - z_0|^q$, with $q < 3$. As in Corollary 2.5, some L^p norm of f will then appear in the estimate of the entropy dissipation, which can be dealt with as in Corollary 2.6 if $q < 2$.

Finally, one can in principle deal with functions ψ which both cancel at a finite number of points, and which have a specific behavior at infinity. When ψ is growing at infinity more than $z \mapsto |z|^2$, we can get results analogous to Corollary 2.7, while if $z \mapsto \psi(z)/|z|^2$ is decaying at infinity, one can get an estimate in which some (decaying) weight appears, and where some L^p norm of f also appears (that is, some mixture of Corollaries 2.2, 2.3 and 2.4 with Corollary 2.5).

All the results presented in Sect. 1 are proven in Sect. 2.

2 Proofs of the Theorems

We begin with the

Proof of Theorem 2: We start as in the proof of the corresponding theorem in [14]. We first observe that (for all $x, y \in \mathbb{R}^3$)

$$y^T\,(|x|^2\,Id - x\otimes x)\,y = \frac{1}{2}\sum_{i,j\in\{1,2,3\}} |x_i\,y_j - x_j\,y_i|^2,$$

so that

$$D_\psi(f) = \frac{1}{4}\sum_{i,j\in\{1,2,3\}} \int\int_{\mathbb{R}^3\times\mathbb{R}^3} f(v)\,f(w)\,\frac{\psi(|v-w|)}{|v-w|^2}\,\left|q^f_{ij}(v,w)\right|^2\,dvdw,$$

where (for $i, j \in \{1,2,3\}$, $i \neq j$),

$$q^f_{ij}(v,w) := (v_i - w_i)\left(\frac{\partial_j f(v)}{f(v)} - \frac{\partial_j f(w)}{f(w)}\right) - (v_j - w_j)\left(\frac{\partial_i f(v)}{f(v)} - \frac{\partial_i f(w)}{f(w)}\right)$$

$$= \left[v_i\,\frac{\partial_j f(v)}{f(v)} - v_j\,\frac{\partial_i f(v)}{f(v)}\right] + w_j\,\frac{\partial_i f(v)}{f(v)} - w_i\,\frac{\partial_j f(v)}{f(v)}$$

$$-v_i\,\frac{\partial_j f(w)}{f(w)} + v_j\,\frac{\partial_i f(w)}{f(w)} + \left[w_i\,\frac{\partial_j f(w)}{f(w)} - w_j\,\frac{\partial_i f(w)}{f(w)}\right].$$

Then, instead of using $w \mapsto \chi(w)\,e^{-\lambda w^2}\,f(w)$, where χ is a polynomial of degree 1 as in [14], we use the functions $w \mapsto \chi(w)\,\phi(|w|^2/2)$, where ϕ is a generic radially symmetric function. Picking $i, j \in \{1,2,3\}$, $i \neq j$, we see that for $\chi(w) = 1$,

$$\int_{\mathbb{R}^3} q^f_{ij}(v,w)\,\phi(|w|^2/2)\,f(w)\,dw = \left[v_i\,\frac{\partial_j f(v)}{f(v)} - v_j\,\frac{\partial_i f(v)}{f(v)}\right]\left(\int_{\mathbb{R}^3}\phi(|w|^2/2)\,f(w)\,dw\right)$$

$$+\left(\int_{\mathbb{R}^3} w_j\,\phi(|w|^2/2)\,f(w)\,dw\right)\frac{\partial_i f(v)}{f(v)} - \left(\int_{\mathbb{R}^3} w_i\,\phi(|w|^2/2)\,f(w)\,dw\right)\frac{\partial_j f(v)}{f(v)}$$

$$+v_i\left(\int_{\mathbb{R}^3} w_j\,\phi'(|w|^2/2)\,f(w)\,dw\right) - v_j\left(\int_{\mathbb{R}^3} w_i\,\phi'(|w|^2/2)\,f(w)\,dw\right).$$

Then, for $\chi(w) = w_i$,

$$\int_{\mathbb{R}^3} q^f_{ij}(v,w)\,w_i\,\phi(|w|^2/2)\,f(w)\,dw = \left[v_i\,\frac{\partial_j f(v)}{f(v)} - v_j\,\frac{\partial_i f(v)}{f(v)}\right]\left(\int_{\mathbb{R}^3} w_i\,\phi(|w|^2/2)\,f(w)\,dw\right)$$

$$+\left(\int_{\mathbb{R}^3} w_j\, w_i\, \phi(|w|^2/2)\, f(w)\, dw\right) \frac{\partial_i f(v)}{f(v)} - \left(\int_{\mathbb{R}^3} w_i^2\, \phi(|w|^2/2)\, f(w)\, dw\right) \frac{\partial_j f(v)}{f(v)}$$

$$+ v_i \left(\int_{\mathbb{R}^3} w_i\, w_j\, \phi'(|w|^2/2)\, f(w)\, dw\right) - v_j \left(\int_{\mathbb{R}^3} (\phi(|w|^2/2) + w_i^2\, \phi'(|w|^2/2))\, f(w)\, dw\right)$$

$$+ \int_{\mathbb{R}^3} w_j\, \phi(|w|^2/2)\, f(w)\, dw.$$

Exchanging i and j (or, equivalently, taking $\chi(w) = w_j$), we get the identity

$$\int_{\mathbb{R}^3} q_{ij}^f(v, w)\, w_j\, \phi(|w|^2/2)\, f(w)\, dw = \left[v_i \frac{\partial_j f(v)}{f(v)} - v_j \frac{\partial_i f(v)}{f(v)}\right] \left(\int_{\mathbb{R}^3} w_j\, \phi(|w|^2/2)\, f(w)\, dw\right)$$

$$-\left(\int_{\mathbb{R}^3} w_j\, w_i\, \phi(|w|^2/2)\, f(w)\, dw\right) \frac{\partial_j f(v)}{f(v)} + \left(\int_{\mathbb{R}^3} w_j^2\, \phi(|w|^2/2)\, f(w)\, dw\right) \frac{\partial_i f(v)}{f(v)}$$

$$- v_j \left(\int_{\mathbb{R}^3} w_i\, w_j\, \phi'(|w|^2/2)\, f(w)\, dw\right) + v_i \left(\int_{\mathbb{R}^3} (\phi(|w|^2/2) + w_j^2\, \phi'(|w|^2/2))\, f(w)\, dw\right)$$

$$- \int_{\mathbb{R}^3} w_i\, \phi(|w|^2/2)\, f(w)\, dw.$$

Considering the above identities as a 3×3 system for the unknowns $v_i \frac{\partial_j f(v)}{f(v)} - v_j \frac{\partial_i f(v)}{f(v)}$, $\frac{\partial_i f(v)}{f(v)}$ and $\frac{\partial_j f(v)}{f(v)}$, and using Cramer's formulas, we end up with the following formula for $\frac{\partial_i f(v)}{f(v)}$:

$$\frac{\partial_i f(v)}{f(v)} = \Delta_\phi(f)^{-1}$$

$$\times\, Det \left(\int_{\mathbb{R}^3} \phi(|w|^2/2)\, f(w) \begin{bmatrix} 1 & w_i & q_{ij}^f(v, w) + P_1(f)(v, w) \\ w_i & w_i^2 & q_{ij}^f(v, w)\, w_i + P_2(f)(v, w) \\ w_j & w_i\, w_j & q_{ij}^f(v, w)\, w_j + P_3(f)(v, w) \end{bmatrix} dw\right),$$

where

$$P_1(f)(v, w) = v_j \frac{w_i\, \phi'(|w|^2/2)}{\phi(|w|^2/2)} - v_i \frac{w_j\, \phi'(|w|^2/2)}{\phi(|w|^2/2)},$$

$$P_2(f)(v, w) = v_j \frac{[\phi(|w|^2/2) + w_i^2\, \phi'(|w|^2/2)]}{\phi(|w|^2/2)} - v_i \frac{w_i\, w_j\, \phi'(|w|^2/2)}{\phi(|w|^2/2)} - w_j,$$

$$P_3(f)(v, w) = v_j \frac{w_i\, w_j\, \phi'(|w|^2/2)}{\phi(|w|^2/2)} - v_i \frac{[\phi(|w|^2/2) + w_j^2\, \phi'(|w|^2/2)]}{\phi(|w|^2/2)} + w_i.$$

Then,

$$\left| \frac{\partial_i f(v)}{f(v)} \right| \le 2\,\Delta_\phi(f)^{-1} \left(\int_{\mathbb{R}^3} \phi(|w|^2/2)\, f(w)\, (1 + |w|^2)\, dw \right)^2$$

$$\times \left(\int_{\mathbb{R}^3} \phi(|w|^2/2)\, f(w) \left[\sum_{k=1}^{3} |P_k(f)(v,w)|\, dw \,+\, |q_{ij}^f(v,w)|\, (1 + |w_i| + |w_j|) \right] dw \right)$$

$$\le 2\,\Delta_\phi(f)^{-1} \left(\int_{\mathbb{R}^3} \phi(|w|^2/2)\, f(w)\, (1 + |w|^2)\, dw \right)^2$$

$$\times \left((1 + |v_i| + |v_j|) \int_{\mathbb{R}^3} f(w) \left[\phi(|w|^2/2)\, (1 + |w_i| + |w_j|) + |\phi'(|w|^2/2)|\, (|w_i| + |w_j| + 2\,|w|^2) \right] dw \right.$$

$$\left. + \int_{\mathbb{R}^3} f(w)\, \phi(|w|^2/2)\, |q_{ij}^f(v,w)|\, (1 + |w_i| + |w_j|)\, dw \right)$$

$$\le 2\,\Delta_\phi(f)^{-1} \left(\int_{\mathbb{R}^3} \phi(|w|^2/2)\, f(w) \,<w>^2 dw \right)^2$$

$$\times \left(\sqrt{3} <v> \int_{\mathbb{R}^3} f(w) \left[\sqrt{3} <w> \phi(|w|^2/2) + 2\sqrt{2} <w>^2 |\phi'|(|w|^2/2) \right] dw \right.$$

$$\left. + \sqrt{3} \int_{\mathbb{R}^3} f(w)\, \phi(|w|^2/2)\, |q_{ij}^f(v,w)| <w> dw \right).$$

Then

$$\int_{\mathbb{R}^3} f(v) \left| \frac{\partial_i f(v)}{f(v)} \right|^2 M(v)\, dv$$

$$\le 4\,\Delta_\phi(f)^{-2} \left(\int_{\mathbb{R}^3} \phi(|w|^2/2)\, f(w) \,<w>^2 dw \right)^4$$

$$\times \left(6 \int_{\mathbb{R}^3} f(v) <v>^2 M(v)\, dv \right.$$

$$\times \left| \int_{\mathbb{R}^3} f(w) \left[\sqrt{3} <w> \phi(|w|^2/2) + 2\sqrt{2} <w>^2 |\phi'|(|w|^2/2) \right] dw \right|^2$$

$$\left. + 6 \int_{\mathbb{R}^3} f(v)\, M(v) \left| \int_{\mathbb{R}^3} f(w)\, \phi(|w|^2/2)\, |q_{ij}^f(v,w)| <w> dw \right|^2 dv \right)$$

$$\le 4\,\Delta_\phi(f)^{-2} \left(\int_{\mathbb{R}^3} \phi(|w|^2/2)\, f(w) \,<w>^2 dw \right)^4$$

$$\times \left\{ 12 \int_{\mathbb{R}^3} f(v) < v >^2 M(v)\, dv \left[3 \left(\int_{\mathbb{R}^3} f(w) < w > \phi(|w|^2/2)\, dw \right)^2 \right.\right.$$

$$\left. + 8 \left(\int_{\mathbb{R}^3} f(w) < w >^2 |\phi'|(|w|^2/2)\, dw \right)^2 \right]$$

$$+ 6 \int_{\mathbb{R}^3} f(v)\, M(v) \left(\int_{\mathbb{R}^3} f(w)\, |q_{ij}^f(v,w)|^2 \frac{\psi(|v-w|)}{|v-w|^2} dw \right)$$

$$\left. \times \left(\int_{\mathbb{R}^3} f(w)\, \phi^2(|w|^2/2) < w >^2 \frac{|v-w|^2}{\psi(|v-w|)} dw \right) \right\}$$

$$\leq 4\, \Delta_\phi(f)^{-2} \left(\int_{\mathbb{R}^3} \phi(|w|^2/2)\, f(w) < w >^2 dw \right)^4$$

$$\times \left\{ 12 \int_{\mathbb{R}^3} f(v) < v >^2 M(v)\, dv \left[3 \left(\int_{\mathbb{R}^3} f(w) < w > \phi(|w|^2/2)\, dw \right)^2 \right.\right.$$

$$\left. + 8 \left(\int_{\mathbb{R}^3} f(w) < w >^2 |\phi'|(|w|^2/2)\, dw \right)^2 \right]$$

$$\left. + 24\, D_\psi(f) \sup_{v \in \mathbb{R}^3} M(v) \left(\int_{\mathbb{R}^3} f(w)\, \phi^2(|w|^2/2) \frac{|v-w|^2}{\psi(|v-w|)} < w >^2 dw \right) \right\}.$$

We conclude the proof of Theorem 2 by noticing that

$$\int_{\mathbb{R}^3} |\nabla \sqrt{f(v)}|^2 M(v)\, dv = \frac{1}{4} \sum_{i=1}^{3} \int_{\mathbb{R}^3} f(v) \left| \frac{\partial_i f(v)}{f(v)} \right|^2 M(v)\, dv.$$

□

We now turn to the proofs of the corollaries of this theorem.

Proof of Corollary 2.1: It is a direct consequence of Theorem 2 and the bounds assumed on M, ϕ and ψ. □

Proof of Corollary 2.2: We recall that $\psi(z) \geq c_0\, |z|^{\gamma+2}$. Using the elementary inequalities

$$\forall x, y, p \in \mathbb{R}_+, \qquad (x+y)^p \leq 2^{\sup(p-1,0)}\, (x^p + y^p),$$

$$\forall x, y, p \in \mathbb{R}_+, \qquad x^p + y^p \leq 2^{\sup(1-p,0)}\, (x+y)^p,$$

we see that (for any $\gamma < 0$, $v, w \in \mathbb{R}^3$)

$$|v-w|^{|\gamma|} \le 2^{\sup(|\gamma|-1,0)} \, (|v|^{|\gamma|} + |w|^{|\gamma|})$$

$$\le 2^{\sup(|\gamma|-1,0)+\sup(2-|\gamma|,0)} \, < v >^{|\gamma|} < w >^{|\gamma|} .$$

Then taking

$$M(v) =< v >^{\gamma}, \qquad \phi(z) = (1+2z)^{\gamma/4},$$

we see that assumption (16) holds provided that $\beta = c_0^{-1}\, 2^{\sup(|\gamma|-1,0)+\sup(2-|\gamma|,0)}$. Noticing that

$$||M||_\infty = 1, \qquad ||\phi||_\infty = 1, \qquad ||\phi'||_\infty = \gamma/2.$$

and using Corollary 2.1, we get Corollary 2.2. □

Proof of Corollary 2.3: We recall that $\psi(z) \ge c_0\, e^{-c_1\,|z|^\delta}$. Still using the elementary inequalities used in the proof of Corollary 2.2, we see that (for all $v, w \in \mathbb{R}^3$)

$$e^{c_1\,|v-w|^\delta} \le e^{\tilde{c_1}\,|v|^\delta}\, e^{\tilde{c_1}\,<w>^\delta},$$

with $\tilde{c_1} = c_1\, 2^{\sup(0,\delta-1)}$, so that taking

$$M(v) = e^{-\tilde{c_1}\,|v|^\delta}, \qquad \phi(z) = e^{-\frac{\tilde{c_1}}{2}\,(1+2z)^\delta},$$

we see that assumption (16) holds provided that $\beta = c_0^{-1}$. We then observe that

$$||M||_\infty = 1, \qquad ||\phi||_\infty = e^{-\frac{\tilde{c_1}}{2}}, \qquad ||\phi'||_\infty \le \sup\left(1, \tilde{c_1}\,\delta\left(\frac{2(\delta-1)}{\tilde{c_1}\,\delta}\right)^{2/\delta} e^{-1+1/\delta}\right).$$

Using Corollary 2.1, we get Corollary 2.3. □

Proof of Corollary 2.4: We recall that $\psi(z) \ge \exp(-e^{|z|})$. Then

$$\exp(e^{|v-w|}) \le \exp(e^{|v|}\, e^{|w|})$$

$$\le 1 + e^{|v|}\, e^{|w|} + \sum_{k=2}^{\infty} \frac{1}{k\,(k-1)} \frac{e^{k|v|}}{[(k-2)!]^{1/2}} \frac{e^{k|w|}}{[(k-2)!]^{1/2}}.$$

If we introduce

$$u_k := \frac{e^{k|v|}}{[(k-2)!]^{1/2}},$$

we see that

$$u_{k+1} \le u_k \qquad \Longleftrightarrow \qquad k \ge 1 + e^{2\,|v|}.$$

Then, the sequence u_k reaches its maximum when $k = [2 + e^{2\,|v|}]$, so that (for all $k \geq 2$)

$$u_k \leq \frac{\exp(|v|\,(2 + e^{2\,|v|}))}{([e^{2\,|v|}]!)^{1/2}},$$

and finally

$$\exp(e^{|v-w|}) \leq 1 + e^{|v|}\,e^{|w|} + \left(\sum_{k=2}^{\infty} \frac{1}{k\,(k-1)}\right) \frac{\exp(|v|\,(2 + e^{2\,|v|}))}{([e^{2\,|v|}]!)^{1/2}} \, \frac{\exp(|w|\,(2 + e^{2\,|w|}))}{([e^{2\,|w|}]!)^{1/2}} \tag{25}$$

$$\leq 1 + e^{|v|}\,e^{|w|} + \exp(|v|\,(2 + e^{2\,|v|}))\,\exp(|w|\,(2 + e^{2\,|w|}))$$

$$\leq 3\,\exp(3\,e^{3\,|v|})\,\exp(3\,e^{3\,<w>}).$$

We then introduce

$$M(v) := \exp(-3\,e^{3\,|v|}), \qquad \phi(z) = \exp\left(-\frac{3}{2}\,e^{3\sqrt{1+2z}}\right),$$

so that

$$\phi'(z) = -\exp\left(-\frac{3}{2}\,e^{3\sqrt{1+2z}}\right)\frac{9}{2}\sqrt{\frac{1}{1+2z}}\,e^{3\sqrt{1+2z}}.$$

We see that

$$||M||_\infty = e^{-3}, \qquad ||\phi||_\infty = e^{-\frac{3}{2}}, \qquad ||\phi'||_\infty \leq \frac{9}{2}.$$

Using estimate (25), we obtain the estimate

$$M(v)\,\phi(|w|^2/2) \leq 3\,\exp(-e^{|v-w|})$$

$$\leq \beta\,\frac{\psi(|v-w|^2)}{|v-w|^2},$$

with $\beta = 3$. Using Corollary 2.1, we end up with the statement of Corollary 2.4. □

Proof of Corollary 2.5: We introduce

$$M(v) = 1, \qquad \phi(z) = (1 + 2z)^{-1/2},$$

so that

$$\phi'(z) = -(1 + 2z)^{-3/2}.$$

Then, using Theorem 2, we see that

$$\int_{\mathbb{R}^3} |\nabla \sqrt{f(v)}|^2 \, dv \le 3\, \Delta_\phi(f)^{-2}\, \mathscr{E}_f^4 \Bigg\{ 132\, \mathscr{E}_f^3$$

$$+ 24\, c_0^{-1}\, D_\psi(f) \sup_{v\in\mathbb{R}^3} \int f(w)\, \sup(1, |v-w|^{-\gamma})\, dw \Bigg\}.$$

We now observe that

$$\sup_{v\in\mathbb{R}^3} \int f(w)\, \sup(1, |v-w|^{-\gamma})\, dw \le \mathscr{E}_f + ||f \,*\, |\cdot|^{-\gamma}\, 1_{\{|\cdot|\le 1\}}||_{L^\infty}.$$

Then, Young's inequality for convolutions ensures that, for any $p > \frac{3}{3-\gamma}$,

$$||f \,*\, |\cdot|^{-\gamma}\, 1_{\{|\cdot|\le 1\}}||_{L^\infty} \le ||f||_{L^p} \left(\frac{4\pi (p-1)}{(3-\gamma)\, p - 3} \right)^{1-1/p}.$$

This concludes the proof of Corollary 2.5. □

Proof of Corollary 2.6: We first write on $\sqrt{f}$ the Sobolev inequality corresponding to the Sobolev embedding $H^1(\mathbb{R}^3) \subset L^6(\mathbb{R}^3)$, that is

$$||f||_{L^3} \le C_s\, ||\nabla \sqrt{f}||_{L^2}^2, \tag{26}$$

where $C_s > 0$ is the (best) constant appearing in the Sobolev inequality.

Denoting

$$a := 396\, \Delta_\phi(f)^{-2}\, \mathscr{E}_f^7,$$

$$b := 72\, \Delta_\phi(f)^{-2}\, \mathscr{E}_f^5\, c_0^{-1},$$

$$c := 72\, \Delta_\phi(f)^{-2}\, \mathscr{E}_f^4\, c_0^{-1} \left(\frac{4\pi\, (p-1)}{3\,(p-1) - \gamma\, p)} \right)^{1-1/p},$$

we see thanks to Corollary 2.5, Hölder's inequality, and the Sobolev inequality (26) that

$$\int_{\mathbb{R}^3} |\nabla \sqrt{f(v)}|^2 \, dv \le a + D_\psi(f)\, (b + c\, ||f||_{L^p})$$

$$\le a + D_\psi(f)\, (b + c\, ||f||_{L^3}^\theta\, \mathscr{E}_f^{1-\theta})$$

$$\le a + D_\psi(f) \left[b + c\, C_s^\theta\, \mathscr{E}_f^{1-\theta} \left(\int_{\mathbb{R}^3} |\nabla \sqrt{f(v)}|^2 \, dv \right)^\theta \right],$$

where $\theta = \frac{3(p-1)}{2p} \in]0,1[$ for some $p \in]\frac{3}{3-\gamma}, 3[$ small enough (remember that $\gamma \in]0,2[$, so that such a choice is possible).

Then, denoting $q = \int_{\mathbb{R}^3} |\nabla \sqrt{f(v)}|^2 \, dv$, we end up with the inequality

$$q \le a + D_\psi(f) \left[b + c\, C_s^\theta \, \mathscr{E}_f^{1-\theta} \, q^\theta \right].$$

Thanks to Young's inequality (applied with conjugate numbers $1/\theta$ and $1/(1-\theta)$), we get for any $d > 0$

$$q^\theta \le \theta \, d \, q + (1-\theta)\, d^{-\theta/(1-\theta)}.$$

As a consequence,

$$q \le a + D_\psi(f)\, b + c\, \mathscr{E}_f^{1-\theta}\, C_s^\theta \, D_\psi(f)\, \theta\, d\, q + c\, \mathscr{E}_f^{1-\theta}\, C_s^\theta\, D_\psi(f)\,(1-\theta)\, d^{-\theta/(1-\theta)}.$$

Selecting $d = \frac{1}{2} c^{-1} \, \mathscr{E}_f^{\theta-1} \, C_s^{-\theta} \, D_\psi(f)^{-1} \, \theta^{-1}$, in such a way that

$$c\, \mathscr{E}_f^{1-\theta}\, C_s^\theta\, D_\psi(f)\, \theta\, d\, q = q/2,$$

we end up with the estimate

$$q \le 2a + 2\, D_\psi(f)\, b + c^{1/(1-\theta)}\, \mathscr{E}_f\, C_s^{\theta/(1-\theta)}\,(1-\theta)\,\theta^{\theta/(1-\theta)}\, 2^{1/(1-\theta)}\, D_\psi(f)^{1/(1-\theta)}.$$

Recalling the definition of a, b, c, q, θ, we obtain Corollary 2.6. □

Proof of Corollary 2.7: We introduce

$$M(v) = < v >^\gamma, \qquad \phi(z) = (1+2z)^{-\frac{\gamma}{4}-\frac{1}{2}},$$

so that

$$||\phi||_\infty \le 1, \qquad ||\phi'||_\infty \le \frac{\gamma}{2} + 1.$$

Then, using Theorem 2, we see that

$$\int |\nabla \sqrt{f(v)}|^2 < v >^\gamma \, dv \le 36\, \Delta_\phi(f)^{-2}\, \mathscr{E}_f^4 \left\{ \left(3 + 8 \left(\frac{\gamma}{2} + 1 \right)^2 \right) \mathscr{E}_f^2 \int f(w) < w >^{2+\gamma} dw \right.$$

$$\left. + 24\, c_0^{-1}\, D_\psi(f) \sup_{v \in \mathbb{R}^3} < v >^\gamma \int f(w) < w >^{-\gamma} |v-w|^{-\gamma} dw \right\}.$$

We now estimate

$$\sup_{v \in \mathbb{R}^3} < v >^\gamma \int f(w) < w >^{-\gamma} |v-w|^{-\gamma} dw$$

$$\le 3^{\gamma/2} \sup_{v\in\mathbb{R}^3} \int f(w) < w >^{\gamma} < w >^{-\gamma} |v-w|^{-\gamma}\, 1_{\{|v-w|\le 1\}}\, dw$$

$$+ \sup_{v\in\mathbb{R}^3} < v >^{\gamma} \int f(w) < w >^{-\gamma} |v-w|^{-\gamma}\, 1_{\{|v-w|\ge 1\}}\, 1_{\{|w|\le |v|/2\}}\, dw$$

$$+ \sup_{v\in\mathbb{R}^3} < v >^{\gamma} \int f(w) < w >^{-\gamma} |v-w|^{-\gamma}\, 1_{\{|v-w|\ge 1\}}\, 1_{\{|w|\ge |v|/2\}}\, dw$$

$$\le 3^{\gamma/2} ||f * |\cdot|^{-\gamma}\, 1_{\{|\cdot|\le 1\}}||_{L^\infty} + \sup_{v\in\mathbb{R}^3} \frac{2^{\gamma} < v >^{\gamma}}{\sup(1, |v|^{\gamma})} ||f||_{L^1}$$

$$+ \sup_{v\in\mathbb{R}^3} < v >^{\gamma} < v/2 >^{-\gamma} ||f||_{L^1}$$

$$\le 3^{\gamma/2} ||f||_{L^p} \left(\frac{4\pi(p-1)}{(3-\gamma)\,p-3}\right)^{1-1/p} + (2^{3\gamma/2} + 2^{\gamma})\, ||f||_{L^1}.$$

Using this estimate and the previous one, we get the statement of Corollary 2.7

Proof of Proposition 3: Observing that $\Delta_\phi(f)$ is a Grad determinant, we see that (for all $\varepsilon > 0,\ R > 0$),

$$\Delta_\phi(f) \ge \left[\inf_{\lambda^2+\mu^2+\nu^2=1} \int_{\mathbb{R}^3} \phi(|v|^2/2)\, f(v)\, |\lambda + \mu\, v_i + \nu\, v_j|^2\, dv\right]^3$$

$$\ge \varepsilon^6 \left[\inf_{\lambda^2+\mu^2+\nu^2=1} \int_{B(0,R)} \phi(|v|^2/2)\, f(v)\, 1_{\{|\lambda+\mu\, v_i+\nu\, v_j|\ge\varepsilon\}}\, dv\right]^3$$

$$\ge \varepsilon^6 \inf_{B(0,R)} \phi\left(\frac{|\cdot|^2}{2}\right)^3 \left(\int_{B(0,R)} f(v)\, dv - \sup_{\lambda^2+\mu^2+\nu^2=1} \int_{B(0,R)} f(v)\, 1_{\{|\lambda+\mu\, v_i+\nu\, v_j|\le\varepsilon\}}\, dv\right)^3.$$

□

Proof of Proposition 4 : Thanks to Proposition 3, we know that for all $R > 1$, $\varepsilon \in\ ]0, 1/2[$, $A > 1$,

$$\Delta_\phi(f) \ge \varepsilon^6 \inf_{B(0,R)} \phi\left(\frac{|\cdot|^2}{2}\right)^3 \left(\int_{B(0,R)} f(v)\, dv - \sup_{\lambda^2+\mu^2+\nu^2=1} \int_{B(0,R)} f(v)\, 1_{\{|\lambda+\mu\, v_i+\nu\, v_j|\le\varepsilon\}}\, dv\right)^3.$$

$$\ge \varepsilon^6 \inf_{B(0,R)} \phi\left(\frac{|\cdot|^2}{2}\right)^3 \left(\int_{\mathbb{R}^3} f(v)\, dv - R^{-2} \int_{\mathbb{R}^3} f(v)\, |v|^2\, dv - \bar{H}\, (\ln A)^{-1} - A \sup_{\lambda^2+\mu^2+\nu^2=1} Y_{\{\lambda,\mu,\nu,R,\varepsilon\}}\right)^3,$$

where $Y_{\{\lambda,\mu,\nu,R,\varepsilon\}}$ is the Lebesgue measure of the set

$$\{v \in \mathbb{R}^3, \quad |\lambda + \mu\, v_i + \nu\, v_j| \le \varepsilon\} \cap B(0, R),$$

and $\bar{H}$ is any constant larger than $\int f(v)\, |\ln f(v)|\, dv$.

Using a rotation, we see that $\sup_{\lambda^2+\mu^2+\nu^2=1} Y_{\{\lambda,\mu,\nu,R,\varepsilon\}} = \sup_{\lambda^2+\mu^2=1} Z_{\{\lambda,\mu,R,\varepsilon\}}$, where $Z_{\{\lambda,\mu,R,\varepsilon\}}$ is the Lebesgue measure of the set

$$\{v \in \mathbb{R}^3, \quad |\lambda + \mu\, v_1| \le \varepsilon\} \cap B(0, R).$$

Then $Z_{\{\lambda,\mu,R,\varepsilon\}} \le 4\, R^2\, W_{\{\lambda,\mu,R,\varepsilon\}}$, where $W_{\{\lambda,\mu,R,\varepsilon\}}$ is the one-dimensional Lebesgue measure of the set

$$\{v_1 \in \mathbb{R}, \quad |\lambda + \mu\, v_1| \le \varepsilon\} \cap B(0, R).$$

As a consequence, for any $\mu_0 > 0$, and $|\mu| \le \mu_0$, $|v_1| \le R$, λ, μ such that $\lambda^2 + \mu^2 = 1$,

$$\begin{aligned}
|\lambda + \mu\, v_1| &\ge |\lambda| - |\mu_0|\, R \\
&\ge \sqrt{1 - \mu_0^2} - \mu_0\, R \\
&\ge 1 - \mu_0 - \mu_0\, R \\
&\ge 1 - 2R\, \mu_0.
\end{aligned}$$

Taking $\mu_0 = \frac{1-\varepsilon}{2R}$, we see that $W_{\{\lambda,\mu,R,\varepsilon\}} = 0$ if $|\mu| \le \mu_0$.

Then, for $|\mu| \ge \mu_0$, $W_{\{\lambda,\mu,R,\varepsilon\}} \le \frac{2\varepsilon}{\mu_0}$, so that finally

$$W_{\{\lambda,\mu,R,\varepsilon\}} \le \frac{4\varepsilon\, R}{1-\varepsilon} \le 8\varepsilon\, R,$$

and

$$Z_{\{\lambda,\mu,R,\varepsilon\}} \le 32\, R^3\, \varepsilon.$$

Taking

$$R = \sup\left(1, 2\left(\frac{\int f(v)\, v^2\, dv}{\int f(v)\, dv}\right)^{1/2}\right),$$

we see that

$$R^{-2} \int f(v)\, v^2\, dv \le \frac{1}{4} \int f(v)\, dv.$$

Then choosing

$$A = \exp\left(\frac{4\, \bar{H}}{\int f(v)\, dv}\right),$$

we also see that

$$\frac{\bar{H}}{\ln A} = \frac{1}{4} \int f(v)\, dv.$$

Finally, considering

$$\varepsilon = \inf \left[2^{-1}, 2^{-7} \int f(v)\, dv \, \exp\left(- \frac{4\bar{H}}{\int f(v)\, dv} \right) \sup\left(1, 2^{-3} \left(\frac{\int f(v)\, dv}{\int f(v)\, v^2\, dv} \right)^{3/2} \right) \right],$$

we obtain the inequality

$$32\, R^3\, \varepsilon\, A \leq \frac{1}{4} \int f(v)\, dv.$$

We end up with estimate (24).

This ends the proof of Proposition 4. □

Acknowledgments The research leading to this paper was funded by the french "ANR blanche" project Kibord: ANR-13-BS01-0004. It was also funded by Université Sorbonne Paris Cité, in the framework of the "Investissements d'Avenir", convention ANR-11-IDEX-0005.

References

1. Alexandre, R.: Sur le taux de dissipation d'entropie sans troncature angulaire. C.R. Acad. Sci. Paris, Série I, **326**, 311–315 (1998)
2. Alexandre, R., Desvillettes, L., Villani, C., Wennberg, B.: Entropy dissipation and long-range interactions. Arch. Ration. Mech. Anal. **152**, 327–355 (2000)
3. Alexandre, R., Morimoto, Y., Ukai, S., Xu, C.J., Yang, T.: Smoothing effect of weak solutions for the spatially homogeneous Boltzmann equation without angular cutoff. Kyoto J. Math. **52**(3), 433–463 (2012)
4. Alexandre, R., Morimoto, Y., Ukai, S., Xu, C.J., Yang, T.: Global well-posedness theory for the spatially inhomogeneous Boltzmann equation without angular cutoff. C.R. Math. **348**(15), 867–871 (2010)
5. Alexandre, R., Villani, C.: On the Landau approximation in plasma physics. Annales Inst. Henri Poincare **21**(1) 61–95 (2004)
6. Bobylev, A., Pulvirenti, M., Saffirio, C.: From particle systems to the Landau equation: a consistency result. Commun. Math. Phys. **319**(3), 683–702 (2013)
7. Bogolyubov, N.N.: Problems of a dynamical theory in statistical physics, State Technical Press (1946) (in Russian). In: de Boer, J., Uhlenbeck, G.E. (eds.) English translation in Studies in Statistical Mechanics I, part A. North-Holland, Amsterdam (1962)
8. Carlen, E.A., Carvalho, M.C.: Strict entropy production bounds and stability of the rate of convergence to equilibrium for the Boltzmann equation. J. Stat. Phys. **67**(3–4), 575–608 (1992)
9. Carlen, E.A., Carvalho, M.C.: Entropy production estimates for Boltzmann equations with physically realistic collision kernels. J. Stat. Phys. **74**(3–4), 743–782 (1994)
10. Carrapatoso, K., Desvillettes, L., He, L.: Estimates for the large time behavior of the Landau equation in the Coulomb case. Work in preparation
11. Chapman, S., Cowling, T.G.: The Mathematical Theory of Non-Uniform Gases. Cambridge University Press, London (1952)
12. Desvillettes, L.: Entropy dissipation rate and convergence in kinetic equations. Commun. Math. Phys. **123**(4), 687–702 (1989)

13. Desvillettes, L.: On asymptotics of the Boltzmann equation when the collisions become grazing. Transp. Theory Stat. Phys. **21**(3), 259–276 (1992)
14. Desvillettes, L.: Entropy dissipation estimates for the Landau equation in the Coulomb case and applications. J. Funct. Anal. **269**, 1359–1403 (2015)
15. Desvillettes, L., Mouhot, C., Villani, C.: Celebrating Cercignani's conjecture for the Boltzmann equation. Kinet. Relat. Models **4**, 277–294 (2011)
16. Desvillettes, L., Villani, C.: On the spatially homogeneous Landau equation for hard potentials. Part I. Existence, uniqueness and smoothness. Commun. Partial Differ. Equ. **25**(1–2), 179–259 (2000)
17. Desvillettes, L., Villani, C.: On the spatially homogeneous Landau equation for hard potentials. Part II. H-Theorem and applications. Commun. Partial Differ. Equ. **25**(1–2), 261–298 (2000)
18. Gressman, P.T., Strain, R.M.: Global classical solutions of the Boltzmann equation with long-range interactions. Proc. Nat. Acad. Sci. U.S.A. **107**(13), 5744–5749 (2010)
19. Gressman, P.T., Strain, R.M.: Global classical solutions of the Boltzmann equation without angular cut-off. J. Am. Math. Soc. **24**(3), 771–847 (2011)
20. Gressman, P.T., Strain, R.M.: Sharp anisotropic estimates for the Boltzmann collision operator and its entropy production. Adv. Math. **227**(6), 2349–2384 (2011)
21. Lifschitz, E.M., Pitaevskii, L.P.: Physical Kinetics. Perg Press, Oxford (1981)
22. Lions, P.L.: Regularity and compactness for Boltzmann collision operators without angular cut-off. C.R. Acad. Sci. Paris, Serie I, **326**(1), 37–41 (1998)
23. Toscani, G., Villani, C.: On the trend to equilibrium for some dissipative systems with slowly increasing a priori bounds. J. Stat. Phys. **98**(5–6), 1279–1309 (2000)
24. Villani, C.: On a new class of weak solutions to the spatially homogeneous Boltzmann and Landau equations. Arch. Ration. Mech. Anal. **143**(3), 273–307 (1998)
25. Villani, C.: Regularity estimates via the entropy dissipation for the spatially homogeneous Boltzmann equation without cut-off. Rev. Matem. Iberoam. **15**(2), 335–352 (1999)
26. Wu, K.-C.: Global in time estimates for the spatially homogeneous Landau equation with soft potentials. J. Funct. Anal. **266**, 3134–3155 (2014)

The Boltzmann Equation over $\mathbb{R}^D$: Dispersion Versus Dissipation

François Golse

Abstract The Boltzmann equation of the kinetic theory of gases involves two competing processes. Dissipation—or entropy production—due to the collisions between gas molecules drives the gas towards local thermodynamic (Maxwellian) equilibrium. If the spatial domain is the Euclidean space $\mathbb{R}^D$, the ballistic transport of gas molecules between collisions results in a dispersion effect which enhances the rarefaction of the gas, and offsets the effect of dissipation. The competition between these two effects leads to a scattering regime for the Boltzmann equation over $\mathbb{R}^D$ with molecular interaction satisfying Grad's angular cutoff assumption. The present paper reports on results in this direction obtained in collaboration with Bardos, Gamba and Levermore [arxiv:1409.1430] and discusses a few open questions related to this work.

Keywords Boltzmann equation · Global Maxwellian · Boltzmann collision integral · Boltzmann H Theorem · Free transport · Large time limit · Scattering operator

1 The Boltzmann Dynamics over $\mathbb{R}^D$: Two Competing Processes

The kinetic theory of gases finds its origins in the works of Maxwell and Boltzmann, published in the second half of the 19th century.

In this theory, the state of a monatomic gas at time t is described by its distribution function, henceforth denoted $F \equiv F(v, x, t) \geq 0$, which is the number density of gas molecules with velocity $v \in \mathbb{R}^D$ at the position $x \in \mathbb{R}^D$ at time t.

An important example of distribution function is

$$\mathbb{M}_{\rho,u,\theta}(v) := \frac{\rho}{(2\pi\theta)^{\frac{D}{2}}} e^{-\frac{|v-u|^2}{2\theta}},$$

F. Golse (✉)
CMLS, École polytechnique, CNRS, Université Paris-Saclay,
91128 Palaiseau Cedex, France
e-mail: francois.golse@polytechnique.edu

© Springer International Publishing Switzerland 2016

P. Gonçalves and A.J. Soares (eds.), *From Particle Systems to Partial Differential Equations III*, Springer Proceedings in Mathematics & Statistics 162,
DOI 10.1007/978-3-319-32144-8_7

known as the Maxwellian with density $\rho \geq 0$, mean velocity $u \in \mathbb{R}^D$ and temperature $\theta > 0$. When the parameters ρ, u, θ are functions of t and x, this distribution function is referred to as a "local Maxwellian", whereas the same distribution is called a "uniform Maxwellian"—or an absolute Maxwellian—if the parameters ρ, u, θ are independent of t and x.

If the effect of external forces (such as gravity) can be neglected, the dynamics of gas molecules involves two different physical processes

- collisions between pairs of gas molecules, and
- inertial motion of gas molecules between collisions.

The first process, i.e. the collisions involving pairs of gas molecules, is purely local and instantaneous, and leads to discontinuities in time in the velocity of gas molecules. It is described by an integral operator, known as the Boltzmann collision integral, acting on the v variable only in the distribution function F. The second process mixes the dependence of the distribution function F in the velocity and space variables, and is described by the advection (or free transport) operator. The evolution of the distribution function F is governed by the Boltzmann equation, which takes the form

$$\underbrace{\partial_t F + v \cdot \nabla_x F}_{\text{Advection}} = \underbrace{\mathcal{B}(F, F)}_{\text{Collision}} .$$

The left hand side of the Boltzmann equation involves the advection (or free transport) operator describing the ballistic transport of gas molecules between collisions, while the right hand side is the Boltzmann collision integral.

1.1 Dissipation Effect of Collisions

In order to focus on the collision process and its dynamical implications on the distribution function, we consider the space homogeneous Boltzmann equation. Equivalenty, we specialize the Boltzmann equation to distribution functions that are independent of the position variable x. In other words, we seek distribution functions of the form $F \equiv F(v, t)$ satisfying

$$\partial_t F = \mathcal{B}(F, F).$$

A first important property of the Boltzmann collision integral is the identity

$$\int_{\mathbb{R}^D} \mathcal{B}(f, f)(v) \begin{pmatrix} 1 \\ v \\ \frac{1}{2}|v|^2 \end{pmatrix} dv = 0$$

satisfied by all measurable functions $f \equiv f(v)$ rapidly decaying as $|v| \to \infty$. Solutions of the space homogeneous Boltzmann equation satisfying this decay assumption must therefore verify the conservation laws of mass, momentum and energy, i.e.

$$\frac{d}{dt}\int_{\mathbb{R}^D}\begin{pmatrix}1\\ v\\ \frac{1}{2}|v|^2\end{pmatrix} F(v,t)\,dv = 0.$$

A second fundamental property of the Boltzmann collision integral is the inequality

$$\int_{\mathbb{R}^D} \mathscr{B}(f,f)(v)\ln f(v)\,dv \le 0$$

satisfied by all measurable functions $f \equiv f(v)$ rapidly decaying as $|v| \to \infty$, and such that $\ln f$ has at most polynomial growth as $|v| \to \infty$. (Any function of the form $f(v) = P(|v|)e^{-|v|^2}$ where P is a polynomial obviously satisfies this assumption.) In addition, the inequality above is an equality if and only if f is a Maxwellian. This second property is known as "Boltzmann's H Theorem" and has the following implication. For each nonnegative measurable function F such that

$$\int_{\mathbb{R}^D} F(v,t)(1+|v|^2)\,dv < \infty,$$

define its H-function:

$$H[F](t) := \int_{\mathbb{R}^D} F(v,t)\ln F(v,t)\,dv \in \mathbb{R}\cup\{+\infty\}.$$

Each solution F of the space homogeneous Boltzmann equation satisfying the decay assumptions mentioned above is such that the function

$$t \mapsto H[F](t) \text{ is nonincreasing.}$$

The dissipation of the H-function associated to solutions of the Boltzmann equation is analogous to the increase of entropy in closed systems, as postulated by the second principle of thermodynamics. Indeed, the entropy of some amount of an ideal gas is equal to minus the H-function of its distribution function. There is however an important difference between the second principle of thermodynamics and Boltzmann's H Theorem. While the former is postulated and of universal bearing, the latter is rigorously derived from the Boltzmann equation, and therefore applies only to the dynamics of ideal monatomic gases. Another difference is that Boltzmann's H Theorem gives an explicit expression for the entropy production rate, at variance with the second principle.

An important consequence of Boltzmann's H Theorem is the convergence of solutions of the space homogeneous Boltzmann equation to a Maxwellian equilibrium in the long time limit, which can be stated as follows. Let $F^{\text{in}} \equiv F^{\text{in}}(v)$ satisfy

$$F^{\text{in}} \geq 0\,, \quad \int_{\mathbb{R}^{\mathrm{D}}} \begin{pmatrix} 1 \\ v \\ \frac{1}{2}|v|^2 \end{pmatrix} F^{\text{in}}(v)\,dv = \begin{pmatrix} \rho^\infty \\ \rho^\infty u^\infty \\ \frac{1}{2}\rho^\infty(|u^\infty|^2 + 3\theta^\infty) \end{pmatrix}$$

where $\rho^\infty \geq 0$, $u^\infty \in \mathbb{R}^{\mathrm{D}}$ and $\theta^\infty > 0$. Then

$$F(v,t) \to \mathbb{M}_{\rho^\infty,u^\infty,\theta^\infty}(v) \quad \text{as } t \to +\infty.$$

In fact, $\mathbb{M}_{\rho^\infty,u^\infty,\theta^\infty}(v)$ is the minimum point of $H[F]$ as F runs through the set of nonnegative measurable functions on $\mathbb{R}^{\mathrm{D}}$ such that

$$\int_{\mathbb{R}^{\mathrm{D}}} \begin{pmatrix} 1 \\ v \\ \frac{1}{2}|v|^2 \end{pmatrix} F(v)\,dv = \begin{pmatrix} \rho^\infty \\ \rho^\infty u^\infty \\ \frac{1}{2}\rho^\infty(|u^\infty|^2 + 3\theta^\infty) \end{pmatrix}.$$

In other words, the dynamics defined by the space homogeneous Boltzmann equation is similar to the gradient flow of $F \mapsto H[F]$ restricted to the set of distribution functions satisfying the conservation laws defined by the initial data. With appropriate decay assumptions on F^{in}, one can prove that $F(v,t)$ converges to $\mathbb{M}_{\rho^\infty,u^\infty,\theta^\infty}(v)$ exponentially fast: see [25].

Consider now the (space inhomogeneous) Boltzmann equation set in the spatial domain $\mathbb{T}^{\mathrm{D}} = \mathbb{R}^{\mathrm{D}}/\mathbb{Z}^{\mathrm{D}}$:

$$(\partial_t + v \cdot \nabla_x)F = \mathscr{B}(F,F)\,, \quad F\Big|_{t=0} = F^{\text{in}}\,, \qquad v \in \mathbb{R}^{\mathrm{D}}\,,\ x \in \mathbb{T}^{\mathrm{D}}.$$

Define

$$\rho^\infty := \iint_{\mathbb{R}^{\mathrm{D}}\times\mathbb{T}^{\mathrm{D}}} F^{\text{in}} dv dx\,, \quad u^\infty := \frac{1}{\rho^\infty}\iint_{\mathbb{R}^{\mathrm{D}}\times\mathbb{T}^{\mathrm{D}}} vF^{\text{in}} dv dx\,,$$

and

$$\theta^\infty := \frac{1}{\rho^\infty}\iint_{\mathbb{R}^{\mathrm{D}}\times\mathbb{T}^{\mathrm{D}}} \tfrac{1}{\mathrm{D}}|v - u^\infty|^2 F^{\text{in}} dv dx.$$

Desvillettes and Villani [10] have proved that, in the limit as $t \to \infty$, and under additional assumptions on the initial distribution function F^{in} and on the solution of the Cauchy problem for the Boltzmann equation, one has

$$\|F(t) - \mathbb{M}_{\rho^\infty,u^\infty,\theta^\infty}\|_{L^1(\mathbb{R}^{\mathrm{D}}\times\mathbb{T}^{\mathrm{D}})} = O(t^{-m}) \quad \text{for all } m \geq 0.$$

1.2 Dispersion Effect of Advection in $\mathbb{R}^D$

The dispersion effect of the advection (or free transport) equation set on a spatial domain that is the Euclidean space $\mathbb{R}^D$ can be formulated as follows—in its simplest possible variant.

Let $f \equiv f(v, x, t)$ be the solution of the advection (free transport) equation

$$\partial_t f + v \cdot \nabla_x f = 0, \qquad f\big|_{t=0} = f^{\text{in}}.$$

Assume that $f^{\text{in}} \in L^1(\mathbb{R}^D{}_x; L^\infty(\mathbb{R}^D{}_v))$: in other words, for some $\phi \in L^1(\mathbb{R}^D)$, one has

$$0 \le f^{\text{in}}(v, x) \le \phi(x) \qquad \text{for a.e. } (v, x) \in \mathbb{R}^D \times \mathbb{R}^D.$$

Then, for each $t > 0$ and a.e. $x \in \mathbb{R}^D$,

$$\begin{aligned} 0 \le \rho(x, t) := \int_{\mathbb{R}^D} f(v, x, t)\, \mathrm{d}v &\le \int_{\mathbb{R}^D} \phi(x - tv)\, \mathrm{d}v \\ &= \frac{1}{t^D} \int_{\mathbb{R}^D} \phi(y)\, \mathrm{d}y = \frac{1}{t^D} \|\phi\|_{L^1}. \end{aligned}$$

1.3 Two Competing Mechanisms

Because of dispersion, the macroscopic density of the gas, i.e.

$$\rho(x, t) := \int_{\mathbb{R}^D} F(v, x, t)\, \mathrm{d}v$$

is expected to decay to 0 as $t \to +\infty$. Thus, dispersion increases the degree of rarefaction of the gas, and therefore offsets the dissipative effect of collisions.

On the other hand, the discussion above suggests that dissipation occurs at a much faster rate—specifically, at an exponential rate for space homogeneous distribution functions—than dispersion. It is therefore conceivable that, in the limit as $t \to +\infty$, solutions of the Boltzmann equation set in the spatial domain $\mathbb{R}^D$ should approach some distribution function $E \equiv E(v, x, t)$ satisfying both

$$(\partial_t + v \cdot \nabla_x) E(v, x, t) = 0 \text{ and } \mathscr{B}(E, E)(v, x, t) = 0 \quad \text{on } \mathbb{R}^D \times \mathbb{R}^D \times \mathbb{R}.$$

The distribution functions satisfying both these conditions are studied in the next section.

The results presented in this work, reported in Sects. 3–5, have been obtained in collaboration with Bardos, Gamba and Levermore [4]. The next section (Sect. 2) summarizes an independent result by Levermore [16] which is used repeatedly in the following sections. Section 6 formulates and discusses in detail four open problems, suggested by the main results reported in Sects. 3–5. These problems offer a new and broader prospective on the results presented here, complementing the conclusions in [4].

2 Global Maxwellians

Definition 2.1 A global Maxwellian is a distribution function $\mathscr{M} \equiv \mathscr{M}(v, x, t)$ such that

$$(\partial_t + v \cdot \nabla_x)\mathscr{M} = \mathscr{B}(\mathscr{M}, \mathscr{M}) = 0, \quad (v, x, t) \in \mathbb{R}^{\mathrm{D}} \times \mathbb{R}^{\mathrm{D}} \times \mathbb{R}.$$

An obvious example of global Maxwellian is

$$\mathscr{M}(v, x, t) := e^{-|x-tv|^2}.$$

Any global Maxwellian $\mathscr{M}$ must be a local Maxwellian, which means that there exist functions $\rho \equiv \rho(x, t) \geq 0$ on $\mathbb{R}^{\mathrm{D}} \times \mathbb{R}$ and $\theta \equiv \theta(x, t) > 0$ on $\mathbb{R}^{\mathrm{D}} \times \mathbb{R}$, and a vector field $u \equiv u(x, t) \in \mathbb{R}^{\mathrm{D}}$ such that

$$\mathscr{M}(v, x, t) = \mathbb{M}_{\rho(x,t), u(x,t), \theta(x,t)}(v).$$

This follows from the equality case in Boltzmann's H Theorem, since

$$\mathscr{B}(\mathscr{M}, \mathscr{M}) = 0 \quad \text{implies that} \quad \int_{\mathbb{R}^{\mathrm{D}}} \mathscr{B}(\mathscr{M}, \mathscr{M}) \ln \mathscr{M} \, \mathrm{d}v = 0.$$

That $(\partial_t + v \cdot \nabla_x)\mathscr{M} = 0$ implies additional constraints on the (x, t) dependence in ρ, u, θ.

Theorem 2.2 (Levermore [16]) *Any global Maxwellians $\mathscr{M}$ such that $\mathscr{M}(v, x, t) \in L^1(\mathbb{R}^{\mathrm{D}} \times \mathbb{R}^{\mathrm{D}})$ for all $t \in \mathbb{R}$ is of the form*

$$\mathscr{M}(v, x, t) = \frac{m}{(2\pi)^{\mathrm{D}}} \sqrt{\det(Q)} \exp(-q(v - v_0, x - tv - x_0)),$$

where $m \geq 0$, and

$$q(w, y) := \tfrac{1}{2}(c|w|^2 + a|y|^2 + 2by \cdot w + 2w \cdot By)$$

with $a, c > 0$, $b \in \mathbb{R}$, *and* $B = -B^T \in M_D(\mathbb{R})$ *such that the symmetric matrix*

$$Q := (ac - b^2)I + B^2 \text{ is definite positive.}$$

Notation: Henceforth Ω designates the set of $(a, b, c, B) \in \mathbb{R} \times \mathbb{R} \times \mathbb{R} \times M_D(\mathbb{R})$ such that

$$a, c > 0, \quad B = -B^T, \quad \text{and } (ac - b^2)I + B^2 > 0.$$

Interestingly, there is a variational characterization of global Maxwellians—in fact, this is the main result in [16].

Theorem 2.3 (Levermore [16]) *For each distribution function* $f \equiv f(v, x)$ *satisfying*

$$f(v, x) \geq 0 \text{ a.e. on } \mathbb{R}^D \times \mathbb{R}^D \quad \text{and} \quad \iint_{\mathbb{R}^D \times \mathbb{R}^D} f(v, x)(1 + |x|^2 + |v|^2)\, dv dx < \infty,$$

there exists a unique global Maxwellian denoted $\mathcal{M}_f$ *such that*

$$\iint_{\mathbb{R}^D \times \mathbb{R}^D} \begin{pmatrix} 1 \\ v \\ |v|^2 \\ x \\ |x|^2 \\ x \cdot v \\ x \wedge v \end{pmatrix} \mathcal{M}_f(v, x, 0)\, dv dx = \iint_{\mathbb{R}^D \times \mathbb{R}^D} \begin{pmatrix} 1 \\ v \\ |v|^2 \\ x \\ |x|^2 \\ x \cdot v \\ x \wedge v \end{pmatrix} f(v, x)\, dv dx,$$

where $a \wedge b := a \otimes b - b \otimes a$ *for each* $a, b \in \mathbb{R}^D$. *Moreover*

$$H[f] := \iint_{\mathbb{R}^D \times \mathbb{R}^D} f(v, x) \ln f(v, x)\, dv dx \geq H[\mathcal{M}_f].$$

The main difficulty in the proof of this theorem is to solve the realizability problem for moments of global Maxwellians (i.e. the first statement). Once the first statement is proved, the inequality involving the Boltzmann H-function is equivalent to the inequality

$$\iint_{\mathbb{R}^D \times \mathbb{R}^D} \left(f(v, x) \ln \frac{f(v, x)}{\mathcal{M}_f(v, x, 0)} - f(v, x) + \mathcal{M}_f(v, x, 0) \right) dv dx \geq 0,$$

which is obvious since the integrand is a.e. nonnegative on $\mathbb{R}^D \times \mathbb{R}^D$, as a consequence of the elementary inequality $\ln z \leq z - 1$, valid for all $z > 0$.

The last statement in the Theorem above can be formulated as follows. Let $\mathscr{M}$ be a global Maxwellian on $\mathbb{R}^D \times \mathbb{R}^D \times \mathbb{R}$ with finite total mass, and set

$$\begin{pmatrix} a_0 \\ a_1 \\ a_2 \\ b_1 \\ b_2 \\ s_2 \\ w_2 \end{pmatrix} := \iint_{\mathbb{R}^D \times \mathbb{R}^D} \begin{pmatrix} 1 \\ v \\ |v|^2 \\ x \\ |x|^2 \\ x \cdot v \\ x \wedge v \end{pmatrix} \mathscr{M}_f(v, x, 0)\, \mathrm{d}v\mathrm{d}x .$$

Then $\mathscr{M}(v, x, 0)$ is the unique minimum point of the variational problem

$$\inf \left\{ H[f] \text{ s.t. } \iint_{\mathbb{R}^D \times \mathbb{R}^D} \begin{pmatrix} 1 \\ v \\ |v|^2 \\ x \\ |x|^2 \\ x \cdot v \\ x \wedge v \end{pmatrix} f(v, x)\, \mathrm{d}v\mathrm{d}x = \begin{pmatrix} a_0 \\ a_1 \\ a_2 \\ b_1 \\ b_2 \\ s_2 \\ w_2 \end{pmatrix} \right\} .$$

See [16] for the proofs of the theorems above.

3 The Cauchy Problem for the Boltzmann Equation in $\mathbb{R}^D$

We briefly recall the main features of the Boltzmann equation, and introduce some elements of notation used in the sequel.

3.1 *The Boltzmann Collision Integral*

The Boltzmann collision integral is given by the formula

$$\mathscr{B}(F, F)(v, x, t) := \iint_{\mathbb{S}^{D-1} \times \mathbb{R}^D} (F' F'_* - F F_*) \mathsf{b}(v - v_*, \omega) \mathrm{d}\omega \mathrm{d}v_* .$$

The notation F, F_*, F' and F'_* designates the values of the distribution function F at the same time and position, but at different values of the velocity variable, corresponding to the pre- or post-collision state of a pair of gas molecules. Specifically,

$$F = F(v, x, t) , \quad F_* = F(v_*, x, t) , \quad F' = F(v', x, t) , \text{ and } F'_* = F(v'_*, x, t) ,$$

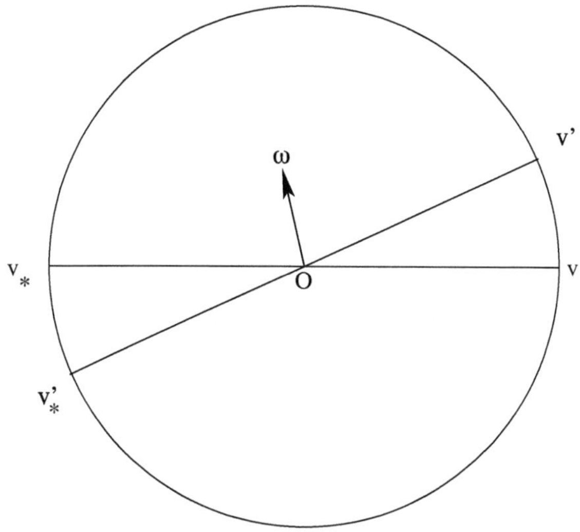

Fig. 1 The velocities v, v_*, v', v'_* and the unit vector ω

where

$$v' \equiv v'(v, v_*, \omega) := v - (v - v_*) \cdot \omega\omega \,, \quad v'_* \equiv v'_*(v, v_*, \omega) := v + (v - v_*) \cdot \omega\omega .$$

As the unit vector ω runs through the unit sphere $\mathbb{S}^{D-1}$, the pair of velocity vectors $(v'(v, v_*, \omega), v'_*(v, v_*, \omega))$ runs through the set of all possible pre-collision velocities leading to the velocities (v, v_*) immediately after an elastic collision between identical point particles. See Fig. 1 for the geometry of pre- and post-collision velocities.

Because the molecular radius is negligeable in the kinetic theory of gases, the unit vector ω, defining the direction of the line joining the centers of two colliding gas molecules, is distributed on the unit sphere instead of being uniquely determined by the positions of the molecules at the instant of collision. The distribution of ω on $\mathbb{S}^{D-1}$ for collisions leading to post-collisions velocities v and v_* is defined in terms of the collision kernel $\mathsf{b}(v - v_*, \omega) > 0$.

We shall henceforth assume that the collision kernel has separated form, i.e. for a.e. $(\omega, z) \in \mathbb{S}^{D-1} \times \mathbb{R}^D$

$$0 < \mathsf{b}(z, \omega) = |z|^\beta \, \hat{\mathsf{b}} \left(\omega \cdot \tfrac{z}{|z|} \right) , \quad -\mathrm{D} < \beta \leq 2 \,, \tag{1}$$

and satisfies the weak angular cutoff condition

$$0 < \overline{\mathsf{b}} := \int_{\mathbb{S}^{D-1}} \hat{\mathsf{b}}(\omega \cdot n) \, d\omega < \infty .$$

An example of collision kernel satisfying these assumptions is

$$\mathsf{b}(z, \omega) = \tfrac{1}{2} d^2 |z \cdot \omega|$$

in the case of elastic collisions between hard spheres of diameter $d > 0$ in space dimension $\mathrm{D} = 3$.

3.2 The Time-Averaged Collision Frequency

Given a distribution function $F \equiv F(v, x, t)$, we denote by $\mathscr{A}(F)$ the corresponding collision frequency, i.e.

$$\mathscr{A}(F)(v, x, t) := \iint_{\mathbb{S}^{\mathrm{D}-1} \times \mathbb{R}^{\mathrm{D}}} F(v_*, x, t) \mathsf{b}(v - v_*, \omega) \, \mathrm{d}\omega \mathrm{d}v_*.$$

Lemma 3.1 *Let* $1 - \mathrm{D} < \beta \le 1$ *and* $\mathscr{M}$ *be a global Maxwellian. Then*

$$\begin{aligned} \nu(\mathscr{M}) := & \sup_{(v,x,t) \in \mathbb{R}^{\mathrm{D}} \times \mathbb{R}^{\mathrm{D}} \times \mathbb{R}} \left| \int_{\mathbb{R}} \mathscr{A}(\mathscr{M})(v, x - tv + sv, s) \, \mathrm{d}s \right| \\ & \le \frac{m \overline{\mathsf{b}}}{(2\pi)^{\mathrm{D}-\frac{1}{2}} \sqrt{a}} \left((2\pi a)^{\mathrm{D}/2} + \frac{|\mathbb{S}^{\mathrm{D}-1}| \sqrt{\det Q}}{\beta + \mathrm{D} - 1} \right). \end{aligned}$$

Because $\mathscr{M}$ is a global Maxwellian, it satisfies the free transport equation

$$(\partial_t + v \cdot \nabla_x) \mathscr{M} = 0 \,,$$

and because $\mathscr{A}(\mathscr{M})$ is an average of $\mathscr{M}$ in the velocity variable, the dispersion effect of advection leads to a gain of one power of $|v|$ by integrating first in the time variable before taking the sup in (v, x, t). This observation is the core of the work of Illner-Shinbrot [14], which gave the first global existence and uniqueness result for the Boltzmann equation in the near-vacuum regime. The lemma above extends the validity of the Illner-Shinbrot estimate to the most general global Maxwellian with finite total mass, including solid rotation (i.e. for $B \neq 0$).

3.3 Existence and Uniqueness for the Cauchy Problem

To each global Maxwellian $\mathscr{M}$, one associates the Banach spaces

$$\begin{aligned} &\mathscr{X}_{\mathscr{M}} := \mathscr{M} L^\infty(\mathbb{R}^{\mathrm{D}} \times \mathbb{R}^{\mathrm{D}} \times \mathbb{R}) && \text{with norm} \|F\|_{\mathscr{M}} := \|F/\mathscr{M}\|_{L^\infty(\mathbb{R}^{\mathrm{D}} \times \mathbb{R}^{\mathrm{D}} \times \mathbb{R})} \\ &\mathscr{Y}_{\mathscr{M}(0)} := \mathscr{M}(0) L^\infty(\mathbb{R}^{\mathrm{D}} \times \mathbb{R}^{\mathrm{D}}) && \text{with norm} \|f\|_{\mathscr{M}(0)} := \|f/\mathscr{M}(0)\|_{L^\infty(\mathbb{R}^{\mathrm{D}} \times \mathbb{R}^{\mathrm{D}})} \end{aligned}$$

Theorem 3.2 (Bardos-Gamba-Golse-Levermore [4]) *Assume $\beta \in (1 - D, 1]$ and pick a global Maxwellian satisfying the condition $\nu(\mathscr{M}) < \frac{1}{4}$.*
(a) For each $F^{in} \in \mathscr{Y}_{\mathscr{M}(0)}$ such that

$$|F^{in} - \mathscr{M}(0)|_{\mathscr{M}(0)} < \tfrac{(1-4\nu(\mathscr{M}))^2}{8\nu(\mathscr{M})},$$

there exists a unique mild solution $F \in \mathscr{X}_{\mathscr{M}}$ of the Boltzmann equation s.t.

$$F(0) = F^{in} \quad \text{and } \|F - \mathscr{M}\|_{\mathscr{M}} \leq r(\mathscr{M}),$$

with

$$r(\mathscr{M}) := \left(\tfrac{1}{4\nu(\mathscr{M})} - 1\right)\left(1 - \sqrt{1 - \tfrac{8\nu(\mathscr{M})|F^{in} - \mathscr{M}(0)|_{\mathscr{M}(0)}}{(1-4\nu(\mathscr{M}))^2}}\right).$$

(b) Moreover, if

$$\tfrac{1}{2} \leq 4\nu(\mathscr{M}) < 1,$$

or if

$$0 < 4\nu(\mathscr{M}) < \tfrac{1}{2} \text{ and } |F^{in} - \mathscr{M}(0)|_{\mathscr{M}(0)} \leq 1 - 6\nu(\mathscr{M}),$$

then $r(\mathscr{M}) \leq 1$, and therefore

$$0 \leq (1-r)\mathscr{M} \leq F \leq (1+r)\mathscr{M} \quad \text{a.e. on } \mathbb{R}^D \times \mathbb{R}^D \times \mathbb{R}.$$

The proof of this theorem is based on a fixed point argument for the integral equation

$$F(v, x, t) = F^{in}(v, x - tv) + \int_0^t \mathscr{B}(F, F)(v, x - (t-s)v, s)\, ds$$

equivalent to the Boltzmann equation (i.e. the definition of mild solutions of the Boltzmann equation) in a suitable closed ball of $\mathscr{X}_{\mathscr{M}}$. This method of construction does not guarantee that solutions of the Boltzmann equation obtained in this way are nonnegative in general. For this reason, the theorem above involves two smallness conditions on the data of the Cauchy problem. The first smallness condition is $\nu(\mathscr{M}) < \frac{1}{4}$ and bears on the global Maxwellian $\mathscr{M}$ only; this condition implies the stability of some appropriate ball in $\mathscr{X}_{\mathscr{M}}$, and the contraction property for the integral equation formulation of the Boltzmann equation. The second smallness condition in part (b) of the theorem above involves both the initial data F^{in} and $\mathscr{M}$, and implies the positivity of the solution F. In other words, the theorem above bears on solutions of the Boltzmann equation which remain uniformly close to a global Maxwellian $\mathscr{M}$ for which $\nu(\mathscr{M})$ is small enough. If the initial data F^{in} is sufficiently close to $\mathscr{M}$, the corresponding solution of the Boltzmann equation remains positive.

Part (a) of the theorem above had been already proved by Hamdache [12] for a less general class of global Maxwellians—specifically, for global Maxwellians without solid rotation, i.e. $B = 0$.

4 Large Time Dynamics for the Boltzmann Equation in $\mathbb{R}^D$

From now on, we denote by A the advection operator $\mathsf{A} = v \cdot \nabla_x$.

We begin with the following elementary observation.

Theorem 4.1 (Bardos-Gamba-Golse-Levemore [4]) *Assume* $\beta \in (1 - \mathrm{D}, 2]$. *Let* $F \equiv F(v, x, t)$ *be a solution of the Boltzmann equation on* $\mathbb{R}^D \times \mathbb{R}^D \times (t_0, +\infty)$ *for some* $t_0 \in \mathbb{R}$. *Assume that, for some global Maxwellian* $\mathscr{M}$,

$$0 \leq F \leq \mathscr{M} \quad \textit{a.e. on } \mathbb{R}^D \times \mathbb{R}^D \times (t_0, +\infty). \tag{2}$$

Then there exists a unique $F^{+\infty} \equiv F^{+\infty}(v, x)$ *s.t.*

$$\|F(t) - e^{t\mathsf{A}} F^{+\infty}\|_{L^1(\mathbb{R}^D \times \mathbb{R}^D)} \to 0 \quad \textit{as } t \to +\infty.$$

Assume for simplicity that $t_0 = 0$. The idea of the proof is to write the integral equation equivalent to the Boltzmann equation

$$F(t) = e^{-t\mathsf{A}} F^{\text{in}} + \int_0^t e^{-(t-s)\mathsf{A}} \mathscr{B}(F, F)(s)\, \mathrm{d}s$$

in the form

$$e^{t\mathsf{A}} F(t) = F^{\text{in}} + \int_0^t e^{s\mathsf{A}} \mathscr{B}(F, F)(s)\, \mathrm{d}s.$$

Then the condition $0 \leq F \leq \mathscr{M}$ implies that

$$|\mathscr{B}(F, F)(s)| \leq 2\mathscr{A}(\mathscr{M})(v, x, s)\mathscr{M}(v, x, s)$$

so that

$$|e^{t_2\mathsf{A}} F(t_2) - e^{t_1\mathsf{A}} F(t_1)| \leq 2 \int_{t_1}^{t_2} \mathscr{A}(\mathscr{M})(v, x + sv, s)\mathscr{M}(v, x + sv, s)\, \mathrm{d}s.$$

Hence for each $t_2 > t_1$, one has

$$\|e^{t_2\mathsf{A}} F(t_2) - e^{t_1\mathsf{A}} F(t_1)\|_{L^1(\mathbb{R}^D \times \mathbb{R}^D)}$$

$$\leq 2 \int_{t_1}^{t_2} \iint_{\mathbb{R}^D \times \mathbb{R}^D} \mathscr{A}(\mathscr{M})(v, x + sv, s)\mathscr{M}(v, x + sv, s)\, \mathrm{d}v\mathrm{d}x\mathrm{d}s$$

$$= 2\int_{t_1}^{t_2}\iint_{\mathbb{R}^D\times\mathbb{R}^D} \mathscr{A}(\mathscr{M})(v,y,s)\mathscr{M}(v,y,s)\,\mathrm{d}v\mathrm{d}y\mathrm{d}s$$

$$\leq \text{Const.}\int_{t_1}^{t_2}\theta(t)^{\frac{D+\beta}{2}}\,\mathrm{d}t,$$

where we recall that $\theta \equiv \theta(t)$ is the temperature corresponding to $\mathscr{M}$. In other words, we recall that

$$\mathscr{M}(v,x,t) = \mathbb{M}_{\rho(x,t),u(x,t),\theta(t)}(v).$$

That the temperature field θ is independent of x follows from elementary computations, together with the fact that

$$\int_{\mathbb{R}} \theta(t)^{\frac{D+\beta}{2}}\,\mathrm{d}t < \infty.$$

The estimate above and Cauchy's criterion show that

$$e^{t\mathsf{A}}F(t) \to F^{+\infty} := F^{\text{in}} + \int_0^{+\infty} e^{s\mathsf{A}}\mathscr{B}(F,F)(s)\,\mathrm{d}s$$

in $L^1(\mathbb{R}^D\times\mathbb{R}^D)$ as $t \to +\infty$.

Remark. By the same token, if F is a solution on $\mathbb{R}^D\times\mathbb{R}^D\times(-\infty,t_0)$, there exists a unique $F^{-\infty}$ s.t. $F(t) - e^{t\mathsf{A}}F^{-\infty} \to 0$ as $t \to -\infty$.

Henceforth, we denote

$$\mathscr{T}^{\pm}F^{\text{in}} := F^{\pm\infty}$$

when either of these limits exists.

By analogy with the large time behavior of solutions of the Boltzmann equation set in the periodic box $\mathbb{T}^D$ (see the work of Desvillettes-Villani [10] recalled above), the result in the theorem above raises the following question.

Question: Is $F^{+\infty}$ of the form $F^{+\infty}(v,x) = \mathscr{M}(v,x,0)$, where $\mathscr{M}$ is a global Maxwellian?

If F^{in} satisfies the assumptions of Theorem 3.2, there exists a unique solution of the Boltzmann equation defined on $\mathbb{R}^D\times\mathbb{R}^D\times\mathbb{R}$, i.e. for all $t\in\mathbb{R}$, and this solution satisfies the domination condition (2). According to Theorem 4.1, there exist two distribution functions $F^{+\infty}$ and $F^{-\infty}$ such that

$$\|F(t) - e^{t\mathsf{A}}F^{\pm\infty}\|_{L^1(\mathbb{R}^D\times\mathbb{R}^D)} \to 0 \quad \text{as } t \to \pm\infty.$$

Since the fundamental conservation laws of the Boltzmann equation imply that

$$\iint_{\mathbb{R}^D\times\mathbb{R}^D} F(v, x+tv, t)\begin{pmatrix}1\\ v\\ |v|^2\\ x\\ |x|^2\\ x\cdot v\\ x\wedge v\end{pmatrix} dvdx = \iint_{\mathbb{R}^D\times\mathbb{R}^D} F^{\text{in}}(v, x)\begin{pmatrix}1\\ v\\ |v|^2\\ x\\ |x|^2\\ x\cdot v\\ x\wedge v\end{pmatrix} dvdx\,,$$

the only possibility for $F^{+\infty}$, were Problem 1 to be answered in the affirmative, would be $F^{+\infty} = \mathscr{M}_{F^{\text{in}}}\big|_{t=0}$. Much less would be known about $F^{-\infty}$: Boltzmann's H Theorem implies that

$$H[F^{-\infty}] \geq H[F^{\text{in}}] \geq H[F^{+\infty}]. \tag{3}$$

Since the limits for $t \to \pm\infty$ appear as quite identical in Theorem 3.2, the lack of information on $F^{-\infty}$ casts some doubt as to whether $F^{+\infty} = \mathscr{M}_{F^{\text{in}}}\big|_{t=0}$.

If the inequalities (3) above are equalities, then

$$H[F](t) = H[F^{\text{in}}] \quad \text{for all } t \in \mathbb{R}\,,$$

so that

$$\int_{\mathbb{R}}\iint_{\mathbb{R}^D\times\mathbb{R}^D} \mathscr{B}(F, F)(v, x, s)\ln F(v, x, s)\, dvdxds = 0.$$

In that case,

$$F^{-\infty} = F^{+\infty} = \mathscr{M}_{F^{\text{in}}}\big|_{t=0}\,, \quad \text{and } F = \mathscr{M} \text{ is a global Maxwellian.}$$

In fact, the question above is answered in the negative. Indeed, $\mathscr{T}^+$ is locally onto in the vicinity of global Maxwellians satisfying the same smallness condition as in Theorem 3.2, as shown by the following theorem.

Theorem 4.2 (Bardos-Gamba-Golse-Levemore [4]) *Assume* $\beta \in (1-\mathrm{D}, 1]$. *Let* $\mathscr{M}$ *be a global Maxwellian s.t.*

$$\nu(\mathscr{M}) < \tfrac{1}{4}.$$

For each $F^{+\infty} \equiv F^{+\infty}(v, x)$ *s.t.*

$$|F^{+\infty} - \mathscr{M}(0)|_{\mathscr{M}(0)} \leq \tfrac{(1-4\nu(\mathscr{M}))^2}{8\nu(\mathscr{M})}\,,$$

there exists a unique F^{in} *satisfying*

$$|F^{\text{in}}_{+} - \mathscr{M}(0)|_{\mathscr{M}(0)} \leq r(\mathscr{M}) \quad \textit{and } \mathscr{T}^+F^{\text{in}} := F^{+\infty}\,,$$

where we recall that

$$r(\mathscr{M}) := \left(\frac{1}{4\nu(\mathscr{M})} - 1\right)\left(1 - \sqrt{1 - \frac{8\nu(\mathscr{M})|F^{\text{in}} - \mathscr{M}(0)|_{\mathscr{M}(0)}}{(1-4\nu(\mathscr{M}))^2}}\right).$$

Observe that the smallness condition on $\mathscr{M}$ is the same in Theorems 3.2 and 4.2. Likewise, the neighborhood of $\mathscr{M}(0)$ included in the range of $\mathscr{T}^+$ according to Theorem 4.2 is exactly the same as the set of initial data F^{in} near $\mathscr{M}(0)$ for which a unique global solution of the Cauchy problem for the Boltzmann equation is obtained according to Theorem 3.2.

In fact, this observation reflects the similarities between the proofs of both Theorems 3.2 and 4.2. In other words, the problem of solving for F^{in} the equation

$$\mathscr{T}^+ F^{\text{in}} = F^{+\infty}$$

is embedded in a Cauchy problem for the Boltzmann equation where the "initial" time is $t = +\infty$. What makes this possible is the fact that the dispersion effect of the advection operator quenches the dissipation effect of collisions in the Boltzmann equation.

5 Scattering Theory for the Boltzmann Equation in $\mathbb{R}^D$

The operators $\mathscr{T}^\pm$ can be thought of as being analogous to the so-called "wave operators" in the Lax-Phillips scattering theory for the wave equation [15]. On the basis of this analogy, we define a scattering operator for the Boltzmann equation in $\mathbb{R}^D$, as follows.

Definition 5.1 The scattering operator for the Boltzmann equation is defined by the formula

$$\mathscr{S} := \mathscr{T}^+ \circ (\mathscr{T}^-)^{-1}.$$

While Theorem 4.2 bears on $\mathscr{T}^+$, it obviously can be adapted without change for $\mathscr{T}^-$. In particular $(\mathscr{T}^-)^{-1}$ is defined near $\mathscr{M}(0)$, provided that $\mathscr{M}$ satisfies the smallness condition $\nu(\mathscr{M}) < \frac{1}{4}$. With this definition of the scattering operator, one sees immediatly that $\mathscr{S}$ is defined on some neighborhood of $\mathscr{M}(0)$.

There is however a slightly more convenient way of thinking of the scattering operator $\mathscr{S}$. If F is a solution of the Boltzmann equation defined on $\mathbb{R}^D \times \mathbb{R}^D \times \mathbb{R}$ and satisfying the domination condition (2) in Theorem 4.1, one has $\mathscr{T}^- F^{\text{in}} = F^{-\infty}$ and $\mathscr{T}^+ F^{\text{in}} = F^{+\infty}$ with $F^{\text{in}} = F\big|_{t=0}$, so that

$$\mathscr{S} F^{-\infty} = \mathscr{T}^+ \mathscr{T}^{-1} F^{-\infty} = \mathscr{T}^+ F^{\text{in}} = F^{+\infty}.$$

Henceforth, we think of $\mathscr{S}$ as the operator mapping $F^{-\infty}$ to $F^{+\infty}$ for each solution of the Boltzmann equation defined on $\mathbb{R}^D \times \mathbb{R}^D \times \mathbb{R}$ for which the limits

$$\lim_{t\to\pm\infty} F(v, x+tv, t) = F^{\pm\infty}(v, x)$$

exist. By the argument as in the remark following Theorem 4.2, one can solve the Cauchy problem for the Boltzmann equation along the characteristic lines of the advection operator, starting from $t = -\infty$ as the initial time, and with initial data $F^{-\infty}$. Then $\mathscr{S}F^{-\infty} = F^{+\infty}$ is obtained as the final state of this Cauchy problem, exactly as in Theorem 4.1.

This alternative definition of the scattering operator has at least one advantage: the set of data $F^{-\infty}$ in the past over which the scattering operator can be defined in this way is larger than the set of $F^{-\infty}$ such that $(\mathscr{T}^-)^{-1}F^{-\infty}$ belongs to the domain of definition of $\mathscr{T}^+$ (see statement (a) in the next theorem).

Theorem 5.2 (Bardos-Gamba-Golse-Levemore [4]) *Assume $\beta \in (1-\mathrm{D}, 1]$, and let $\mathscr{M}$ be a global Maxwellian s.t. $\nu(\mathscr{M}) < \frac{1}{4}$. Set*

$$\eta := \tfrac{(1-4\nu(\mathscr{M}))^2}{8\nu(\mathscr{M})}, \quad r := \left(\tfrac{1}{4\nu(\mathscr{M})} - 1\right)\left(1 - \sqrt{1 - \tfrac{|F^{\mathrm{in}} - \mathscr{M}(0)|_{\mathscr{M}(0)}}{\eta}}\right).$$

(a) The scattering operator $\mathscr{S} : \overline{B_{\mathscr{Y}_{\mathscr{M}(0)}}(\mathscr{M}(0), \eta)} \to \overline{B_{\mathscr{Y}_{\mathscr{M}(0)}}(\mathscr{M}(0), r)}$ is Lipschitz continuous in $\mathscr{Y}_{\mathscr{M}(0)}$, with Lipschitz constant

$$L = \frac{1}{1 - 8\nu(2r+1)}.$$

(b) One has

$$\mathscr{S}\mathscr{M}(0) = \mathscr{M}(0).$$

(c) The scattering operator $\mathscr{S}$ satisfies the global conservation laws

$$\iint_{\mathbb{R}^D\times\mathbb{R}^D} \begin{pmatrix} 1 \\ v \\ |v|^2 \\ x \\ |x|^2 \\ x\cdot v \\ x\wedge v \end{pmatrix} \mathscr{S}f(v,x)\,\mathrm{d}v\mathrm{d}x = \iint_{\mathbb{R}^D\times\mathbb{R}^D} \begin{pmatrix} 1 \\ v \\ |v|^2 \\ x \\ |x|^2 \\ x\cdot v \\ x\wedge v \end{pmatrix} f(v,x)\mathrm{d}v\mathrm{d}x.$$

(d) The scattering operator $\mathscr{S}$ decreases the Boltzmann H function:

$$\iint_{\mathbb{R}^D\times\mathbb{R}^D} \mathscr{S}f(v,x)\ln\mathscr{S}f(v,x)\mathrm{d}x\mathrm{d}v \le \iint_{\mathbb{R}^D\times\mathbb{R}^D} f(v,x)\ln f(v,x)\mathrm{d}x\mathrm{d}v\,,$$

with equality iff there exists a global Maxwellian $\mathscr{M}$ s.t.

$$f = \mathscr{M}(0).$$

Several comments are in order after stating Theorem 5.2.

First, the existence of the "wave operators" $\mathscr{T}^{\pm}$ was already known for discrete velocity models of the Boltzmann equation: see the works of Tartar [21], Beale [6, 7], and of Bony [8]. However, the case of discrete velocity models is noticeably different from the Boltzmann equation itself. For instance, the wave operators are known to be discontinuous in the discrete velocity case (see Sect. 5.6 in [8]), whereas it is rather elementary to prove that the operators $\mathscr{T}^{\pm}$ are locally Lipschitz continuous in $\mathscr{Y}_{\mathscr{M}(0)}$ near $\mathscr{M}(0)$ (see Theorem 2.7 in [4]). Another important difference between the Boltzmann equation and the discrete velocity models is that there is no analogue of the notion of global Maxwellian in the discrete velocity case. Finally, the existence of wave operators has been obtained by Hamdache (Theorem 4.3 in [12]), in the vicinity of global Maxwellians without solid rotation.

6 Some Open Problems

We conclude this brief presentation of a scattering regime for the Boltzmann equation in $\mathbb{R}^D$ in the case of cutoff molecular interaction with a few comments and a list of open problems.

6.1 *Problem 1: The BGK Model*

The BGK equation is a simple relaxation model for the Boltzmann equation introduced by Bhatnagar, Gross and Krook in [5]. It has many essential features in common with the Boltzmann equation, especially the local conservation laws of mass, momentum and energy and the H Theorem. It has the same equilibrium solutions as the Boltzmann equation, and essentially the same fluid dynamic limits. The BGK equation is

$$(\partial_t + v \cdot \nabla_x) F = \nu[F](\mathbb{M}_F - F)\,,$$

where

$$\mathbb{M}_F := \mathbb{M}[\rho_F, u_F, \theta_F]\,,$$

with

$$\rho_F := \int_{\mathbb{R}^D} F \mathrm{d}v\,, \quad u_F := \frac{1}{\rho_F}\int_{\mathbb{R}^D} vF\,\mathrm{d}v\,, \quad \theta_F := \frac{1}{\rho_F}\int_{\mathbb{R}^D} \tfrac{1}{D}|v - u_F|^2 F\,\mathrm{d}v.$$

The collision frequency $\nu[F]$ in the original BGK model is

$$\nu[F] = \nu_0 \rho_F , \qquad \text{where } \nu_0 > 0 \text{ is a constant.}$$

With this choice, the "BGK collision operator" $F \mapsto \nu[F](\mathbb{M}_F - F)$ is homogeneous of degree 2 in F, exactly as the map $F \mapsto \mathscr{B}(F, F)$, where $\mathscr{B}(F)$ is the Boltzmann collision integral.

In the mathematical literature, one finds very often the alternate choice $\nu[F] = \nu_0$ (see [19, 20]). With this choice, the existence and uniqueness of solutions of the Cauchy problem for the BGK equation is known: see [19, 20]. However, the choice of a constant collision frequency is physically questionable. On the contrary, one would expect the collision frequency to increase with the density of the gas. The choice $\nu[F] = \nu_0 \rho_F$ is therefore more satisfying on physical grounds. Unfortunately, this choice seems to complicate seriously the analysis of the Cauchy problem for the BGK model, and we are not aware of mathematical results in this direction at the time of this writing.

6.1.1 Dispersion and Global in Time Solutions of the BGK Equation

The analysis in the present paper suggests the idea that the dispersion effect of the advection operator should offset the dependence of the collision frequency in the local density ρ_F. In other words, can one use dispersion in order to prove the existence of a solution of the Cauchy problem for the BGK model with collision frequency $\nu[F] = \nu_0 \rho_F$ in the vicinity of global Maxwellians satisfying some appropriate smallness condition? One obvious difficulty in this problem is that the "gain part" of the BGK collision operator, i.e. $F \mapsto \mathbb{M}_F$, is not order-preserving in general—in other words, $0 < F(v) \leq G(v)$ a.e. on $\mathbb{R}^D$ does not imply in general that $\mathbb{M}_F \leq \mathbb{M}_G$ a.e. on $\mathbb{R}^D$. This is a striking difference with the gain part of the Boltzmann collision integral

$$\mathscr{B}_+(F, F)(v) := \iint_{\mathbb{S}^{D-1} \times \mathbb{R}^D} F' F'_* b(v - v_*, \omega) \, d\omega dv_* ,$$

which is obviously order-preserving since $b > 0$ a.e. on $\mathbb{S}^{D-1} \times \mathbb{R}^D$.

6.2 *Problem 2: Scattering and Cercignani's Conjecture*

By Theorem 5.2(c), (d), any (measurable) distribution function $f \equiv f(v, x) \geq 0$ a.e. with finite H-function and second order moments, i.e. such that

$$\iint_{\mathbb{R}^D \times \mathbb{R}^D} (1 + |x|^2 + |v|^2 + |\ln f(v, x)|) f(v, x) \, dv dx < \infty ,$$

satisfies

$$0 \le H(f) - H(\mathscr{S} f) \le H(f) - H(\mathscr{M}_f(0)).$$

Moreover, if any one of these two inequalities is an equality, then $f = \mathscr{M}_f(0)$ a.e., and both inequalities are equalities.

This suggests that the quantities $H(f) - H(\mathscr{S} f)$ and $H(f) - H(\mathscr{M}_f(0))$ appearing in the inequality above may in fact be comparable in some sense. More precisely, does there exists $C > 0$ and $\alpha \ge 1$ such that

$$C(H(f) - H(\mathscr{M}_f(0))^\alpha \le H(f) - H(\mathscr{S} f)$$

for all distribution functions $f \equiv f(v, x)$ satisfying the assumptions above and in the vicinity of $\mathscr{M}_f(0)$?

This question can be seen as the analogue of Cercignani's conjecture [9] on entropy production in the context of the Boltzmann equation set over $\mathbb{R}^D$ and in the scattering regime. Cercignani's conjecture on entropy production for the Boltzmann equation is the following statement: there exists a positive constant C such that

$$C(H[f] - H[\mathbb{M}[1, 0, 1]]) \le -\int_{\mathbb{R}^D} \mathscr{B}(f, f) \ln f \, dv$$

for each $f \equiv f(v) \ge 0$ a.e. on $\mathbb{R}^D$ such that

$$\int_{\mathbb{R}^D} f(v)\, dv = 1\,, \quad \int_{\mathbb{R}^D} v f(v)\, dv = 0\,, \quad \int_{\mathbb{R}^D} |v|^2 f(v)\, dv = D.$$

See [23] for a discussion of the conjecture (which, in Villani's own words, is "always almost true and sometimes true").

The sought inequality falls in the class of "entropy-entropy production inequalities" [24], of the form

$$\Theta(H(f) - H(\mathscr{M}_f(0))) \le -\int_{\mathbb{R}} \iint_{\mathbb{R}^D \times \mathbb{R}^D} \mathscr{B}(F, F) \ln F(v, x, t)\, dv dx dt$$

for each solution F of the Boltzmann equation such that

$$\mathscr{T}^- \left(F\big|_{t=0}\right) = f\,,$$

where $\Theta : \mathbb{R}_+ \to \mathbb{R}_+$ is a continuous increasing function (independent of F) satisfying the condition $\Theta(0) = 0$.

A natural way of approaching this problem is to consider the linearization of the Boltzmann equation about some global Maxwellian. In this regime, the problem stated above can be reduced to the existence of a spectral gap near the eigenvalue 1 for the linearized scattering operator. This question has been studied by Zakrevskiy [26].

6.3 Problem 3: Non Cutoff Molecular Interactions

The analysis in the present paper relies very heavily on the weak angular cutoff assumption on the collision kernel. However, we feel that some of the questions considered here should be of interest in the case of collision kernels having the same separated form as before in space dimension $\mathrm{D} = 3$, with

$$\frac{1}{C}\vartheta^{-1-2s} \le \hat{\mathfrak{b}}(\cos\vartheta)\sin\vartheta \le C\vartheta^{-1-2s}, \quad \vartheta \in (0, \tfrac{\pi}{2})$$

for some constant $C > 1$ and $s \in (0, 1)$, while the exponent β in formula (1) satisfies $\beta > -1 - 2s$. These conditions hold for all of the collision kernels arising from an inverse power law intermolecular potential.

Existence and uniqueness of the solution of the Cauchy problem near uniform Maxwellians for the solution of the Boltzmann equation with such molecular interactions has been studied in [1, 11], and it would be useful to check whether the same proof applies in the vicinity of global, non-uniform Maxwellians.

In the affirmative, the asymptotic behavior of such solutions as $t \to +\infty$ would be an interesting object of study, and should lead to results rather different from the ones which hold in the cutoff case. For instance, the proof of Theorem 4.1 relies on the decomposition of the Boltzmann collision integral in its gain term $\mathscr{B}_+$ and its loss term $\mathscr{B}_- = \mathscr{B}_+ - \mathscr{B}$, and on the fact that both $\mathscr{B}_+$ and $\mathscr{B}_-$ are order-preserving. This decomposition is known to be impossible in the case of non cutoff molecular interactions.

Another striking difference lies with Theorem 4.2. In the case of a non cutoff molecular interaction, the evolution semigroup associated to the Boltzmann equation, defined in the vicinity of uniform Maxwellians (see Theorem 1 in [11]) has regularizing properties: see for instance [17] and Sect. 3 in [1]. These regularizing properties suggest the following question: does one have

$$\mathscr{T}^+ F^{\text{in}} \in C^\infty(\mathbb{R}^\mathrm{D} \times \mathbb{R}^\mathrm{D})$$

for all initial data F^{in} such that $\mathscr{T}^+ F^{\text{in}}$ exists? If so, it is clear that not all distribution functions in the vicinity of $\mathscr{M}(0)$ in the space $\mathscr{Y}_{\mathscr{M}(0)}$ are asymptotic states $F^{+\infty}$ of solutions F of the Boltzmann equation in the case of a non cutoff molecular interaction.

In short, it seems dubious that a scattering regime can be established for the Boltzmann equation in the case of a non cutoff molecular interaction. Studying the linearization of the non cutoff Boltzmann equation about a global Maxwellian might shed some light on this problem.

6.4 Problem 4: Scattering and Boltzmann-Grad Limit

Consider a set of N identical balls moving freely in the Euclidean space $\mathbb{R}^3$, and interacting through elastic collisions. Sinai asked whether the total number of collisions in such a system is finite over the infinite time interval. This question has been answered in the affirmative by Vasershtein [22]—see also [13] for a simplified proof of this result. In other words, there exists some finite time T such that the motion of the N balls is free over the time interval $(T, +\infty)$.

Since a system of N balls interacting through elastic collisions is mechanically reversible, by the same token there exists some finite time $T' < T$ such that all the collisions take place in the time interval $[T', T]$, and the motion of the N balls is free for all times belonging to $(-\infty, T') \cup (T, +\infty)$. In particular, there is a scattering regime for such a system of N balls in $\mathbb{R}^3$ interacting through elastic collisions.

This suggests the following question: is the scattering regime for the Boltzmann equation described in the present paper the limit of the scattering regime for a system of N hard balls of radius r in the Boltzmann-Grad scaling, i.e. for $N \to \infty$ and $r \to 0$ with Nr^2 converging to some positive constant?

Vasershtein's result is based on an inequality showing that collisions tend to speed up the alignment of the vector of the N positions $(x_1(t), \ldots, x_N(t))$ of the ball centers with the vector $(v_1(t), \ldots, v_N(t))$ of their N velocities in $(\mathbb{R}^3)^N$: see Lemma 1.4 in [13]. It would be interesting to understand whether there is some analogous property at the level of the Boltzmann equation.

References

1. Alexandre, R., Morimoto, Y., Ukai, S., Xu, C.-J., Yang, T.: The Boltzmann equation without angular cutoff in the whole space: qualitative properties of solutions. Arch. Ration. Mech. Anal. **202**, 599–661 (2011)
2. Alonso, R., Gamba, I.M.: Distributional and classical solutions to the Cauchy Boltzmann problem for soft potentials with integrable angular cross section. J. Stat. Phys. **137**, 1147–1165 (2009)
3. Arsenio, D.: On the global existence of mild solutions to the Boltzmann equation for small data in L^D. Commun. Math. Phys. **302**, 453–476 (2011)
4. Bardos, C., Gamba, I.M., Golse, F., Levermore, C.D.: Global Solutions of the Boltzmann Equation over $\mathbb{R}^D$ near Global Maxwellians with Small Mass. arXiv:1409.1430
5. Bhatnagar, P.L., Gross, E.P., Krook, M.: A model for collision processes in gases. I. Small amplitude processes in charged and neutral one-component systems. Phys. Rev. **94**(3), 511–525
6. Beale, J.T.: Large-time behavior of the Broadwell model of a discrete velocity gas. Commun. Math. Phys. **102**, 217–235 (1985)
7. Beale, J.T.: Large-time behavior of discrete velocity Boltzmann equations. Commun. Math. Phys. **106**, 659–678 (1986)
8. Bony, J.-M.: Existence globale et diffusion pour les modèles discrets de la cinétique des gaz. In: Joseph, A., Mignot, F., Murat, F., Prum, B., Rentschler, R. (eds.): First European Congress of Mathematics. Progress in Mathematics 119, Paris, 1992, vol. I, pp. 391–410. Birkhäuser, Basel (1994)

9. Cercignani, C.: H-theorem and trend to equilibrium in the kinetic theory of gases. Arch. Mech. **34**, 231–241 (1982)
10. Desvillettes, L., Villani, C.: On the trend to global equilibrium for spatially inhomogeneous kinetic systems: the Boltzmann equation. Invent. Math. **159**, 245–316 (2005)
11. Gressman, P.T., Strain, R.M.: Global classical solutions of the Boltzmann equation with long-range interactions. Proc. Nat. Acad. Sci. **107**, 5744–5749 (2010)
12. Hamdache, K.: Existence in the large and asymptotic behaviour for the Boltzmann equation. Jpn. J. Appl. Math. **2**, 1–15 (1985)
13. Illner, R.: On the number of collisions in a hard sphere particle system in all space. Transp. Theor. Stat. Phys. **18**, 71–86 (1989)
14. Illner, R., Shinbrot, M.: The Boltzmann equation: global existence for a rare gas in an infinite vacuum. Commun. Math. Phys. **95**, 217–226 (1984)
15. Lax, P.D., Phillips, R.S.: Scattering Theory, Revised edn. Academic Press, San Diego (1989)
16. Levermore, C.D.: Global Maxwellians over All Space and Their Relation to Conserved Quantites of Classical Kinetic Equations
17. Lions, P.-L.: On Boltzmann and Landau equations. Phil. Trans. R. Soc. Lond. A **346**, 191–204 (1994)
18. Lu, X.: Spatial decay of solutions of the Boltzmann equation: converse properties of long time limiting behavior. SIAM J. Math. Anal. **30**, 1151–1174 (1999)
19. Perthame, B.: Global existence to the BGK model of Boltzmann equation. J. Diff. Equ. **82**, 191–205 (1989)
20. Perthame, B., Pulvirenti, M.: Weighted L^∞ bounds and uniqueness for the Boltzmann BGK model. Arch. Ration. Mech. Anal. **125**, 289–295 (1993)
21. Tartar, L.: From Hyperbolic Systems to Kinetic Theory. Springer, Berlin (2008)
22. Vaserstein, L.N.: On systems of particles with finite range and/or repulsive interactions. Commun. Math. Phys. **69**, 31–56 (1979)
23. Villani, C.: Cercignani's conjecture is sometimes true and always almost true. Commun. Math. Phys. **234**, 455–490 (2003)
24. Villani, C.: Entropy production and convergence to equilibrium. In: Golse, F., Olla, S. (eds.) Entropy Methods for the Boltzmann Equation. Lecture Notes in Mathematics 1916, Springer, Berlin (2008)
25. Wennberg, B.: Stability and exponential convergence for the Boltzmann equation. Arch. Ration. Mech. Anal. **130**, 103–144 (1995)
26. Zakrevskiy, T.: Global solutions of the linearised Boltzmann equation over $\mathbb{R}^D$ near global Maxwellians with small mass. Chapter 4, Ph.D. thesis, Ecole polytechnique (2015)

The Gradient Flow Approach to Hydrodynamic Limits for the Simple Exclusion Process

Max Fathi and Marielle Simon

Abstract We present a new approach to prove the macroscopic hydrodynamic behaviour for interacting particle systems, and as an example we treat the well-known case of the symmetric simple exclusion process (SSEP). More precisely, we characterize any possible limit of its empirical density measures as solutions to the heat equation by passing to the limit in the gradient flow structure of the particle system.

Keywords Hydrodynamic limits · Gradient flows structure · Diffusive heat equation · Reversible markov chains

1 Introduction

The aim of this work is to show how one can use gradient flow structures to prove convergence to the hydrodynamic limit for interacting particle systems. The exposition is focused on the case of the symmetric simple exclusion process on the discrete d-dimensional torus, but the strategy can be adapted to other reversible particle systems, such as zero-range processes (see [11] for the definitions of these models). Gradient flows are ordinary differential equations of the form

$$\dot{x}(t) = -\nabla V(x(t)),$$

where ∇V denotes the gradient of the function V. De Giorgi and his collaborators showed in [5] how to give a meaning to solutions to such equations in the setting of

M. Fathi
Department of Mathematics, University of California, Berkeley, USA
e-mail: mfathi@phare.normalesup.org

M. Simon (✉)
Équipe MEPHYSTO, INRIA Lille Nord Europe, 40 avenue du Halley,
59650 Villeneuve d'Ascq, Lille, France
e-mail: marielle.simon@inria.fr

© Springer International Publishing Switzerland 2016

P. Gonçalves and A.J. Soares (eds.), *From Particle Systems to Partial Differential Equations III*, Springer Proceedings in Mathematics & Statistics 162,
DOI 10.1007/978-3-319-32144-8_8

metric spaces: these solutions are called minimizing-movement solutions, or curves of maximal slope. When considering the case of spaces of probability measures, one can use this notion to rewrite the partial differential equations governing the time evolution of the laws of diffusion processes, such as the heat equation, as gradient flows, for which V is the entropy with respect to the optimal transport (or Wasserstein) distance. We refer to [3] for more details. This framework can be adapted to the case of reversible Markov chains on finite spaces. This was proven independently by Maas [13] and Mielke [16], who both developed a discrete counterpart to the Lott-Sturm-Villani theory of lower bounds on Ricci curvature for metric spaces.

These gradient flow structures are a powerful tool to study convergence of sequences of dynamics to some limit. Two main strategies have already been developed. One of them consists in using the discrete (in time) approximation schemes suggested by the gradient flow structure (see for example [4]). The second one, which we shall use here, consists in characterizing gradient flows in terms of a relation between the energy function and its variations, and passing to the limit in this characterization. It was first developed by Sandier and Serfaty in [18], and then generalized in [19]. This strategy can be combined with the gradient flow structure of [13, 16] to prove convergence to some scaling limit for interacting particle systems. This was recently done for chemical reaction equations in [14] and mean-field interacting particle systems on graphs in [6].

Gradient flow structures are also related to large deviations, at least when considering diffusion processes, see [1, 8]. While we only present here the case of the SSEP, the technique is fairly general, and can be adapted to other reversible interacting particle systems. For example, the adaptation of the proof to the case of a zero-range process on the lattice (with nice rates) is quite straightforward. It would be very interesting to apply this method to obtain other PDEs, more degenerate than the heat equation, as hydrodynamic limits of some interacting particle system: for instance, porous medium and fast diffusion equations (see [17, 20]) also have a gradient flow structure, and are not directly solvable by standard techniques.

The plan of the sequel is as follows: in Sect. 2, we present the gradient flow framework for Markov chains on discrete spaces developed in [13, 16]. In Sect. 3, we expose the setup for proving convergence of gradient flows. Finally, in Sect. 4, we investigate the symmetric simple exclusion process, and reprove the convergence to its hydrodynamic limit.

2 Gradient Flow Structure for Reversible Markov Chains

2.1 Framework

We start by describing the gradient flow framework for Markov chains on discrete spaces. The presentation we use here is the one of [13]. We consider an irreducible continuous time reversible Markov chain on a finite space $\mathscr{X}$ with kernel $K : \mathscr{X} \times$

$\mathscr{X} \to \mathbb{R}_+$ and invariant probability measure ν. Let $\mathscr{P}(\mathscr{X})$ (resp. $\mathscr{P}_+(\mathscr{X})$) be the set of probability densities (resp. positive) with respect to ν. The probability law $\rho_t \nu$ of the Markov chain at time t satisfies the evolution equation

$$\dot{\rho}_t(x) + \sum_{y \in \mathscr{X}} (\rho_t(x) - \rho_t(y))K(x, y) = 0, \quad \text{for all } x \in \mathscr{X}. \tag{1}$$

Hereafter we denote by $\dot{\rho}_t(x)$ the derivative with respect to time of the function $(t, x) \mapsto \rho_t(x)$. Given a function $\psi : \mathscr{X} \to \mathbb{R}$, we define $\nabla\psi(x, y) := \psi(y) - \psi(x)$. The discrete divergence of a function $\Phi : \mathscr{X} \times \mathscr{X} \longrightarrow \mathbb{R}$ is defined as

$$\operatorname{div}(\Phi)(x) := \frac{1}{2} \sum_{y \in \mathscr{X}} (\Phi(x, y) - \Phi(y, x))K(x, y).$$

With these definitions, we have the integration by parts formula

$$\sum_{x,y \in \mathscr{X}} \nabla\psi(x, y)\Phi(x, y)K(x, y)\nu(x) = -\sum_{x \in \mathscr{X}} \psi(x) \operatorname{div}(\Phi)(x)\nu(x). \tag{2}$$

Let us introduce three notions we shall use to define the gradient flow structure:

Definition 2.1 1. The *relative entropy with respect to* ν is defined as

$$\operatorname{Ent}_\nu(\rho) := \sum_{x \in \mathscr{X}} \nu(x)\rho(x) \log \rho(x), \quad \text{for } \rho \in \mathscr{P}(\mathscr{X}),$$

with the convention that $\rho(x) \log \rho(x) = 0$ if $\rho(x) = 0$. We sometimes denote $\mathscr{H}(\rho) := \operatorname{Ent}_\nu(\rho)$, whenever ν is fixed and no confusion arises.

2. The *symmetric Dirichlet form* is given for two real-valued functions ϕ, ψ by

$$\mathscr{E}(\phi, \psi) := \frac{1}{2} \sum_{x,y \in \mathscr{X}} (\phi(y) - \phi(x))(\psi(y) - \psi(x))K(x, y)\nu(x),$$

3. The *Fisher information* (or entropy production) writes as $\mathscr{I}(\rho) := \mathscr{E}(\rho, \log \rho)$.

Notice that Ent_ν is the *mathematical* entropy, and not the physical entropy. It decreases along solutions of (1), so in physical terms it plays the role of a free energy. We call $\mathscr{I}$ the entropy production since along solutions of (1) we have

$$\frac{d}{dt} \operatorname{Ent}_\nu(\rho_t) = -\mathscr{I}(\rho_t).$$

2.2 Continuity Equation

We introduce the *logarithmic mean* $\Lambda(a, b)$ of two non-negative numbers a, b as

$$\Lambda(a,b) = \int_0^1 a^s b^{1-s} ds = \frac{b-a}{\log(b) - \log(a)} \quad \text{if } a \neq b, a > 0, b > 0,$$

and $\Lambda(a, a) = a$, and also $\Lambda(a, b) = 0$ if $a = 0$ or $b = 0$. The mean Λ satisfies:

$$\sqrt{ab} \leq \Lambda(a,b) \leq (a+b)/2, \qquad \text{for all } a \geq 0, b \geq 0. \tag{3}$$

Let us now define, for $\rho \in \mathscr{P}_+(\mathscr{X})$, its *logarithmic mean* $\hat{\rho}$ defined on $\mathscr{X} \times \mathscr{X}$ as

$$\hat{\rho}(x,y) := \Lambda(\rho(x), \rho(y)). \tag{4}$$

In order to define a suitable metric on $\mathscr{P}(\mathscr{X})$, we need a representation of curves as solving a continuity equation:

Lemma 2.1 *Given a smooth flow of positive probability densities $\{\rho_t\}_{t\geq 0}$ on $\mathscr{X}$, there exists a function $(t, x) \mapsto \psi_t(x)$ such that the following continuity equation holds for any $t \geq 0$ and $x \in \mathscr{X}$:*

$$\dot{\rho}_t(x) + \sum_{y\in\mathscr{X}} (\psi_t(y) - \psi_t(x)) K(x,y) \hat{\rho}_t(x,y) = 0. \tag{5}$$

Moreover, for any $t \geq 0$, $\psi_t(\cdot)$ is unique up to an additive constant.

We refer to [13, Sect. 3] for the proof.

Definition 2.2 Given $\rho \in \mathscr{P}(\mathscr{X})$ and $\psi : \mathscr{X} \to \mathbb{R}$, we define *the action*

$$\mathscr{A}(\rho,\psi) := \frac{1}{2} \sum_{x,y\in\mathscr{X}} (\psi(y) - \psi(x))^2 \, \hat{\rho}(x,y) K(x,y) \nu(x) \geq 0.$$

A distance between two probability densities (ρ_0, ρ_1) could then be defined as the infimum of the action of all curves $\{\rho_t, \psi_t\}_{t\in[0,1]}$ linking these densities, as was done in [13]. However, we do not need to introduce that metric here, since we shall only use the formulation of gradient flows as minimizing-movement curves, as follows:

Proposition 2.2 *Let $\{\rho_t\}_{t\geq 0}$ be a smooth flow of probability densities on $\mathscr{X}$, and let $\{\psi_t\}_{t\geq 0}$ be such that the continuity equation* (5) *holds. Then, for any $T > 0$,*

$$\mathrm{Ent}_\nu(\rho_T) - \mathrm{Ent}_\nu(\rho_0) + \frac{1}{2}\int_0^T \mathscr{I}(\rho_t) dt + \frac{1}{2}\int_0^T \mathscr{A}(\rho_t, \psi_t) dt \geq 0, \tag{6}$$

with equality if and only if $\{\rho_t\}_{t\geq 0}$ is the flow of the Markov process on $\mathscr{X}$ with kernel K and invariant measure ν, solution to (1).

This is the analogue of the characterization of solutions to $\dot{x}_t = -\nabla V(x_t)$ on $\mathbb{R}^d$ as the only curves for which the non-negative functional

$$V(x_T) - V(x_0) + \frac{1}{2}\int_0^T |\nabla V(x_t)|^2 dt + \frac{1}{2}\int_0^T |\dot{x}_t|^2 dt$$

cancels. Hence in the framework of Markov chains, Ent_ν plays the role of V, and the entropy production $\mathscr{I}$ plays the role of $|\nabla V|^2$.

Proof (*of Proposition* 2.2) Denote $\mathscr{H}(\rho) = \mathrm{Ent}_\nu(\rho)$. We have

$$\mathscr{H}(\rho_T) - \mathscr{H}(\rho_0) = \int_0^T \frac{d}{dt}\big(\mathscr{H}(\rho_t)\big)dt = \int_0^T \sum_{x\in\mathscr{X}} \nu(x)\frac{d}{dt}\big(\rho_t(x)\log\rho_t(x)\big)dt$$
$$= \int_0^T \sum_{x\in\mathscr{X}} \nu(x)\dot{\rho}_t(x)\log\rho_t(x)dt.$$

Using the reversibility of the invariant measure ν, we write

$$\mathscr{H}(\rho_T) - \mathscr{H}(\rho_0) = -\int_0^T \sum_{x,y\in\mathscr{X}} \nu(x)(\psi_t(y) - \psi_t(x))K(x,y)\hat{\rho}_t(x,y)\log\rho_t(x)\,dt$$
$$= \int_0^T \frac{1}{2}\sum_{x,y\in\mathscr{X}} (\psi_t(y) - \psi_t(x))(\log\rho_t(y) - \log\rho_t(x))\hat{\rho}_t(x,y)K(x,y)\nu(x)\,dt$$
$$\geq -\frac{1}{4}\int_0^T \sum_{x,y\in\mathscr{X}} (\psi_t(x) - \psi_t(y))^2\hat{\rho}_t(x,y)K(x,y)\nu(x)\,dt$$
$$-\frac{1}{4}\int_0^T \sum_{x,y\in\mathscr{X}} (\log\rho_t(x) - \log\rho_t(y))^2\hat{\rho}_t(x,y)K(x,y)\nu(x)\,dt$$
$$= -\frac{1}{2}\int_0^T \mathscr{A}(\rho_t,\psi_t)\,dt - \frac{1}{2}\int_0^T \mathscr{I}(\rho_t)\,dt$$

with equality if and only if, for all $x, y \in \mathscr{X}$ and almost every $t \in [0, T]$, we have

$$\psi_t(x) - \psi_t(y) = \log\rho_t(y) - \log\rho_t(x)$$

which is equivalent to saying that for almost every t and for every x we have

$$\dot{\rho}_t(x) + \sum_{y\in\mathscr{X}} (\log\rho_t(x) - \log\rho_t(y))\hat{\rho}_t(x,y)K(x,y) = 0.$$

3 Scaling Limits and Gradient Flows

With the formulation of Proposition 2.2, we can use the approach of Sandier and Serfaty [19] to study convergence of sequences of Markov chains to a scaling limit. Let (K_n) be a sequence of reversible Markov kernels on finite spaces $\mathscr{X}_n$, and let (ν_n) be the sequence of invariant measures on $\mathscr{X}_n$. Since we wish to investigate the asymptotic behaviour of the sequence of random processes, it is much more convenient to work in a single space $\mathscr{X}$ that contains all the $\mathscr{X}_n$. Hence we shall assume that we are given a space $\mathscr{X}$ and a collection of embeddings $\mathbf{p}_n : \mathscr{X}_n \longrightarrow \mathscr{X}$. In practice, the choice of $\mathscr{X}$ and $\mathbf{p}_n$ is suggested by the model under investigation. In the next section, which is focused on the simple exclusion process on the torus, the embeddings will map a configuration η onto the associated empirical measure $\pi^n(\eta)$ (see (12), Sect. 4.1). Such embeddings immediately define embeddings of $\mathscr{P}(\mathscr{X}_n)$ into $\mathscr{P}(\mathscr{X})$.

In order to simplify the exposition below, we adopt the following convention: whenever we say that a sequence (x_n) of elements of $\mathscr{X}_n$ converges to $x \in \mathscr{X}$, we shall mean that $\mathbf{p}_n(x_n) \longrightarrow x$ as n goes to infinity. In particular, the topology used for convergence is implicitly the topology of $\mathscr{X}$, which is assumed to be a separable complete metric space. The strategy is to characterize possible candidates for the limit as gradient flows. For that purpose we give a definition of minimizing-movement curves in the metric setting:

Definition 3.1 Let $(\mathscr{X}, d)$ be a complete metric space. The *gradient flows* of an energy functional $\mathscr{H} : \mathscr{X} \to \mathbb{R}$ with respect to the metric d are the curves $\{m_t\}$ s.t.

$$\mathscr{H}(m_T) - \mathscr{H}(m_0) + \frac{1}{2}\int_0^T g(m_t)dt + \frac{1}{2}\int_0^T |\dot{m}_t|^2 dt = 0,$$

where g is the local slope for $\mathscr{H}$, defined as

$$g(m) := \limsup_{\tilde{m}\to m} \frac{\mathscr{H}(m) - \mathscr{H}(\tilde{m})}{d(m, \tilde{m})} \quad \text{and} \quad |\dot{m}_t| = \limsup_{h\to 0} \frac{d(m_t, m_{t+h})}{h}.$$

Remark 3.1 This is not a complete definition. To make it correct, we should introduce the notion of *absolutely continuous curves*, whose slopes are well defined. This is not a real issue here, since we shall only use it for reversible Markov chains (for which the notions have already been well defined previously for curves of strictly positive probability measures) and the heat equation, for which smooth curves of strictly positive functions do not cause any issue (see the next example). We refer to [3, 19] for a more rigorous discussion of the issues in the metric setting.

Example 3.1 (Heat equation) Let us consider the parabolic PDE

$$\frac{\partial m}{\partial t} = \frac{\partial^2 m}{\partial \theta^2}, \qquad t \geq 0,\ \theta \in (0, 1). \tag{7}$$

We know from [2] that (7) is associated to a gradient flow, since we have:

$$\int_0^1 h(m(T,\theta))d\theta - \int_0^1 h(m(0,\theta))d\theta$$
$$+ \frac{1}{2}\int_0^T \int_0^1 m(1-m)\Big(\frac{\partial(h'(m))}{\partial\theta}\Big)^2 d\theta dt + \frac{1}{2}\int_0^T \Big\|\frac{\partial m}{\partial t}\Big\|_{-1,m}^2 dt = 0,$$

with $h(x) = x\log x + (1-x)\log(1-x)$ and, given $u : [0,1] \to \mathbb{R}$,

$$||u||_{-1,m}^2 := \sup_J \left\{2\int_0^1 J(\theta)u(\theta)d\theta - \int_0^1 m(1-m)\big(J'(\theta)\big)^2 d\theta\right\},$$

where the supremum is over all smooth test functions J.

We now state the main result of that section. Hereafter, when we assert that a sequence of curves of probability measures (in $\mathscr{P}(\mathscr{X}_n) \hookrightarrow \mathscr{P}(\mathscr{X})$) converges to a deterministic curve $\{m_t\}$ (in $\mathscr{X}$), we mean that it converges to a curve of Dirac measures $\{\delta_{m_t}\}$. Definition 4.1 below gives a more precise meaning in the case of particle systems. It is important to stress that we study convergence of probability measures, which are deterministic objects.

Theorem 3.1 *Let (a_n) be an increasing diverging sequence of positive numbers. We first assume that the topology on $\mathscr{P}(\mathscr{X})$ has the following property:*

(**P**) *For any sequence $(\rho_t^n \nu_n)$ of smooth curves of positive probability measures that converges to some deterministic curve $\{m_t\}$, the following inequalities hold:*

$$\liminf_{n\to\infty} \frac{1}{a_n} \operatorname{Ent}_{\nu_n}(\rho_T^n) \ge \mathscr{H}(m_T) \tag{8}$$

$$\liminf_{n\to\infty} \frac{1}{a_n} \int_0^T \mathscr{I}_n(\rho_t^n)dt \ge \int_0^T g(m_t)dt \tag{9}$$

$$\liminf_{n\to\infty} \frac{1}{a_n} \int_0^T \mathscr{A}_n(\rho_t^n, \psi_t^n)dt \ge \int_0^T |\dot{m}_t|^2 dt, \tag{10}$$

where ψ_t^n is such that (ρ_t^n, ψ_t^n) solves (5).

Now consider a sequence $(\rho_t^n \nu_n)$ of gradient flows (so that there is equality in (6)*), assume that the initial sequence $(\rho_0^n \nu_n)$ does converge in distribution to some m_0, and that moreover*

$$\lim_{n\to\infty} \frac{1}{a_n} \operatorname{Ent}_{\nu_n}(\rho_0^n) = \mathscr{H}(m_0).$$

Then, any possible weak limit $\{m_t\}$ of $(\rho_t^n \nu_n)$ is almost surely a gradient flow of the energy $\mathscr{H}$, starting from m_0. In particular, if gradient flows starting from a given initial data are unique, $(\rho_t^n \nu_n)$ weakly converges to a Dirac measure concentrated on the unique gradient flow of $\mathscr{H}$ starting from m_0.

Moreover, for any $t \in [0, T]$, we have

$$\frac{1}{a_n} \operatorname{Ent}_{\nu_n}(\rho_t^n) \xrightarrow[n\to\infty]{} \mathscr{H}(m_t).$$

Above (a_n) is a sequence of weights that corresponds to the correct scaling of the system. For particle systems on the discrete torus of length n in dimension d under diffusive scaling, we would take $a_n = n^d$. This result is a slight variation of the abstract method developed in [19], to which we refer for more details. The main difference (apart from the setting which is restricted to gradient flows in spaces of probability measures arising from reversible Markov chains) is that we consider curves of probability measures that converge to a deterministic curve, rather than any possible limit.

One of the interesting features of this technique is that it does not require an assumption of uniform semi-convexity on the sequence of relative entropies, which can be hard to establish for interacting particle systems (see [7] for the general theory of geodesic convexity of the entropy for Markov chains, and [9] for the study of this property for interacting particle systems on the complete graph). Such an assumption of semi-convexity is known as a lower bound on Ricci curvature for the Markov chain, by analogy with the situation for Brownian motion on a Riemannian manifold. For the simple exclusion on the discrete torus, it seems reasonable to conjecture that curvature is non-negative, but this is still an unsolved problem.

Proof First of all, for any weak limit $\mathscr{Q}$ of the laws of the trajectories, we also have

$$\liminf_{n\to\infty} \frac{1}{a_n} \int_0^T \mathscr{E}(\rho_t^n, \log \rho_t^n)dt \geq \mathscr{Q}\left[\int_0^T g(m_t)dt\right]$$

and

$$\liminf_{n\to\infty} \frac{1}{a_n} \int_0^T \mathscr{A}(\rho_t^n, \psi_t^n)dt \geq \mathscr{Q}\left[\int_0^T |\dot{m}_t|^2 dt\right],$$

where we denote by $\{m_t\}$ a random trajectory with law $\mathscr{Q}$. This is a direct consequence of the following lemma (whose proof is given below):

Lemma 3.2 *Let (f_n) be a sequence of real-valued, non-negative functions on a space $(\Omega, \mathbb{P})$, and assume that there exists a function f such that for any sequence of random variables (X_n) that converges in law to a deterministic limit x, we have*

$$\liminf_{n\to\infty} \mathbb{E}[f_n(X_n)] \geq f(x).$$

Then, for any sequence (X_n) of random variables that converges in law to a random variable X_∞, we have

$$\liminf_{n\to\infty} \mathbb{E}[f_n(X_n)] \geq \mathbb{E}[f(X_\infty)].$$

We now use Proposition 2.2 with the gradient flows $\{\rho_t^n \nu_n\}$, and pass to the limit in

$$\frac{1}{a_n}\left(\mathrm{Ent}_{\nu_n}(\rho_T^n) - \mathrm{Ent}_{\nu_n}(\rho_0^n) + \frac{1}{2}\int_0^T \mathscr{E}(\rho_t^n, \log \rho_t^n)dt + \frac{1}{2}\int_0^T \mathscr{A}(\rho_t^n, \psi_t^n)dt\right) = 0,$$

and therefore

$$\mathscr{Q}[\mathscr{H}(m_T)] - \mathscr{H}(m_0) + \frac{1}{2}\mathscr{Q}\left[\int_0^T g(m_t)dt\right] + \frac{1}{2}\mathscr{Q}\left[\int_0^T |\dot{m}_t|^2 dt\right] \leq 0. \quad (11)$$

Since the above quantity is an expectation of a non-negative functional, we see that

$$\mathscr{H}(m_T) - \mathscr{H}(m_0) + \frac{1}{2}\int_0^T g(m_t)dt + \frac{1}{2}\int_0^T |\dot{m}_t|^2 dt = 0, \quad \mathscr{Q}\text{-almost surely.}$$

This means that $\mathscr{Q}$-almost surely, $\{m_t\}$ is a gradient flow of $\mathscr{H}$. If uniqueness of gradient flows with initial condition m_0 holds, convergence immediately follows.

Convergence of the relative entropy at time T necessarily holds, since otherwise it would contradict (11). Finally, it also holds at any other time $t \in [0, T]$, since one can rewrite the same result on the time-interval $[0, t]$.

We still have to prove Lemma 3.2. This proof is taken from [6].

Proof (*of Lemma* 3.2) Consider a sequence (X_n) that converges in law to a random variable X_∞. Using the almost-sure representation theorem, there exists a sequence (Y_n) such that for any n, Y_n has the same law as X_n, and (Y_n) almost surely converges to Y_∞. If we condition the whole sequence on the event $\{Y_\infty = y\}$, then (Y_n) almost surely converges to y. Then we have, using Fatou's lemma

$$\begin{aligned}\liminf_{n\to\infty} \mathbb{E}[f_n(X_n)] &= \liminf_{n\to\infty} \mathbb{E}[f_n(Y_n)] = \liminf_{n\to\infty} \mathbb{E}_{Y_\infty}\left[\mathbb{E}[f_n(Y_n)|Y_\infty]\right] \\ &\geq \mathbb{E}_{Y_\infty}\left[\liminf_{n\to\infty} \mathbb{E}[f_n(Y_n)|Y_\infty]\right] \geq \mathbb{E}_{Y_\infty}[f(Y_\infty)] = \mathbb{E}[f(X_\infty)].\end{aligned}$$

4 Symmetric Simple Exclusion Process (SSEP)

4.1 Model: Definitions and Notations

To make notations easier we consider the interacting particle systems on the one-dimensional torus $\mathbb{T}_n = \{0, \ldots, n-1\}$, but the result is valid in any dimension

$d \geq 1$. Let us define $\mathscr{X}_n := \{0, 1\}^{\mathbb{T}_n}$, $\mathscr{X} := \{0, 1\}^{\mathbb{Z}}$, and $\mathbb{T} = [0, 1)$ the continuous torus, We create a Markov process $\{\eta_t^n\ ;\ t \geq 0\}$ on the state space $\mathscr{X}_n$, which satisfies for any $\eta \in \mathscr{X}_n$:

- $\eta(i) = 1$ if there is a particle at site $i \in \mathbb{T}_n$,
- $\eta(i) = 0$ if the site i is empty,
- any particle waits independently an exponential time and then jumps to one of its neighbouring sites with probability $1/2$, provided that the chosen site is empty.

We are looking at the evolution of the Markov process in the diffusive time scale, meaning that time is accelerated by n^2. The generator is given for $f : \mathscr{X}_n \to \mathbb{R}$ by

$$\mathscr{L}_n(f)(\eta) := n^2 \sum_{i \in \mathbb{T}_n} \eta(i)(1 - \eta(i+1))(f(\eta^{i,i+1}) - f(\eta)),$$

where $\eta^{i,j}$ is the configuration obtained from η exchanging the occupation variables $\eta(i)$ and $\eta(j)$. The hydrodynamics behavior of the SSEP is well-known, and we refer the reader to [11] for a survey. Let ν_α^n be the Bernoulli product measure of parameter $\alpha \in (0, 1)$, the invariant measures for the dynamics. Under ν_α^n, the variables $\{\eta(i)\}_{i \in \mathbb{T}_n}$ are independent with marginals given by

$$\nu_\alpha^n\{\eta(i) = 1\} = \alpha = 1 - \nu_\alpha^n\{\eta(i) = 0\}.$$

Let us fix once and for all $\alpha \in (0, 1)$ and denote by ρ_t^n the probability density of the law of η_t^n (whose time evolution is generated by $n^2\mathscr{L}_n$) with respect to ν_α^n.

To prove convergence, we need to embed our particle configurations in a single metric space. For each configuration $\eta \in \mathscr{X}_n$, we construct a measure on $\mathscr{X}$ associated to η, denoted by $\pi^n(\eta)$. We do it here through the *empirical measures*:

$$\pi_t^n(d\theta) := \pi^n(\eta_t^n)(d\theta) = n^{-1} \sum_{i \in \mathbb{T}_n} \eta_t^n(i)\delta_{i/n}(d\theta), \tag{12}$$

where δ_θ stands for the Dirac measure concentrated on $\theta \in \mathbb{T}$. Let us denote by $\mathscr{M}_+ = \mathscr{M}_+(\mathscr{X})$ the space of finite positive measures on $\mathscr{X}$ endowed with the weak topology. Assume moreover that, for each n, $\pi^n : \mathscr{X}_n \to \mathscr{M}_+$ is a continuous function. Our goal is to prove the convergence of the flow of measures $(\pi^n(\eta_t^n))$. In particular, π^n inherits the Markov property from η^n.

We start by defining properly two notions of convergence. For any function $G : \mathscr{X} \to \mathbb{R}$ and any measure π on $\mathscr{X}$, we denote by $\langle \pi, G \rangle$ the integral of G with respect to the measure π. In the following $T > 0$ is fixed.

Definition 4.1 Let (π_t^n) be a sequence of flows of measures, each element belonging to the Skorokhod space $\mathscr{D}([0, T], \mathscr{M}_+)$. For each n, let $\mathscr{Q}_n$ be the probability measure on $\mathscr{D}([0, T], \mathscr{M}_+)$ corresponding to $\{\pi_t^n\ ;\ t \in [0, T]\}$.

1. We say that the sequence (π_t^n) *converges to the deterministic flow* $\{\pi_t\}$ if the probability measure $\mathcal{Q}_n$ converges to the Dirac probability measure concentrated on the deterministic flow $\{\pi_t\}$.
2. Fix $t \in [0, T]$. We say that (π_t^n) *converges in probability* to the deterministic measure $\pi_t \in \mathcal{M}_+$ if, for all smooth test functions $G : \mathcal{X} \to \mathbb{R}$, and all $\delta > 0$,

$$\mathcal{Q}_n\Big[\big|\langle \pi_t^n, G\rangle - \langle \pi_t, G\rangle\big| > \delta\Big] \xrightarrow[n\to\infty]{} 0. \tag{13}$$

The next proposition gives the equivalence between the two notions above.

Proposition 4.1 *Let (π_t^n) be a sequence of flows of measures which converges to a deterministic flow $\{\pi_t\}$. Assume that $t \in [0, T] \mapsto \pi_t \in \mathcal{M}_+$ is continuous (with respect to the weak topology). Then, for any $t \in [0, T]$ fixed, (π_t^n) converges in probability to $\pi_t \in \mathcal{M}_+$.*

Proof By assumption, the limiting probability measure on $\mathcal{D}([0, T], \mathcal{M}_+)$ is concentrated on weakly continuous trajectories. Therefore, the limiting flow is almost surely continuous, and the map $\{\pi_t \ ; \ t \in [0, T]\} \mapsto \pi_t$ is continuous from $\mathcal{D}([0, T], \mathcal{M}_+)$ to $\mathcal{M}_+$. Then, for $t \in [0, T]$, (π_t^n) converges in distribution to π_t. Since the latter is deterministic, this induces convergence in probability.

We recall here the main result, that we are going to prove in a different way. Recall that π_t^n is the empirical measure defined in (12) and $\mathcal{Q}_n$ is the probability measure on $\mathcal{D}([0, T], \mathcal{M}_+)$ corresponding to the flow $\{\pi_t^n\}$.

Theorem 4.2 (Hydrodynamic limits for the SSEP) *Fix a density profile $m_0 : \mathbb{T} \to [0, 1]$ and let (μ^n) be a sequence of probability measures such that, under μ^n, the sequence $(\pi_0^n(d\theta))$ converges in probability to $m_0(\theta)d\theta$. In other words,*

$$\limsup_{n\to\infty} \mu^n\bigg[\Big|n^{-1}\sum_{i\in\mathbb{T}_n} G(i/n)\eta(i) - \int_{\mathbb{T}} G(\theta)m_0(\theta)d\theta\Big| > \delta\bigg] = 0,$$

for any $\delta > 0$ and any smooth function $G : \mathbb{T} \to \mathbb{R}$. Assume moreover that this initial data is well-prepared, *in the sense that:*

$$\lim_{n\to\infty} \frac{1}{n} \operatorname{Ent}_{\nu_\alpha^n}(\mu^n) = \int_{\mathbb{T}} h(m_0(\theta))d\theta - h\Big(\int_{\mathbb{T}} m_0(\theta)d\theta\Big), \tag{14}$$

where h has been defined in Example 3.1. Then, for any $t > 0$, the sequence $\{\pi_t^n\}_{n\in\mathbb{N}}$ converges in probability to the deterministic measure $\pi_t(d\theta) = m(t,\theta)d\theta$ where m is solution to the heat equation (7) *on $\mathbb{R}_+ \times \mathbb{T}$. The entropy also converges:*

$$\lim_{n\to\infty} \frac{1}{n} \operatorname{Ent}_{\nu_\alpha^n}(\mu_t^n) = \int_{\mathbb{T}} h(m(t,\theta))d\theta - h\Big(\int_{\mathbb{T}} m(t,\theta)d\theta\Big).$$

Note that $\int m(t,\theta)d\theta$ is actually a constant, given by the fixed density of particles. The convergence of the entropy is equivalent to the local Gibbs behavior (see [12]). Hence, the assumptions and conclusions are those obtained with the relative entropy method of [21]. However, the techniques and restrictions are the same as for the entropy method of [10]: we do not use smoothness of solutions to the hydrodynamic PDE, but we use the replacement lemma (see Sect. 4.3), which relies on the two-block estimate, rather than the one-block estimate alone as in [21].

4.2 The Gradient Flow Approach to Theorem 4.2

We are going to apply Theorem 3.1 to obtain Theorem 4.2. The main steps are as follows:

1. We first need to prove that the sequence $(\mathscr{Q}_n)$ is relatively compact, so that there exists a converging subsequence. Such an argument was already part of the entropy method of [10]. We refer to [11][Chap. 4, Sect. 2] for the proof in the context of the simple exclusion process.
2. In order to prove (8), we have to investigate the convergence of the relative entropy with respect to the invariant measure ν_α^n towards the free energy associated to the limiting PDE (7), which in our case reads as

$$\mathscr{H}(m) = \int_{\mathbb{T}} h(m(\theta))d\theta - h\Big(\int_{\mathbb{T}} m(\theta)d\theta\Big).$$

   This result is actually equivalent to the large deviation principle for ν_α^n (see for example [15]), and is standard (see [11]). Moreover, if our initial data is close (in relative entropy) to a slowly varying Bernoulli product measure[1] associated to m, which satisfies $\nu_{\rho(\cdot)}^n\{\eta(i)=1\} = m(i/n)$, then its relative entropy with respect to ν_α^n converges to the limiting free energy, so that we can easily have (14).
3. We prove the lower bound for the entropy production along curves (9) and the lower bound for the slopes (10) in Sect. 4.3.
4. When passing to the limit, we obtain that for any weak limit $\mathscr{Q}$ of $(\mathscr{Q}_n)$,

$$\mathscr{Q}\Big[\int_{\mathbb{T}} h(m(T,\theta))d\theta - \int_{\mathbb{T}} h(m(0,\theta))d\theta + \frac{1}{2}\int_0^T\int_{\mathbb{T}} m(1-m)\Big(\frac{\partial(h'(m))}{\partial\theta}\Big)^2 d\theta dt + \frac{1}{2}\int_0^T \|\dot{m}_t\|_{-1,m}^2\, dt\Big] \le 0.$$

[1] This is also the assumption used to make Yau's relative entropy method work, see [21].

Since the expression inside the expectation is the characterization of solutions to the heat equation as minimizing-movement curves, it is non-negative, and almost surely m is a solution to the heat equation. Uniqueness of solutions starting from m_0 allows us to conclude.

4.3 Bounds and Convergence

Here we prove that (9) and (10) are satisfied for the density ρ_t^n of the SSEP accelerated in time, assuming that the empirical measure $(\pi_t^n(d\theta))$ converges to a deterministic curve $m_t(\theta)d\theta$. Let us start with (10). The argument is based on a duality argument (Proposition 4.3) and on the replacement lemma (Lemma 4.4) which is commonly used in the literature (see for example [11]).

Proposition 4.3 *Consider a couple (ρ_t, ψ_t) satisfying the continuity equation* (5) *for almost every $t \geq 0$. For any smooth (in time) function $J : [0, T] \times \mathscr{X} \to \mathbb{R}$,*

$$\int_0^T \mathscr{A}(\rho_t, \psi_t)dt \geq 2\sum_{x\in\mathscr{X}} J(T,x)\rho_T(x)\nu(x) - 2\sum_{x\in\mathscr{X}} J(0,x)\rho_0(x)\nu(x)$$

$$-2\int_0^T \sum_{x\in\mathscr{X}} \partial_t J(t,x)\rho_t(x)\nu(x)dt - \int_0^T \sum_{x,y}(J(t,x)-J(t,y))^2\hat{\rho}_t(x,y)K(x,y)\nu(x)dt.$$

Proof From the continuity equation (5), we have

$$\sum_{x\in\mathscr{X}} J(T,x)\rho_T(x)\nu(x) - \sum_{x\in\mathscr{X}} J(0,x)\rho_0(x)\nu(x)$$

$$= \int_0^T \sum_{x\in\mathscr{X}} \partial_t J(t,x)\rho_t(x)\nu(x) + J(t,x)\dot{\rho}_t(x)\nu(x)dt = \int_0^T \sum_{x\in\mathscr{X}} \partial_t J(t,x)\rho_t(x)\nu(x)dt$$

$$- \int_0^T \sum_{x,y\in\mathscr{X}} J(t,x)(\psi_t(y)-\psi_t(x))\hat{\rho}_t(x,y)K(x,y)\nu(x)dt.$$

We symmetrize in x and y the last term, and get

$$2\sum_{x\in\mathscr{X}} J(T,x)\rho_T(x)\nu(x) - 2\sum_{x\in\mathscr{X}} J(0,x)\rho_0(x)\nu(x) = 2\int_0^T \sum_{x\in\mathscr{X}} \partial_t J(t,x)\rho_t(x)\nu(x)dt$$

$$- \int_0^T \sum_{x,y\in\mathscr{X}} (J(t,x)-J(t,y))(\psi_t(y)-\psi_t(x))\hat{\rho}_t(x,y)K(x,y)\nu(x)dt,$$

and therefore

$$\frac{1}{2n}\int_0^T \mathscr{A}(\rho_t,\psi_t)dt \geq \frac{1}{n}\sum_{x\in\mathscr{X}} J(T,x)\rho_T(x)\nu(x) - \frac{1}{n}\sum_{x\in\mathscr{X}} J(0,x)\rho_0(x)\nu(x)$$

$$-\frac{1}{n}\int_0^T \sum_{x\in\mathscr{X}} \partial_t J(t,x)\rho_t(x)\nu(x)dt - \frac{1}{2n}\int_0^T \sum_{x,y}(J(t,x)-J(t,y))^2\hat{\rho}_t(x,y)K(x,y)\nu(x)dt.$$

To apply Proposition 4.3 to the SSEP, we consider observables of the form

$$J(t,\eta) = \sum_{i\in\mathbb{T}_n} G\Big(t,\frac{i}{n}\Big)\eta(i) \tag{15}$$

for smooth functions $G : [0,T]\times\mathbb{T}\to\mathbb{R}$. For any $\ell\in\mathbb{N}$ and $i\in\mathbb{T}_n$, we denote by $\eta^\ell(i)$ the empirical density of particles in a box of size $2\ell+1$ centered at i:

$$\eta^\ell(i) := \frac{1}{2\ell+1}\sum_{|j-i|\leq\ell}\eta(i).$$

Hereafter we also denote by τ_x the translated operator that acts on local functions $g : \{0,1\}^{\mathbb{Z}}\to\mathbb{R}$ as $(\tau_x g)(\eta) := g(\tau_x\eta)$, and $\tau_x\eta$ is the configuration obtained from η by shifting: $(\tau_x\eta)_y = \eta_{x+y}$. The main tool that we are going to use is the well-known replacement lemma, which is a consequence of the averaging properties of the SSEP. We recall the main statement and refer the reader to [10, 11] for a proof:

Lemma 4.4 (Replacement Lemma) *Denote by $\mathbb{P}_{\mu^n}$ the probability measure on the Skorokhod space $\mathscr{D}([0,T],\mathscr{X}_n)$ induced by the Markov process $\{\eta_t^n\}_{t\geq 0}$ starting from μ^n. Then, for every $\delta>0$ and every local function g,*

$$\limsup_{\varepsilon\to 0}\limsup_{n\to\infty}\mathbb{P}_{\mu^n}\left[\int_0^T n^{-1}\sum_{x\in\mathbb{T}_n}\tau_x V_{\varepsilon n}(\eta_s)ds \geq \delta\right] = 0,$$

where

$$V_\ell(\eta) = \left|\frac{1}{2\ell+1}\sum_{|y|\leq\ell}\tau_y g(\eta) - \tilde{g}(\eta^\ell(0))\right|$$

and $\tilde{g} : (0,1)\to\mathbb{R}$ corresponds to the expected value: $\tilde{g}(\alpha) := \int g(\eta)d\nu_\alpha(\eta)$.

We are now able to conclude the proof. We treat separately the terms in the right-hand side of Proposition 4.3, taking J as in (15). Since, for any fixed t, $(\pi_t^n(d\theta))$ converges in probability to $\pi_t(d\theta) = m_t(\theta)d\theta$ we have

$$\frac{1}{n}\sum_{\eta\in\mathscr{X}_n} J(T,\eta)\rho_T^n(\eta)\nu_\alpha^n(\eta) = \mathscr{Q}_n\big[\langle\pi_T^n, G\rangle\big] \xrightarrow[n\to\infty]{} \int_{\mathbb{T}} G(T,\theta)m_T(\theta)d\theta.$$

And the same happens at initial time for ρ_0^n. Similarly,

$$\frac{1}{n}\int_0^T \sum_{\eta\in\mathscr{X}_n} \frac{\partial J}{\partial t}(t,\eta)\rho_t^n(\eta)\nu_\alpha^n(\eta)dt \longrightarrow \int_0^T\int_{\mathbb{T}} \frac{\partial G}{\partial t}(t,\theta)m_t(\theta)d\theta dt.$$

Then, we write

$$\int_0^T \frac{1}{n} \sum_{\eta,\eta'\in\mathscr{X}_n} (J(t,\eta)-J(t,\eta'))^2\, \widehat{\rho_t^n}(\eta,\eta')K_n(\eta,\eta')\nu_\alpha^n(\eta)dt$$
$$= \frac{n^2}{n}\int_0^T \sum_{\eta,i}\Big[G\Big(t,\frac{i}{n}\Big)-G\Big(t,\frac{i+1}{n}\Big)\Big]^2 \eta(i)(1-\eta(i+1))\, \widehat{\rho_t^n}(\eta,\eta^{i,i+1})\nu_\alpha^n(\eta)dt.$$

We now use the logarithmic inequality (3) and write that the latter is smaller than

$$n\int_0^T \sum_{\eta,i}\Big[G\Big(t,\frac{i}{n}\Big)-G\Big(t,\frac{i+1}{n}\Big)\Big]^2 \eta(i)(1-\eta(i+1))\, \frac{\rho_t^n(\eta)+\rho_t^n(\eta^{i,i+1})}{2}\nu_\alpha^n(\eta)dt.$$

From the invariance property of ν_α^n with respect to the change of variables $\eta\to\eta^{i,i+1}$, and from the smoothness of G we get that the above quantity is equal to

$$\frac{1}{n}\int_0^T \sum_{\eta\in\mathscr{X}_n}\sum_{i\in\mathbb{T}_n}\Big[G'\Big(t,\frac{i}{n}\Big)\Big]^2 \eta(i)(1-\eta(i+1))\, \rho_t^n(\eta)\nu_\alpha^n(\eta)dt + o\Big(\frac{1}{n}\Big), \quad (16)$$

where G' denotes the space derivative of G.

Above we want to replace $\eta(i)(1-\eta(i+1))$ by $m(i/n)(1-m(i/n))$. For $\varepsilon>0$ we define the approximation of the identity $i_\varepsilon(u)=(2\varepsilon)^{-1}\mathbf{1}\{|u|\le\varepsilon\}$. With that notation, $\eta_t^{\varepsilon n}(0)$ is very close to $\langle\pi_t^n, i_\varepsilon\rangle$. Let us denote $h(\eta):=\eta(0)(1-\eta(1))$. Since G is a smooth function, (16) equals

$$\frac{1}{n}\int_0^T \sum_{\eta\in\mathscr{X}_n}\sum_{i\in\mathbb{T}_n}\frac{1}{2\varepsilon n+1}\sum_{|j-i|\le\varepsilon n}\Big[G'\Big(t,\frac{j}{n}\Big)\Big]^2 \tau_i h(\eta)\rho_t^n(\eta)\nu_\alpha^n(\eta)dt + O(\varepsilon^2).$$

A summation by parts shows that the previous term can be written as

$$\frac{1}{n}\int_0^T \sum_{\eta\in\mathscr{X}_n}\sum_{i\in\mathbb{T}_n}\Big[G'\Big(t,\frac{i}{n}\Big)\Big]^2\frac{1}{2\varepsilon n+1}\sum_{|j-i|\le\varepsilon n} \tau_j h(\eta)\rho_t^n(\eta)\nu_\alpha^n(\eta)dt + O(\varepsilon^2).$$

By Lemma 4.4, this expression is then equal to

$$\frac{1}{n}\int_0^T \sum_{\eta\in\mathscr{X}_n}\sum_{i\in\mathbb{T}_n}\Big[G'\Big(t,\frac{i}{n}\Big)\Big]^2\tau_i\tilde{h}\big(\langle\pi_t^n(\eta),i_\varepsilon\rangle\big)\nu_\alpha^n(\eta)dt + R_{n,\varepsilon,T},$$

where $R_{n,\varepsilon,T}$ vanishes in probability as n goes to infinity and then ε goes to 0. From the convergence in probability of (π_t^n), the last expression converges to

$$\int_0^T \int_{\mathbb{T}} m_t(\theta)(1 - m_t(\theta))(G'(t,\theta))^2 \, d\theta dt.$$

As a result, since the convergences above are valid for any smooth function G,

$$\begin{aligned}\liminf_{n\to\infty} \frac{1}{n} \int_0^T \mathscr{A}(\rho_t^n, \psi_t^n) dt &\geq \int_0^T \sup_G \left\{ 2 \int_{\mathbb{T}} G \, \dot{m}_t \, d\theta - \int_{\mathbb{T}} m_t(1 - m_t)(G')^2 d\theta \right\} dt \\ &= \int_0^T \|\dot{m}_t\|_{-1,m}^2 \, dt.\end{aligned}$$

In the same way, we need to prove

$$\liminf_{n\to\infty} \frac{1}{n} \int_0^T \mathscr{E}(\rho_t^n, \log \rho_t^n) dt \geq \int_0^T \int_{\mathbb{T}} m(1-m) \Big(\frac{\partial (h'(m))}{\partial \theta} \Big)^2 d\theta dt.$$

Since the arguments are essentially the same as for the slopes, we shall be more brief in the exposition. We denote

$$\nabla_n G\Big(t, \frac{i}{n}\Big) = n\Big[G\Big(t, \frac{i+1}{n}\Big) - G\Big(t, \frac{i}{n}\Big)\Big].$$

By duality, we have

$$\begin{aligned}
&\frac{1}{2n} \int_0^T \mathscr{E}(\rho_t^n, \log \rho_t^n) dt \\
&\geq \int_0^T \sum_{\eta,i} (\log \rho_t^n(\eta^{i,i+1}) - \log \rho_t^n(\eta)) \nabla_n G\Big(t, \frac{i}{n}\Big) \eta_i (1 - \eta_{i+1}) \widehat{\rho_t^n}(\eta, \eta^{i,i+1}) \nu_\alpha^n(\eta) dt \\
&\quad - \frac{1}{2n} \int_0^T \sum_{\eta,i} \Big[\nabla_n G\Big(t, \frac{i}{n}\Big)\Big]^2 \eta_i (1 - \eta_{i+1}) \widehat{\rho_t^n}(\eta, \eta^{i,i+1}) \nu_\alpha^n(\eta) dt \\
&\geq \int_0^T \sum_{\eta,i} (\rho_t^n(\eta^{i,i+1}) - \rho_t^n(\eta)) \nabla_n G\Big(t, \frac{i}{n}\Big) \eta_i (1 - \eta_{i+1}) \nu_\alpha^n(\eta) dt \\
&\quad - \frac{1}{2n} \int_0^T \sum_{\eta,i} \Big[\nabla_n G\Big(t, \frac{i}{n}\Big)\Big]^2 \eta_i (1 - \eta_{i+1}) \rho_t^n(\eta) \nu_\alpha^n(\eta) dt \\
&= - \int_0^T \sum_{\eta,i} \rho_t^n(\eta) \Big[\nabla_n G\Big(t, \frac{i}{n}\Big) - \nabla_n G\Big(t, \frac{i-1}{n}\Big)\Big] \eta_i \nu_\alpha^n(\eta) dt \\
&\quad - \frac{1}{2n} \int_0^T \sum_{\eta,i} \Big[\nabla_n G\Big(t, \frac{i}{n}\Big)\Big]^2 \eta_i (1 - \eta_{i+1}) \rho_t^n(\eta) \nu_\alpha^n(\eta) dt
\end{aligned}$$

Using the replacement lemma, passing to the supremum in G, and to the limit,

$$\begin{aligned}
&\liminf_{n\to\infty} \frac{1}{n}\int_0^T \mathscr{E}(\rho_t^n, \log \rho_t^n)dt \\
&\geq \sup_G\left\{ -2\int_0^T\int_{\mathbb{T}} G'' m_t \, d\theta dt - \int_0^T\int_{\mathbb{T}} m_t(1-m_t)(G')^2 d\theta dt \right\} \\
&= \sup_G\left\{ -2\int_0^T\int_{\mathbb{T}} G'' m_t(1-m_t)h''(m_t) d\theta dt - \int_0^T\int_{\mathbb{T}} m_t(1-m_t)(G')^2 d\theta dt \right\} \\
&= \int_0^T\int_{\mathbb{T}} m_t(1-m_t)\left(\frac{\partial h'(m_t)}{\partial\theta}\right)^2 d\theta dt,
\end{aligned}$$

and this is exactly what we were seeking to prove.

Acknowledgments M.F. : I would like to thank Hong Duong, Matthias Erbar, Vaios Laschos and André Schlichting for discussions on convergence of gradient flows. Part of this work was done while I was staying at the Hausdorff Institute for Mathematics in Bonn, whose support is gratefully acknowledged. I also benefited from funding from GDR MOMAS and from NSF FRG grant DMS-1361185. M.S.: This work has been supported by the French Ministry of Education through the grant ANR (EDNHS), and also by CAPES (Brazil) and IMPA (Instituto de Matematica Pura e Aplicada, Rio de Janeiro) through a post-doctoral fellowship.

References

1. Adams, S., Dirr, N., Peletier, M.A., Zimmer, J.: From a large-deviations principle to the Wasserstein gradient flow: a new micro-macro passage. Commun. Math. Phys. **307**, 791–815 (2011)
2. Adams, S., Dirr, N., Peletier, M. A., Zimmer, J.: Large deviations and gradient flows. Phil. Trans. R. Soc. A **371**(2005), 0341 (2013)
3. Ambrosio, L., Gigli, N., Savaré, G.: Gradient Flows in Metric Spaces and in the Space of Probability Measures. Lectures in Mathematics ETH Zurich, 2nd edn. Birkhauser Verlag, Basel (2008)
4. Ambrosio, L., Savaré, G., Zambotti, L.: Existence and stability for Fokker-Planck equations with log-concave reference measure. Probab. Theory Relat. Fields **145**, 517–564 (2009)
5. De Giorgi, E., Marino, A., Tosques, M.: Problems of evolution in metric spaces and maximal decreasing curve. Att. Acc. Naz. Linc. R. Cl. Sci. Fis. Mat. Nat. **8(68)**(3), 180–187 (1980)
6. Erbar, M., Fathi, M., Laschos, V., Schlichting, A.: Gradient flow structure for McKean-Vlasov equations on discrete spaces. arXiv:1601.08098
7. Erbar, M., Maas, J.: Ricci curvature of finite Markov chains via convexity of the entropy. Arch. Ration. Mech. Anal. **206**(3), 997–1038 (2012)
8. Fathi, M.: A gradient flow approach to large deviations for diffusion processes. Journal de Mathématiques Pures et Appliquées (2014). arXiv:1405.3910
9. Fathi, M., Maas, J.: Entropic Ricci curvature bounds for discrete interacting systems. Ann. Appl. Probab. (2015)
10. Guo, M.Z., Papanicolaou, G.C., Varadhan, S.R.S.: Nonlinear diffusion limit for a system with nearest neighbor interactions. Commun. Math. Phys. **118**, 31–59 (1988)
11. Kipnis, C., Landim, C.: Scaling Limits of Interacting Particle Systems. Grundlehren der Mathematischen Wissenschaften [Fundamental Principles of Mathematical Sciences], vol. 320. Springer, Berlin (1999)

12. Kosygina, E.: The behavior of the specific entropy in the hydrodynamic scaling limit. Ann. Probab. **29**(3), 1086–1110 (2001)
13. Maas, J.: Gradient flows of the entropy for finite Markov chains. J. Funct. Anal. **261**(8), 2250–2292 (2011)
14. Maas, J., Mielke, A.: Gradient structures for chemical reactions with detailed balance: I. Modeling and large-volume limit (2015)
15. Mariani, M.: A Gamma-convergence approach to large deviations. Preprint arXiv:1204.0640v1 (2012)
16. Mielke, A.: Geodesic convexity of the relative entropy in reversible Markov chains. Calc. Var. Partial. Differ. Equ. **48**(1–2), 1–31 (2013)
17. Otto, F.: The geometry of dissipative evolution equations: the porous medium equation. Commun. Partial. Differ. Equ. **26**(1–2), 101–174 (2001)
18. Sandier, E., Serfaty, S.: Gamma-convergence of gradient flows with applications to Ginzburg-Landau. Commun. Pure Appl. Math **57**(12), 1627–1672 (2004)
19. Serfaty, S.: Gamma-convergence of gradient flows on Hilbert and metric spaces and applications. Discret. Contin. Dyn. Syst. **A 31**(4), 1427–1451 (2011)
20. Vazquez, J.L.: The Porous Medium Equation: Mathematical Theory. Oxford Mathematical Monographs. A Clarendon Press Publication, Oxford (2006)
21. Yau, H.T.: Relative entropy and hydrodynamics of Ginzburg-Landau models. Lett. Math. Phys. **22**, 63–80 (1991)

Symmetries and Martingales in a Stochastic Model for the Navier-Stokes Equation

Rémi Lassalle and Ana Bela Cruzeiro

Abstract A stochastic description of solutions of the Navier-Stokes equation is investigated. These solutions are represented by laws of finite dimensional semi-martingales and characterized by a weak Euler-Lagrange condition. A least action principle, related to the relative entropy, is provided. Within this stochastic framework, by assuming further symmetries, the corresponding invariances are expressed by martingales, stemming from a weak Noether's theorem.

Keywords Stochastic analysis · Stochastic control · Navier-Stokes equation

Mathematics Subject Classification: 93E20 · 60H30

Several stochastic models for the *Navier-Stokes* equation have been proposed in the literature. Some refer to random perturbations of the fluid velocity. This is not the case here: we are interested in stochastic Lagrangian paths whose (mean) velocity, or drift, represent the deterministic velocity of the fluid. Different studies of stochastic Lagrangian paths in fluid dynamics and in particular in turbulence can be found in a collection of works, from which we refer to [2, 15] and [16] as examples. Also representation formulae in terms of different random processes were given in [5, 6, 8], among others.

Concerning the derivation of solutions of Navier-Stokes equations from (stochastic) variational principles, after the early articles [17] and [20], such principles were developed in [3] and subsequent works, for instance see [4]. We mention also [11] and [9] for different, unrelated approaches to the same kind of problems.

R. Lassalle
CEREMADE, Université Paris-Dauphine, Place du Maréchal de Lattre de Tassigny, 75016 Paris, France
e-mail: lassalle@ceremade.dauphine.fr

A.B. Cruzeiro (✉)
GFMUL and Departamento Matemática Instituto Superior Técnico, Universidade de Lisboa, Av. Rovisco Pais, 1049-001 Lisbon, Portugal
e-mail: abcruz@math.tecnico.ulisboa.pt

© Springer International Publishing Switzerland 2016

P. Gonçalves and A.J. Soares (eds.), *From Particle Systems to Partial Differential Equations III*, Springer Proceedings in Mathematics & Statistics 162, DOI 10.1007/978-3-319-32144-8_9

In [7] and [13] a weak description of a stochastic deformation of mechanics has been investigated: the Euler-Lagrange condition extends as a condition on laws of stochastic processes [13]. The associated calculus of variations is developed in [7]. As a by product, the latter provides the underlying geometry to primal dynamical Schrödinger problems of [19] and [21]. It also applies to stochastic kinematics unrelated to martingales ; specific martingales merely appearing in the dynamics. In particular, it allows to define consistent perturbations of statistical mixture of paths, the associated variations of functionals, and Euler-Lagrange conditions, by considering convex combinations of Dirac measures concentrated on smooth enough trajectories. The main originality of this approach, with respect to other weak deformations of mechanics, is to handle problems in a functional-analytic framework, in a full consistency with the deterministic case.

In this paper, within the framework of [7] and [13], we develop a specific case, from an example of [7], which provides a *stochastic description* for solutions to the *Navier-Stokes equation.* Then we investigate several *symmetries*, whose associated invariances correspond in this setup to *martingales.*

Section 1 fixes the framework and notations of the paper; the *weak Euler-Lagrange condition* of [13] and [7] is recalled. Under conditions, in Sect. 2, a map

$$P : u \to P_u,$$

associates laws of $\mathbb{R}^d$-valued *semi-martingales* to *divergence free vector fields*. Solutions of the *Navier-Stokes equation* are shown to be divergence free vector fields u, whose associated probability P_u satisfies a weak *Euler-Lagrange condition* (Proposition 1); Proposition 2 characterizes those solutions as *critical points* of the *stochastic action*

$$\mathscr{S}^p(\nu) := E_\nu \left[\int_0^1 \left(\frac{|v_t^\nu|^2}{2} - p(1-t, W_t) \right) dt \right],$$

where ν denotes the *law* of specific continuous *semi-martingales*, and (v_t^ν) the characteristic drift of ν, as stated accurately below. The function p is a smooth *pressure field* which is assumed to be *given*.

The action functional above is related to the relative *entropy* with respect to a reference law μ_p induced by the pressure field.

Finally, within this stochastic model, Sect. 3 investigates *invariances*, stemming from *symmetries*, by the weak *Noether's theorem* of [7]; within this stochastic framework *martingales* on the canonical space play the rôle of constants of motion in classical mechanics.

1 The Weak Stochastic Euler-Lagrange Condition

The *weak* stochastic *Euler-Lagrange condition*, recalled below, was introduced in [7, 13]. It *embeds* in probability measures, specifically in a set of *laws* of semi-martingales, the classical condition. Thus, it provides a *functional analytic* approach to tackle *stochastic variational problems*; in particular, in contrast with usual diffusion approaches, it canonically extends the classical, not-stochastic, approach. In this context the extension of *Noether's theorem* becomes natural. Moreover, as stated in [7], another specificity of this framework is that it provides critical conditions to *semi-martingale optimal transportation problems*. The latter, introduced in [18], correspond to a relaxation of a specific *dynamical Schrödinger problems* (see [14]) by allowing, in particular, the characteristic dispersion to be not-trivial. Finally, as it is expected of optimization over a subset of Borel probabilities on a Polish space, one crucial advantage of this framework is that *compactness* is rather simple to obtain.

1.1 Admissible Trajectories

Trajectories of infinitely small passive tracers in fluids can be described by elements in the space $W := C([0,1], \mathbb{R}^d)$, namely the set of continuous $\mathbb{R}^d$-valued paths (where we consider the norm $|.|_W$ of uniform convergence), endowed with the Borel sigma-field $\mathscr{B}(W)$. In particular, trajectories of finite energy can be described by the subset

$$H := \left\{ h \in W, h := \int_0^{\cdot} \dot{h}_s ds, \int_0^1 |\dot{h}_s|^2_{\mathbb{R}^d} ds < \infty \right\},$$

of absolutely continuous paths with square integrable derivatives.

Let $(W_t)_{t\in[0,1]}$ denote the *evaluation process*

$$(t, \omega) \in [0,1] \times W \to W_t(\omega) := \omega(t) \in \mathbb{R}^d.$$

We consider $(\mathscr{F}_t^0)$ the natural (past) filtration and denote by $\mathscr{P}_W$ the set of Borel probabilities on W.

1.2 A Weak Description of Random Trajectories

In order to avoid measurability issues, we consider $(\mathscr{F}_t^\nu)$, the ν-usual augmentation of the filtration $(\mathscr{F}_t^0)$ under $\nu \in \mathscr{P}_W$. The latter naturally models random trajectories, since any $\nu \in \mathscr{P}_W$ is the law of the evaluation process, on the completed probability space $(W, \mathscr{B}(W)^\nu, \nu)$. We define $\mathbb{S}$ to be the subset of $\nu \in \mathscr{P}_W$ such that there exists a $(\mathscr{F}_t^\nu)$-martingale (M_t^ν) on $(W, \mathscr{B}(W)^\nu, \nu)$, which satisfies

$$W_t = W_0 + M_t^\nu + \int_0^t v_s^\nu ds,$$

for all $t \in [0, 1]$, $\nu - a.s.$, where (v_s^ν) is a $(\mathscr{F}_t^\nu)$-*predictable process* on the same space, and where the *predicable covariation process* of (M^ν) is of the specific form

$$< (M_t^\nu)^i, (M_t^\nu)^j >= \int_0^\cdot (\alpha_s^\nu)^{ij} ds,$$

for a predictable process (α_s^ν); subsequently, by abuse of language, we refer to (v_t^ν, α_t^ν) as the *characteristics* of ν. In the whole paper, notations are those of [7].

1.3 A Weak Euler-Lagrange Condition

Given a *smooth Lagrangian* function

$$\mathscr{L} : (t, x, v, a) \in [0, 1] \times \mathbb{R}^d \times \mathbb{R}^d \times (\mathbb{R}^d \otimes \mathbb{R}^d) \to \mathscr{L}_t(x, v, a) \in \mathbb{R}, \quad (1)$$

the classical *Euler-Lagrange condition* naturally extends to $\mathbb{S}$ (see [13]). A semi-martingale $\nu \in \mathbb{S}$ satisfies the Euler-Lagrange condition if there exists a $(\mathscr{F}_t^\nu)$ *càd-làg martingale* (N_t^ν), such that

$$\partial_v \mathscr{L}_t(W_t, v_t^\nu, \alpha_t^\nu) - \int_0^t \partial_q \mathscr{L}_s(W_s, v_s^\nu, \alpha_s^\nu) ds = N_t^\nu \quad \lambda \otimes \nu - a.e., \quad (2)$$

$\partial_q \mathscr{L}$ and $\partial_v \mathscr{L}_t$ denoting the respective gradients (in the first and in the second variables, respectively).

Remark 1 Similar conditions were considered in [1] for arbitrary semi-martingales U on abstract stochastic basis $(\Omega, \mathscr{A}, (\mathscr{A}_t), \mathscr{P})$. On the contrary, condition (2) imposes constraints on laws of processes. A semi-martingale U on an arbitrary stochastic basis whose law satisfies (2) exhibits very precise properties, depending on the Lagrangian; for instance, in a specific case, it is associated to systems of coupled stochastic differential equations; the latter are not satisfied, in general, when U verifies the critical condition of [1]. Moreover, the associated variational principles of [1] do not contain the optimum criticality for semi-martingale optimal transportation problems either.

2 Navier-Stokes Equation and the Weak Euler-Lagrange Condition

Henceforth, and *until the end of the paper, p denotes a smooth map*

$$p : (t, x) \in [0, 1] \times \mathbb{R}^d \to p(t, x) \in \mathbb{R}^+, \tag{3}$$

which is further assumed to be bounded, with bounded derivatives; p models the *pressure field*. This map being given, we provide a *stochastic model* for solutions of the equation

$$\partial_t u + (u.\nabla)u = -\nabla p + \frac{\Delta u}{2} \ ; \ div\, u = 0. \tag{4}$$

For sake of clarity, we focus on the case where the *divergence free velocity vector field*, involved in the Navier-Stokes equation, belongs to

$$C^{1,2}_{b,div}([0,1] \times \mathbb{R}^d) := \{u \in C^{1,2}([0,1] \times \mathbb{R}^d; \mathbb{R}^d) \cap C_b([0,1] \times \mathbb{R}^d; \mathbb{R}^d) : div\, u(t,.) = 0, \ \textit{for all } t \in [0,1]\}.$$

2.1 Description of Dissipative Flows by Laws of Semi-martingales

Given $u \in C^{1,2}_{b,div}([0,1] \times \mathbb{R}^d)$, we define P_u to be the probability measure, which is equivalent with respect to the *Wiener measure* $\mu \in \mathscr{P}_W$ (the law of *standard Brownian motion*), with density defined by

$$\frac{dP_u}{d\mu} := \exp\left(-\int_0^1 u(1-t, W_t)dW_t - \frac{1}{2}\int_0^1 |u(1-t, W_t)|^2 dt\right).$$

By the *Girsanov theorem* (see for example [10]), we obtain a map

$$P : u \in C^{1,2}_{b,div}([0,1] \times \mathbb{R}^d) \to P_u \in \mathbb{S}, \tag{5}$$

such that $\lambda \otimes P_u$ *a.e.*, $\alpha_s^{P_u} = I_{\mathbb{R}^d}$, and $v_t^{P_u} = -u(1-t, W_t)$, λ denoting the Lebesgue measure. *Itô's formula* yields the following:

Proposition 1 *A time-dependent vector field $u \in C^{1,2}_{b,div}([0,1] \times \mathbb{R}^d)$ satisfies the Navier-Stokes equation (4) if and only if P_u satisfies (2) for*

$$\mathscr{L}_t^p(x, v, a) := \frac{|v|^2}{2} - p(1-t, x). \tag{6}$$

2.2 Stochastic Action and Relative Entropy

Define the *stochastic action* associated to the Lagrangian $\mathscr{L}^p$ of (6) by

$$\mathscr{S}^p : \nu \in \mathbb{S} \to \mathscr{S}^p(\nu) := E_\nu \left[\int_0^1 \mathscr{L}_s^p(W_s, v_s^\nu, \alpha_s^\nu) ds \right] \in \mathbb{R} \cup \{+\infty\}. \quad (7)$$

As $\mathscr{L}^p$ and P_u satisfy the assumptions of Theorem 5.1 of [7], for all p as above and $u \in C_{b,div}^{1,2}([0,1] \times \mathbb{R}^d)$, the $\mathbb{S}$-functional $\mathscr{S}^p$ is differentiable in the sense considered in [7]. Thus, we obtain the following result:

Proposition 2 *A vector field $u \in C_{b,div}^{1,2}([0,1] \times \mathbb{R}^d)$ is a solution of the Eq. (4) if and only if*

$$\delta \mathscr{S}_{P_u}[h] = 0,$$

for all $(\mathscr{F}_t^{P_u})$-adapted process (h_t) of finite energy, such that

$$h_0 = h_1 = 0 \ P_u - a.s.,$$

where $P_u \in \mathbb{S}$ is given by (5) and $\delta \mathscr{S}_{P_u}$ denotes the $\mathbb{S}$-differential of [7] at P_u.

Remark 2 In particular, taking the pullback with the map (5), a stochastic action on $C_{b,div}^{1,2}([0,1] \times \mathbb{R}^d)$ is given by

$$S : u \in C_{b,div}^{1,2}([0,1] \times \mathbb{R}^d) \to S_u = E_{P_u} \left[\int_0^1 \frac{|u(1-t, W_t)|^2}{2} - p(1-t, W_t) dt \right]. \quad (8)$$

2.3 Least Action Principle and Relative Entropy

Subsequently, assuming $\nu \in \mathscr{P}_W$ to be absolutely continuous with respect to a reference law $\eta \in \mathscr{P}_W$, the relative *entropy* of ν w.r.t. η is defined as

$$\mathscr{H}(\nu|\eta) := E_\nu \left[\ln \frac{d\nu}{d\eta} \right].$$

By the representation formula in [12], we obtain

$$\mathscr{S}^p(P_u) = \mathscr{H}(P_u|\mu_p) + \ln Z_p, \quad (9)$$

for all $u \in C^{1,2}([0,1] \times \mathbb{R}^d)$, where P_u is defined by (5), and where μ_p is the absolutely continuous probability with respect to the *standard Wiener measure*

($W_0 = 0$, $\mu - a.s.$), whose density is given by

$$\frac{d\mu_p}{d\mu} := \frac{\exp\left(\int_0^1 p(1-s, W_s)ds\right)}{Z_p},$$

with Z_p a normalization constant. Whence we obtain the following ersatz of Proposition 2:

Proposition 3 *A time-dependent vector field $u \in C^{1,2}_{b,div}([0,1] \times \mathbb{R}^d)$ is a solution of the Eq. (4) if and only if*

$$\delta \mathscr{H}(.|\mu_p)_{P_u}[h] = 0$$

for all $(\mathscr{F}_t^\nu)$-adapted process (h_t) of finite energy, such that

$$h_0 = h_1 = 0 \;\; P_u - a.s..$$

3 Invariances and the Stochastic Noether Theorem

Within this section we consider the case $d = 3$ and we denote by (e_1, e_2, e_3) the canonical orthogonal basis of $\mathbb{R}^3$. We further assume that $u \in C^{1,2}_{b,div}([0,1] \times \mathbb{R}^3)$ is a solution of (4) for a given smooth function p. By Proposition 1, the associated law of the continuous semi-martingale P_u (c.f. (5)) satisfies (2) for $\mathscr{L}^p$ defined by (6).

The next subsections investigate different *symmetries* and compute the related local *martingales*, stemming from the *weak Noether Theorem* 6.1. of [7]. In each particular case considered below *the symmetries are expressed through a condition on the pressure field p*. Given the associated family of transformations (h^ε), subsequently, the *symmetry condition* on p yields that (h^ε) is a smooth family of $\mathbb{S}$-invariant transformations for $\mathscr{L}^p$, in the sense considered in [7].

We recall this *symmetry condition* on $\mathbb{S}$. First, by setting

$$\Gamma^\varepsilon : \omega \in W \to \Gamma^\varepsilon(\omega) \in W,$$

where

$$\Gamma_t^\varepsilon(\omega) := h^\varepsilon(t, \omega(t)),$$

for all $t \in [0,1]$, $\omega \in W$, (h^ε) induces a family (Γ^ε) of transformations of W. Given $\eta \in \mathbb{S}$, for all ε, (Γ_t^ε) defines a stochastic process on the probability space $(W, \mathscr{B}(W)^\eta, \eta)$. Thus, by *Itô's formula* on the probability space $(W, \mathscr{B}(W)^\eta, \eta)$, for all $\varepsilon \in \mathbb{R}$, *the transformation h^ε of the state space $\mathbb{R}^3$ is lifted to a transformation*

$$\eta \in \mathbb{S} \to \Gamma_\star^\varepsilon \eta \in \mathbb{S}$$

of $\mathbb{S}$, by pushforward. The *symmetry condition* considered in [7] consists in the relation

$$\mathscr{L}_t^p(W_t, v_t^\eta, \alpha_t^\eta) = \mathscr{L}_t^p(\Gamma_t^\varepsilon, v_t^{\Gamma_\star^\varepsilon \eta} \circ \Gamma^\varepsilon, \alpha_t^{\Gamma_\star^\varepsilon \eta} \circ \Gamma^\varepsilon), \tag{10}$$

holding a.e., for all $\eta \in \mathbb{S}$ in the domain of the map defined in (7), and for all $\varepsilon \in \mathbb{R}$. Here $\circ$ denotes the pullback of the ($\Gamma_\star^\varepsilon \eta$—equivalence class of) map(s) $v_t^{\Gamma_\star^\varepsilon \eta} : W \to \mathbb{R}^3$ with the (η—equivalence class of) map(s) $\Gamma^\varepsilon : W \to W$.

Consider a smooth Lagrangian $\mathscr{L}$ and assume that $\nu \in \mathbb{S}$ satisfies the weak Euler-Lagrange condition for this Lagrangian. The stochastic weak Noether's Theorem in [7] associates to a family (h_ε) of $\mathbb{S}$-invariant transformations of $\mathscr{L}$ local martingales on the probability space $(W, \mathscr{B}(W)^\nu, \nu)$. These local martingales, that we denote by $(\mathscr{I}_t)_{t\in[0,1)}$, are in fact explicitly given:

$$\mathscr{I}_t := < \frac{d}{d\varepsilon}|_{\varepsilon=0} h_t^\varepsilon(W_t), \phi_t^\nu >_{\mathbb{R}^d} - \sum_i \left[\frac{d}{d\varepsilon}|_{\varepsilon=0} h_\cdot^\varepsilon(W_\cdot)^i, \phi_\cdot^{\nu i}\right]_t + \int_0^t \theta_s ds \tag{11}$$

where $[.,.]$ stands for the quadratic co-variation process of *càd-làg* semi-martingales, (ϕ_t^ν) denotes a *càdlàg* modification of the process $\partial_v \mathscr{L}_t(W_t, v_t^\nu, \alpha_t^\nu)$, and

$$\theta_s := \sum_{i,j} \kappa_s^{i,j} \frac{\partial \mathscr{L}}{\partial \alpha_{i,j}}(W_s, v_s^\nu, \alpha_s^\nu), \tag{12}$$

where $(\kappa_s(\omega))$ is the $\mathscr{M}_d(\mathbb{R})$-valued process defined by

$$\kappa_s(\omega) := \alpha_s^\nu . \left(\left(\nabla \frac{d}{d\varepsilon} h^\varepsilon|_{\varepsilon=0}\right)(s, W_s)\right)^\dagger + \left(\left(\nabla \frac{d}{d\varepsilon} h^\varepsilon|_{\varepsilon=0}\right)(s, W_s)\right)\alpha_s^\nu . \tag{13}$$

3.1 Symmetry by Translation and the Momentum Process

Assume the *symmetry by translation* along e_3 of the pressure, that is

$$p(t, x + ae_3) = p(t, x),$$

for all $a \in \mathbb{R}, t \in [0,1], x \in \mathbb{R}^3$. To check that Noether's theorem yields the expected result, set

$$h^\varepsilon : (t, x) \in [0,1] \times \mathbb{R}^3 \to h^\varepsilon(t, x) := x + \varepsilon e_3 \in \mathbb{R}^3.$$

By proposition 3.2. of [7], (10) is trivially satisfied, so that, by Theorem 6.1 of [7], we obtain $(< v_t^{P_u}, e_z >)$, as the related $(\mathscr{F}_t^{P_u})$-*local martingale*.

3.2 Symmetry by Rotation and the Kinetic Momentum Process

Assume the *symmetry by rotation* along the axis e_3 of the pressure; that is,

$$p(t, R^\varepsilon x) = p(t, x), \tag{14}$$

for all $(t, x) \in [0, 1] \times \mathbb{R}^3$ and $\varepsilon \in \mathbb{R}$, where $R^\varepsilon : \mathbb{R}^3 \to \mathbb{R}^3$ denotes the operator of rotation, along the axis e_3, with angle ε. We consider the family of space transformations

$$h^\varepsilon : (t, x) \in [0, 1] \times \mathbb{R}^3 \to h^\varepsilon(t, x) := R^\varepsilon x \in \mathbb{R}^3. \tag{15}$$

Applying Lemma 3.2. of [7], with h^ε given by (15), we compute the characteristics of $\Gamma^\varepsilon_\star \eta$ and we obtain the relation

$$|\nu^{\Gamma^\varepsilon_\star \eta} \circ \Gamma^\varepsilon|_{\mathbb{R}^3} = |\nu^\eta|_{\mathbb{R}^3} \quad \lambda \otimes \eta - a.e.,$$

for any $\varepsilon \in \mathbb{R}$. Whence, the *symmetry condition* (10) is satisfied, from (14). Define the stochastic process (l_t), on the complete probability space $(W, \mathscr{B}(W)^{P_u}, P_u)$, by

$$l_t := < W_t, e_1 >_{\mathbb{R}^3} < v_t^{P_u}, e_2 >_{\mathbb{R}^3} - < W_t, e_2 >_{\mathbb{R}^3} < v_t^{P_u}, e_1 >_{\mathbb{R}^3}$$

for $t \in [0, 1]$, the stochastic counterpart to the *kinetic momentum* along e_3. Further denoting by

$$rot\ u_t : \mathbb{R}^3 \to \mathbb{R}^3,$$

the rotational of $u(t, .)$, the weak Noether Theorem 6.1 of [7] implies that the corresponding process $(\mathscr{I}_t)$, defined by

$$\mathscr{I}_t := l_t + \int_0^t < rot\ u_{1-s}(W_s), e_3 > ds,$$

is the *local martingale* associated to this *symmetry by rotation*. The latter expresses the *dissipation*, modeled through the martingale part of P_u, which is involved in the covariation process in the expression of $\mathscr{I}_t$.

References

1. Bismut, J.-M.: Mécanique aléatoire. Lecture Notes in Mathematics. Springer, New York (1981)
2. Bossy, M., Jabir, J.F., Talay, D.: On conditional McKean Lagrangian stochastic models. Probab. Theory Relat. Fields **151**, 319–351 (2011)
3. Cruzeiro, A.B., Cipriano, F.: Navier-Stokes equation and diffusions on the group of homeomorphisms of the torus. Commun. Math. Phys. **275**, 255–269 (2007)

4. Cruzeiro, A.B., Lassalle, R.: On the least action principle for the Navier-Stokes equation. In: Stochastic Analysis and Applications 2014. Springer Proceedings in Mathematics and Statistics, vol. 100. Springer (2014)
5. Constantin, P.: An Eulerian-Lagrangian approach to the Navier-Stokes equations. Commun. Math. Phys. **216**(3), 663–686 (2001)
6. Constantin, P., Iyer, G.: A stochastic Lagrangian representation of the three-dimensional incompressible Navier-Stokes equations. Commun. Pure Appl. Math. Phys. **61**(3), 330–345 (2008)
7. Cruzeiro, A.B., Lassalle, R.: Weak calculus of variations for functionals of laws of semi-martingales (2015)
8. Cruzeiro, A.B., Shamarova, E.: Navier-Stokes equations and forward-backward SDEs on the group of diffeomorphisms of a torus. Stoch. Process. Appl. **119**(12), 4034–4060 (2009)
9. Gomes, D.A.: A variational formulation for the Navier-Stokes equation. Commun. Math. Phys. **257**(1), 227–234 (2005)
10. Ikeda, N., Watanabe, S.: Stochastic Differential Equations and Diffusion Processes. North Holland, Amsterdam (Kodansha Ltd., Tokyo) (1981)
11. Inoue, A., Funaki, T.: A new derivation of the Navier-Stokes equation. Commun. Math. Phys. **65**, 83–90 (1979)
12. Föllmer, H.: Random fields and diffusion processes. In: Ecole d' Été de Saint Flour XV–XVII (1988)
13. Lassalle, R., Zambrini, J.C.: A weak approach to the stochastic deformation of classical mechanics. J. Geom. Mech. (2016)
14. Leonard, C.: A survey of the Schrödinger problem and some of its connections with optimal transport. Discrete Continu. Dyn. Syst. A **34**(4), 1533 (2014)
15. Mikulevicius, R., Rozovskii, B.L.: Stochastic Navier-Stokes equations for turbulent flows. SIAM J. Math. Anal. **35**(5), 1250–1310 (2004)
16. Mikulevicius, R., Rozovskii, B.L.: On equations of stochastic fluid mechanics. In: Hida, T., Karandikar, R., Kunita, H., Rajput, P., Watanabe, S., Xiong, J. (eds.) Stochastics in Finite and Infinite Dimensions, pp. 285–302. Birkhauser Boston, Boston, MA, (2001)
17. Nakagomi, T., Yasue, K., Zambrini, J.C.: Stochastic variational derivations of the Navier-Stokes equation. Lett. Math. Phys. **5**(6), 545–552 (1981)
18. Tan, X., Touzi, N.: Optimal transportation under controlled stochastic dynamics. Ann. Prob. (2012)
19. Thieullen, M., Zambrini, J.C.: Probability and quantum symmetries I, the theorem of Noether in Schrödinger's Euclidean quantum mechanics. Ann. Inst. H.Poincaré, Phys. Theo. **67**(3), 297 (1997)
20. Yasue, K.: A variational principle for the Navier-Stokes equation. J. Funct. Anal. **51**(2), 133–141 (1983)
21. Zambrini, J.C.: Stochastic mechanics according to E. Schrödinger. Phys. Rev. A. **33**(3), 1532–1548 (1986)

Convergence of Diffusion-Drift Many Particle Systems in Probability Under a Sobolev Norm

Jian-Guo Liu and Yuan Zhang

Abstract In this paper we develop a new martingale method to show the convergence of the regularized empirical measure of many particle systems in probability under a Sobolev norm to the corresponding mean field PDE. Our method works well for the simple case of Fokker Planck equation and we can estimate a lower bound of the rate of convergence. This method can be generalized to more complicated systems with interactions.

Keywords Many particle system · Martingale method · Energy-dissipation inequality

MSC 2010 subject classifications: 35Q70 · 65M75

1 Introduction

We are considering the stochastic processes $\{X_i(t)\}_{i=1}^N$ in $\mathbb{R}^d$ of the following SDE

$$dX_i(t) = \mathbf{F}(X_i(t), t)\, dt + dB_i(t), \quad i = 1, 2, \ldots, N \tag{1}$$

with an initial condition $X_i(0)$ and a sequence of independent d-dimensional standard Brownian motions $\{B_i(t)\}_{i=1}^N$. We show the system converges to the following Fokker-Planck equation

J.-G. Liu (✉)
Department of Physics and Department of Mathematics,
Duke University, 90320, Durham, NC 27708-0320, USA
e-mail: jliu@math.duke.edu

Y. Zhang
Department of Mathematics, University of California, Los Angeles,
951555, Los Angeles, CA 90095-1555, USA
e-mail: yuanzhang@math.ucla.edu

© Springer International Publishing Switzerland 2016

P. Gonçalves and A.J. Soares (eds.), *From Particle Systems to Partial Differential Equations III*, Springer Proceedings in Mathematics & Statistics 162,
DOI 10.1007/978-3-319-32144-8_10

$$\frac{\partial \rho}{\partial t}(x,t) = \frac{1}{2}\Delta\rho(x,t) - \nabla\cdot(\rho(x,t)\mathbf{F}(x,t)) \tag{2}$$

with boundary condition $\rho(x,0) \in L^2(R^d)$, in the $L^\infty(L^2)\cap L^2(H^1)$ norm and in probability, and we estimate the convergence rate. The motivation of developing this kind of estimates is hoping that it can be adapted in the analysis of the propagation of chaos and the mean field limit for some interacting many particle systems

$$dX_i(t) = \frac{1}{N}\sum_{j\neq i}^{N}\mathbf{F}_0\big(X_i(t) - X_j(t)\big)dt + \sigma dB_t^i, \quad i = 1,\dots,N. \tag{3}$$

For the simplified model (1) and (2), we show in Theorem 1 that the regularized empirical measure:

$$\rho_{\varepsilon,N}(x,t) = \frac{1}{N}\sum_{i=1}^{N}\varphi_\varepsilon\left(x - X_i(t)\right), \quad \varphi_\varepsilon(x) = \frac{1}{\varepsilon^d}\varphi(\frac{x}{\varepsilon}), \quad \varepsilon = N^{-1/3d} \tag{4}$$

has a rate of convergence to the mean field solution ρ in the Sobolev space $L^\infty\big(0,T;L^2(\mathbb{R}^d)\big)\cap L^2\big(0,T;H^1(\mathbb{R}^d)\big)$, for any $T>0$,

$$\begin{aligned} P\Big(\sup_{s\in[0,T]}\big(\|(\rho-\rho_{\varepsilon,N})(\cdot,s)\|^2 + \int_0^T \|\nabla(\rho-\rho_{\varepsilon,N})(\cdot,s)\|^2\,ds\big) \\ \geq C_T\big(\|(\rho-\rho_{\varepsilon,N})(\cdot,0)\|^2 + N^{-1/6d}\big)\Big) \leq C_T N^{-1/6d} \end{aligned} \tag{5}$$

for some function C_T that depends only on T, $||\rho_0||_{L^2}$ and the Lipschitz constant of $\mathbf{F}$. Thus the free energy-dissipation inequality for the difference between the regularized empirical measure $\rho_{\varepsilon,N}$ and the mean field density ρ holds with high probability. The second term in (5) is the dissipation term and it is contributed from the Brownian motion. In the usual coupling method, the contribution of the Brownian motion is removed in the estimation and hence the dissipation term is lost. When the interaction kernel $\mathbf{F}(x)$ is Lipschitz continuous, the propagation of chaos can be directly justified by the McKean's coupling method [13, 17]. In some physically important systems such as two dimensional Navier-Stokes equation [6, 7, 11, 12, 15] and the Keller-Segel equation [10, 16], the interaction is given by the gradient or the curl of the Newtonian potential. The free energy-dissipation inequality plays a curial role in analyze such kind of systems, particularly, mathematical justifications of the propagation of chaos and mean field limit.

Recently, we have shown that the new martingale method developed in this paper can be generalized to prove the convergence under first order Sobolev norm when interactions are introduced in the particle system. Let $\{X_i(t)\}_{i=1}^N$ be the N-particles system defined in (3), where $\mathbf{F}_0$ is a bounded function and Lipschitz continuous against x with Lipschitz constant L_F for all $t\geq 0$, and the deterministic PDE model be as follows:

$$\begin{cases} \dfrac{\partial \rho}{\partial t}(x,t) = \dfrac{1}{2}\Delta\rho - \nabla \cdot (\rho \mathbf{F}(x,t)) \\ \mathbf{F}(x,t) = \displaystyle\int_{R^d} \mathbf{F}_0(x-y)\rho(y,t)dy. \end{cases} \tag{6}$$

To show the regularized empirical measure converges to the solution of the PDE above, we first introduce an intermediate self-consistent system $\{\hat{X}_i(t)\}_{i=1}^N$, which is defined by the following SDE

$$d\hat{X}_i(t) = \mathbf{F}\left(\hat{X}_i(t), t\right) dt + dB_i(t), \quad i = 1, 2, \ldots, N \tag{7}$$

where

$$\mathbf{F}(x,t) = \int_{R^d} \mathbf{F}_0(x-y)\rho(y,t)dy. \tag{8}$$

According to previous study, we are able to control the distance between $\hat{X}_i(t)$ and $X_i(t)$. And we can use the similar martingale method as in this paper to control the distance between the self-consistent system $\{\hat{X}_i(t)\}_{i=1}^N$ and the deterministic PDE model. The details of the proof will be presented in a separate paper.

The use of the regularized empirical measure is important in computation and the regularized kernel φ is known as a blob function in the vortex method. Pioneered by Chorin in 1973 [2], the random vertex blob method is one of the most successful computational methods for fluid dynamics and other related fields. The success of the method is exemplified by the accurate computation of flow past a cylinder at the Reynolds numbers up to 9500 in the 1990s [9]. The convergence analysis for the random vortex method for the Navier-Stokes equation is given by [7, 11, 12] in the 1980s. We refer to the book [3] for theoretical and practical use of vortex methods, refer to Goodman [7] and Long [11] for the convergence analysis of the random vortex method to the Navier-Stokes equation. We also hoped that the estimation (5) can be adapted to do numerical analysis.

2 Convergence in $L^\infty(L^2) \cap L^2(H^1)$ Norm

As described in Introduction, in this section we will show that the regularized empirical measure of the many particle system defined in (1) converge to the solution of the Fokker Planck equation (2) in probability under $L^\infty(L^2) \cap L^2(H^1)$ norm. In this paper, we will denote the $L^2(\mathbb{R}^d)$ norm as $\|\cdot\|$. To show this, we need to make some assumption of the initial state to make sure that the particles are not too close to each other. In the following work, we show that our convergence result holds under either of the following 2 assumptions:

Assumption 1 There is some constant $C < \infty$ independent to N, such that for any N and $i \neq j, m, n \leq N$, and $\delta_{i,j} = X_i(0) - X_j(0)$, we always have $|\delta_{i,j}| \geq 2CN^{-1/d}$.

Assumption 2 $(X_1(0), \ldots, X_N(0))$ has a joint distribution such that there is some constant $C < \infty$ independent to N, such that for any N and $i \neq j, i, j \leq N$, $\delta_{i,j}$ has a density function $f_{i,j,N}$ and it satisfies $\|f_{i,j,N}\|_\infty \leq C$.

Remark A special case of Assumption 2 is that $X_1(0), X_2(0), \ldots, X_N(0)$ are i.i.d. with density $\rho(x, 0)$, since that for any i, j, $X_i(0) - X_j(0)$ has density function

$$f_{i,j,N}(a) = \int_{\mathbb{R}^d} \rho(x, 0)\rho(x + a, 0)\, dx \leq \|\rho(x, 0)\|^2 < \infty.$$

For any N, we consider the regularized empirical measure:

$$\rho_{\varepsilon,N}(x, t) = \frac{1}{N} \sum_{i=1}^{N} \varphi_\varepsilon (x - X_i(t)). \tag{9}$$

To show that $\rho_{\varepsilon,N}(x, t)$ converges to ρ, we have theorem as follows:

Theorem 1 *Let $X_i(t), i = 1, \ldots, N$ be solutions of stochastic differential equation (1) with initial data $X_i(0)$ satisfying either Assumption 1 or Assumption 2 and $\rho_{\varepsilon,N}$ be the constructed regularized empirical measure (9) with regularized parameter $\varepsilon_N = N^{-1/3d}$. Let ρ be the solution of the corresponding mean field equation (2) with initial density $\rho_0 \in L^2(\mathbb{R}^d)$. Then, there is a positive function $c(t), t > 0$ (will be specified in (83)) dependent only on t, φ, and $||\rho_0||$, such that*

$$P\Big(\sup_{s\in[0,t]} \big(\|(\rho - \rho_{\varepsilon_N,N})(\cdot, s)\|^2 + \int_0^s \|\nabla(\rho - \rho_{\varepsilon_N,N})(\cdot, h)\|^2\, dh\big) \\ < 2e^{C_0 t}\big(\|(\rho - \rho_{\varepsilon_N,N})(\cdot, 0)\|^2 + c(t)\, N^{-1/6d}\big)\Big) \geq 1 - c(t)\, N^{-1/6d}. \tag{10}$$

where $C_0 = 2dL_F$ and L_F is the Lipschitz constant of $\mathbf{F}$.

The proof of this theorem is divided into one proposition (Proposition 1) and four lemmas (Lemmas 1–4). We will give the proof of this theorem after the proof of these preliminary results.

Proposition 1 *For the difference between the PDE density ρ and the empirical measure $\rho_{\varepsilon,N}$, we have*

$$\|(\rho - \rho_{\varepsilon,N})(\cdot, t)\|^2 = \|(\rho - \rho_{\varepsilon,N})(\cdot, 0)\|^2 - \int_0^t \|\nabla(\rho - \rho_{\varepsilon,N})(\cdot, s)\|^2\, ds \\ - \int_0^t \int_{\mathbb{R}^d} \nabla \cdot \mathbf{F}(x, s) \left((\rho - \rho_{\varepsilon,N})(x, s)\right)^2 dxds \\ + Res(t) + \tilde{M}_t + M_t + \frac{t}{N}\|\nabla\varphi_\varepsilon\|_2^2 \tag{11}$$

where M_t is defined by $M_t = \sum_{i=1}^N M_t^i$ with

$$M_t^i = \frac{2}{N}\int_0^t \int_{\mathbb{R}^d} \rho(x,s)\nabla\varphi_\varepsilon(x - X_i(s))\,dx\cdot dB_i(s), \tag{12}$$

and

$$\tilde{M}_t = \sum_{n=1}^N \tilde{M}_t^i \tag{13}$$

with $\tilde{M}_t^i$ equals to

$$\frac{2}{N^2}\int_0^t \int_{\mathbb{R}^d} \varphi_\varepsilon(x)\left(\sum_{j=1}^{i-1}\nabla\varphi_\varepsilon\left(x + B_i(s) - B_j(s)\right) - \sum_{j=i+1}^{N}\nabla\varphi_\varepsilon\left(x + B_j(s) - B_i(s)\right)\right) dx\cdot dB_i(s) \tag{14}$$

and

$$Res(t) = 2\int_0^t \int_{\mathbb{R}^d} \frac{1}{N}\sum_{i=1}^N \varphi_\varepsilon(x - X_i(s))\left(\mathbf{F}(x,s) - \mathbf{F}(X_i(s),s)\right)\cdot\nabla(\rho - \rho_{\varepsilon,N})(x,s)\,dxds. \tag{15}$$

Proof To prove the proposition, first note that for any ε and N,

$$\|(\rho - \rho_{\varepsilon,N})(\cdot,t)\|^2 = \|\rho(\cdot,t)\|^2 - 2\int_{\mathbb{R}^d}\rho(x,t)\rho_{\varepsilon,N}(x,t)\,dx + \|\rho_{\varepsilon,N}(\cdot,t)\|^2.$$

First for the deterministic part of $\|\rho(\cdot,t)\|^2$, we have

$$\begin{aligned}\int_{\mathbb{R}^d}\rho(x,t)^2\,dx &= \int_{\mathbb{R}^d}\rho(x,0)^2\,dx + \int_0^t\int_{\mathbb{R}^d}\rho(x,s)\left(\Delta\rho(x,s) - 2\nabla\cdot(\rho\mathbf{F})(x,s)\right)\,dxds\\ &= \int_{\mathbb{R}^d}\rho(x,0)^2\,dx - \int_0^t \|\nabla\rho\|^2\,ds + 2\int_0^t\int_{\mathbb{R}^d}\rho(x,s)\mathbf{F}(x,s)\cdot\nabla\rho(x,s)\,dxds.\end{aligned} \tag{16}$$

Then for the second part which equals to

$$-\frac{2}{N}\int_{\mathbb{R}^d}\rho(x,t)\sum_{i=1}^N\varphi_\varepsilon(x - X_i(t))\,dx,$$

note that for each i, by Ito's formula, we have

$$
\begin{aligned}
\rho(x,t)\varphi_\varepsilon(x-X_i(t)) = \rho(x,0)\varphi_\varepsilon(x-X_n(0)) &+ \int_0^t \frac{\partial\rho(x,s)}{\partial t}\varphi_\varepsilon(x-X_i(s))\,ds \\
&- \int_0^t \rho(x,s)\nabla\varphi_\varepsilon(x-X_i(s))\cdot\mathbf{F}(X_i(s),s)\,ds \\
&- \int_0^t \rho(x,s)\nabla\varphi_\varepsilon(x-X_i(s))\cdot dB_i(s) \\
&+ \frac{1}{2}\int_0^t \rho(x,s)\Delta\varphi_\varepsilon(x-X_i(s))\,ds.
\end{aligned}
\tag{17}
$$

Note that for the second term of the sum above, according to the definition of the Fokker-Planck's PDE,

$$
\int_0^t \frac{\partial\rho(x,s)}{\partial t}\varphi_\varepsilon(x-X_i(s))\,ds = \int_0^t \left(\frac{1}{2}\Delta\rho(x,s) - \nabla\cdot\big(\rho(x,s)\mathbf{F}(x,s)\big)\right)\varphi_\varepsilon(x-X_i(s))\,ds.
$$

Then integrate it over $x \in R^d$, we have

$$
\begin{aligned}
&\int_0^t\int_{\mathbb{R}^d} \frac{\partial\rho(x,s)}{\partial t}\varphi_\varepsilon(x-X_i(s))\,dxds \\
&\qquad = -\frac{1}{2}\int_0^t\int_{\mathbb{R}^d} \nabla\rho(x,s)\cdot\nabla\varphi_\varepsilon(x-X_i(s))\,dxds \\
&\qquad\quad + \int_0^t\int_{\mathbb{R}^d} \rho(x,s)\mathbf{F}(x,s)\cdot\nabla\varphi_\varepsilon(x-X_i(s))\,dxds.
\end{aligned}
\tag{18}
$$

□

Remark Here we can also apply Ito's formula on the integration itself and it is more rigorous since we do not need to change the order of integrations. However, since all the calculations are the same, we will use the followings notations for simplicity.

Then integrating the third term in (17) over $x \in R^d$ we have by divergence theorem that

$$
\begin{aligned}
&-\int_0^t\int_{\mathbb{R}^d} \rho(x,s)\nabla\varphi_\varepsilon(x-X_i(s))\cdot\mathbf{F}(X_i(s),s)\,dxds \\
&\qquad = \int_0^t\int_{\mathbb{R}^d} \varphi_\varepsilon(x-X_i(s))\,\mathbf{F}(X_i(s),s)\cdot\nabla\rho(x,s)\,dxds.
\end{aligned}
\tag{19}
$$

Combining (17), (18) and (19) we have

$$\begin{aligned}\int_{\mathbb{R}^d} \rho(x,t)\varphi_\varepsilon(x - X_i(t))\,dx \\ &= \int_{\mathbb{R}^d} \rho(x,0)\varphi_\varepsilon(x - X_i(0))\,dx \\ &\quad - \int_0^t \int_{\mathbb{R}^d} \nabla\rho(x,s)\cdot\nabla\varphi_\varepsilon(x - X_i(s))\,dxds \\ &\quad + \int_0^t \int_{\mathbb{R}^d} \rho(x,s)\mathbf{F}(x,s)\cdot\nabla\varphi_\varepsilon(x - X_i(s))\,dxds \\ &\quad + \int_0^t \int_{\mathbb{R}^d} \varphi_\varepsilon(x - X_i(s))\,\mathbf{F}(X_i(s),s)\cdot\nabla\rho(x,s)dxds - \frac{N}{2}M_t^i\end{aligned} \tag{20}$$

where M_t^i is

$$M_t^i = \frac{2}{N}\int_0^t \int_{\mathbb{R}^d} \rho(x,s)\nabla\varphi_\varepsilon(x - X_i(s))\,dx\cdot dB_i(s).$$

Summing up over $i = 1, 2, \ldots, N$ we have

$$\begin{aligned}-2\int_{\mathbb{R}^d} \rho_{\varepsilon,N}(x,t)\rho(x,t)\,dx \\ &= -2\int_{\mathbb{R}^d} \rho_{\varepsilon,N}(x,0)\rho(x,0)\,dx + 2\int_0^t \int_{\mathbb{R}^d} \nabla\rho(x,s)\cdot\nabla\rho_{\varepsilon,N}(x,s)\,dx\,ds \\ &\quad -2\int_0^t \int_{\mathbb{R}^d} \rho(x,s)\mathbf{F}(x,s)\cdot\nabla\rho_{\varepsilon,N}(x,s)\,dx\,ds \\ &\quad -2\int_0^t \int_{\mathbb{R}^d} \frac{1}{N}\sum_{i=1}^N \varphi_\varepsilon(x - X_i(s))\,\mathbf{F}(X_i(s),s)\cdot\nabla\rho(x,s)\,dx\,ds \\ &\quad + M_t\end{aligned} \tag{21}$$

where M_t is the first martingale term in (11) defined in (12). I.e.,

$$M_t = 2\int_0^t \int_{\mathbb{R}^d} \rho(x,s)\frac{1}{N}\sum_{i=1}^N \nabla\varphi_\varepsilon(x - X_i(s))\,dx\cdot dB_i(s).$$

Lastly, for the part of $\int_{\mathbb{R}^d} \rho_{\varepsilon,N}(x,t)\rho_{\varepsilon,N}(x,t)\,dx$ which equals to

$$\begin{aligned}\frac{1}{N^2}\sum_{i=1}^N \int_{\mathbb{R}^d} \varphi_\varepsilon(x - X_i(t))^2\,dx + \frac{2}{N^2}\sum_{i<j}\int_{\mathbb{R}^d} \varphi_\varepsilon\left(x - X_j(t)\right)\varphi_\varepsilon(x - X_i(t))\,dx \\ = \frac{1}{N}\|\varphi_\varepsilon\|_2^2 + \frac{2}{N^2}\sum_{i<j}\int_{\mathbb{R}^d} \varphi_\varepsilon\left(x - X_j(t)\right)\varphi_\varepsilon(x - X_i(t))\,dx.\end{aligned} \tag{22}$$

And for each $i < j$,

$$\int_{\mathbb{R}^d} \varphi_\varepsilon\left(x - X_j(t)\right) \varphi_\varepsilon\left(x - X_i(t)\right) dx = \int_{\mathbb{R}^d} \varphi_\varepsilon(x)\, \varphi_\varepsilon\left(x + X_j(t) - X_i(t)\right) dx.$$

Then we can again apply the Ito's formula on $\varphi_\varepsilon\left(x + X_j(t) - X_i(t)\right)$:

$$\begin{aligned}\varphi_\varepsilon\big(x + X_j(0) - X_i(0)\big) &+ \int_0^t \Delta\varphi_\varepsilon\left(x + X_j(s) - X_i(s)\right) ds \\ &+ \int_0^t \nabla\varphi_\varepsilon\left(x + X_j(s) - X_i(s)\right) \cdot \left(\mathbf{F}\left(X_j(s), s\right) - \mathbf{F}\left(X_i(s), s\right)\right) ds \\ &+ \int_0^t \nabla\varphi_\varepsilon\left(x + X_j(s) - X_i(s)\right) \cdot \left(dB_j(t) - dB_i(t)\right).\end{aligned} \tag{23}$$

Integrating the first and second terms over x, we have

$$\int_{\mathbb{R}^d} \varphi_\varepsilon(x)\varphi_\varepsilon\left(x + X_j(0) - X_i(0)\right) dx = \int_{\mathbb{R}^d} \varphi_\varepsilon\left(x - X_j(0)\right) \varphi_\varepsilon\left(x - X_i(0)\right) dx$$

and

$$\int_0^t \int_{\mathbb{R}^d} \varphi_\varepsilon(x)\, \Delta\varphi_\varepsilon\left(x + X_j(s) - X_i(s)\right) dxds = -\int_0^t \int_{\mathbb{R}^d} \nabla\varphi_\varepsilon\left(x - X_j(s)\right) \cdot \nabla\varphi_\varepsilon\left(x - X_i(s)\right) dxds.$$

Moreover for the third term, we have

$$\begin{aligned}&\int_0^t \int_{\mathbb{R}^d} \varphi_\varepsilon(x)\nabla\varphi_\varepsilon\left(x + X_j(s) - X_i(s)\right) \cdot \mathbf{F}\left(X_j(s), s\right) dxdt \\ &\qquad = \int_0^t \int_{\mathbb{R}^d} \varphi_\varepsilon\left(x - X_j(s)\right) F\left(X_j(s), s\right) \cdot \nabla\varphi_\varepsilon\left(x - X_i(s)\right) dxds\end{aligned}$$

and

$$\begin{aligned}&\int_0^t \int_{\mathbb{R}^d} \varphi_\varepsilon(x)\nabla\varphi_\varepsilon\left(x + X_j(s) - X_i(s)\right) \cdot \mathbf{F}\left(X_i(s), s\right) dxdt \\ &\qquad = -\int_0^t \int_{\mathbb{R}^d} \varphi_\varepsilon\left(x - X_i(s)\right) F\left(X_i(s), s\right) \cdot \nabla\varphi_\varepsilon\left(x - X_j(s)\right) dxds.\end{aligned}$$

So after we sum up over all m, n, we have

$$
\begin{aligned}
\|\rho_{\varepsilon,N}(\cdot,t)\|^2 =& \|\rho_{\varepsilon,N}(\cdot,0)\|^2 - \int_0^t \|\nabla \rho_{\varepsilon,N}(\cdot,s)\|^2 ds \\
&+ 2\int_0^t \int_{\mathbb{R}^d} \left[\frac{1}{N}\sum_{i=1}^N \varphi_\varepsilon\left(x - X_i(s)\right) F\left(X_i(s), s\right)\right] \cdot \nabla \rho_{\varepsilon,N}(x,s)\, dxds \\
&+ \tilde{M}_t + \frac{t}{N}\|\nabla\varphi_\varepsilon\|_2^2
\end{aligned}
\tag{24}
$$

where $\tilde{M}_t$ is the second martingale term in (11), which is defined in (13) and (14). I.e.,

$$
\tilde{M}_t = \sum_{i=1}^N \tilde{M}_t^i
$$

with $\tilde{M}_t^i$ equals to

$$
\frac{2}{N^2}\int_0^t \int_{\mathbb{R}^d} \varphi_\varepsilon(x) \left(\sum_{j=1}^{i-1} \nabla\varphi_\varepsilon\left(x + B_i(s) - B_j(s)\right) - \sum_{j=i+1}^{N} \nabla\varphi_\varepsilon\left(x + B_j(s) - B_i(s)\right)\right) dx \cdot dB_i(s).
$$

Then combine (16), (21) and (24) we have

$$
\begin{aligned}
\|(\rho-\rho_{\varepsilon,N})(\cdot,t)\|^2 =& \|(\rho-\rho_{\varepsilon,N})(\cdot,0)\|^2 - \int_0^t \|\nabla(\rho-\rho_{\varepsilon,N})(\cdot,s)\|^2 ds \\
&+ 2\int_0^t \int_{\mathbb{R}^d} \left(\rho(x,s)\mathbf{F}(x,s) - \frac{1}{N}\sum_{i=1}^N \varphi_\varepsilon\left(x - X_i(s)\right) \mathbf{F}\left(X_i(s), s\right)\right) \cdot \nabla(\rho-\rho_{\varepsilon,N})(x,s)\, dxds \\
&+ \tilde{M}_t + M_t + \frac{t}{N}\|\nabla\varphi_\varepsilon\|_2^2.
\end{aligned}
\tag{25}
$$

Then plus and minus the term of $-\frac{1}{N}\sum_{i=1}^N \varphi_\varepsilon\left(x - X_i(s)\right) \mathbf{F}\left(x,s\right) \cdot \nabla(\rho-\rho_{\varepsilon,N})(x,s)$ we have

$$
\begin{aligned}
\|(\rho-\rho_{\varepsilon,N})(\cdot,t)\|^2 =& \|(\rho-\rho_{\varepsilon,N})(\cdot,0)\|^2 - \int_0^t \|\nabla(\rho-\rho_{\varepsilon,N})(\cdot,s)\|^2 ds \\
&+ 2\int_0^t \int_{\mathbb{R}^d} (\rho-\rho_{\varepsilon,N})(x,s)\mathbf{F}(x,s) \cdot \nabla(\rho-\rho_{\varepsilon,N})(x,s)\, dxds \\
&+ 2\int_0^t \int_{\mathbb{R}^d} \frac{1}{N}\sum_{i=1}^N \varphi_\varepsilon\left(x - X_i(s)\right) \left(\mathbf{F}(x,s) - \mathbf{F}\left(X_i(s), s\right)\right) \cdot \nabla(\rho-\rho_{\varepsilon,N})(x,s)\, dxds \\
&+ \tilde{M}_t + M_t + \frac{t}{N}\|\nabla\varphi_\varepsilon\|_2^2.
\end{aligned}
\tag{26}
$$

Applying Green's identity on the third term of Eq. (26) and recalling the definition of Res(t) in (15) we have

$$\begin{aligned}\|(\rho-\rho_{\varepsilon,N})(\cdot,t)\|^2 &= \|(\rho-\rho_{\varepsilon,N})(\cdot,0)\|^2 - \int_0^t \|\nabla(\rho-\rho_{\varepsilon,N})(\cdot,s)\|^2\,ds \\ &\quad - \int_0^t\int_{\mathbb{R}^d} \nabla\cdot\mathbf{F}(x,s)\left((\rho-\rho_{\varepsilon,N})(x,s)\right)^2\,dxds \\ &\quad + \mathrm{Res}(t) + \tilde{M}_t + M_t + \frac{t}{N}\|\nabla\varphi_\varepsilon\|_2^2.\end{aligned} \tag{27}$$

Thus, the proof of Proposition 1 is complete.

Lemma 1 *For all $t \geq 0$, we have the second moment control*

$$E\left(M_t^2\right) \leq \frac{4}{N\varepsilon^{d+2}}\|\nabla\varphi\|^2 \int_0^t \|\rho(x,s)\|^2 ds. \tag{28}$$

Proof Here and in Lemma 2, we will use the natural filtration $\mathscr{F}_t^N$, which is generated by the Brownian motions $B_1(t), \dots B_N(t)$. Note that $M_t = \sum_{i=1}^N M_t^i$ where

$$M_t^i = \frac{2}{N}\int_0^t\int_{\mathbb{R}^d} \rho(x,s)\nabla\varphi_\varepsilon\left(x - X_i(s)\right)dx \cdot dB_i(s) = \sum_{k=1}^d M_t^{i,k},$$

and

$$M_t^{i,k} = \frac{2}{N}\int_0^t\int_{\mathbb{R}^d} \rho(x,s)\frac{\partial\varphi_\varepsilon\left(x - X_i(s)\right)}{\partial x_k}dx\, dB_i^{(k)}(s).$$

The $B_i^{(k)}(s)$ in the equation above is the kth coordinate of the Brownian motion $B_i(t)$ and it is itself a one dimension Brownian motion and a square integrable martingale under filtration $\mathscr{F}_t^N$ noting that $B_i^{(k)}(s)$ is independent to $B_j^{(h)}(s)$ for all $h \neq k$, or $i \neq j$. For each i and k we have the integrand

$$Y_{i,k}(s) = \int_{\mathbb{R}^d} \rho(x,s)\frac{\partial\varphi_\varepsilon\left(x - X_i(s)\right)}{\partial x_k}dx$$

continuous and adapted to filtration $\mathscr{F}_s^N$. Moreover

$$|Y_{i,k}(s)| = \left|\int_{\mathbb{R}^d} \rho(x,s)\frac{\partial\varphi_\varepsilon\left(x - X_i(s)\right)}{\partial x_k}dx\right| \leq \left\|\frac{\partial\varphi_\varepsilon}{\partial x_k}\right\| \times \|\rho(\cdot,s)\| < \infty.$$

Thus by Theorem 5.2.3 in [5], for all $i \in \{1, 2, \dots, N\}$ and $k \in \{1, 2, \dots, d\}$, $M_t^{i,k}$ is a square integrable martingale with

$$E[(M_t^{i,k})^2] = \frac{4}{N^2}E\left(\int_0^t Y_{i,k}(s)^2 ds\right) \le \frac{4}{N^2}\left\|\frac{\partial \varphi_\varepsilon}{\partial x_k}\right\|^2 \int_0^t \|\rho(\cdot, s)\|^2 ds. \tag{29}$$

And for all $(i, k) \neq (j, h)$ we have that

$$\begin{aligned}\langle M_t^{i,k}, M_t^{j,h}\rangle &= \langle Y_{i,k} \cdot B_i^{(k)}(t), Y_{i,h} \cdot B_j^{(h)}(t)\rangle \\ &= \int_0^t Y_{i,k}(s) \cdot Y_{j,h}(s)\, d\langle B_i^{(k)}(t), B_j^{(h)}(t)\rangle \\ &= \int_0^t Y_{i,k}(s) \cdot Y_{j,h}(s)\, d0 = 0,\end{aligned} \tag{30}$$

since $\langle B_i^{(k)}(t), B_j^{(h)}(t)\rangle \equiv 0$ for two independent Brownian motions, where $\langle X_t, Y_t\rangle$ is the quadratic covariance between the two processes X_t and Y_t, defined by

$$\langle X_t, Y_t\rangle = \frac{1}{2}(\langle X_t + Y_t\rangle - \langle X_t\rangle - \langle Y_t\rangle).$$

Noting that $M_t^{i,k}$ and $M_t^{j,h}$ are both square integrable martingales, (30) implies that

$$E\left(M_t^{i,k} M_t^{j,h}\right) \equiv 0. \tag{31}$$

Combining (29) and (31) immediately gives us

$$E[(M_t^i)^2] = \sum_{k=1}^d E[(M_t^{i,k})^2] \le \frac{4}{N^2}\|\nabla \varphi_\varepsilon\|^2 \int_0^t \|\rho(\cdot, s)\|^2\, ds.$$

and

$$E(M_t^i M_t^j) = 0$$

which implies that

$$E\big((M_t)^2\big) = \sum_{i=1}^N E[(M_t^i)^2] \le \frac{4}{N}\|\nabla \varphi_\varepsilon\|^2 \int_0^t \|\rho(\cdot, s)\|^2\, ds = \frac{4}{N\varepsilon^{d+2}}\|\nabla \varphi\|^2 \int_0^t \|\rho(\cdot, s)\|^2\, ds. \tag{32}$$

□

Lemma 2 *For all $t \ge 0$, we have the second moment control*

$$E\big((\tilde{M}_t)^2\big) \le \frac{4}{N\varepsilon^{2d+2}}\|\varphi\|^2 \|\nabla \varphi\|^2 t. \tag{33}$$

Proof Again note that $\tilde{M}_t = \sum_{i=1}^N \tilde{M}_t^i$ with

$$\tilde{M}_t^i = \sum_{k=1}^d \tilde{M}_t^{i,k}$$

where

$$\tilde{M}_t^{i,k} = \int_0^t Z_{i,k}(s) dB_i^{(k)}(s)$$

and

$$\begin{aligned} Z_{i,k}(t) =& \frac{2}{N^2} \int_{\mathbb{R}^d} \varphi_\varepsilon(x) \sum_{j=1}^{i-1} \frac{\partial \varphi_\varepsilon\left(x + B_i(s) - B_j(s)\right)}{\partial x_k} dx \\ & - \frac{2}{N^2} \int_{\mathbb{R}^d} \varphi_\varepsilon(x) \sum_{j=i+1}^{N} \frac{\partial \varphi_\varepsilon\left(x + B_j(s) - B_i(s)\right)}{\partial x_k} dx. \end{aligned} \tag{34}$$

It is easy to see that the integrand $Z_{i,k}(t)$ is continuous and adapted to $\mathscr{F}_t^N$ and that

$$|Z_{i,k}(t)| \leq \frac{2}{N} \|\varphi_\varepsilon\| \cdot \left\| \frac{\partial \varphi_\varepsilon}{\partial x_k} \right\| \tag{35}$$

Then again accordion to Theorem 5.2.3 in [5] we have for all $i \in \{1, 2, \ldots, N\}$ and $k \in \{1, 2, \ldots, d\}$, $M_t^{i,k}$ is a square integrable martingale with that

$$E[(\tilde{M}_t^{i,k})^2] = E\left(\int_0^t Z_{i,k}(s)^2 ds\right) \leq \frac{4t}{N^2} \|\varphi_\varepsilon\|^2 \cdot \left\| \frac{\partial \varphi_\varepsilon}{\partial x_k} \right\|^2 \tag{36}$$

and that for all $(i, k) \neq (j, h)$ we have that

$$\begin{aligned} \langle \tilde{M}_t^{i,k}, \tilde{M}_t^{j,h} \rangle &= \langle Z_{i,k} \cdot B_i^{(k)}(t), Z_{i,h} \cdot B_j^{(h)}(t) \rangle \\ &= \int_0^t Z_{i,k}(s) \cdot Z_{j,h}(s) \, d\langle B_i^{(k)}(t), B_j^{(h)}(t) \rangle \\ &= \int_0^t Z_{i,k}(s) \cdot Z_{j,h}(s) \, d0 = 0, \end{aligned} \tag{37}$$

which implies that

$$E\left(\tilde{M}_t^{i,k} \tilde{M}_t^{j,h}\right) \equiv 0. \tag{38}$$

Thus we immediately have

$$E[(\tilde{M}_t^i)^2] \leq \frac{4t}{N^2} \|\varphi_\varepsilon\|^2 \cdot \|\nabla \varphi_\varepsilon\|^2$$

and $E\big(\tilde{M}_t^i \tilde{M}_t^j\big) = 0$ for all $i \neq j$. Thus

$$E\big((\tilde{M}_t)^2\big) = \sum_{i=1}^{N} E\big((\tilde{M}_t^i)^2\big) \leq \frac{4t}{N} \|\varphi_\varepsilon\|^2 \|\nabla\varphi_\varepsilon\|^2 = \frac{4t}{N\varepsilon^{2d+2}} \|\varphi\|^2 \|\nabla\varphi\|^2. \tag{39}$$

□

With the martingale parts M_t and $\tilde{M}_t$ both controlled, our last step is to bound the "residue" part $\text{Res}(t)$. First, we again apply Cauchy Schwarz inequality and have

$$\begin{aligned} |\text{Res}(t)| \leq & \frac{1}{2}\int_0^t \|\nabla(\rho - \rho_{\varepsilon,N})(\cdot, s)\|^2 ds \\ & + \frac{2}{N^2}\int_0^t \int_{\mathbb{R}^d} \left| \sum_{i=1}^{N} \varphi_\varepsilon\left(x - X_i(s)\right)\left(\mathbf{F}(x,s) - \mathbf{F}\left(X_i(s), s\right)\right)\right|^2 dxds. \end{aligned} \tag{40}$$

It is easy to see that we can rewrite the integrand in the second term as follows:

$$\left| \sum_{i=1}^{N} \varphi_\varepsilon\left(x - X_i(s)\right)\left(\mathbf{F}(x,s) - \mathbf{F}\left(X_i(s), s\right)\right)\right|^2 = \sum_{i,j \leq N} R_{i,j}(x,s)$$

where

$$R_{i,j}(x,s) = \varphi_\varepsilon\left(x - X_i(s)\right)\varphi_\varepsilon\left(x - X_j(s)\right)\left(\mathbf{F}(x,s) - \mathbf{F}\left(X_i(s), s\right)\right)\cdot\left(\mathbf{F}(x,s) - \mathbf{F}\left(X_j(s), s\right)\right).$$

Note that for any $i, j \leq N$, $R_{i,j}(x,s) = 0$ when $|X_j(s) - X_i(s)| > 2\varepsilon$. And when $|X_i(s) - X_j(s)| \leq 2\varepsilon$, noting that F is Lipschitz continuous with the Lipschitz constant less than or equal to L_F,

$$\left|R_{i,j}(x,s)\right| \leq L_F^2 \varepsilon^2 \left|\varphi_\varepsilon\left(x - X_i(s)\right)\varphi_\varepsilon\left(x - X_j(s)\right)\right|.$$

Thus for all $i, j \leq N$, we have the spatial integral

$$\int_{\mathbb{R}^d} |R_{i,j}(x,s)|\, dx \leq \varepsilon^{2-d} L_F^2 \|\varphi\|^2 \mathbb{1}_{|X_i(s)-X_j(s)|\leq 2\varepsilon}. \tag{41}$$

Combining (40) and (41) we have

$$|\text{Res}(t)| \leq \frac{1}{2}\int_0^t \|\nabla(\rho - \rho_{\varepsilon,N})(\cdot, s)\|^2\, ds + R^*(t) \tag{42}$$

where

$$R^*(t) = \frac{2L_F^2 \|\varphi\|^2}{N^2 \varepsilon^{d-2}} \sum_{i,j \leq N} \int_0^t \mathbb{1}_{|X_i(s)-X_j(s)|\leq 2\varepsilon}\, ds. \tag{43}$$

By definition, $R^*(t)$ is monotonically increasing over t. So if we want to control $\sup_{s\le t}|\mathrm{Res}(s)|$, it is sufficient to control $R^*(t)$. Noting that $\mathbb{1}_{|X_i-X_j|\le 2\varepsilon}\equiv 1$, we can take expectation on (43) and have

$$E[R^*(t)] = \frac{2L_F^2\|\varphi\|^2 t}{N\varepsilon^{d-2}} + \frac{2L_F^2\|\varphi\|^2}{N^2\varepsilon^{d-2}} \sum_{i,j\le N: i\ne j} E\left(\int_0^t \mathbb{1}_{|X_i(s)-X_j(s)|\le 2\varepsilon}\, ds\right). \tag{44}$$

Noting that $X_i(s)-X_j(s)$ is continuous and adaptable to $\mathscr{F}_t^N$ (which implies progressive), $\mathbb{1}_{|X_i(s)-X_j(s)|\le 2\varepsilon}\times \mathbb{1}_{0\le s\le t}$ is measurable on $[0,t]\times\Omega$ and bounded and thus integrable. By Fubini's Theorem,

$$E[R^*(t)] = \frac{2L_F^2\|\varphi\|^2 t}{N\varepsilon^{d-2}} + \frac{2L_F^2\|\varphi\|^2}{N^2\varepsilon^{d-2}} \sum_{i,j\le N: i\ne j} \int_0^t P\big(|X_i(s)-X_j(s)|\le 2\varepsilon\big) ds. \tag{45}$$

At this point, we have reduced the problem of controlling $|\mathrm{Res}(s)|$ to controlling the upper bounds for the probabilities of $P(|X_i(s)-X_j(s)|\le 2\varepsilon)$, $i\ne j\le N$. We show this under Assumption 1 and 2 respectively. However, since the proofs are similar, we will only show the proof of the more complicated case under Assumption 1. One can proof the same result under Assumption 2 followings exact the same steps but the calculations are easier since the sums will be replaced by integrals in that case.

Under Assumption 1, for $j=1,\ldots,N$, let

$$E_j(t) = \frac{1}{N}\sum_{i\le N: i\ne j}\int_0^t P(|X_i(s)-X_j(s)|\le 2\varepsilon)\, ds \tag{46}$$

which implies that

$$E[R^*(t)] = \frac{2L_F^2\|\varphi\|^2 t}{N\varepsilon^{d-2}} + \frac{2L_F^2\|\varphi\|^2}{\varepsilon^{d-2}}\left(\frac{1}{N}\sum_{j=1}^N E_j(t)\right).$$

We have the following Lemma.

Lemma 3 *Under Assumption 1, for any $t\ge 0$, there exist some constant $C_1(t)$ and $C_2(t)$ depends only on t such that*

$$E_j(t)\le C_1(t)\varepsilon^{d-1} + C_2(t)\frac{1}{N\varepsilon} \tag{47}$$

for all $j=1,2,\ldots,N$, when ε is sufficiently small.

Proof For any N and j. Fix $i\ne j$, $i\le N$, and let $\{\Omega,\mathscr{F}_t^{i,j},P\}$ be our probability measure space where , $\mathscr{F}_t^{i,j}$ is the natural filtration generated by $B_{i,j}^*(t)=[B_i(t),B_j(t)]$, which is a $2d$-dimensional Brownian motion. Let $\theta_{i,j}(s) = -\big(F(X_i(s),s),F(X_j(s),s)\big)$

be the integrand and consider the adapted measurable process

$$\Gamma_s = \int_0^s \theta_{i,j}(h) \cdot dB^*_{i,j}(h). \tag{48}$$

Note that for any $s \geq 0$,

$$|\theta_{i,j}(s)|^2 \leq 2d \, \|F\|_\infty^2. \tag{49}$$

Thus the Novikov condition (see page 198 of [14] for details) is satisfied, i.e.,

$$E\left[\exp\left(\frac{1}{2}\int_0^s |\theta_{i,j}(h)|^2 \, dh\right)\right] \leq \exp(sd \, \|F\|_\infty^2) < \infty,$$

by Girsanov Theorem (see Theorem 3.5.1 of [14]) we can define a probability measure Q in our probability space with Radon-Nikodym derivative

$$\frac{dQ_{i,j}}{dP}\bigg| \mathscr{F}_t = \mathscr{E}_t = \exp\left[\Gamma_t - \frac{1}{2}\int_0^s |\theta_{i,j}(h)|^2 \, dh\right]. \tag{50}$$

Then we have

$$\begin{bmatrix} X_i(t) - X_i(0) \\ X_j(t) - X_j(0) \end{bmatrix} = \begin{bmatrix} B_i(t) + \int_0^t F(X_i(s), s)ds \\ B_j(t) + \int_0^t F(X_j(s), s)ds \end{bmatrix} = B^*_{i,j}(t) - \langle \Gamma, B^*_{i,j} \rangle_t$$

is a standard $2d$-dimensional Brownian motion under probability measure $Q_{i,j}$, where $\langle \Gamma, B^*_{i,j} \rangle_t$ is again the quadratic covariance between Γ_t and $B^*_{i,j}(t)$. Thus by Radon-Nikodym Theorem we have

$$\int_{|X_i(t)-X_j(t)|\leq 2\varepsilon} \mathscr{E}_t dP = P\left(|B_i(t) - B_j(t) + \delta_{i,j}| \leq 2\varepsilon\right). \tag{51}$$

Moreover,

$$P\left(|X_i(t) - X_j(t)| \leq 2\varepsilon\right) \leq P(\mathscr{E}_t < \varepsilon^{1/2}) + P\left(|X_i(t) - X_j(t)| \leq 2\varepsilon \cap \mathscr{E}_t \geq \varepsilon^{1/2}\right)$$

and for the first part we have,

$$P(\mathscr{E}_t < \varepsilon^{1/2}) \leq P\left(\exp\left[\int_0^t (-\theta_{i,j}(s)) \cdot dB^*_{i,j}(s)\right] > \varepsilon^{-1/2}\exp(-td \, \|F\|_\infty^2)\right). \tag{52}$$

To control the right had side of the inequality above, we consider the L^{4d} norm:

$$E\left[\left(\exp\left[\int_0^t (-\theta_{i,j}(s)) \cdot dB^*_{i,j}(s)\right]\right)^{4d}\right] = E\left(\exp\left[\int_0^t (-4d\,\theta_{i,j}(s)) \cdot dB^*_{i,j}(s)\right]\right) \tag{53}$$

and note that again by Girsanov Theorem,

$$\mathcal{E}_t' = \exp\left[\int_0^t (-4d\theta_{i,j}(s)) \cdot dB_{i,j}^*(s)\right] \exp\left(-8d^2 \int_0^t |\theta_{i,j}(s)|^2 \, ds\right)$$

is again a Radon-Nikodym derivative. Thus we have

$$E(\mathcal{E}_t') = 1$$

which combining with (49), implies

$$E\left(\exp\left[\left[\int_0^t (-4d\theta_{i,j}(s)) \cdot dB_{i,j}^*(s)\right]\right]\right) \le \exp(16d^3 t \, \|F\|_\infty^2) < \infty. \tag{54}$$

Combining (52), (54) and Chebyshev's Inequality gives us

$$P(\mathcal{E}_t < \varepsilon^{1/2}) \le \varepsilon^{2d} \exp\left((4d^2 + 16d^3)t \, \|F\|_\infty^2\right). \tag{55}$$

Then for the second part, according to (51) we have

$$\int_{|X_i(t)-X_j(t)|\le 2\varepsilon \cap \mathcal{E}_t \ge \varepsilon^{1/2}} \mathcal{E}_t dP \le P\left(|B_i(t) - B_j(t) + \delta_{i,j}| \le 2\varepsilon\right).$$

and thus

$$\begin{aligned} &P\left(|X_i(t) - X_j(t)| \le 2\varepsilon \cap \mathcal{E}_t \ge \varepsilon^{1/2}\right) \\ &\le \varepsilon^{-1/2} P\left(|B_i(t) - B_j(t) + \delta_{i,j}| \le 2\varepsilon\right). \end{aligned} \tag{56}$$

Combining the two inequalities above, we have

$$\begin{aligned} P\left(|X_i(t) - X_j(t)| \le 2\varepsilon\right) \le &\varepsilon^{2d} \exp\left((4d^2 + 16d^3)t \, \|F\|_\infty^2\right) \\ &+ \varepsilon^{-1/2} P\left(|B_i(t) - B_j(t) + \delta_{i,j}| \le 2\varepsilon\right) \end{aligned} \tag{57}$$

for any $t \ge 0$. Integrating (57) on $[0, t]$ and averaging over all $i \ne j$, $i \le N$, we have

$$\begin{aligned} E_j(t) \le &\frac{\varepsilon^{2d}}{(4d^2 + 16d^3) \, \|F\|_\infty^2} \exp\left((4d^2 + 16d^3)t \, \|F\|_\infty^2\right) \\ &+ \frac{1}{N\varepsilon^{1/2}} \sum_{i:i\ne j, i\le N} \int_0^t P\left(|B_i(s) - B_j(s) + \delta_{i,j}| \le 2\varepsilon\right) ds. \end{aligned} \tag{58}$$

According to (58) to proof this lemma it is sufficient to have the following lemma for standard Brownian motions: □

Lemma 4 *Under Assumption 1, for any $t \geq 0$, there is some constant $C_1^*(t)$ and $C_2^*(t)$ such that*

$$\frac{1}{N} \sum_{i:i \neq j, i \leq N} \left[\int_0^t P\left(|B_i(s) - B_j(s) + \delta_{i,j}| \leq 2\varepsilon \right) ds \right] \leq C_1^*(t)\varepsilon^d + C_2^*(t)\frac{1}{N}. \tag{59}$$

Proof We first note that for any s and m, n, $B_i(s) - B_j(s) + \delta_{i,j}$ has a d-dimensional normal distribution with mean $\delta_{i,j}$ and variance $2s$. So we have

$$\begin{aligned} &\int_0^t P\left(|B_i(s) - B_j(s) + \delta_{i,j}| \leq 2\varepsilon \right) ds \\ &= \int_0^t \int_{|x| \leq 2\varepsilon} \frac{1}{(4\pi s)^{d/2}} \exp\left(-\frac{|\delta_{i,j} - x|^2}{4s} \right) dxds \\ &= \int_{|x| \leq 2\varepsilon} \int_0^t \frac{1}{(4\pi s)^{d/2}} \exp\left(-\frac{|\delta_{i,j} - x|^2}{4s} \right) dsdx. \end{aligned} \tag{60}$$

To deal with equation (60), we need to separate the case of $d = 1, d = 2$ and $d \geq 3$.

Case 1: $d = 1$. In this case we simply use the bound

$$\begin{aligned} &\int_0^t P\left(|B_i(s) - B_j(s) + \delta_{i,j}| \leq 2\varepsilon \right) ds \\ &\leq \int_{-2\varepsilon}^{2\varepsilon} \int_0^t s^{-1/2}\, dsdx = 8\varepsilon\sqrt{t}. \end{aligned}$$

Averaging over m gives us the desired result.

Case 2: $d = 2$. In this case we have

$$\begin{aligned} &\int_0^t P\left(|B_i(s) - B_j(s) + \delta_{i,j}| \leq 2\varepsilon \right) ds \\ &= \int_{|x| \leq 2\varepsilon} \int_0^t \frac{1}{4\pi s} \exp\left(-\frac{|\delta_{i,j} - x|^2}{4s} \right) dsdx. \end{aligned}$$

If $\delta_{i,j} \geq 1$, then for all $\varepsilon < 1/4$ and $x < 2\varepsilon$ we have

$$\begin{aligned} &\int_{|x| \leq 2\varepsilon} \int_0^t \frac{1}{4\pi s} \exp\left(-\frac{|\delta_{i,j} - x|^2}{4s} \right) dsdx \\ &\leq \int_{|x| \leq 2\varepsilon} \int_0^t \frac{1}{s} \exp\left(-\frac{1}{16s} \right) dsdx \\ &\leq 16\varepsilon^2 \int_0^t \frac{1}{s} \exp\left(-\frac{1}{16s} \right) ds \end{aligned} \tag{61}$$

When $\delta_{i,j} < 1$ taking $h = \frac{|\delta_{i,j}-x|^2}{4s}$, we have

$$\int_0^t P\left(|B_i(s) - B_j(s) + \delta_{i,j}| \leq 2\varepsilon\right) ds$$
$$= \int_{|x|\leq 2\varepsilon} \int_0^t \frac{1}{4\pi s} \exp\left(-\frac{|\delta_{i,j} - x|^2}{4s}\right) dsdx$$
$$= \int_{|x|\leq 2\varepsilon} \int_{|\delta_{i,j}-x|^2/4t}^{\infty} \frac{1}{4\pi h} \exp(-h)\, dhdx.$$

Note that $h^{-1}\exp(-h) < h^{-1}$ and $h^{-1}\exp(-h) \leq \exp(-h)$ when $h \geq 1$. We have

$$\int_{|\delta_{i,j}-x|^2/4t}^{\infty} h^{-1}\exp(-h)\, dh \leq \int_{|\delta_{i,j}-x|^2/4t}^{1} h^{-1} dh + \int_1^\infty e^{-h} dh$$
$$\leq 2|\log(|\delta_{i,j} - x|)| + |\log t| + 1 + \log 4. \tag{62}$$

Moreover, let $\delta_1 = CN^{-1/2}$, where C is the constant in Assumption 1, $\delta_2 = \delta_1 + 4\varepsilon$ and $M = [\delta_2^{-1}] + 1$. For all $k = 0, 1, \ldots, M$ consider the following sets

$$A_k := \left\{i : k\delta_2 \leq |\delta_{i,j}| < (k+1)\delta_2\right\}. \tag{63}$$

By definition, it is easy to see that when N is large and ε is small

$$\bigcup_{k=0}^{M} A_k \supset \left\{i : |\delta_{i,j}| < 1\right\}. \tag{64}$$

If we first look at A_0, according to Assumption 1, the little balls $\{N(\delta_{i,j}, \delta)\}_{i\leq N, i\neq j}$ (where $N(x, y)$ is the neighborhood of x with radius y) have no intersections with each other. And for all $i \in A_0$,

$$N(\delta_{i,j}, \delta_1) \subset N(0, \delta_1 + \delta_2)$$

This immediately implies that

$$\mathrm{card}(A_0) \leq \left(\frac{\delta_1 + \delta_2}{\delta_1}\right)^2 = \left(2 + \frac{4\varepsilon}{C} N^{1/2}\right)^2 \leq 8 + \frac{32\varepsilon^2}{C^2} N,$$

since the sum of areas of disjoint disks with radius δ_1 in A_0 cannot be larger than the area of A_0 itself. Thus we have

$$\frac{1}{N} \sum_{i\in A_0} \int_0^t P\left(|B_i(s) - B_j(s) + \delta_{i,j}| \leq 2\varepsilon\right) ds \leq \frac{t}{N}\mathrm{card}(A_0) \leq \frac{8t}{N} + \frac{32\varepsilon^2 t}{C^2} \tag{65}$$

Similarly, for each $k \geq 1$ and $i \in A_k$,

$$N(\delta_{i,j}, \delta_1) \subset \{y : (k-1)\delta_2 \leq |x| < (k+2)\delta_2\}$$

which implies that

$$\text{card}(A_k) \leq \frac{[(k+2)^2 - (k-1)^2]\delta_2^2}{\delta_1^2} \leq 9k\left(1 + \frac{4\varepsilon N^{1/2}}{C}\right)^2.$$

Noting that for all $i \in A_k$ and $|x| \leq 2\varepsilon$

$$|\log(|\delta_{i,j} - x|)| \leq \max\{\log 2, |\log(|k\delta_2 - 2\varepsilon|)|\} \leq \log 2 + |\log(k\delta_2)|.$$

Thus according to (62) and the inequality above

$$\begin{aligned}
&\frac{1}{N}\sum_{i\in A_k}\int_0^t P\left(|B_i(s) - B_j(s) + \delta_{i,j}| \leq 2\varepsilon\right) ds \\
&\leq \frac{1}{N}\sum_{i\in A_k}\left[\int_{|x|\leq 2\varepsilon} |\log t| + \log 4 + 1 + 2|\log(|\delta_{i,j} - x|)|\, dx\right] \\
&\leq \frac{1}{N}\sum_{i\in A_k}\left[\int_{|x|\leq 2\varepsilon} |\log t| + \log 16 + 1 + 2|\log(k\delta_2)|\, dx\right] \\
&\leq \left[\frac{1}{N}\sum_{i\in A_k} 16(|\log t| + \log 16 + 1)\varepsilon^2\right] + \frac{288\varepsilon^2 k\left(1 + \frac{4\varepsilon N^{1/2}}{C}\right)^2}{N}|\log(k\delta_2)| \\
&= \left[\frac{1}{N}\sum_{i\in A_k} 16(|\log t| + \log 16 + 1)\varepsilon^2\right] + \frac{288\varepsilon^2}{C^2}\delta_2\left[k\delta_2|\log(k\delta_2)|\right]
\end{aligned} \tag{66}$$

Summing over $k = 0, 1, \ldots, M$ we have

$$\begin{aligned}
&\frac{1}{N}\sum_{i:\delta_{i,j}<1}\int_0^t P\left(|B_i(s) - B_j(s) + \delta_{i,j}| \leq 2\varepsilon\right) ds \\
&\leq \frac{8t}{N} + \frac{32\varepsilon^2 t}{C^2} + 16(|\log t| + \log 16 + 1)\varepsilon^2 + \frac{288\varepsilon^2}{C^2}\sum_{k=1}^{[\delta_2^{-1}]+1}\delta_2\left[k\delta_2|\log(k\delta_2)|\right]
\end{aligned} \tag{67}$$

Note that the last term in the inequality above is a Riemann sum of function $x|\log x|$ and the fact that $x|\log x| \leq \max\{\log 2, e^{-1}\} < 1$ on $[0, 2]$.

$$\sum_{k=1}^{[\delta_2^{-1}]+1} \delta_2 \left[k\delta_2 |\log(k\delta_2)|\right] \le \int_0^2 dt = 2.$$

So we have

$$\begin{aligned} &\frac{1}{N}\sum_{i:\delta_{i,j}<1}\int_0^t P\left(|B_i(s) - B_j(s) + \delta_{i,j}| \le 2\varepsilon\right) ds \\ &\le \frac{8t}{N} + \frac{32\varepsilon^2 t}{C^2} + 16(|\log t| + \log 16 + 1)\varepsilon^2 + \frac{576\varepsilon^2}{C^2} \end{aligned} \tag{68}$$

Combining (68) and (61), and letting

$$\begin{aligned} C_1^*(t) &:= 16\int_0^t \frac{1}{s}\exp\left(-1/16s\right) ds + \frac{32t}{C^2} + 16(|\log t| + \log 16 + 1) + \frac{576}{C^2} \\ C_2^*(t) &:= 8t \end{aligned} \tag{69}$$

we finally get

$$\frac{1}{N}\sum_{i:i\neq j, i\le N}\int_0^t P\left(|B_i(s) - B_j(s) + \delta_{i,j}| \le 2\varepsilon\right) ds \le C_1^*(t)\varepsilon^d + C_2^*(t)\frac{1}{N}$$

when $d = 2$, and the proof for case 2 is complete.

Case 3: $d \ge 3$. The proof in this case is similar but simpler than the case of $d = 2$. Again we have

$$\begin{aligned} &\int_0^t P\left(|B_i(s) - B_j(s) + \delta_{i,j}| \le 2\varepsilon\right) ds \\ &= \int_{|x|\le 2\varepsilon}\int_0^t \frac{1}{(4\pi s)^{d/2}} \exp\left(-\frac{|\delta_{i,j} - x|^2}{4s}\right) dsdx. \end{aligned}$$

If $\delta_{i,j} \ge 1$, then for all $\varepsilon < 1/4$ and $|x| < 2\varepsilon$ we have

$$\begin{aligned} &\int_{|x|\le 2\varepsilon}\int_0^t \frac{1}{(4\pi s)^{d/2}} \exp\left(-\frac{|\delta_{i,j} - x|^2}{4s}\right) dsdx \\ &\le \int_{|x|\le 2\varepsilon}\int_0^t \frac{1}{s^{d/2}} \exp\left(-\frac{1}{16s}\right) dsdx \\ &\le 2^{2d}\varepsilon^d \int_0^t \frac{1}{s^{d/2}} \exp\left(-\frac{1}{16s}\right) ds \end{aligned} \tag{70}$$

When $\delta_{i,j} < 1$ taking $h = \frac{|\delta_{i,j}-x|^2}{4s}$, we have

$$\begin{aligned}
&\int_0^t P\left(|B_i(s) - B_j(s) + \delta_{i,j}| \le 2\varepsilon\right) ds \\
&= \int_{|x|\le 2\varepsilon}\int_0^t \frac{1}{(4\pi s)^{d/2}} \exp\left(-\frac{|\delta_{i,j} - x|^2}{4s}\right) dsdx \\
&< C_d \int_{|x|\le 2\varepsilon} |\delta_{i,j} - x|^{-d+2} dx.
\end{aligned} \tag{71}$$

where constant

$$C_d := 2^{4d} \int_0^\infty h^{-2+d/2} \exp(-h) dh.$$

Then again we can define $\delta_1 = CN^{-1/d}$, where C is the constant in Assumption 1, $\delta_2 = \delta_1 + 4\varepsilon$ and $M = [\delta_2^{-1}] + 1$. For all $k = 0, 1, \ldots, M$ consider the following sets

$$A_k := \left\{i : k\delta_2 \le |\delta_{i,j}| < (k+1)\delta_2\right\}. \tag{72}$$

such that

$$\bigcup_{k=0}^{M} A_k \supset \left\{i : |\delta_{i,j}| < 1\right\}.$$

Then similarly, we have

$$\text{card}(A_0) \le \left(\frac{\delta_1 + \delta_2}{\delta_1}\right)^d = \left(2 + \frac{4\varepsilon}{C} N^{1/d}\right)^d \le 2^{2d-1} + \frac{2^{3d-1}\varepsilon^d}{C^d} N$$

and

$$\text{card}(A_k) \le \frac{[(k+2)^d - (k-1)^d]\delta_2^d}{\delta_1^d} \le 3^d k^{d-1} \left(1 + \frac{4\varepsilon N^{1/d}}{C}\right)^d.$$

Thus

$$\frac{1}{N}\sum_{i\in A_0} \int_0^t P\left(|B_i(s) - B_j(s) + \delta_{i,j}| \le 2\varepsilon\right) ds \le \frac{t}{N}\text{card}(A_0) \le \frac{2^{2d-1}t}{N} + \frac{2^{3d-1}t\varepsilon^d}{C^d} \tag{73}$$

and

$$\begin{aligned}&\frac{1}{N}\sum_{i\in A_k}\int_0^t P\left(|B_i(s)-B_j(s)+\delta_{i,j}|\le 2\varepsilon\right)\,ds\\&\le \frac{C_d}{N}\sum_{m\in A_k}\int_{|x|\le 2\varepsilon}|\delta_{i,j}-x|^{-d+2}dx\\&\le \frac{C_d}{N}3^d k^{d-1}\left(1+\frac{4\varepsilon N^{1/d}}{C}\right)^d(2^{2d}\varepsilon^d)\times\left(2^d[k(CN^{-1/d}+4\varepsilon)]^{-d+2}\right)\end{aligned} \tag{74}$$

Summing over $k=0,1,\ldots,M$,

$$\begin{aligned}&\frac{1}{N}\sum_{i:\delta_{i,j}<1}\int_0^t P\left(|B_i(s)-B_j(s)+\delta_{i,j}|\le 2\varepsilon\right)\,ds\\&\le \frac{2^{2d-1}t}{N}+\frac{2^{3d-1}t\varepsilon^d}{C^d}+C_d\left(\frac{24}{C}\right)^d\varepsilon^d\sum_{k=1}^{[\delta_2^{-1}]+1}[(k\delta_2)\delta_2].\end{aligned} \tag{75}$$

Again for the last term we have

$$\sum_{k=1}^{[\delta_2^{-1}]+1}[(k\delta_2)\delta_2]\le\int_0^2 tdt=2.$$

Thus

$$\begin{aligned}&\frac{1}{N}\sum_{i:\delta_{i,j}<1}\int_0^t P\left(|B_i(s)-B_j(s)+\delta_{i,j}|\le 2\varepsilon\right)\,ds\\&\le \frac{2^{2d-1}t}{N}+\frac{2^{3d-1}t\varepsilon^d}{C^d}+2C_d\left(\frac{24}{C}\right)^d\varepsilon^d\end{aligned} \tag{76}$$

Then combining (70) and (76), and letting

$$\begin{aligned}C_1^*(t)&:=2^{2d}\int_0^t\frac{1}{s^{d/2}}\exp\left(-\frac{1}{16s}\right)\,ds+\frac{2^{3d-1}t}{C^d}+2C_d\left(\frac{24}{C}\right)^d\\C_2^*(t)&:=2^{2d-1}t\end{aligned} \tag{77}$$

We complete the proof of case 3. With the Lemma 4 proved, the proof of Lemma 3 is complete. □

Plugging the result of this lemma into (45),

$$E[R^*(t)]\le\frac{2\|\varphi\|^2t}{N\varepsilon^{d-2}}+2\|\varphi\|^2C_1(t)\varepsilon+\frac{2\|\varphi\|^2}{N\varepsilon^{d-1}}C_2(t). \tag{78}$$

Proof of Theorem 1 Combining (28), (33) and (78), noting that $\varepsilon_N = N^{-1/3d} \to 0$, then for any $t \geq 0$, letting

$$\delta_1(t) := \left(4N^{-1/2d}\, \|\nabla\varphi\|^2 \int_0^t \|\rho(x,s)\|^2 ds\right)^{1/3}$$
$$\delta_2(t) := \left(4N^{-1/2d}\, \|\nabla\varphi\|^2 \|\varphi\|^2\right)^{1/3}$$
$$\delta_3(t) := \left(2N^{-1/3d}\, \|\varphi\|^2 (t + C_1(t) + C_2(t))\right)^{1/2}$$

by Doob and Chebyshev inequality, inequality (78) and the fact that R^* is nonnegative, it is easy to see that as $N \to \infty$ the probability of the following events:

$$\begin{aligned} P(A_N) &:= P\big(\sup_{s\leq t} |M_t| < \delta_1(t)\big) \geq 1 - \delta_1(t) \\ P(B_N) &:= P\big(\sup_{s\leq t} |\tilde{M}_t| < \delta_2(t)\big) \geq 1 - \delta_2(t) \\ P(C_N) &:= P\big(R^*(t) < \delta_3(t)\big) \geq 1 - \delta_3(t). \end{aligned} \tag{79}$$

Note that for the constant terms in Proposition 1, we have

$$\frac{t}{N}\|\nabla\varphi_{\varepsilon_N}\|_2^2 = \frac{t\varepsilon_N^{d+2}}{N}\|\nabla\varphi\|^2 < tN^{-1/3d}$$

when N is sufficiently large. And under event C_N, for any $s \leq t$, by (42)

$$\begin{aligned} \mathrm{Res}(s) &\leq \frac{1}{2}\int_0^s \|\nabla(\rho - \rho_{\varepsilon_N,N})(\cdot,h)\|^2\, dh + R^*(s) \\ &\leq \frac{1}{2}\int_0^s \|\nabla(\rho - \rho_{\varepsilon_N,N})(\cdot,h))\|^2\, dh + R^*(t) \\ &\leq \frac{1}{2}\int_0^s \|\nabla(\rho - \rho_{\varepsilon_N,N})(\cdot,h)\|^2\, dh + \delta_3(t), \end{aligned}$$

which implies that

$$\sup_{s\leq t}\left(\mathrm{Res}(s) - \frac{1}{2}\int_0^s \|\nabla(\rho - \rho_{\varepsilon_N,N})(\cdot,h)\|^2\, dh\right) \leq \delta_3(t). \tag{80}$$

So under the event

$$P(A_N \cap B_N \cap C_N) \geq 1 - \delta_1(t) - \delta_2(t) - \delta_3(t)$$

we have that for all $s \leq t$

$$
\begin{aligned}
\|(\rho-\rho_{\varepsilon,N})(\cdot,s)\|^2 \le & \|(\rho-\rho_{\varepsilon,N})(\cdot,0)\|^2 - \frac{1}{2}\int_0^s \|\nabla(\rho-\rho_{\varepsilon_N,N})(\cdot,h)\|^2\,dh \\
& - \int_0^s\int_{\mathbb{R}^d} \nabla\cdot\mathbf{F}(x,h)\left((\rho-\rho_{\varepsilon,N})(x,h)\right)^2 dxdh \\
& + \delta_1(t)+\delta_2(t)+\delta_3(t)+N^{-1/3d}\,t
\end{aligned} \tag{81}
$$

which implies

$$
\begin{aligned}
\sup_{s\in[0,t]} & \left(2\|(\rho-\rho_{\varepsilon,N})(\cdot,s)\|^2 + \int_0^s \|\nabla(\rho-\rho_{\varepsilon_N,N})(\cdot,h)\|^2\,dh\right) \\
& \le 2\|(\rho-\rho_{\varepsilon,N})(\cdot,0)\|^2 - 2\int_0^s\int_{\mathbb{R}^d} \nabla\cdot\mathbf{F}(x,h)\left(\rho(x,h)-\rho_{\varepsilon,N}(x,h)\right)^2 dxdh \\
& + 2\left(\delta_1(t)+\delta_2(t)+\delta_3(t)+N^{-1/3d}t\right).
\end{aligned} \tag{82}
$$

Since F is Lipschitz continuous from $R^d \to R^d$, it is also differentiable almost everywhere by Rademacher's theorem, see Theorem 3.1.6 of [8]. Thus for any x such that F is differentiable, we have $||\nabla\cdot\mathbf{F}||_{L^\infty} \le dL_F$, which implies that $||\rho(\cdot,s)||^2 \le e^{C_0 s}||\rho_0||^2$ where C_0 is the constant defend in the statement of Theorem 1. I.e., $C_0 = 2dL_F \ge 2||\nabla\cdot\mathbf{F}||_{L^\infty}$. Let

$$
\begin{aligned}
c(t) = & \left(\frac{4e^{C_0 t}}{C_0}\|\nabla\varphi\|^2\|\rho_0\|^2\right)^{1/3} + \left(4\|\nabla\varphi\|^2\|\varphi\|^2\right)^{1/3} \\
& + \left(2\|\varphi\|^2[t+C_1(t)+C_2(t)]\right)^{1/2} + t.
\end{aligned} \tag{83}
$$

Gronwall's inequality finishes the proof of Theorem 1.

3 Discussion About Higher Order Sobolev Norm

In the previous discussions, we proved the convergence under H^1 norm. We hope what this method can be generalized to prove the convergence in the higher order Sobolev norms. However, currently we are only able to prove the simple case when $\mathbf{F} \equiv 0$. I.e., the limit PDE is now the heat equation

$$
\frac{\partial\rho}{\partial t}(x,t) = \frac{1}{2}\Delta\rho(x,t)
$$

and the paths of the particles are i.i.d. standard Brownian motions $B_i(t)$, $i = 1, 2, \ldots, N$ and

$$
\rho_{\varepsilon,N}(x,t) = \frac{1}{N}\sum_{i=1}^N \varphi_\varepsilon\left(x - B_i(t)\right). \tag{84}
$$

Again for any t we consider

$$\|D^{\alpha}(\rho-\rho_{\varepsilon,N})(\cdot,t)\|^2=\|D^{\alpha}\rho(\cdot,t)\|^2-2\int_{\mathbb{R}^d}D^{\alpha}\rho(x,t)\,D^{\alpha}\rho_{\varepsilon,N}(x,t)dx+\|D^{\alpha}\rho_{\varepsilon,N}(\cdot,t)\|^2. \tag{85}$$

For the first term we have

$$\begin{aligned}\|D^{\alpha}\rho(\cdot,t)\|^2&=\|D^{\alpha}\rho(\cdot,0)\|^2+\int_{\mathbb{R}^d}\int_0^t 2D^{\alpha}\rho(x,s)\,D^{\alpha}\frac{\partial\rho}{\partial t}(x,s)dsdx\\&=\|D^{\alpha}\rho(\cdot,0)\|^2-\int_0^t\|\nabla D^{\alpha}\rho(\cdot,s)\|^2ds.\end{aligned} \tag{86}$$

Then for the second term we have

$$\int_{\mathbb{R}^d}D^{\alpha}\rho(x,t)\,D^{\alpha}\rho_{\varepsilon,N}(x,t)dx=\frac{1}{N}\sum_{i=1}^{N}\int_{\mathbb{R}^d}D^{\alpha}\rho(x,t)\,D^{\alpha}\varphi_{\varepsilon}\left(x-B_i(t)\right)dx$$

and for each $i\leq N$, and for all $x\in R^d$, we have by Ito's formula,

$$\begin{aligned}D^{\alpha}\rho(x,t)\,D^{\alpha}\varphi_{\varepsilon}\left(x-B_i(t)\right)&=D^{\alpha}\rho(x,0)\,D^{\alpha}\varphi_{\varepsilon}\left(x-B_i(0)\right)\\&+\int_0^t D^{\alpha}\frac{\partial\rho}{\partial t}(x,s)\,D^{\alpha}\varphi_{\varepsilon}\left(x-B_i(s)\right)ds\\&-\int_0^t D^{\alpha}\rho(x,s)\nabla D^{\alpha}\varphi_{\varepsilon}\left(x-B_i(s)\right)\cdot dB_i(s)\\&+\frac{1}{2}\int_0^t D^{\alpha}\rho(x,s)\Delta D^{\alpha}\varphi_{\varepsilon}\left(x-B_i(s)\right)ds.\end{aligned} \tag{87}$$

integrating over R^d and use integration by parts, we have

$$\begin{aligned}\int_{\mathbb{R}^d}D^{\alpha}\rho(x,t)\,D^{\alpha}\varphi_{\varepsilon}\left(x-B_i(t)\right)dx&=\int_{\mathbb{R}^d}D^{\alpha}\rho(x,0)\,D^{\alpha}\varphi_{\varepsilon}\left(x-B_i(0)\right)dx\\&+\int_0^t\int_{\mathbb{R}^d}D^{\alpha}\frac{\partial\rho}{\partial t}(x,s)\,D^{\alpha}\varphi_{\varepsilon}\left(x-B_i(t)\right)dxds\\&-\frac{1}{2}\int_0^t\int_{\mathbb{R}^d}\nabla D^{\alpha}\rho(x,s)\cdot\nabla D^{\alpha}\varphi_{\varepsilon}\left(x-B_i(t)\right)dxds\\&-NM_t^i\end{aligned}$$

where

$$M_t^i=\frac{1}{N}\int_0^t\int_{\mathbb{R}^d}\left[D^{\alpha}\rho(x,s)\,\nabla D^{\alpha}\varphi_{\varepsilon}\left(x-B_i(s)\right)\right]dx\cdot dB_i(s)$$

Averaging over $i = 1, 2, \ldots, N$ we have

$$\begin{aligned}\int_{\mathbb{R}^d} D^\alpha \rho(x,t)\, D^\alpha \rho_{\varepsilon,N}\,(x - B_n(t))\, dx = &\int_{\mathbb{R}^d} D^\alpha \rho(x,0)\, D^\alpha \rho_{\varepsilon,N}\,(x - B_n(0))\, dx \\ &- \int_0^t \int_{\mathbb{R}^d} \nabla D^\alpha \rho(x,s) \cdot \nabla D^\alpha \rho_{\varepsilon,N}\,(x - B_n(t))\, dxds \\ &- M_t\end{aligned} \tag{88}$$

where $M_t = \sum_{i=1}^N M_t^i$. And by the same argument in Lemma 1, we have

$$E\big((M_t)^2\big) \le \frac{1}{N\varepsilon^{d+2|\alpha|+2}} \|\nabla D^\alpha \varphi\|^2 \int_0^t \|D^\alpha \rho(x,s)\|^2 ds \tag{89}$$

Finally for the last term $\|D^\alpha \rho_{\varepsilon,N}(x,t)\|^2$, we have

$$\|D^\alpha \rho_{\varepsilon,N}(x,t)\|^2 = \frac{1}{N}\|D^\alpha \varphi_\varepsilon\| + \frac{2}{N^2}\sum_{j>i}\int_{\mathbb{R}^d} D^\alpha \varphi_\varepsilon\left(x - B_j(t)\right) D^\alpha \varphi_\varepsilon\,(x - B_i(t))\, dx. \tag{90}$$

Then for each $j > i$, by change of variables equals to

$$\int_{\mathbb{R}^d} D^\alpha \varphi_\varepsilon\left(x - B_j(t)\right) D^\alpha \varphi_\varepsilon\,(x - B_i(t))\, dx = \int_{\mathbb{R}^d} D^\alpha \varphi_\varepsilon\,(x)\, D^\alpha \varphi_\varepsilon\left(x + B_j(t) - B_i(t)\right) dx.$$

Again, we use Ito's formula, for any $j > i$ and $x \in R^d$,

$$\begin{aligned}D^\alpha \varphi_\varepsilon\left(x + B_j(t) - B_i(t)\right) =& D^\alpha \varphi_\varepsilon\left(x + B_j(0) - B_i(0)\right) \\ &+ \int_0^t \nabla D^\alpha \varphi_\varepsilon\left(x + B_j(s) - B_i(s)\right) \cdot dB_j(s) \\ &- \int_0^t \nabla D^\alpha \varphi_\varepsilon\left(x + B_j(s) - B_i(s)\right) \cdot dB_i(s) \\ &+ \int_0^t \Delta D^\alpha \varphi_\varepsilon\left(x + B_j(s) - B_i(s)\right) ds.\end{aligned} \tag{91}$$

Integrating over R^d we have $\int_{\mathbb{R}^d} D^\alpha \varphi_\varepsilon\big(x - B_j(t)\big) D^\alpha \varphi_\varepsilon\,(x - B_i(t))\, dx$ equals to

$$\begin{aligned}&\int_{\mathbb{R}^d} D^\alpha \varphi_\varepsilon\big(x - B_j(0)\big) D^\alpha \varphi_\varepsilon\,(x - B_i(0))\, dx \\ &\qquad - \int_0^t \int_{\mathbb{R}^d} \nabla D^\alpha \varphi_\varepsilon\left(x - B_j(s)\right) \nabla D^\alpha \varphi_\varepsilon\,(x - B_i(s))\, dxds \\ &\qquad + \int_0^t \int_{\mathbb{R}^d} D^\alpha \varphi_\varepsilon\,(x)\, \nabla D^\alpha \varphi_\varepsilon\left(x + B_j(s) - B_i(s)\right) dx \cdot dB_j(s) \\ &\qquad - \int_0^t \int_{\mathbb{R}^d} D^\alpha \varphi_\varepsilon\,(x)\, \nabla D^\alpha \varphi_\varepsilon\left(x + B_j(s) - B_i(s)\right) dx \cdot dB_i(s).\end{aligned}$$

Then summing over all $j > i$, we have

$$\|D^\alpha \rho_{\varepsilon,N}(\cdot,t)\|^2 = \|D^\alpha \rho_{\varepsilon,N}(\cdot,0)\|^2 - \int_0^t \|\nabla D^\alpha \rho_{\varepsilon,N}(\cdot,s)\|^2 ds + \tilde{M}_t + \frac{t}{N}\|\nabla D^\alpha \varphi_\varepsilon\|^2 \tag{92}$$

where

$$\tilde{M}_t = \sum_{i=1}^N \tilde{M}_t^i,$$

with

$$\begin{aligned}\tilde{M}_t^i =& \frac{2}{N^2}\int_0^t \int_{\mathbb{R}^d} \left[\sum_{j=1}^{i-1} D^\alpha \varphi_\varepsilon(x)\, \nabla D^\alpha \varphi_\varepsilon\big(x + B_i(s) - B_j(s)\big)\right] dx \cdot dB_i(s)\\ &- \frac{2}{N^2}\int_0^t \int_{\mathbb{R}^d} \left[\sum_{j=i+1}^{N} D^\alpha \varphi_\varepsilon(x)\, \nabla D^\alpha \varphi_\varepsilon\big(x + B_j(s) - B_i(s)\big)\right] dx \cdot dB_i(s).\end{aligned}$$

Then according to the same argument of Lemma 2, we have

$$E\big((\tilde{M}_t)^2\big) \le \frac{4t}{N\varepsilon^{2d+4|\alpha|+2}}\|D^\alpha \varphi\|^2 \|\nabla D^\alpha \varphi\|^2. \tag{93}$$

Combining (86), (88) and (92) we have

$$\begin{aligned}\|D^\alpha(\rho - \rho_{\varepsilon,N})(\cdot,t)\|^2 =& \|D^\alpha(\rho - \rho_{\varepsilon,N})(x,0)\|^2 - \int_0^t \|\nabla D^\alpha(\rho - \rho_{\varepsilon,N})(\cdot,s)\|^2 ds\\ &+ 2M_t + \tilde{M}_t + \frac{t}{N}\|\nabla D^\alpha \varphi_\varepsilon\|^2.\end{aligned} \tag{94}$$

With inequality (89) and (93), again we let $\varepsilon_N = N^{-1/3(d+2|\alpha|)} \to 0$ as $N \to \infty$. Then for any $t \ge 0$, letting

$$\begin{aligned}\delta_1^*(t) &:= \left(4N^{-1/2(d+2|\alpha|)}\|\nabla D^\alpha \varphi\|^2 \int_0^t \|D^\alpha \rho(x,s)\|^2 ds\right)^{1/3}\\ \delta_2^*(t) &:= \left(4N^{-1/2(d+2|\alpha|)}\|\nabla D^\alpha \varphi\|^2 \|D^\alpha \varphi\|^2\right)^{1/3}\end{aligned}$$

by Doob and Chebyshev inequality, it is easy to see that as $N \to \infty$ the probability of the following events:

$$\begin{aligned}P(A_N^*) &:= P(\sup_{s\le t}|M_t| < \delta_1(t)) \ge 1 - \delta_1(t)\\ P(B_N^*) &:= P(\sup_{s\le t}|\tilde{M}_t| < \delta_2(t)) \ge 1 - \delta_2(t)\end{aligned}$$

Noting that for the the constant term, we have

$$\frac{t}{N}\|\nabla D^\alpha \varphi_{\varepsilon_N}\|_2^2 < N^{-1/3(d+2|\alpha|)}t$$

when N is sufficiently large, so under the event

$$P(A_N^* \cap B_N^*) \geq 1 - \delta_1^*(t) - \delta_2^*(t)$$

we have that for all $s \leq t$

$$\|D^\alpha(\rho - \rho_{\varepsilon,N})(\cdot,s)\|^2 \leq \|D^\alpha(\rho - \rho_{\varepsilon,N})(\cdot,0)\|^2 - \int_0^s \|\nabla D^\alpha(\rho - \rho_{\varepsilon,N})(\cdot,h)\|^2\,dh \\ + \delta_1^*(t) + \delta_2^*(t) + N^{-1/3(d+2|\alpha|)}t. \tag{95}$$

Let

$$c_\alpha(t) = \left(4t\|\nabla D^\alpha\varphi\|^2\|D^\alpha\rho_0\|^2\right)^{1/3} + \left(4\|\nabla D^\alpha\varphi\|^2\|D^\alpha\varphi\|^2\right)^{1/3} + t \tag{96}$$

In above we have used the fact that $\|D^\alpha\rho(\cdot,s)\| \leq \|D^\alpha\rho_0\|$. Then for $\varepsilon_N = N^{-1/3(d+2|\alpha|)}$ that for and any $t \geq 0$, and the $c_\alpha(t)$ defined above that depends only on t and $\rho(x,s)$, we have

$$P\Big(\sup_{s\leq t}\big(\|D^\alpha(\rho - \rho_{\varepsilon_N,N})(\cdot,s)\|^2 + \int_0^s \|\nabla D^\alpha(\rho - \rho_{\varepsilon_N,N})\|^2 dh\big) \\ < \|D^\alpha(\rho - \rho_{\varepsilon_N,N})(\cdot,0)\|^2 + c_\alpha(t)N^{-1/6(d+2|\alpha|)}\Big) \geq 1 - c_\alpha(t)N^{-1/6(d+2|\alpha|)}. \tag{97}$$

when N is sufficiently large. So we have

Theorem 2 *Let ρ be the solution of the heat equation with initial density $\rho_0 \in H^k(\mathbb{R}^d)$, $B_i(t)$ be independent copies of Brownian Motions with initial data $B_i(0)$, and $\rho_{\varepsilon_N,N}$ be the constructed regularized empirical measure defined in (84), with regularized parameter $\varepsilon_N = N^{-1/3(d+2k)}$. For any $t \geq 0$, $|\alpha| \leq k$, we have*

$$P\Big(\sup_{s\leq t}\big(\|D^\alpha(\rho - \rho_{\varepsilon_N,N})(\cdot,s)\|^2 + \int_0^t \|\nabla D^\alpha(\rho - \rho_{\varepsilon_N,N})\|^2 dh\big) \\ < \|D^\alpha(\rho - \rho_{\varepsilon_N,N})(\cdot,0)\|^2 + c_\alpha(t)N^{-1/6(d+2|\alpha|)}\Big) \geq 1 - c_\alpha(t)N^{-1/6(d+2|\alpha|)}. \tag{98}$$

where $c_\alpha(t)$, $t > 0$ is positive function dependent only on t, φ and $\|\rho_0\|_{H^k}$ and is defined in (96).

Acknowledgments The research was partially supported by KI-Net NSF RNMS Grant No. 1107291, and NSF DMS Grant No. 1514826.

References

1. Bolley, F., Guillin, A., Villani, C.: Quantitative concentration inequalities for empirical measures on non-compact spaces. Prob. Theory Rel. Fields **137**, 541–593 (2007)
2. Chorin, A.J.: Numerical study of slightly viscous flow. J. Fluid Mech. **57**, 785–796 (1973)
3. Cottet, G.H., Koumoutsakos, P.: Vortex Methods: Theory and Practice, Cambridge University Press (2000)
4. Durrett, R.T.: Stochastic Calculus: A Practical Introduction, CRC Press (1996)
5. Ethier, S.N., Kurtz, R.G.: Markov Processes: Characterization and Convergence. Wiley (2005)
6. Fournier, N., Hauray, M., Mischler, S.: Propagation of chaos for the 2D viscous vortex model. J. Eur. Math. Soc. **16**, 1425–1466 (2014)
7. Goodman, J.: Convergence of the random vortex method. Comm. Pure Appl. Math. **40**, 189–220 (1987)
8. Federer, H.: Geometric Measure Theory. Springer, New York (1969)
9. Koumoutsakos, P., Leonard, A.: High-resolution simulations of the flow around an impulsively started cylinder using vortex methods. J. Fluid Mech. **96**, 1–38 (1995)
10. Liu, J.-G., Yang, R.: A random particle blob method for the Keller-Segel equation and convergence analysis. Math. Comp., to appear
11. Long, D.G.: Convergence of the random vortex method in two dimensions. J. Amer. Math. Soc **1**, 779–804 (1988)
12. Marchioro, C., Pulvirenti, M.: Hydrodynamics in two dimensions and vortex theory. Commun. Math. Phys. **84**, 483–504 (1982)
13. McKean, H.P.: Propagation of chaos for a class of non-linear parabolic equation. In: Lecture Series in Differential Equations, session 7, pp. 177–194, Catholic University (1967)
14. Karatzas, I., Shreve, S.: Brownian Motion and Stochastic Calculus, 2nd edn. Springer (1991)
15. Osada, H.: Propagation of chaos for the two dimensional Navier-Stokes equation. In: Probabilistic Methods in Mathematical Physics (Katata/Kyoto, 1985), pp. 303–334. Academic Press, Boston, MA (1987)
16. Stevens, A.: The derivation of chemotaxis equations as limit dynamics of moderately interacting stochastic many-particle systems. SIAM J. Appl. Math. **61**, 183–212 (2000)
17. Sznitman, A.-S.: Topics in propagation of chaos, In: Ecole d'Eté de Probabilités de Saint-Flour XIX-1989. Lecture Notes in Mathematics, vol. 1464. Springer, Berlin (1991)

References

[illegible]

From Market Data to Agent-Based Models and Stochastic Differential Equations

R. Vilela Mendes

Abstract A survey of results from 2008 to 2014 on the construction of a stochastic market model, from the empirical data to its modelling interpretation and proof of mathematical consistency (no-arbitrage and completeness).

Keywords Fractional volatility · Arbitrage · Completeness · Agent-based models

1 Introduction

After 2008, excessive reliance on mathematical models was blamed as the cause for the subprime collapse and the following global crisis. This criticism soon reached the popular level on newspaper articles and cartoons

It is true that a closed mathematical model of the economy may be an impossible dream, as is also true that some models use unrealistic assumptions, trading empirical evidence for mathematical beauty or simplicity. On the other hand, there are also cases where the mathematics is there, providing sound results, but nobody pays attention,

R.V. Mendes (✉)
Centro de Matemática e Aplicações Fundamentais, University of Lisbon,
Lisbon, Portugal
e-mail: rvmendes@fc.ul.pt; rvilela.mendes@gmail.com

© Springer International Publishing Switzerland 2016

P. Gonçalves and A.J. Soares (eds.), *From Particle Systems to Partial Differential Equations III*, Springer Proceedings in Mathematics & Statistics 162,
DOI 10.1007/978-3-319-32144-8_11

because sometimes not paying attention is more profitable in the short term. An example:

A theorem by Föllmer and Schied [1] states that if M is the set of all probability measures on a finite space of scenarios Ω, ρ will be a convex risk measure iff there is a penalty function $\alpha : M \to (-\infty, \infty]$ such that

$$\rho(X) = \sup_{Q \in P}(E_Q[-X] - \alpha(Q)) \quad (1)$$

with α convex, lower semicontinuous and $\alpha(Q) \geq -\rho(0)$.

In such a convex risk measure, the first term represents the **maximal expected loss in the scenario** Q and $\alpha(Q)$ accounts for the **probability of the scenario**. Whereas the calculation of $E_Q[-X]$ is a simple exercise in stochastic analysis, estimation of $\alpha(Q)$ involves many factors which are frequently not taken into account. *For example, if the historical data that is being used does not contain unfavorable events, it is tempting (or profitable) to say that meltdowns are improbable.*

Intrigued by why the well-qualified experts of the rating agencies had rated AAA the "toxic" products and had not predicted the 2008 crisis, 4 economists of the Federal Reserve of Atlanta, made an extensive analysis of their reports in the years before the crisis. The conclusion was that most experts reported that a small fall in the price of the houses would lead to disaster, but assigned a very small (*penalty*) probability to that event [2]. In the Atlanta paper, instead of the language of convex measures, the authors consider the probability of foreclosures decomposed into

$$\frac{df}{dt} = \frac{df}{dp}\frac{dp}{dt} \quad (2)$$

$\frac{df}{dp}$ being the sensitivity of foreclosures to price (HPA) and $\frac{dp}{dt}$ the time variation of the price of houses. Their conclusion is that the estimation of $\frac{df}{dp}$ was correct but not of $\frac{dp}{dt}$ (which is equivalent to the effect of the penalty term). However looking at the housing bubble it should have been clear after 2001 that the probability of a downturn in HPA was high. In addition, how could inflation-adjusted prices continue to rise when real incomes of most Americans, especially at the bottom, continued to fall?

Why was a small value assigned to the *penalty term?* Conflict of interests is a possible reason. The SEC recognition of the main agencies (Fitch, Moody's, S&P and Dominion) together with the recommendations of Basel II, put them in the center of the financial world. However their clients were exactly the creators of the securities. If one agency does not provide favorable reports, look elsewhere. A recent event lends credibility to this hypothesis: In February 2011, Redwood Trust put in the market the second (since the crisis) mortgage-based private-label security in the USA. Redwood asked both Fitch and Moody's for ratings, but only published Fitch's report. Later on, perhaps to assert its credibility, Moody's published its report which was quite negative. On the other hand if the client were the buyer of the securities, a conflict of interest of opposite sign might also occur. A recent proposal is the creation of government-sponsored clearing agencies. Will it ever work?

Another example is the Bernard Madoff affair. Already in 1999 and later in 2005 Markopoulos' letters to the SEC had shown, using simple mathematics that Madoff's strategy could not generate the 12 % average annual return unless he was either using **insider-trading** or a **Ponzi scheme.** Although some large brokerage institutions (Goldman Sachs, for example) stayed away from any deals with Madoff, he continued to attract a lot of investment in Europe and the USA. Once again, the mathematics was there but very few were paying attention.

The work that is reported in this paper concerns another aspect of the relation of mathematics to economic life, namely the modelization of the price $S(t)$ fluctuations in stock exchanges. The basic (stylized) facts provided by the empirical data are:

1. The **returns** ($r(t, \Delta) = \frac{S(t+\Delta)-S(t)}{S(t)}$) have nearly no autocorrelation;
2. The **autocorrelations of** $|r(t, \Delta)|$ decline slowly with increasing lag Δ, a long memory effect;
3. **Leptokurtosis**: asset returns have distributions with fat tails and excess peakedness at the mean;
4. **Autocorrelations of sign** $r(t, \Delta)$ are insignificant;
5. **Volatility clustering**: there is a tendency of large changes to follow large changes and small changes to follow small changes. Volatility occurs in bursts;
6. **Volatility is mean-reversing** and the distribution is close to lognormal or inverse gamma;
7. **Leverage effect**: volatility tends to rise more following a large price fall than following a price rise.

Geometrical Brownian motion (GBM)

$$\frac{dS_t}{S_t} = \mu dt + \sigma dB(t) \tag{3}$$

is a basis for most of mathematical finance (Black-Scholes, etc.). Is it consistent with the empirical data? No. In GMB price changes would be log-normal. No leptokurtosis and scaling properties $E\left|\frac{S(t+\Delta)-S(t)}{S(t)}\right| \approx \Delta^{1/2}$ which is not born out by the data. In addition, the volatility σ being constant, there is no volatility clustering nor leverage effect. One of the most famous consequences of GBM is the Black-Scholes formula for option pricing. When the historical volatility σ is used in the formula, the resulting price is quite distinct from the one that is actually practiced in the market. Nevertheless the Black-Scholes continues to be used in the following way: The market price of liquid options is used to infer what would be the (implied) volatility σ_{imp} that would lead to that price. σ_{imp} is then used to compute the price of less traded options. Black-Scholes is used as an invertible mapping between price and σ_{imp} and not for its theoretical value.

In conclusion: the wide use of GBM is a case where oversimplification of the mathematics leads to important departures from the empirical evidence. This was the motivation that led to the present attempt to construct a market model, which although preserving some of the nice features of GBM, would be consistent with (and directly inspired by) the empirical data.

2 A Data-Reconstructed Market Model [3]

The basic hypothesis for the model construction were:

- (**H1**) *The log-price process* $\log S_t$ *belongs to a probability space* $\Omega \otimes \Omega'$, where the first one, Ω, is the Wiener space W and the second, Ω', is a probability space to be empirically reconstructed.
  Denote $\log S_t(\omega, \omega')$ with $\omega \in \Omega$, $\omega' \in \Omega'$, and $\mathscr{F}_t$, $\mathscr{F}_t'$ are the $\sigma-$algebras in Ω and Ω' generated by the processes up to time t.
- (**H2**) The second hypothesis is stronger: *Assume that for each fixed* ω', $\log S_t(\bullet, \omega')$ *is a square integrable random variable in* Ω

From (H2) it follows [4] that, for each fixed ω',

$$\frac{dS_t}{S_t}(\bullet, \omega') = \mu_t(\bullet, \omega')dt + \sigma_t(\bullet, \omega')dB(t) \tag{4}$$

with $\mu_t(\bullet, \omega')$ and $\sigma_t(\bullet, \omega')$ well-defined processes in Ω.

If $\{X_t, \mathscr{F}_t\}$ is a process such that

$$dX_t = \mu_t dt + \sigma_t dB(t) \tag{5}$$

with μ_t and σ_t being $\mathscr{F}_t-$adapted processes, then

$$\begin{aligned} \mu_t &= \lim_{\varepsilon \to 0} \tfrac{1}{\varepsilon} \{E(X_{t+\varepsilon} - X_t) | \mathscr{F}_t\} \\ \sigma_t^2 &= \lim_{\varepsilon \to 0} \tfrac{1}{\varepsilon} \left\{E(X_{t+\varepsilon} - X_t)^2 \middle| \mathscr{F}_t\right\} \end{aligned} \tag{6}$$

The process associated to the probability space Ω' is now inferred from the data. For each fixed ω' realization in Ω' one has

$$\sigma_t^2(\bullet, \omega') = \lim_{\varepsilon \to 0} \frac{1}{\varepsilon} \left\{E(\log S_{t+\varepsilon} - \log S_t)^2\right\} \tag{7}$$

Because each set of market data corresponds to a particular realization ω', the σ_t^2 process may indeed be reconstructed from the data. The question is how to construct a mathematical model for this *induced volatility* process. For this purpose we looked for scaling properties of the data, namely

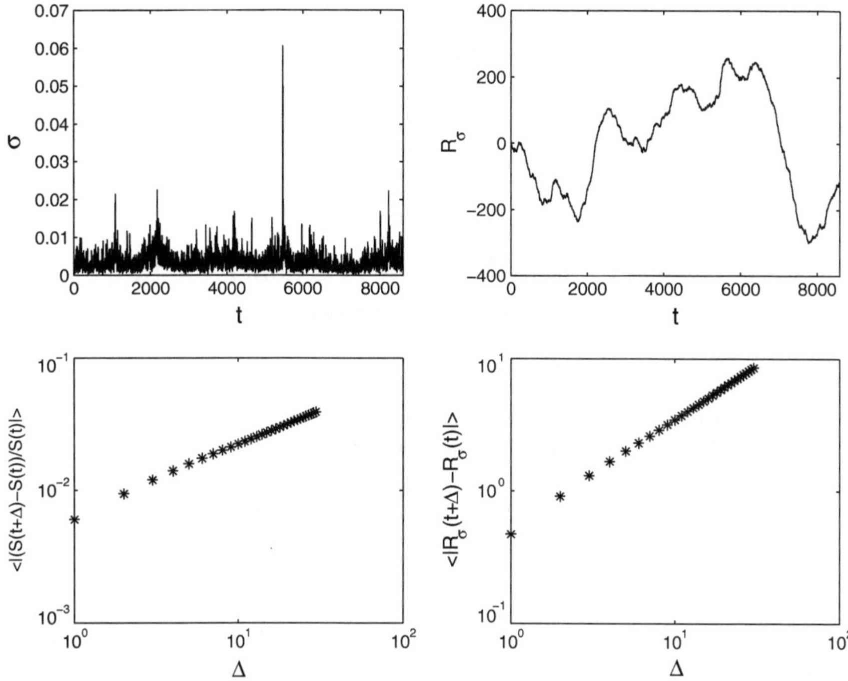

Fig. 1 The R_σ process and its scaling properties

$$E\left|\sigma(t+\Delta)-\sigma(t)\right| \sim \Delta^H \qquad\qquad E\left|\frac{\sigma(t+\Delta)-\sigma(t)}{\sigma(t)}\right| \sim \Delta^H \tag{8}$$

but neither of these hold. By contrast, the empirical integrated log-volatility is well represented by a relation of the form $\sum_{n=0}^{t/\delta} \log\sigma(n\delta) = \beta t + R_\sigma(t)$ with the $R_\sigma(t)$ process displaying very accurate self-similar properties (Fig. 1).

If a nondegenerate process X_t has finite variance, stationary increments and is self-similar

$$Law(X_{at}) = Law(a^H X_t)$$

then it necessarily has covariance

$$Cov(X_s, X_t) = \frac{1}{2}(|s|^{2H} + |t|^{2H} - |s-t|^{2H})E(X_1^2) \tag{9}$$

and the simplest process with these properties is a Gaussian process called *fractional Brownian motion*. Therefore the simplest model compatible with the data is

$$\boxed{\begin{array}{ll} dS_t & = \mu S_t dt + \sigma_t S_t dB(t) \\ \log\sigma_t & = \beta + \frac{k}{\delta}\left\{B_H(t) - B_H(t-\delta)\right\} \end{array}} \tag{10}$$

which has been called the *fractional volatility model* (FVM). δ is the observation time scale and, from the data, H is found to be in the range 0.8–0.9. From (10) it follows that the volatility (at resolution δ) is

$$\sigma(t) = \theta e^{\frac{k}{\delta}\{B_H(t) - B_H(t-\delta)\} - \frac{1}{2}\left(\frac{k}{\delta}\right)^2 \delta^{2H}} \tag{11}$$

3 Mathematical Consistency of the Fractional Volatility Model: Arbitrage and Completeness [5]

The main consistency check for any market model is the no-arbitrage principle. In addition completeness or incompleteness of the model should also be checked.

The no-arbitrage principle: *A (perfect) market does not allow for risk-free profits with no initial investment or, equivalently, to profits without any risk.*

A self-financing portfolio $V(t) = \sum_{i=1}^{n} h_t^i S_t^i$ is an *arbitrage portfolio* if

$$V(0) = 0; \;\; P(V(T) > 0) > 0 \tag{12}$$

for $T > 0$, P being the probability measure on the market scenarios. *A market is arbitrage free if and only if there is an equivalent martingale measure Q for the discounted price processes* [1, 6].

On the other hand completeness is related to the possibility of hedging portfolios. H is said to be *an hedge* for the portfolio X (or to *replicate* X) if

$$H \text{ is self-financing; } V_H(T) = X(T), \;\; P - \text{almost surely} \tag{13}$$

A market is *complete* if all X can be hedged. Furthermore, *a market is complete if and only if the martingale measure Q is unique.*

3.1 No-Arbitrage

In the fractional volatility model (FVM) one has two probability spaces $(\Omega_1, \mathcal{F}_1, P_1)_W$ and $(\Omega_2, \mathcal{F}_2, P_2)_{B_H}$ and the product space $(\overline{\Omega}, \overline{\mathcal{F}}, \overline{P})$ with π_1 and π_2 the projections of $\overline{\Omega}$ onto Ω_1 and Ω_2.

Consider a risky asset with price S_t and a risk-free asset with dynamics

$$dA_t = rA_t \, dt \qquad A_0 = 1 \tag{14}$$

The volatility σ_t of the risky asset is a measurable $\overline{\mathbb{F}}$–adapted process satisfying for $0 \le t < \infty$

$$\mathbb{E}_{\overline{P}}\left[\int_0^t \sigma_s^2 ds\right] = \int_0^t \theta^2 e^{-(\frac{k}{\delta})^2\delta^{2H}} \mathbb{E}_{\overline{P}}\left[e^{\frac{2k}{\delta}\{B_H(s)-B_H(s-\delta)\}}\right] ds$$
$$= \theta^2 \exp\left\{(\frac{k}{\delta})^2\delta^{2H}\right\} t < \infty \tag{15}$$

by Fubini's theorem and the moment generating function of the Gaussian random variable $B_H(s) - B_H(s-\delta)$.

$\int_0^t |\mu_s|\, ds$ being finite $\overline{P}$–almost surely for $0 \le t < \infty$, application of Itô's formula yields

$$S_t = S_0 \exp\left\{\int_0^t (\mu_s - \frac{1}{2}\sigma_s^2) ds + \int_0^t \sigma_s dB_s\right\} \tag{16}$$

Lemma 1 *Consider the measurable process*

$$\gamma_t = \frac{r - \mu_t}{\sigma_t}, \qquad 0 \le t < \infty \tag{17}$$

with $\mu \in L^\infty([0, T] \times \overline{\Omega})$. *Then, for a continuous version of* B_H

$$\exp\left[\frac{1}{2}\int_0^T \gamma_s^2(\omega_2) ds\right] < A(\omega_2) < \infty \tag{18}$$

P_2*–almost all* $\omega_2 \in \Omega_2$.

The proof uses the fact that P_2–almost surely, a continuous version of fractional Brownian motion is Hölder continuous of any order $\alpha \ge 0$ less than H, that is, there is a random variable $C_\alpha > 0$ such that for P_2–almost all $\omega_2 \in \Omega_2$ $|B_H(t) - B_H(s)| \le C_\alpha(\omega_2)\, |t - s|^\alpha$

Theorem 1 *The market* (A_t, S_t, σ_t) *is free of arbitrage*

Proof Restricting the process to a particular path ω_2 of the B_H–process, construct the stochastic exponential of $\int_0^t \gamma_s(\omega_2) dB_s$,

$$\eta_t(\omega_2) = \exp\left\{\int_0^t \gamma_s(\omega_2) dB_s - \frac{1}{2}\int_0^t \gamma_s^2(\omega_2) ds\right\} \tag{19}$$

The bound in Lemma 1 is the Kallianpur condition that insures

$$\mathbb{E}_{P_1}\left[\eta_t(\omega_2)\right] = 1 \qquad \omega_2 - a.s. \tag{20}$$

Hence, we are in the framework of Girsanov theorem and $\eta_t(\omega_2)$ is a true P_1- martingale. We can define for each $0 \leq T < \infty$ a new probability measure $Q_T(\omega_2)$ on $\mathscr{F}_1$ by

$$\frac{dQ_T(\omega_2)}{dP_1} = \eta_T(\omega_2), \qquad P_1 - a.s. \tag{21}$$

By the Cameron-Martin-Girsanov theorem, for each $T \in [0, \infty)$, the process

$$B_t^* = B_t - \int_0^t \frac{r - \mu_s}{\sigma_s(\omega_2)} ds \qquad 0 \leq t \leq T \tag{22}$$

is a Brownian motion on the probability space $(\Omega, \mathscr{F}_1, Q_T(\omega_2))$.

Under the new probability measure $Q_T(\omega_2)$ (equivalent to P_1 on $\mathscr{F}_1$) the discounted price process, $Z_t = \frac{S_t}{A_t} \quad 0 \leq t \leq T$ with dynamical law

$$Z_t(\omega_2) = Z_0 + \int_0^t \sigma_s(\omega_2) Z_s(\omega_2)\, dB_s^* \tag{23}$$

is a martingale in the probability space $(\Omega_1, \mathscr{F}_1, Q_T(\omega_2))$. *By the fundamental theorem of asset pricing* [1, 6], *the existence of an equivalent martingale measure for Z_t implies that there are no arbitrages*, that is, $\mathbb{E}_{Q_T(\omega_2)}\left[Z_t(\omega_2)|\mathscr{F}_{1,s}\right] = Z_s(\omega_2)$ for $0 \leq s < t \leq T$.

This proves that there are no arbitrages for P_2-almost all ω_2 trajectories of the B_H process. Because this process is independent from the B process it follows that the no-arbitrage result is also valid in the product space. □

3.2 *Incompleteness*

In this financial model, trading takes place only in the risky asset and in the money market. As a consequence the volatility risk cannot be hedged. Having more sources of risk than tradable assets, suggests that the market is incomplete

Theorem 2 *The market defined by (A_t, S_t, σ_t) is incomplete*

Proof Use an integral representation for the fractional Brownian motion

$$B_H(t) = \int_0^t K_H(t, s) dW_s \tag{24}$$

W_t being a Brownian motion independent from B_t and K_H is the square integrable kernel

$$K_H(t, s) = C_H s^{\frac{1}{2}-H} \int_s^t (u - s)^{H-\frac{3}{2}} u^{H-\frac{1}{2}}\, du, \quad s < t \tag{25}$$

$(H > 1/2)$. Then the process

$$\eta'_t = \exp(W_t - \frac{1}{2}t) \tag{26}$$

is a square-integrable P_2–martingale. Now, define a standard bi-dimensional Brownian motion, $W_t^* = (B_t, W_t)$ and the process $\eta_t^*(\omega_2) = \eta_t \eta'_t(\omega_2)$

$$\eta_t^*(\omega_2) = \exp\left\{\int_0^t \Gamma_s(\omega_2) \bullet dW_t^* - \frac{1}{2}\int_0^t \|\Gamma_s(\omega_2)\|^2 ds\right\} \tag{27}$$

where, by the Lemma 1, $\Gamma(\omega_2) = (\gamma(\omega_2), 1)$ is also a P_1–martingale. Then, by the Cameron-Martin-Girsanov theorem, the process $\widetilde{W}_t^* = (\widetilde{W}_t^{*(1)}, \widetilde{W}_t^{*(2)})$ defined by

$$\widetilde{W}_t^{*(1)} = B_t - \int_0^t \gamma_s(\omega_2)ds; \; \widetilde{W}_t^{*(2)} = W_t - t \tag{28}$$

is a bi-dimensional Brownian motion on the probability space $(\Omega_1, \mathscr{F}_1, Q_T^*(\omega_2))$, where $Q_T^*(\omega_2)$ is the probability measure $\frac{dQ_T^*(\omega_2)}{dP_1} = \eta_T^*(\omega_2)$. Moreover, the discounted price process Z remains a martingale with respect to the new measure $Q_T^*(\omega_2)$. $Q_T^*(\omega_2)$ being an equivalent martingale measure distinct from $Q_T(\omega_2)$, the market is incomplete. □

3.3 Leverage, a Modified Model and Completeness [5, 7]

The following nonlinear correlation of the returns

$$L(\tau) = \left\langle |r(t+\tau)|^2 r(t)\right\rangle - \left\langle |r(t+\tau)|^2\right\rangle \langle r(t)\rangle \tag{29}$$

is called *leverage* and the *leverage effect* is the fact that, for $\tau > 0$, $L(\tau)$ starts from a negative value whose modulus decays to zero whereas for $\tau < 0$ it has almost negligible values (see Fig. 2 which shows a typical behavior in the NYSE data).

As expressed in (10) the fractional volatility model has the volatility process σ_t acting on the log-price, but not conversely. Therefore, in its simplest form, the *fractional volatility model contains no leverage effect*.

However, leverage may be implemented by a small modification of the model. Use a (truncated) representation for fractional Brownian motion as a stochastic integral over Brownian motion and *identify the random generator of the log-price process with the stochastic integrator of the volatility*. A leverage effect is then obtained.

$$\mathscr{H}(t) = \Pi^{(M)}\left[C_H\left\{\begin{array}{c}\int_{-\infty}^0((t-u)^{H-\frac{1}{2}} - (-u)^{H-\frac{1}{2}})dW_u \\ + \int_0^t (t-u)^{H-\frac{1}{2}} dW_u\end{array}\right\}\right] \tag{30}$$

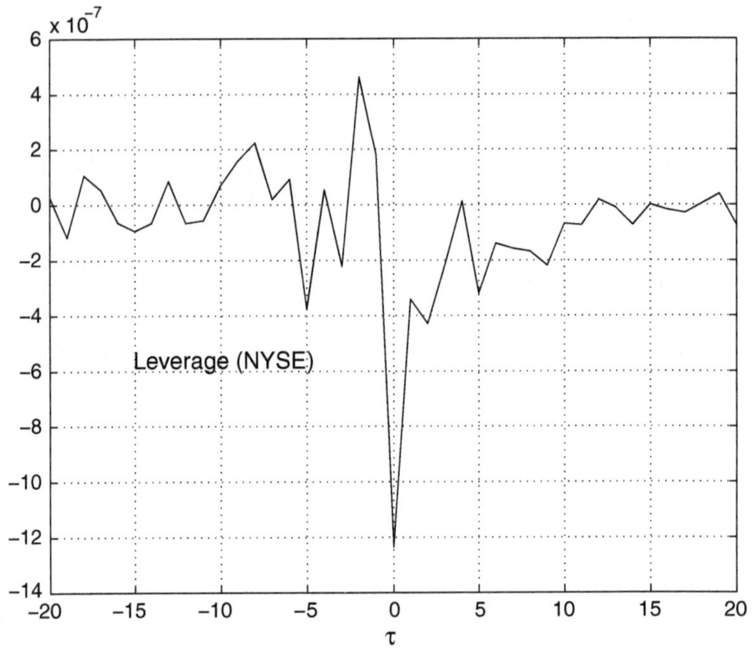

Fig. 2 Example of the leverage effect in NYSE data

$\Pi^{(M)}$ meaning the truncation of the representation to an interval $[-M, M]$ with M arbitrarily large.

Now, instead of two, there is only one source of risk and the new fractional volatility model would be

$$\begin{aligned} dS'_t &= \mu_t S'_t dt + \sigma_t S'_t dW_t \\ \log \sigma'_t &= \beta + \frac{k'}{\delta} \{\mathscr{H}(t) - \mathscr{H}(t-\delta)\} \end{aligned} \tag{31}$$

Theorem 3 *The market (A_t, S'_t, σ'_t) is free of arbitrage and complete.*

Proof Because the two processes are not independent we cannot use the same argument as before to obtain the Kallianpur condition. However, with the truncation, the Hölder condition is trivially verified for all the truncated paths of σ'_t and the construction of an equivalent martingale measure follows the same steps as in Theorem 2. Hence we have a P_1-martingale with respect to $(\mathscr{F}_{1,t})_{0 \leq t < T}$

$$\eta_t = \exp \left\{ \int_0^t \frac{r - \mu_s}{\sigma_s} dW_s - \frac{1}{2} \int_o^t (\frac{r - \mu_s}{\sigma_s})^2 ds \right\} \tag{32}$$

and Q_T, defined by $\frac{dQ_T}{dP_1} = \eta_T$ is an equivalent martingale measure.

The set of equivalent local martingale measures being non-empty, let Q^* be an element in this set. By the Girsanov converse there is a $\mathbb{R}$-valued process ϕ such that the Radon-Nikodym density of Q^* is

$$\frac{dQ_T^*}{dP_1} = \exp\left\{\int_0^T \phi_s dW_s - \frac{1}{2}\int_0^T \phi_s^2 ds\right\} \tag{33}$$

Moreover the process W_t^* given by

$$W_t^* = W_t - \int_0^t \phi_s ds \tag{34}$$

is a standard Q^*–Brownian motion and the discounted price process Z' satisfies the following stochastic differential equation

$$dZ_t' = (\mu_t - r + \sigma_t'\phi_t)Z_t'dt + \sigma_t Z_t' dW_t^* \tag{35}$$

Because Z_t' is a Q^*–martingale, then it must be $\mu(t, \omega) - r + \sigma'(t, \omega)\phi(t, \omega) = 0$ almost everywhere w.r.t. $dt \times P$ in $[0, T] \times \Omega$. It implies

$$\phi(t, \omega) = \frac{r - \mu(t, \omega)}{\sigma'(t, \omega)} \tag{36}$$

a. e. $(t, \omega) \in [0, T] \times \Omega_1$. Hence $Q_T^* = Q_T$, that is, Q_T is the unique equivalent martingale measure. This market model is complete. □

Figure 3 compares the leverage effect in the two models, that is, the original FVM (10) and the modified one (31).

3.4 A Remark on Long Memory and Fractional Brownian Motion

In the past, several authors had already tried to describe long memory effects in the market data by replacing in the price process Brownian motion by fractional Brownian motion with $H > 1/2$. However it was soon realized [8–11] that this replacement implied the existence of arbitrage. These results might be avoided either by restricting the class of trading strategies [12], introducing transaction costs [13] or replacing pathwise integration by a different type of integration [14, 15]. However this is not free of problems because the Skorohod integral approach requires the use of a Wick product either on the portfolio or on the self-financing condition, leading to unreasonable situations from the economic point of view (for example positive portfolio with negative Wick value, etc.) [16].

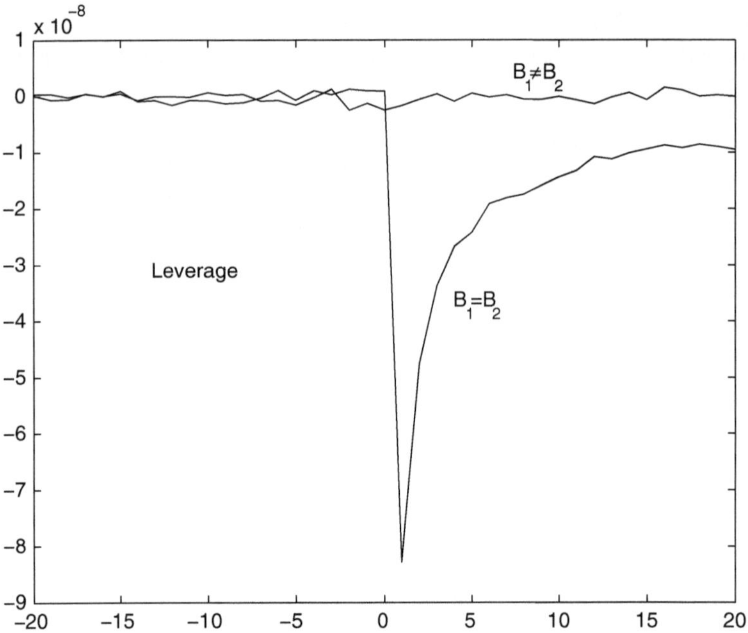

Fig. 3 Comparison of leverage in the original and the modified model

The fractional volatility model in Eq. (10) is not affected by these considerations, because it is the volatility process that is driven by fractional noise, not the price process and, as shown, a no-arbitrage result may be proven. This is no surprise because the requirement (H2) that, for each sample path $\omega_2 \in \Omega_2$, $\log S_t(\cdot, \omega_2)$ is a square integrable random variable in Ω_1 already implies that $\int \sigma_t dB_t$ is a martingale. The square integrability is also essential to guarantee the possibility of reconstruction of the σ process from the data.

4 Agent-Based Interpretation of the Fractional Volatility Model

In [17] two agent-based models were considered:

- In the first the traders strategies play a determinant role.
- In the second the determinant effect is the limit-order book dynamics, the agents having a random nature.

4.1 A Market Model with Self-adapted or Fixed Strategies

A set of investors is playing *against* the market. In addition to the impact of this group of investors, other factors are represented by a stochastic process η_t

$$z_{t+1} = f(z_t, \omega_t) + \eta_t \tag{37}$$

($z_t = \log S_t$) and ω_t is the total investment made by the group of traders. After r time steps, s agents copy the strategy of the s best performers and, at the same time, have some probability to mutate that strategy.

The model was run with different initial conditions and with or without evolution of the strategies. When the model is run with evolution the asymptotic steady-state behavior depends on the initial conditions. Different types of return statistics corresponded to the relative importance of either "value investors" or "technical traders". The occurrence of market bubbles and fat tails corresponds to situations where technical trader strategies were well represented. A situation where there are 50 % of fundamental (value-investing) strategies and 50 % trend-following, was chosen to compare its statistical properties with those of the FVM. The squared volatility $\sigma_t^2 = \frac{1}{|T_0-T_1|} var(\log p_t)$ and the parameters in $\sum_{n=0}^{t/\delta} \log \sigma(n\delta) = \beta t + R_\sigma(t)$ and $|R_\sigma(t+\Delta) - R_\sigma(t)|$ were estimated from the simulations. Figure 4 shows the results.

Notice the lack of scaling behavior of $R_\sigma(t)$ with an asymptotic exponent 0.55, denoting the lack of memory of the volatility process. This might already be evident from the time behavior of $R_\sigma(t)$ in the lower left plot. Also, although the returns have fat tails in this case, they are of different shape from those observed in the market data. Similar conclusions are obtained with other combinations of agent strategies.

In conclusion: *It seems that the features of the fractional volatility model (which are also those of the bulk market data) are not easily captured by a choice of strategies in an agent-based model.*

Agents' reactions and strategies are very probably determinant during market crisis and market bubbles but not in business-as-usual days.

4.2 A Limit-Order Book Market Model

In this model, *asks* and *bids* arrive at random on a window $[S_t - w, S_t + w]$ around the current price S_t. Every time a *buy* order arrives it is fulfilled by the closest non-empty ask slot, the new current price being determined by the value of the ask that fulfills it. If no ask exists when a buy order arrives it goes to a cumulative register to wait to be fulfilled. The symmetric process occurs when a *sell* order arrives, the new price being the bid that buys it. Sell and buy orders, asks and bids all arrive at random. Because the window around the current price moves up and down, asks and bids that are too far away from the current price are automatically eliminated.

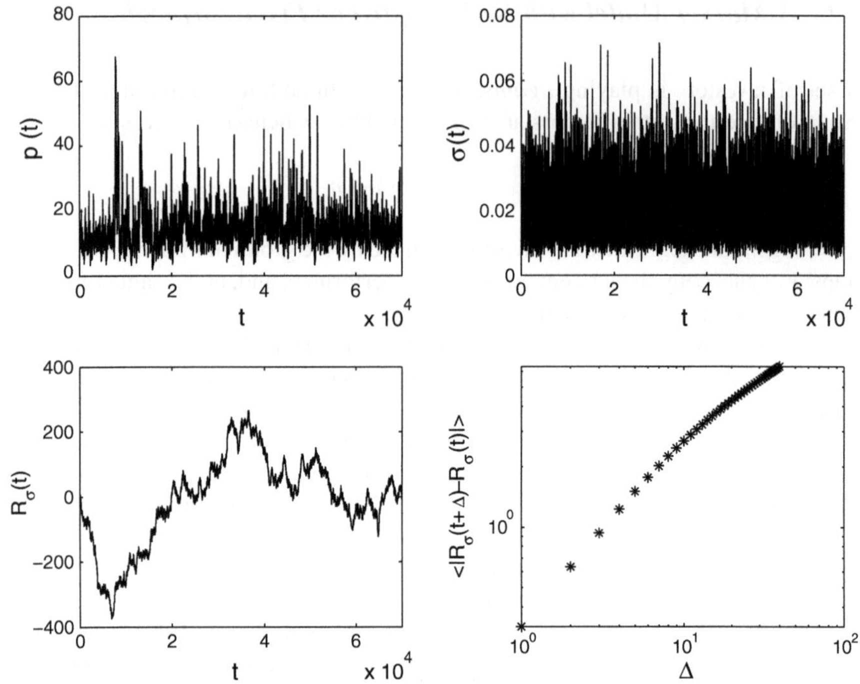

Fig. 4 Statistical properties of the model with agent strategies

The only parameters of the model are the width w of the limit-order book and the size n of the asks and bids, the sell and buy orders being normalized to one.

The model was run for different widths w and liquidities n. Although the exact values of the statistical parameters depend on w and n, the statistical nature of the results is essentially the same. In the Fig. 5 ($n = 2$) the limit-order book is divided into $2w + 1 = 21$ discrete price slots with $\Delta p = 0.1$. The scaling properties of $R_\sigma(t)$ are quite evident from the lower right plot in the figure, the Hurst coefficient being 0.96.

Conclusion: *the main statistical properties of the market data (fast decay of the linear correlation of the returns, non-Gaussianity and volatility memory) are already generated by the dynamics of the limit-order book with random behavior of the agents. A large part of the market statistical properties (in normal business-as-usual days) depends more on the nature of the price fixing financial institutions than on particular investor strategies.*

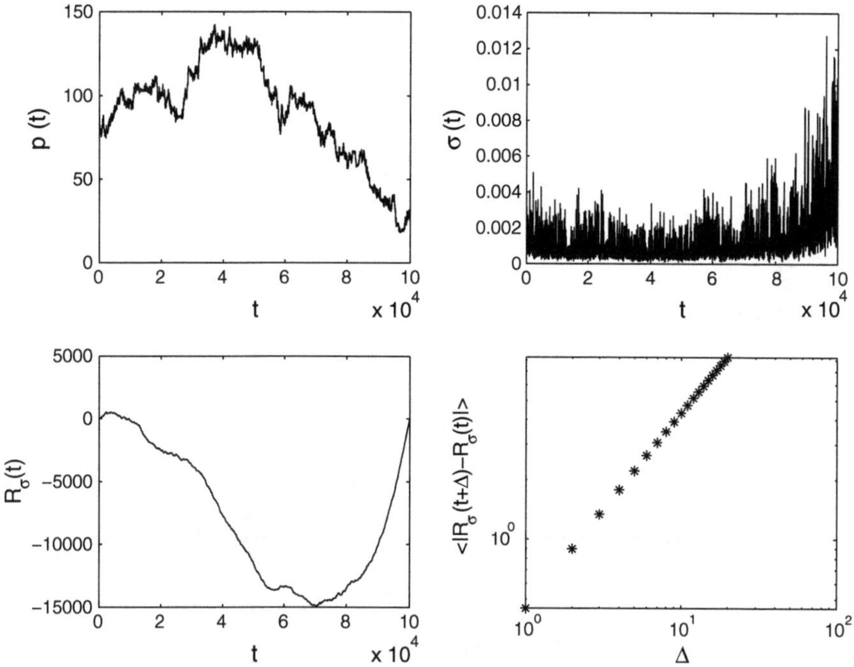

Fig. 5 Statistical properties of the limit order book model

5 Further Properties of the Fractional Volatility Model

In the FVM the statistics of returns is obtained in closed form. From

$$P_\delta(r(\Delta)) = \frac{1}{4\pi\theta k\delta^{H-1}\sqrt{\Delta}} \int_0^\infty dx x^{-\frac{1}{2}} e^{-\frac{1}{C}(\log x)^2} e^{-\lambda x} \tag{38}$$

$$r(\Delta) = \log S_{t+\Delta} - \log S_t \ , \ \theta = e^\beta, \ \lambda = \frac{(r(\Delta) - r_0)^2}{2\Delta\theta^2} \tag{39}$$

$$r_0 = \left(\mu - \frac{\sigma^2}{2}\right)\Delta \ , \ C = 8k^2\delta^{2H-2} \tag{40}$$

one obtains

$$\boxed{P_\delta(r(\Delta)) = \frac{1}{4\pi\theta k\delta^{H-1}\sqrt{\Delta}} \frac{1}{\sqrt{\lambda}} \left(e^{-\frac{1}{C}(\log\lambda - \frac{d}{dz})^2}\Gamma(z)\right)\Big|_{z=\frac{1}{2}}} \tag{41}$$

with asymptotic behavior, for large returns

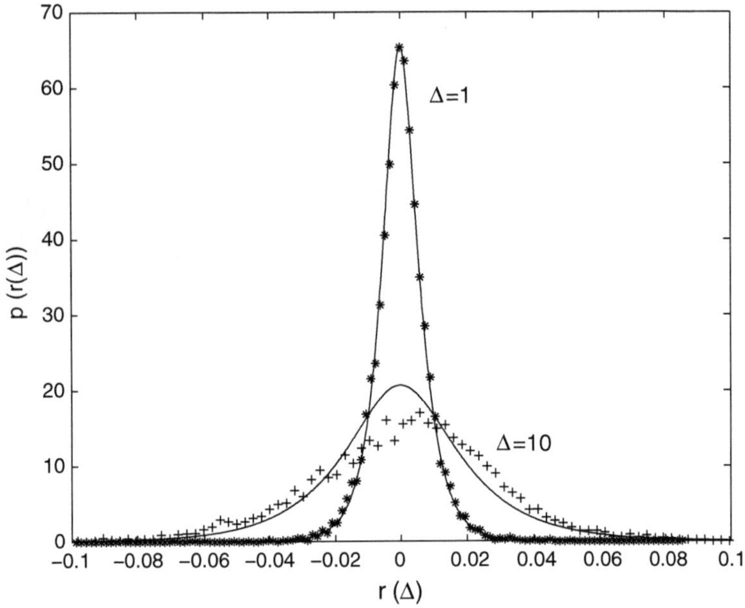

Fig. 6 Statistics of returns in the FVM compared with NYSE data

$$\boxed{P_\delta(r(\Delta)) \sim \frac{1}{\sqrt{\Delta\lambda}} e^{-\frac{1}{c}\log^2\lambda}} \tag{42}$$

This form provides a good fit of the empirical data. Figure 6 compares NYSE data with (41) for $H = 0.83$, $k = 0.59$, $\beta = -5$, $\delta = 1$, $\Delta = 1$ and $\Delta = 10$.

That NYSE returns are well described by (41) is no wonder because the FVM itself was reconstructed from that data. What seemed stranger, at first sight, was the fact that, once the parameters of the model are fixed, a simple change of Δ would predict the returns in a quite different market. This is shown in Fig. 7 where the same parameters as above were used, simply changing to $\Delta = \frac{1}{440}$ (1 min). The prediction of the model is compared with 1-min data of USDollar-Euro market for a couple of months in 2001. The result may be surprising, because one would not expect the volatility parametrization to carry over to such a different time scale and also because one is dealing with a different market. However, if the conclusion from the agent-based models is correct, that in business-as-usual days the statistics of the data depends more on the price fixation process than on agent strategies or other market features, then this result is no longer a surprise.

Using a simple risk neutrality argument, a new option pricing was also obtained, namely

$$\boxed{\begin{aligned} V(S_t, \sigma_t, t) = \{ & S_t \left[aM(\alpha, a, b) + bM(\alpha, b, a) \right] \\ & -Ke^{-r(T-t)} \left[aM(\alpha, a, -b) - bM(\alpha, -b, a) \right] \} \end{aligned}} \tag{43}$$

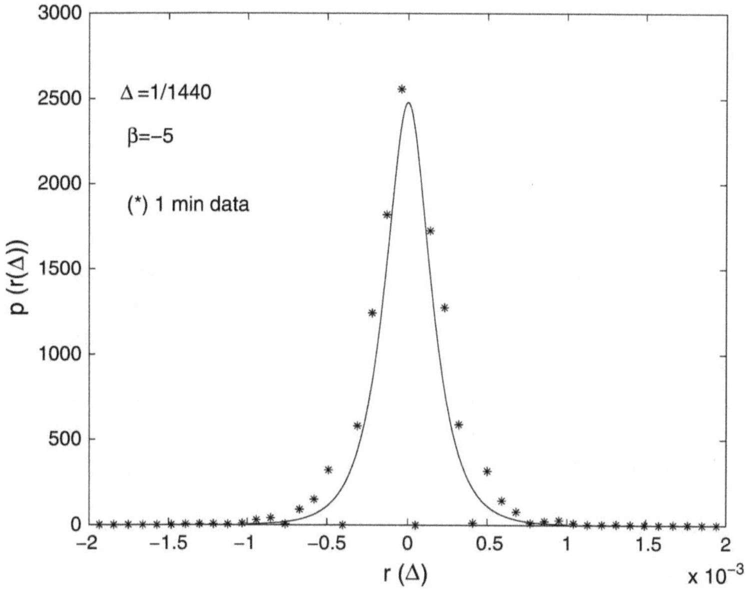

Fig. 7 Statistics of returns for 1 min exchange data

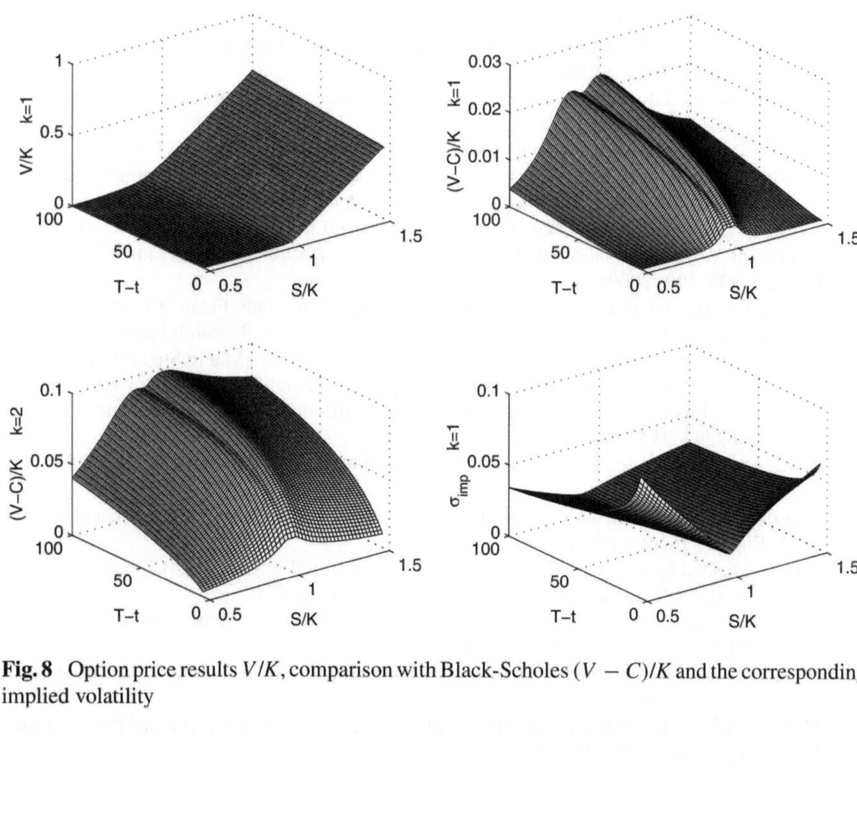

Fig. 8 Option price results V/K, comparison with Black-Scholes $(V - C)/K$ and the corresponding implied volatility

with

$$M(\alpha, a, b) = \frac{1}{2\pi\alpha} \int_{-1}^{\infty} dy \int_{0}^{\infty} dx e^{-\frac{\log^2 x}{2\alpha^2}} e^{-\frac{y^2}{2}(ax+\frac{b}{x})^2}$$
$$= \frac{1}{4\alpha} \sqrt{\frac{2}{\pi}} \int_{0}^{\infty} dx \frac{e^{-\frac{\log^2 x}{2\alpha^2}}}{ax+\frac{b}{x}} erfc\left(-\frac{ax}{\sqrt{2}} - \frac{b}{\sqrt{2}x}\right) \quad (44)$$

K is the strike price and T the maturity time. In Fig. 8 is plotted $V(S_t, \sigma_t, t)$ in the range $T - t \in [5, 100]$ with $S/K \in [0.5, 1.5]$ as well as $(V(S_t, \sigma_t, t) - C(S_t, \sigma_t, t))$ $/K$ for $k = 1$ and $k = 2$. $C(S_t, \sigma_t, t)$ is the Black-Scholes result. Other parameters are fixed at $\sigma = 0.01$, $r = 0.001$, $\delta = 1$, $H = 0.8$. To compare with Black-Scholes (BS), the implied volatility, that would reproduce the same results, was also computed. The implied volatility surface corresponding to $V(S_t, \sigma_t, t)$ is shown for for $k = 1$. It predicts a smile effect with the smile increasing as maturity approaches.

References

1. Föllmer, H., Schied, A.: Stochastic Finance: An Introduction in Discrete Time. Walter de Gruyter, New York (2004)
2. Gerardi, G.S., Lehnert, S., Sherland, S.M., Willen, P.S.: Making sense of the subprime crisis, Working paper 2009-2, Federal Bank of Atlanta
3. Mendes, R.V., Oliveira, M.J.: A data-reconstructed fractional volatility model, E-Economics/discussion paper/2008-22
4. Nualart, D.: The Malliavin Calculus and Related Topics. Springer, Berlin (2006)
5. Mendes, R.V., Oliveira, M.J., Rodrigues, A.M.: No-arbitrage, leverage and completeness in a fractional volatility model. Phys. A: Stat. Mech. Appl. **419**, 470–478 (2015)
6. Björk, T.: Arbitrage Theory in Continuous Time. Oxford University Press, New York (2009)
7. Mendes, R.V.: A fractional calculus interpretation of the fractional volatility model. Nonlinear Dyn. **55**, 395–399 (2009)
8. Rogers, L.C.G.: Arbitrage with fractional Brownian motion. Math. Finance **7**, 95–105 (1997)
9. Shiryaev, A.N.: On arbitrage and replication for fractal models, Research Report 20. Department of Mathematical Sciences, University of Aarhus, Denmark, MaPhySto (1998)
10. Salopek, D.M.: Tolerance to arbitrage. Stochast. Process. Appl. **76**, 217–230 (1998)
11. Sottinen, T.: Fractional Brownian motion, random walks and binary market models. Finance Stochast. **5**, 343–355 (2001)
12. Cheridito, P.: Arbitrage in fractional Brownian motion models. Finance Stochast. **7**, 533–553 (2003)
13. Guasoni, P.: No arbitrage under transaction costs, with fractional Brownian motion and beyond. Math. Finance **16**, 569–582 (2006)
14. Hu, Y., Øksendal, B.: Fractional white noise calculus and applications to finance. Infinite Dimensional Anal. Quantum Probab. Relat. Top. **6**, 1–32 (2003)
15. Elliott, R.J., van der Hoek, J.: A general fractional white noise theory and applications to finance. Math. Finance **13**, 301–330 (2003)
16. Björk, T., Hult, H.: A note on Wick products and the fractional Black-Scholes model. Finance Stochast. **9**, 197–209 (2005)
17. Mendes, R.V.: The fractional volatility model: an agent-based interpretation. Phys. A: Stat. Mech. Appl. **387**, 3987–3994 (2008)

Global Asymptotic Stability of a General Nonautonomous Cohen-Grossberg Model with Unbounded Amplification Functions

José J. Oliveira

Abstract For a class of nonautonomous differential equations with infinite delay, we give sufficient conditions for the global asymptotic stability of an equilibrium point. This class is general enough to include, as particular cases, the most of famous neural network models such as Cohen-Grossberg, Hopfield, and bidirectional associative memory. It is relevant to notice that here we obtain global stability criteria without assuming bounded amplification functions. As illustrations, results are applied to several concrete models studied in some earlier publications and new global stability criteria are given.

Keywords Cohen-Grossberg neural networks · Unbounded time-varying coefficients · Unbounded distributed delays · Unbounded amplification functions · Global asymptotic stability

1 Introduction

The Cohen-Grossberg neural network models, first proposed and studied by Cohen and Grossberg [4] in 1983, have been the subject of an active research due to their extensive applications in various engineering and scientific areas such as neural-biology, population biology, and computing technology. The neural network model in [4] can be described by the following system of ordinary differential equations

$$x_i'(t) = -a_i(x_i(t)) \left[b_i(x_i(t)) - \sum_{j=1}^{n} c_{ij} f_j(x_j(t)) + I_i \right], \quad t \geq 0,\ i = 1, \ldots, n, \tag{1}$$

J.J. Oliveira (✉)
Centro de Matemática (CMAT), Departamento de Matemática e Aplicações, Universidade do Minho, Campus de Gualtar, 4710-057 Braga, Portugal
e-mail: jjoliveira@math.uminho.pt

© Springer International Publishing Switzerland 2016

P. Gonçalves and A.J. Soares (eds.), *From Particle Systems to Partial Differential Equations III*, Springer Proceedings in Mathematics & Statistics 162,
DOI 10.1007/978-3-319-32144-8_12

where n is the number of neurons, $x_i(t)$ is the state variable of the ith neuron at time t, f_j are activation functions, b_i are self-inhibition term, a_i are amplification functions, c_{ij} are connection weights, and I_i are external inputs. Very close to the model (1), other neural network type models appeared such as Hopfield neural networks [10] and bidirectional associative memory neural networks (BAM), [14].

In order to be more realistic, differential equations describing neural networks should incorporate time delays to take into account the synaptic transmission time among neurons, or, in artificial neural networks, the communication time among amplifiers. In 1989, Marcus and Westervelt [16] introduced for the first time a discrete delay in a neural network model, and they observed that the delay can destabilize the system. In fact, the delays can affect the dynamic behavior of neural network models [1] and, for this reason, stability of delayed neural networks has been investigated extensively ([2, 3, 5, 6, 9, 11–13, 15, 17–21], and the references therein). Since neural networks have a spatial nature due to the presence of a multitude of parallel pathways with a variety of axon sizes and lengths, it is also fundamental to consider distributed delays thus, neural network models become more realistic if both discrete time-varying delays and distributed delays are taken into account (see [18]). Many interesting results on stability of autonomous Cohen-Grossberg neural network models with delays have been reported in the literature [3, 6, 11, 17], however, nonautonomous phenomena often occur in many realistic situations. Thus, neural network models with temporal structure of neural activities should be introduced and investigated (see [2, 19]). In this paper we study a class of functional differential equations general enough to include nonautonomous neural network models with both discrete time-varying delays and distributed delays.

Since the global stability of delayed neural network models plays an important role in their applications, it is of significance and necessary to study the global asymptotic stability of an equilibrium point of Cohen-Grossberg neural network models. However, in most of the existing results in the literature, it has been assumed directly or indirectly that the amplification functions are bounded and uniformly positive i.e.,

$$\underline{a}_i < a_i(u) < \overline{a}_i, \quad \forall u \in \mathbb{R}, \quad \forall i \in \{1, \ldots, n\}, \tag{2}$$

for some $\underline{a}_i, \overline{a}_i > 0$. To the best of our knowledge, in [17], for the first time the global asymptotic stability of an equilibrium point of an autonomous Cohen-Grossberg neural network was proved, without assuming condition (2). The main propose of this paper is to achieve the same stability result for a general class of nonautonomous functional differential equation.

After the introduction, the present paper is divided into three sections. In Sect. 2 some notation and definition are presented, the studied functional differential equation is described and its phase space is given. In Sect. 3, we present the main results about global asymptotic stability of an equilibrium point of a general class of nonautonomous functional differential equations with infinite distributed delays or with mixed delays (that is both discrete time-varying delays and infinite distributed delays). Finally, in Sect. 4, we illustrate the results with well-known neural network models and we compare our results with the literature given new stability criteria.

2 Preliminaries

For $n \in \mathbb{N}$, we denote by $BC = BC((-\infty, 0]; \mathbb{R}^n)$ the space of bounded and continuous functions, $\varphi : (-\infty, 0] \to \mathbb{R}^n$, equipped with the norm $||\varphi|| = \sup_{s\leq 0} |\varphi(s)|$, where $|\cdot|$ is the maximum norm in $\mathbb{R}^n$, i.e. $|x| = \max\{|x_i| : i = 1, \ldots, n\}$ for $x = (x_1, \ldots, x_n) \in \mathbb{R}^n$. For $c \in \mathbb{R}^n$, we also use c to denote the constant function $\varphi(s) = c$ in BC. For a vector $c = (c_1, \ldots, c_n) \in \mathbb{R}^n$ and $\varphi = (\varphi_1, \ldots, \varphi_n) \in BC$, we denote $c \cdot \varphi = (c_1\varphi_1, \ldots, c_n\varphi_n) \in BC$, and the vector c is said to be positive if $c_i > 0$ for all $i \in \{1, \ldots, n\}$ and in this case we write $c > 0$. A function $\gamma : [a, +\infty) \to \mathbb{R}$, $a \in \mathbb{R}$, is said to be *eventually monotone* if there is $t^* > a$ such that γ is non-decreasing (or non-increasing) on $[t^*, +\infty)$. For a real sequence $(u_n)_{n\in\mathbb{N}}$ and $a \in \mathbb{R} \cup \{+\infty\}$, we write $u_n \nearrow a$ to say that $(u_n)_{n\in\mathbb{N}}$ is an increasing sequence such that $\lim_{n\to+\infty} u_n = a$.

For an open set $D \subseteq BC$ and $f : [0, +\infty) \times D \to \mathbb{R}^n$ a continuous function, consider the functional differential equation (FDE) given in general setting by

$$x'(t) = f(t, x_t), \quad t \geq 0, \tag{3}$$

where, as usual, x_t denotes the function $x_t : (-\infty, 0] \to \mathbb{R}^n$ defined by $x_t(s) = x(t + s)$ for $s \leq 0$. By a solution of (3) on an interval $I \subseteq \mathbb{R}$, we mean a function $x : (-\infty, \sup I) \to \mathbb{R}^n$ such that $x_t \in D$, $x(t)$ is continuous differentiable, and (3) holds for all $t \in I$ (see [8]).

It is well-known that the Banach space BC is not an admissible phase space for (3), in the sense of [7], thus the standard existence, uniqueness, continuous dependence type results are not available. Instead of BC, we consider the admissible Banach space

$$UC_g = \left\{\phi \in C((-\infty, 0]; \mathbb{R}^n) : \sup_{s\leq 0} \frac{|\phi(s)|}{g(s)} < \infty, \right.$$
$$\left. \frac{\phi(s)}{g(s)} \text{ is uniformly continuous on } (-\infty, 0]\right\},$$

equipped with the norm $||\phi||_g = \sup_{s\leq 0} \dfrac{|\phi(s)|}{g(s)}$, where $g : (-\infty, 0] \to [1, \infty)$ is a function satisfying:

(g1) g is a non-increasing continuous function and $g(0) = 1$;

(g2) $\lim_{u \to 0^-} \frac{g(s+u)}{g(s)} = 1$ uniformly on $(-\infty, 0]$;

(g3) $g(s) \to +\infty$ as $s \to -\infty$.

See [8] for more details.

As $BC \subseteq UC_g$, then BC is a subspace of UC_g, and we denote by BC_g the space BC with the norm $|| \cdot ||_g$. When $n = 1$, we denote the spaces UC_g and BC by UC_g^1 and BC^1 respectively.

As UC_g is an admissible Banach space, we consider the FDE (3) in the phase space UC_g, for a convenient function g, and we assume that f has enough smooth properties to ensure the existence and uniqueness of solution for the initial value problem (see [8]). The solution of (3) with initial condition $x_{t_0} = \varphi$, for $t_0 \geq 0$ and $\varphi \in UC_g$, is denoted by $x(t, t_0, \varphi)$. It is relevant to emphasize that, from [8], the solution $x(t, t_0, \varphi)$ is continuous, once the continuity of the solutions plays an important role in the proof of main results.

In view of our applications to neural network systems, we restrict our attention to initial bounded conditions, i.e.,

$$x_{t_0} = \varphi, \qquad \text{with} \qquad \varphi \in BC, \tag{4}$$

for some $t_0 \geq 0$. From [8], if f maps closed bounded subsets of its domain into bounded sets of $\mathbb{R}^n$, then the solution of (3)–(4) is extensible to $(-\infty, a]$, with $a > t_0$, whenever it is bounded.

Definition 2.1 A solution $x^*(t)$ of (3) is said to be:

(i) *uniformly stable* if for any $\varepsilon > 0$, there is $\delta = \delta(\varepsilon) > 0$ such that, for all $t_0 \geq 0$ and $\varphi \in BC$,

$$\|\varphi - x^*_{t_0}\|_g < \delta \Rightarrow \|x_t(\cdot, t_0, \varphi) - x^*_t\|_g < \varepsilon, \textit{ for all } t \geq t_0;$$

(ii) *globally attractive* if

$$\lim_{t \to +\infty} \big(x(t, t_0, \varphi) - x^*(t)\big) = 0,$$

for any $t_0 \geq 0$ and $\varphi \in BC$;

(iii) *globally asymptotically stable* if it is uniform stable and globally attractive.

It should be mentioned that it is usual in the literature on neural networks with delays to define an equilibrium point as globally asymptotic stable if it is global attractive (see [6, 12]).

In [6], the following two lemmas were established.

Lemma 2.1 *[6] Consider system* (3) *in* UC_g, *and assume that* f *transforms closed bounded sets of* $[0, +\infty) \times D$ *into bounded sets of* $\mathbb{R}^n$. *Assume also that*

(H1) for all $t \geq 0$ *and* $\varphi \in BC_g$ *such that* $\frac{|\varphi(s)|}{g(s)} < |\varphi(0)|$, *for* $s \in (-\infty, 0)$, *then* $\varphi_i(0)f_i(t, \varphi) < 0$ *for some* $i \in \{1, \ldots, n\}$ *such that* $|\varphi(0)| = |\varphi_i(0)|$.

Then, the solutions $x(t) = x(t, 0, \varphi)$, $\varphi \in BC_g$, *of* (3) *are defined on* $\mathbb{R}$ *and satisty* $|x(t, 0, \varphi)| \leq \|\varphi\|_g$ *for all* $t \geq 0$.

Lemma 2.2 *[6] Consider system* (3) *in* UC_g, *and assume that* f *transforms closed bounded sets of* $[0, +\infty) \times D$ *into bounded sets of* $\mathbb{R}^n$. *Assume also that*

(H2) for all $t \geq 0$ *and* $\varphi \in BC$ *such that* $|\varphi(s)| < |\varphi(0)|$, *for* $s \in (-\infty, 0)$, *then* $\varphi_i(0)f_i(t, \varphi) < 0$ *for some* $i \in \{1, \ldots, n\}$ *such that* $|\varphi(0)| = |\varphi_i(0)|$.

Then, the solutions $x(t) = x(t, 0, \varphi)$, $\varphi \in BC$, *of* (3)–(4) *is defined on* $\mathbb{R}$ *and satisty* $|x(t, 0, \varphi)| \leq \|\varphi\|$ *for all* $t \geq 0$.

Remark 1 We note that, assuming (H1), Lemma 2.1 assures that the solution of (3)–(4) is defined and bounded on $\mathbb{R}$ and the zero solution of (3) is uniformly stable. If hypothesis (H2) holds instead of (H1), then, from Lemma 2.2, the solution of (3)–(4) is also defined and bounded on $\mathbb{R}$ but we can not to conclude that the zero solution of (3) is uniformly stable.

In the phase space UC_g, with $g : (-\infty, 0] \to [1, +\infty)$ being a function satisfying (g1)–(g3), consider the following general nonautonomous class of FDEs with infinite distributed delays

$$x_i'(t) = -a_i(t, x_t)\left[b_i(t, x_i(t)) + f_i(t, x_t)\right], \quad t \geq 0, \ i = 1, \ldots, n, \tag{5}$$

where $a_i : [0, +\infty) \times UC_g \to (0, +\infty)$, $b_i : [0, +\infty) \times \mathbb{R} \to \mathbb{R}$, and $f_i : [0, +\infty) \times UC_g \to \mathbb{R}$ are continuous functions. This model is a generalization of a model introduced in [6] and it is particularly relevant in terms of applications, since it includes different types of neural network models with delays, such as Hopfield, Cohen-Grossberg, and BAM.

For (5) the following hypotheses will be considered:

(A1) there exists $x^* = (x_1^*, \ldots, x_n^*) \in \mathbb{R}^n$ an equilibrium point of (5);

(A2) for each $i \in \{1, \ldots, n\}$ and for any $M > 0$,

$$\inf \{a_i(t, \varphi) : \varphi \in BC, \quad \|\varphi\| \leq M, \quad t \geq 0\} > 0;$$

(A3) for each $i \in \{1, \ldots, n\}$, there exists a function $\beta_i : [0, +\infty) \to (0, +\infty)$ such that

$$\frac{b_i(t, u) - b_i(t, v)}{u - v} \geq \beta_i(t), \quad \forall t \geq 0, \ \forall u, v \in \mathbb{R}, u \neq v;$$

(A4) for each $i \in \{1, \ldots, n\}$, there exists a function $F_i : [0, +\infty) \to [0, +\infty)$ such that

$$|f_i(t, \varphi) - f_i(t, \psi)| \leq F_i(t)\|\varphi - \psi\|_g, \qquad \forall t \geq 0, \ \forall \varphi, \psi \in BC;$$

(A5) there exists $\alpha > 0$ such that, for each $i \in \{1, \ldots, n\}$,

$$\beta_i(t) - F_i(t) > \alpha, \ \forall t \geq 0, \tag{6}$$

and

$$\limsup_{t \to +\infty} \frac{F_i(t)}{\beta_i(t)} < 1. \tag{7}$$

We remark that hypothesis (A2) does not imply the uniformly positiveness of the amplification functions i.e., it does not imply that $a_i(t, \varphi) > \underline{a}$ for some real positive constant $\underline{a}$.

We also remark that the hypothesis (A3) says that, for each $t \geq 0$ and $i \in \{1, \ldots, n\}$, the function $u \mapsto b_i(t, u)$ is increasing with a growth rate at least $\beta_i(t)$. For example, (A3) is trivially satisfied if $b_i(t, u) = \beta_i(t)u$ for all $t \geq 0$ and $u \in \mathbb{R}$.

Finally, we should say that, in general, one does not expect having an equilibrium point in a nonautonomous system, (5). However, in view of our applications to neural network models, such as (1), very often the self-inhibition terms b_i are linear on the second variable and the activation function f_j are sigmoid functions, for example $f_j(x) = \tanh x$, $f_j(x) = |x + 1| - |x - 1|$, or $f_j(x) = \arctan x$ (see examples in [9, 20, 21]). Consequently, if we have no external inputs, then zero is an equilibrium point.

3 Asymptotic Stability

Before to study the global stability of the equilibrium point x^* of (5), we need to prove that it is unique.

Lemma 3.1 *Assume (A3)–(A5) hold.*

If the system (5) *has an equilibrium point, then it is unique.*

Proof Assume that $x^* = (x_1^*, \ldots, x_n^*) \in \mathbb{R}^n$ and $y^* = (y_1^*, \ldots, y_n^*) \in \mathbb{R}^n$ are two different equilibrium points of (5) and choose $i \in \{1, \ldots, n\}$ such that $|x^* - y^*| = |x_i^* - y_i^*| > 0$. As $a_i(t, \varphi) > 0$ for all $t \geq 0$ and $\varphi \in UC_g$, then

$$b_i(t, x_i^*) + f_i(t, x^*) = 0 = b_i(t, y_i^*) + f_i(t, y^*), \quad \forall t \geq 0.$$

From the hypotheses (A3), (A4), and (A5) and assuming that $x_i^* - y_i^* > 0$ (the situation $x_i^* - y_i^* < 0$ is analogous), we obtain

$$\begin{aligned} 0 &= b_i(t, x_i^*) - b_i(t, y_i^*) + f_i(t, x^*) - f_i(t, y^*) \\ &\geq \beta_i(t)(x_i^* - y_i^*) - F_i(t)|x^* - y^*| \\ &= (\beta_i(t) - F_i(t))(x_i^* - y_i^*) > 0, \end{aligned} \tag{8}$$

which is a contradiction. Thus $x^* = y^*$ and the result follows.

Now, we are in position to prove the uniform stability of the equilibrium point of the FDE (5).

Theorem 3.2 *For* (5) *assume (A1)–(A5) hold. Then the equilibrium point of* (5) *is uniformly stable.*

Proof Let $x^* = (x_1^*, \dots, x_n^*) \in \mathbb{R}^n$ the equilibrium point of (5). Translating x^* to the origin by the change $\bar{x}(t) = x(t) - x^*$, the system (5) becomes

$$\bar{x}_i'(t) = -\bar{a}_i(t, \bar{x}_t)\left[\bar{b}_i(t, \bar{x}_i(t)) + \bar{f}_i(t, \bar{x}_t)\right], \quad t \geq 0, \ i = 1, \dots, n, \tag{9}$$

where $\bar{a}_i(t, \varphi) = a_i(t, \varphi + x^*)$, $\bar{b}_i(t, u) = b_i(t, u + x_i^*)$, and $\bar{f}_i(t, \varphi) = f_i(t, \varphi + x^*)$, with zero as the unique equilibrium point, i.e. $\bar{b}_j(t, 0) + \bar{f}_j(t, 0) = 0$ for all $j \in \{1, \dots, n\}$ and $t \geq 0$. Clearly a_i, b_i, and f_i satisfy (A2)–(A4) if and only if $\bar{a}_i$, $\bar{b}_i$, and $\bar{f}_i$ satisfy (A2)–(A4), with the same functions β_i and F_i. Hence, we consider (9), where, for simplicity, we drop the bars.

Let $\varphi \in BC_g$ be such that $\|\varphi\|_g = |\varphi(0)| > 0$ and consider $i \in \{1, \dots, n\}$ such that $|\varphi_i(0)| = \|\varphi\|_g$. If $\varphi_i(0) > 0$ (the situation $\varphi_i(0) < 0$ is analogous), then $\|\varphi\|_g = \varphi_i(0)$ and from the hypotheses we conclude that

$$\begin{aligned} a_i(t, \varphi)[b_i(t, \varphi_i(0)) + f_i(t, \varphi)] &= \\ &= a_i(t, \varphi)[(b_i(t, \varphi_i(0)) - b_i(t, 0)) + (f_i(t, \varphi) - f_i(t, 0))] \\ &\geq a_i(t, \varphi)(\beta_i(t) - F_i(t))\|\varphi\|_g > 0. \end{aligned} \tag{10}$$

In particular, (H1) holds and from Lemma 2.1 we deduce that all solutions are defined and bounded on $\mathbb{R}$, and that $x = 0$ is uniformly stable. This means that the equilibrium point of (5) is uniformly stable.

Theorem 3.3 *For* (5) *assume (A1)–(A5) hold. Then the equilibrium point of* (5) *is globally asymptotically stable.*

Proof From Theorem 3.2, it remains to prove that the equilibrium point is globally attractive. As in the above proof, after a translation, we may assume that the equilibrium point is zero, i.e., $b_i(t, 0) + f_i(t, 0) = 0$ for all $i \in \{1, \dots, n\}$ and $t \geq 0$. For $x(t) = x(t, t_0, \varphi) = (x_1(t), \dots, x_n(t))$ the solution of initial value problem (5)–(4), from Theorem 3.2 we conclude that $x(t)$ is a bounded function and we define the limits

$$-v_i = \liminf_{t \to +\infty} x_i(t), \quad u_i = \limsup_{t \to +\infty} x_i(t), \quad i \in \{1, \dots, n\}, \tag{11}$$

and

$$v = \max_i \{v_i\}, \quad u = \max_i \{u_i\}. \tag{12}$$

Note that $u, v \in \mathbb{R}$ and $-v \leq u$. It is sufficient to prove that $\max\{u, v\} = 0$. Assume e.g. that $|v| \leq u$, so that $\max\{u, v\} = u$. (The situation is analogous for $|u| \leq v$.)

Let $i \in \{1, \ldots, n\}$ such that $u_i = u$. Now we prove that there is a sequence $(t_k)_{k\in\mathbb{N}}$ such that

$$t_k \nearrow +\infty, \quad x_i(t_k) \to u, \quad \text{and} \quad x_i'(t_k) \to 0, \quad \text{as} \quad k \to +\infty. \tag{13}$$

Case 1. Assume that $x_i(t)$ is eventually monotone. In this case, $\lim_{t\to+\infty} x_i(t) = u$ and, for t large, $x_i(t)$ is a differentiable, monotone, and bounded real function. Hence there is a sequence $(t_k)_{k\in\mathbb{N}}$ such that $t_k \nearrow +\infty$ and $x_i'(t_k) \to 0$.

Case 2. Assume that $x_i(t)$ is not eventually monotone. In this case there is a sequence $(t_k)_{k\in\mathbb{N}}$ such that $t_k \nearrow +\infty$, $x_i'(t_k) = 0$ and $x_i(t_k) \to u$, as $k \to +\infty$, and (13) holds.

Now we have to show that $u = 0$, hence $v = 0$ as well. For the sake of contradiction, assume that $u > 0$.

Fix $\varepsilon > 0$ and let $T = T(\varepsilon) > t_0$ be such that $\varepsilon < u$ and, for all $t \geq T$, $\dfrac{\|\varphi\|}{g(t_0 - t)} < u + \varepsilon$ and $|x(t)| < u + \varepsilon$. Defining $\mathscr{X} := \max_{s\in[t_0,T]} |x(s)|$, from (g3) we conclude that there is $s_0 < 0$ such that $\dfrac{\mathscr{X}}{g(s)} < u + \varepsilon$ for all $s < s_0$. Thus, for $t > T - s_0$, we have

$$\begin{aligned}
\sup_{s\in[t_0-t,0]} \frac{|x(t+s)|}{g(s)} &\leq \max\left\{ \sup_{s\in[t_0-t,T-t]} \frac{|x(t+s)|}{g(s)}, \sup_{s\in[T-t,0]} \frac{|x(t+s)|}{g(s)} \right\} \\
&\leq \max\left\{ \sup_{s\in[t_0-t,T-t]} \frac{\mathscr{X}}{g(s)}, \sup_{s\in[T-t,0]} \frac{u+\varepsilon}{g(s)} \right\} \leq u + \varepsilon.
\end{aligned}$$

Now, from the hypotheses (A3) and (A4) we have, for $t_k > T - s_0$,

$$\begin{aligned}
x_i'(t_k) &= -a_i(t_k, x_{t_k}) \left[b_i(t_k, x_i(t_k)) + f_i(t_k, x_{t_k}) \right] \\
&= -a_i(t_k, x_{t_k}) \left[b_i(t_k, x_i(t_k)) - b_i(t_k, 0) + f_i(t_k, x_{t_k}) - f_i(t_k, 0) \right] \\
&\leq -a_i(t_k, x_{t_k}) \left[\beta_i(t_k) x_i(t_k) - F_i(t_k) \|x_{t_k}\|_g \right] \\
&\leq -a_i(t_k, x_{t_k}) \left[\beta_i(t_k) x_i(t_k) - F_i(t_k) \max\left\{ \frac{\|\varphi\|}{g(t_0 - t_k)}, \sup_{s\in[t_0-t_k,0]} \frac{|x(t_k+s)|}{g(s)} \right\} \right] \\
&\leq -a_i(t_k, x_{t_k}) \left[\beta_i(t_k) x_i(t_k) - F_i(t_k)(u+\varepsilon) \right],
\end{aligned}$$

which implies that

$$x_i(t_k) \leq -\frac{x_i'(t_k)}{a_i(t_k, x_{t_k})\beta_i(t_k)} + \frac{F_i(t_k)}{\beta_i(t_k)}(u+\varepsilon). \tag{14}$$

As $x(t)$ is bounded, from hypotheses (A2) and (A5) we have $\liminf_{k\to+\infty} a_i(t_k, x_{t_k})\beta_i(t_k) > 0$. Letting $\varepsilon \to 0$ and $k \to +\infty$, from (13) and (A5), we get

$$u \le \left(\limsup_{t\to+\infty} \frac{F_i(t)}{\beta_i(t)}\right) u < u,$$

which is a contradiction.

It is easy to see that, if $F_i(t)$ are bounded functions then condition (6) implies condition (7) and the following result holds:

Corollary 3.4 *Assume (A1)–(A4) hold, where $F_i(t)$ are bounded functions. If* (6) *holds, then the equilibrium point of* (5) *is globally asymptotically stable.*

Remark 2 In [6], the global asymptotic stability of the equilibrium point of (5) was proved assuming constant functions $\beta_i(t)$ and $F_i(t)$ i.e., $F_i(t) = F_i$ and $\beta_i(T) = \beta_i$ with $\beta_i \in (0, +\infty)$ and $F_i \in [0, +\infty)$. In this scenery, condition (6) reads as $\beta_i - F_i > 0$ and Corollary 3.4 improves the stability result in [6].

Now, we consider the following nonautonomous FDE with distributed time-varying delays:

$$x_i'(t) = -a_i(t, x_t)\left[b_i(t, x_i(t)) + h_i(t, x_t) + f_i(t, x_t)\right], \quad t \ge 0, \ i = 1, \dots, n, \quad (15)$$

where the functions a_i, b_i, and f_i are as in the model (5) and $h_i : [0, +\infty) \times UC_g \to \mathbb{R}$ are continuous functions such that, for each $i \in \{1, \dots, n\}$ there exist $H_i, \tau_i : [0, +\infty) \to [0, +\infty)$ such that

$$|h_i(t, \varphi) - h_i(t, \psi)| \le H_i(t)\left(\sup_{\theta \ge -\tau_i(t)} |\varphi(\theta) - \psi(\theta)|\right), \forall t \ge 0, \ \forall \varphi, \psi \in BC \quad (16)$$

and

$$t - \tau_i(t) \to +\infty \quad \text{as} \quad t \to +\infty. \quad (17)$$

We note that condition (16) implies the Lipschitz condition in the space BC i.e.,

$$|h_i(t, \varphi) - h_i(t, \psi)| \le H_i(t)\|\varphi - \psi\|, \quad \forall t \ge 0, \quad \forall \varphi, \psi \in BC,$$

but (16) does not imply the Lipschitz condition in the space BC_g i.e., relative to the norm $\|\cdot\|_g$. This is the reason why the following result only establishes the global attractiveness of the equilibrium point of (15) instead of its global asymptotic stability.

Theorem 3.5 *Assume that* (15) *has an equilibrium point and the hypotheses (A2), (A3), (A4),* (16), *and* (17) *hold.*

If there exists $\alpha > 0$ *such that*

$$\beta_i(t) - [H_i(t) + F_i(t)] > \alpha, \ \forall t \geq 0, \tag{18}$$

and

$$\limsup_{t\to+\infty} \frac{H_i(t) + F_i(t)}{\beta_i(t)} < 1, \tag{19}$$

then the equilibrium point of (15) *is globally attractive.*

Proof Following the same ideas in the proof of Lemma 3.1, we conclude that (15) has at most one equilibrium point.

As we done in the proof of Theorem 3.2, without losing generality, we may assume that the equilibrium point is zero, i.e., $b_i(t, 0) + h_i(t, 0) + f_i(t, 0) = 0$ for all $i \in \{1, \dots, n\}$ and $t \geq 0$.

First, we prove that all solutions of (15), with bounded initial condition, are defined and bounded on $\mathbb{R}$. Let $\varphi \in BC$ be such that $\|\varphi\| = |\varphi(0)| > 0$ and consider $i \in \{1, \dots, n\}$ such that $|\varphi_i(0)| = \|\varphi\|$. If $\varphi_i(0) > 0$ (the situation $\varphi_i(0) < 0$ is analogous), then $\|\varphi\| = \varphi_i(0)$ and from the hypotheses we conclude that

$$\begin{aligned}
a_i(t, \varphi)&[b_i(t, \varphi_i(0)) + h_i(t, \varphi) + f_i(t, \varphi)] = \\
&= a_i(t, \varphi)[(b_i(t, \varphi_i(0)) - b_i(t, 0)) + (h_i(t, \varphi) - h_i(t, 0)) + (f_i(t, \varphi) - f_i(t, 0))] \\
&\geq a_i(t, \varphi)\left[\beta_i(t)\varphi_i(0) - H_i(t)\left(\sup_{\theta\geq-\tau_i(t)} |\varphi(\theta)|\right) - F_i(t)\|\varphi\|_g\right] \\
&\geq a_i(t, \varphi)\,[\beta_i(t) - H_i(t) - F_i(t)]\,\|\varphi\| > 0.
\end{aligned} \tag{20}$$

In particular, (H2) holds and from Lemma 2.2 we deduce that all solutions, with bounded initial conditions, are defined and bounded on $\mathbb{R}$.

Let $t_0 > 0$, $\varphi \in BC$, and $x(t) = x(t, t_0, \varphi)$ be the solution of (15). As $x(t)$ is bounded we define u and v as in (12) and, for the situation $|v| \leq u$ (the situation $|u| \leq v$ is analogous), we choose $i \in \{1, \dots, n\}$ such that $u_i = u = \max\{u, v\}$. For the sake of contradiction, assume that $u > 0$.

Following the same ideas in the proof of Theorem 3.3, we obtain a sequence $(t_k)_{k\in\mathbb{N}}$ satisfying (13) and, for any $\varepsilon \in (0, u)$, there are $T = T(\varepsilon) > t_0$ and $s_0 < 0$ such that, for all $t > T - s_0$, $\|x_t\|_g \leq u + \varepsilon$.

As $|x(t)| < u + \varepsilon$ for all $t \geq T$ and $\lim_{t\to+\infty} (t - \tau_i(t)) = +\infty$, then there is $S > 0$ such that $\sup_{s\in[-\tau_i(t),0]} |x(t + s)| < u + \varepsilon$ for all $t \geq S$. Thus, for $t_k > \max\{T - s_0, S\}$, we have

$$\begin{aligned}
x_i'(t_k) &= -a_i(t_k, x_{t_k})\left[b_i(t_k, x_i(t_k)) + h_i(t_k, x_{t_k}) + f_i(t_k, x_{t_k})\right] \\
&\leq -a_i(t_k, x_{t_k})\left[\beta_i(t_k)x_i(t_k) - H_i(t_k)\left(\sup_{s\in[-\tau_i(t_k),0]} |x(t_k + s)|\right) - F_i(t_k)\|x_{t_k}\|_g\right] \\
&\leq -a_i(t_k, x_{t_k})\,[\beta_i(t_k)x_i(t_k) - (H_i(t_k) + F_i(t_k))(u + \varepsilon)]\,,
\end{aligned}$$

which implies that

$$x_i(t_k) \leq -\frac{x_i'(t_k)}{a_i(t_k, x_{t_k})\beta_i(t_k)} + \frac{H_i(t_k) + F_i(t_k)}{\beta_i(t_k)}(u + \varepsilon). \tag{21}$$

As $x(t)$ is bounded, from hypotheses (A2) and (18) we have $\liminf_{k \to +\infty} a_i(t_k, x_{t_k})\beta_i(t_k) > 0$. Letting $\varepsilon \to 0$ and $k \to +\infty$ we get $x_i(t_k) \to u$, $x_i'(t_k) \to 0$, and

$$u \leq \left(\limsup_{t \to +\infty} \frac{H_i(t) + F_i(t)}{\beta_i(t)} \right) u,$$

which is impossible from (19). Consequently $u = 0$ and the equilibrium point of (15) is globally attractive.

4 Applications

In this section, we apply the previous results to the following general nonautonomous Cohen-Grossberg neural network model with both discrete time-varying delays and distributed delays

$$x_i'(t) = -a_i(t, x_t) \left[b_i(t, x_i(t)) + \sum_{j=1}^{n} h_{ij}(t, x_j(t - \tau_{ij1}(t)), \ldots, x_j(t - \tau_{ijP}(t))) + \sum_{j=1}^{n} f_{ij}(t, x_{j,t}) \right], \quad t \geq 0, \; i = 1, \ldots, n, \tag{22}$$

where $P \in \mathbb{N}$ and, for some function $g : (-\infty, 0] \to [1, +\infty)$ satisfying (g1)-(g3), $a_i : [0, +\infty) \times UC_g \to (0, +\infty)$, $b_i : [0, +\infty) \times \mathbb{R} \to \mathbb{R}$, $\tau_{ijp} : [0, +\infty) \to [0, +\infty)$, $h_{ij} : [0, +\infty) \times \mathbb{R}^P \to \mathbb{R}$, and $f_{ij} : [0, +\infty) \times UC_g^1 \to \mathbb{R}$ are continuous functions such that a_i satisfy (A2), b_i satisfy (A3), $\lim_{t \to +\infty} \left(t - \tau_{ijp}(t)\right) = +\infty$, and h_{ij} and f_{ij} are Lipschitz function on the second variable, i.e., there exist $H_{ij} : [0, +\infty) \to [0, +\infty)$ and $F_{ij} : [0, +\infty) \to [0, +\infty)$ such that

$$|h_{ij}(t, u) - h_{ij}(t, v)| \leq H_{ij}(t)|u - v|, \quad t \geq 0, \quad u, v \in \mathbb{R}^P, \tag{23}$$

$$|f_{ij}(t, \varphi) - f_{ij}(t, \psi)| \leq F_{ij}(t)\|\varphi - \psi\|_g, \quad t \geq 0, \quad \varphi, \psi \in BC_g^1, \tag{24}$$

for $i, j \in \{1, \ldots, n\}, p \in \{1, \ldots, P\}$. The global stability of (22) was recently studied in [5] assuming bounded amplification functions.

Theorem 4.1 *Assume that* (22) *has an equilibrium point, the hypotheses (A2), (A3),* (23)*, and* (24) *hold,* $\lim_{t\to+\infty}\big(t-\tau_{ijp}(t)\big)=+\infty$*, and there exist* $\alpha>0$ *and* $d=(d_1,\ldots,d_n)>0$ *such that, for each* $i\in\{1,\ldots,n\}$,

$$\beta_i(t)-\sum_{j=1}^{n}\frac{d_j}{d_i}(H_{ij}(t)+F_{ij}(t))>\alpha,\ \ \forall t\geq 0, \tag{25}$$

and

$$\limsup_{t\to+\infty}\sum_{j=1}^{n}\frac{d_j(H_{ij}(t)+F_{ij}(t))}{d_i\beta_i(t)}<1. \tag{26}$$

Then the equilibrium point of (22) *is globally attractive.*

If $h_{ij}(t,u)=0$*, for all* $i,j\in\{1,\ldots,n\}$*,* $t\geq 0$*, and* $u\in\mathbb{R}^P$*, then the equilibrium point of* (22) *is globally asymptotically stable.*

Proof As in Lemma 3.1, we conclude that the equilibrium point of (22) is unique and, by translation, we may assume that it is zero that is,

$$b_i(t,0)+\sum_{j=1}^{n}(h_{ij}(t,0)+f_{ij}(t,0))=0,\quad t\geq 0,\ i=1,\ldots,n.$$

The change of variables $y_i(t)=d_i^{-1}x_i(t)$ transforms (22) into

$$\begin{aligned} y_i'(t)=-a_i(t,d\cdot y_t)d_i^{-1}\Bigg[&b_i(t,d_iy_i(t))\\ &+\sum_{j=1}^{n}h_{ij}(t,d_jy_j(t-\tau_{ij1}(t)),\ldots,d_jy_j(t-\tau_{ijP}(t)))+\sum_{j=1}^{n}f_{ij}(t,d_jy_{j,t})\Bigg], \end{aligned} \tag{27}$$

for $t\geq 0$ and $i=1,\ldots,n$. Defining, for each $i\in\{1,\ldots,n\}$, $\tilde{a}_i(t,\varphi):=a_i(t,d\cdot\varphi)$, $\tilde{b}_i(t,u):=d_i^{-1}b_i(t,d_iu)$, $\tilde{h}_i(t,\varphi):=d_i^{-1}\sum_{j=1}^{n}h_{ij}(t,d_j\varphi_j(-\tau_{ij1}(t)),\ldots,d_j\varphi_j(-\tau_{ijP}(t)))$, and $\tilde{f}_i(t,\varphi):=d_i^{-1}\sum_{j=1}^{n}f_{ij}(t,d_j\varphi_j)$, for all $t\geq 0$, $u\in\mathbb{R}$, $\varphi=(\varphi_1,\ldots,\varphi_n)\in UC_g$, the system (27) has the form

$$y_i'(t)=-\tilde{a}_i(t,y_t)\Big[\tilde{b}_i(t,y_i(t))+\tilde{h}_i(t,y_t)+\tilde{f}_i(t,y_t)\Big],\quad t\geq 0,\ i=1,\ldots,n, \tag{28}$$

where zero is the equilibrium point, $\tilde{a}_i$ trivially satisfy (A2), and $\tilde{b}_i$ satisfy (A3) with $\tilde{\beta}_i(t) = \beta_i(t)$. Denoting $\tau_i(t) = \max_{j,p}\{\tau_{ijp}(t)\}$, from (23), for $\varphi = (\varphi_1, \dots, \varphi_n)$, $\psi = (\psi_1, \dots, \psi_n) \in BC_g$ we have

$$
\begin{aligned}
|\tilde{h}_i(t,\varphi) - \tilde{h}_i(t,\psi)| &\le d_i^{-1} \sum_{j=1}^{n} |h_{ij}(t, d_j\varphi_j(-\tau_{ij1}(t)), \dots, d_j\varphi_j(-\tau_{ijP}(t))) \\
&\qquad - h_{ij}(t, d_j\psi_j(-\tau_{ij1}(t)), \dots, d_j\psi_j(-\tau_{ijP}(t)))| \\
&\le d_i^{-1} \sum_{j=1}^{n} H_{ij}(t) d_j |(\varphi_j(-\tau_{ij1}(t)), \dots, \varphi_j(-\tau_{ijP}(t))) \\
&\qquad - (\psi_j(-\tau_{ij1}(t)), \dots, \psi_j(-\tau_{ijP}(t)))| \\
&\le \sum_{j=1}^{n} \frac{d_j}{d_i} H_{ij}(t) \left(\sup_{\theta \in [-\tau_i(t), 0]} |\varphi_j(\theta) - \psi_j(\theta)| \right)
\end{aligned}
$$

which means that

$$
|\tilde{h}_i(t,\varphi) - \tilde{h}_i(t,\psi)| \le \left(\sum_{j=1}^{n} \frac{d_j}{d_i} H_{ij}(t) \right) \sup_{\theta \in [-\tau_i(t), 0]} |\varphi(\theta) - \psi(\theta)|
$$

and consequently $\tilde{h}_i$ satisfies (16) with the function $H_i(t) = \sum_{j=1}^{n} \frac{d_j}{d_i} H_{ij}(t)$ for each $i \in \{1, \dots, n\}$. Similarly, from (24) we have

$$
\begin{aligned}
|\tilde{f}_i(t,\varphi) - \tilde{f}_i(t,\psi)| &\le d_i^{-1} \sum_{j=1}^{n} |f_{ij}(t, d_j\varphi_j) - f_{ij}(t, d_j\psi_j)| \\
&\le d_i^{-1} \sum_{j=1}^{n} F_{ij}(t) \|d_j\varphi_j - d_j\psi_j\|_g \\
&\le \left(\sum_{j=1}^{n} \frac{d_j}{d_i} F_{ij}(t) \right) \|\varphi - \psi\|_g, \qquad (29)
\end{aligned}
$$

which implies that $\tilde{f}_i$ satisfies (A4) with the function $F_i(t) := \sum_{j=1}^{n} \frac{d_j}{d_i} F_{ij}(t)$, for each $i \in \{1, \dots, n\}$. Finally, conditions (18) and (19) follow from (25) and (26), and the conclusions follow from Theorems 3.5 and 3.3.

Remark 3 For the situation that (22) has not an equilibrium point, Esteves and Oliveira [5, Theorem 3.7] showed that

$$\lim_{t\to+\infty}\big(x(t,t_0,\varphi)-x(t,t_0,\psi)\big)=0,\quad \forall t_0\geq 0,\quad \forall\varphi,\psi\in BC,$$

assuming the uniformly positiveness of the amplification functions (recall (2)) and the stronger conditions

$$\underline{a}_i\beta_i(t)-\sum_{j=1}^{n}\overline{a}_j\frac{d_j}{d_i}(H_{ij}(t)+F_{ij}(t))>\alpha,\ \ \forall t\geq 0,$$

and

$$\limsup_{t\to+\infty}\sum_{j=1}^{n}\frac{\overline{a}_j}{\underline{a}_i}\frac{d_j(H_{ij}(t)+F_{ij}(t))}{d_i\beta_i(t)}<1.$$

instead of (25) and (26).

Now, we shall apply the previous stability criterion to several nonautonomous neural network models with infinite delays. The broad framework of our result (Theorem 4.1) allows us to treat most of the neural network models considered in the literature as particular cases of the system (22).

Before to study the stability of several neural network models, we need an auxiliary lemma, established in [6], to define a function g satisfying (g1)–(g3) required to build the convenient phase space.

Lemma 4.2 *[6] Consider $\eta_i:(-\infty,0]\to\mathbb{R}$, $i=1,\ldots,m$, non-decreasing and bounded functions, and $\alpha>0$ such that*

$$\int_{-\infty}^{0}d\eta_i(s)<\alpha,\qquad i=1,\ldots,m.$$

Then, there is a continuous function $g:(-\infty,0]\to[1,+\infty)$ satisfying (g1)–(g3), and such that

$$\int_{-\infty}^{0}g(s)d\eta_i(s)<\alpha,\qquad i=1,\ldots,m.$$

Example 1 Consider the following nonautonomous Cohen-Grossberg neural network model with distributed delays:

$$\begin{aligned}x_i'(t)=-a_i(x_i(t))\Bigg[b_i(t,x_i(t))-\sum_{j=1}^{n}\Bigg(c_{ij}(t)\int_{-\tau_{ij}}^{0}f_j(x_j(t+s))d\varsigma_{ij}(s)\\+d_{ij}(t)\int_{-\infty}^{0}g_j(x_j(t+s))d\eta_{ij}(s)\Bigg)+I_i(t)\Bigg],\ t\geq 0,\ i=1,\ldots,n,\qquad(30)\end{aligned}$$

where $\tau_{ij} \geq 0$, $a_i : \mathbb{R} \to (0, +\infty)$, $b_i : [0, +\infty) \times \mathbb{R} \to \mathbb{R}$, $c_{ij}, d_{ij}, I_i : [0, +\infty) \to \mathbb{R}$, and $f_j, g_j : \mathbb{R} \to \mathbb{R}$ are continuous functions, and $\varsigma_{ij} : [-\tau_{ij}, 0] \to \mathbb{R}$ and $\eta_{ij} : (-\infty, 0] \to \mathbb{R}$ are non-decreasing, bounded, and normalized, i.e. $\varsigma_{ij}(0) - \varsigma_{ij}(-\tau_{ij}) = 1$ and $\eta_{ij}(0) - \eta_{ij}(-\infty) = 1$, for all $i, j \in \{1, \ldots, n\}$. Applying Theorem 4.1 to this system, we have:

Corollary 4.3 *Assume that the model* (30) *has an equilibrium point, $f_j, g_j : \mathbb{R} \to \mathbb{R}$ are Lipschitz functions with Lipschitz constants μ_j, σ_j respectively, and b_i satisfy (A3).*

If there exist $\alpha > 0$ and $d = (d_1, \ldots, d_n) > 0$ such that, for each $i \in \{1, \ldots, n\}$,

$$d_i\beta_i(t) - \sum_{j=1}^{n} d_j\big(|c_{ij}(t)|\mu_j + |d_{ij}(t)|\sigma_j\big) > \alpha, \qquad \forall t \geq 0, \tag{31}$$

$$\limsup_{t\to+\infty} \sum_{j=1}^{n} \frac{d_j\big(|c_{ij}(t)|\mu_j + |d_{ij}(t)|\sigma_j\big)}{d_i\beta_i(t)} < 1, \tag{32}$$

then the equilibrium point of (30) *is globally asymptotically stable.*

Proof As $\alpha > 0$, from (31) and (32), we conclude that there are $\gamma > 0$ and $\delta > 0$ (small enough) such that, for each $i \in \{1, \ldots, n\}$,

$$d_i\beta_i(t) - \sum_{j=1}^{n} d_j\left(|c_{ij}(t)|\mu_j + |d_{ij}(t)|\sigma_j\right)(1+\delta) > \gamma, \quad \forall t \geq 0, \tag{33}$$

and

$$\limsup_{t\to+\infty} \sum_{j=1}^{n} \frac{d_j\big(|c_{ij}(t)|\mu_j + |d_{ij}(t)|\sigma_j\big)(1+\delta)}{d_i\beta_i(t)} < 1. \tag{34}$$

Since $\int_{-\tau_{ij}}^{0} d\varsigma_{ij}(s) < 1+\delta$ and $\int_{-\infty}^{0} d\eta_{ij}(s) < 1+\delta$ for all $i, j \in \{1, \ldots, n\}$, from Lemma 4.2, there is $g : (-\infty, 0] \to [1, \infty)$ satisfying (g1)–(g3) such that

$$\int_{-\tau_{ij}}^{0} g(s)d\varsigma_{ij}(s) < 1+\delta, \quad \text{and} \quad \int_{-\infty}^{0} g(s)\,d\eta_{ij}(s) < 1+\delta, \quad \forall i, j \in \{1, \ldots, n\},$$

and we consider UC_g as the phase space of (30). Defining, for each $i, j \in \{1, \ldots, n\}$, $h_{ij}(t, u) = 0$ and

$$f_{ij}(t, \varphi_j) := c_{ij}(t)\int_{-\tau_{ij}}^{0} f_j(\varphi_j(s))\,d\varsigma_{ij}(s) + d_{ij}(t)\int_{-\infty}^{0} g_j(\varphi_j(s))\,d\eta_{ij}(s) + \frac{I_i(t)}{n},$$

with $u \in \mathbb{R}^P$, $\varphi = (\varphi_1, \ldots, \varphi_n) \in UC_g$, and $t \geq 0$, the system (30) has the form (22).

For $\phi, \varphi \in BC^1$ and $t \geq 0$, since f_j, g_j are Lipschitz functions and ς_{ij}, η_{ij} are non-decreasing, we have, for each $i, j \in \{1, \ldots, n\}$,

$$\begin{aligned}
\left|f_{ij}(t,\phi) - f_{ij}(t,\varphi)\right| &\leq |c_{ij}(t)| \int_{-\tau_{ij}}^{0} \left|f_j(\phi(s)) - f_j(\varphi(s))\right| d\varsigma_{ij}(s) \\
&\quad + |d_{ij}(t)| \int_{-\infty}^{0} \left|g_j(\phi(s)) - g_j(\varphi(s))\right| d\eta_{ij}(s) \\
&\leq |c_{ij}(t)|\mu_j \int_{-\tau_{ij}}^{0} g(s) \frac{|(\phi-\varphi)(s)|}{g(s)} d\varsigma_{ij}(s) \\
&\quad + |d_{ij}(t)|\sigma_j \int_{-\infty}^{0} g(s) \frac{|(\phi-\varphi)(s)|}{g(s)} d\eta_{ij}(s) \\
&\leq \left(|c_{ij}(t)|\mu_j \int_{-\tau_{ij}}^{0} g(s) d\varsigma_{ij}(s) + |d_{ij}(t)|\sigma_j \int_{-\infty}^{0} g(s) d\eta_{ij}(s) \right) \|\phi - \varphi\|_g.
\end{aligned}$$

This means that

$$|f_{ij}(t,\phi) - f_{ij}(t,\varphi)| \leq \left(\left(|c_{ij}(t)|\mu_j + |d_{ij}(t)|\sigma_j\right)(1+\delta) \right) \|\phi - \varphi\|_g,$$

and, together with (33) and (34), the hypotheses (A3), (24), (25) and (26) hold. Now, the conclusion follows from Theorem 4.1.

Remark 4 The global asymptotic stability of the Cohen-Grossberg neural network model (30) was studied in [15] under a different set of hypotheses, including bounded coefficients $c_{ij}(t)$ and $d_{ij}(t)$ and bounded amplification functions $a_i(u)$. More recently, the same model was also studied in [5], where the global asymptotic stability was obtained assuming bounded amplification functions $a_i(u)$ and stronger conditions than (31) and (32). However, in [5] the existence of an equilibrium point was not assumed.

Example 2 As second example, consider the following Cohen-Grossberg neural network model with both discrete time-varying delays and distributed delays

$$\begin{aligned}
x_i'(t) = -a_i(t, x_i(t)) \Bigg[& b_i(t, x_i(t)) + \sum_{j=1}^{n} w_{ij}(t) f_j(x_j(t)) + \sum_{j=1}^{n} v_{ij}(t) g_j(x_j(t - \tau_{ij}(t))) \\
& + \sum_{j=1}^{n} q_{ij}(t) \int_{-\infty}^{0} k_{ij}(-s) p_j(x_j(t+s)) ds + I_i(t) \Bigg],
\end{aligned} \tag{35}$$

$t \geq 0, i = 1, \ldots, n$, where $a_i : [0, +\infty) \times \mathbb{R} \to (0, +\infty)$, $b_i : [0, +\infty) \times \mathbb{R} \to \mathbb{R}$, $w_{ij}, v_{ij}, q_{ij}, I_i : [0, +\infty) \to \mathbb{R}$, $\tau_{ij} : [0, +\infty) \to [0, +\infty)$, $f_j, g_j, p_j : \mathbb{R} \to \mathbb{R}$ are continuous functions, and the delay kernel functions $k_{ij} : [0, +\infty) \to [0, +\infty)$ are piecewise continuous such that

$$\int_0^{+\infty} k_{ij}(s)ds = 1. \tag{36}$$

System (35) also arises as a particular case of the model (22) when we consider $P = 2$, $a_i(t, \varphi) = a_i(t, \varphi_i(0))$, $h_{ij}(t, u) = w_{ij}(t)f_j(u_1) + v_{ij}(t)g_j(u_2)$, $\tau_{ij1}(t) = 0$, $\tau_{ij2}(t) = \tau_{ij}(t)$, $f_{ij}(t, \varphi_j) = q_{ij}(t) \int_{-\infty}^0 p_j(\varphi_j(t+s))d\eta_{ij}(s) + \frac{I_i(t)}{n}$ for $t \geq 0$, $\varphi = (\varphi_1, \dots, \varphi_n) \in BC$, and $u = (u_1, u_2) \in \mathbb{R}^2$, and the functions η_{ij} are defined by

$$\eta_{ij}(s) = \int_{-\infty}^{s} k_{ij}(-u)du, \quad s \in (-\infty, 0], \quad i, j \in \{1, \dots, n\}.$$

Since η_{ij} are non-decreasing, bounded, and $\int_{-\infty}^0 d\eta_{ij}(s) = \int_{-\infty}^0 k_{ij}(-s)ds = 1$ for $i, j \in \{1, \dots, n\}$, from Lemma 4.2, we conclude the existence of a function $g : (-\infty, 0] \to [1, +\infty)$ satisfying (g1)–(g3) and we consider UC_g as the phase space of (35). Now, following the same ideas in the proof of Corollary 4.3, we obtain the following stability criterion.

Corollary 4.4 *Assume that the model* (35) *has an equilibrium point, $f_j, g_j, p_j : \mathbb{R} \to \mathbb{R}$ are Lipschitz functions with Lipschitz constants μ_j, σ_j, ρ_j respectively, a_i satisfy (A2), and b_i satisfy (A3).*

If there exist $\alpha > 0$ and $d = (d_1, \dots, d_n) > 0$ such that, for each $i \in \{1, \dots, n\}$,

$$d_i\beta_i(t) - \sum_{j=1}^{n} d_j\big(|w_{ij}(t)|\mu_j + |v_{ij}(t)|\sigma_j + |q_{ij}(t)|\rho_j\big) > \alpha, \qquad \forall t \geq 0, \tag{37}$$

$$\limsup_{t\to+\infty} \sum_{j=1}^{n} \frac{d_j\big(|w_{ij}(t)|\mu_j + |v_{ij}(t)|\sigma_j + |q_{ij}(t)|\rho_j\big)}{d_i\beta_i(t)} < 1, \tag{38}$$

then the equilibrium point of (35) *is globally attractive.*

Remark 5 In [9], the authors studied the global asymptotic stability of a periodic solution of the model (35), assuming finite time delays and periodic coefficients. This means that τ_{ij} are bounded, the delay kernel functions k_{ij} are defined on $[-\nu_{ij}, 0]$, for some $\nu_{ij} \in \mathbb{R}^+$, and there is $\omega \in \mathbb{R}^+$ such that $w_{ij}(t)$, $v_{ij}(t)$, and $q_{ij}(t)$ are ω-periodic and $a_i(t, x)$, $b_i(t, x)$ are ω-periodic with respect to their first argument. As a final remark, we note that $a_i(t, x)$ being continuous and periodic with respect to the first argument imply hypothesis (A2) holds.

To the best of our knowledge, in [9], for the first time the stability of a nonautonomous Cohen-Grossberg neural network model was studied without assuming bounded amplification functions. Here, we do not assume bounded amplification functions either, but the hypotheses set considered is different. To illustrate this, consider the following 2π-periodic system

$$\begin{cases} x_1'(t) = -(|x_1(t)|+1)\left[\sqrt{2}x_1(t) + \frac{3\sin t}{4}\tanh(x_1(t-1)) \right. \\ \qquad\qquad \left. + \frac{3\cos t}{4}\tanh(x_2(t-2))\right] \\ x_2'(t) = -(|x_2(t)|+2)\left[\sqrt{2}x_2(t) + \frac{3\sin t}{4}\tanh(x_1(t-2)) \right. \\ \qquad\qquad \left. + \frac{3\cos t}{4}\tanh(x_2(t-1))\right] \end{cases}. \tag{39}$$

Although (39) is a periodic system, the Theorem 2 in [9] can not be used to get the stability of the equilibrium point $(x_1, x_2) = (0, 0)$, because

$$-\mathscr{H} = \begin{bmatrix} \sqrt{2} - \sup_{t\geq 0}\left|\frac{3\sin t}{4}\right| & -\sup_{t\geq 0}\left|\frac{3\cos t}{4}\right| \\ -\sup_{t\geq 0}\left|\frac{3\sin t}{4}\right| & \sqrt{2} - \sup_{t\geq 0}\left|\frac{3\cos t}{4}\right| \end{bmatrix} = \begin{bmatrix} \sqrt{2} - \frac{3}{4} & -\frac{3}{4} \\ -\frac{3}{4} & \sqrt{2} - \frac{3}{4} \end{bmatrix},$$

is not a nonsingular M-matrix (in fact, $\sqrt{2} - \frac{3}{2}$ is a negative eigenvalue of $-\mathscr{H}$, see [9, Proposition 2]). However, it is easy to see that conditions (37) and (38) hold and from Corollary 4.4 we conclude that the equilibrium point of (39) is globally attractive.

Example 3 Finally, consider the following nonautonomous BAM neural network model with both discrete time-varying delays and distributed delays

$$\begin{cases} x_i'(t) = -a_i(x_i(t))\left[b_i(t, x_i(t)) + \sum_{j=1}^{m} \widetilde{c}_{ij}(t)\widetilde{f}_{ij}(y_j(t-\tau_{ij}(t)))\right], \\ y_j'(t) = -\widetilde{a}_j(y_j(t))\left[\widetilde{b}_j(t, y_j(t)) + \sum_{i=1}^{n} c_{ji}(t)\int_{-\infty}^{0} k_{ji}(-s)f_{ji}(x_i(t+s))ds\right], \end{cases} \tag{40}$$

$t \geq 0, i = 1, \ldots, n,$ and $j = 1, \ldots, m,$ where $a_i, \widetilde{a}_j : \mathbb{R} \to (0, +\infty), b_i, \widetilde{b}_j : [0, +\infty) \times \mathbb{R} \to \mathbb{R}$, $c_{ji}, \widetilde{c}_{ij} : [0, +\infty) \to \mathbb{R}$, and $\tau_{ij} : [0, +\infty) \to [0, +\infty)$ are continuous functions, $f_{ji}, \widetilde{f}_{ij} : \mathbb{R} \to \mathbb{R}$ are Lipschitz functions with Lipschitz constants μ_{ji}, $\widetilde{\mu}_{ij}$ respectively, and the delay kernel functions $k_{ji} : [0, +\infty) \to [0, +\infty)$ are piecewise continuous such that

$$\int_{0}^{+\infty} k_{ji}(s)ds = 1. \tag{41}$$

As in the previous examples, it is easy to see that system (40) is also a particular case of model (22), and the next corollary is obtained following the same ideas presented in the previous Examples 1 and 2.

Corollary 4.5 *Assume that the model* (40) *has an equilibrium point,* $f_{ji}, \widetilde{f}_{ij} : \mathbb{R} \to \mathbb{R}$ *are Lipschitz functions with Lipschitz constants* $\mu_{ji}, \widetilde{\mu}_{ij}$ *respectively,* $a_i, \widetilde{a}_j : \mathbb{R} \to (0, +\infty)$, $b_i, \widetilde{b}_j : [0 + \infty) \times \mathbb{R} \to \mathbb{R}$, $\widetilde{c}_{ij}, c_{ji} : [0, +\infty) \to \mathbb{R}$, $\tau_{ij} : [0, +\infty) \to [0, +\infty)$ *are continuous functions such that* b_i *and* $\widetilde{b}_j$ *satisfy (A3) with the functions* $\beta_i(t)$ *and* $\widetilde{\beta}_j(t)$ *respectively, and the delay kernel functions* $k_{ji} : [0, +\infty) \to [0, +\infty)$ *are piecewise continuous functions satisfying* (41).

If there exist $d = (d_1, \ldots, d_m) > 0$, $\widetilde{d} = (\widetilde{d}_1, \ldots, \widetilde{d}_n) > 0$, *and* $\alpha > 0$ *such that*

$$d_i\beta_i(t) - \sum_{j=1}^{m} \widetilde{d}_j |\widetilde{c}_{ij}(t)| \widetilde{\mu}_{ij} > \alpha, \quad \forall t \geq 0, \ i = 1, \ldots, n, \tag{42}$$

$$\widetilde{d}_j\widetilde{\beta}_j(t) - \sum_{i=1}^{n} d_i |c_{ji}(t)| \mu_{ji} > \alpha, \quad \forall t \geq 0, \ j = 1, \ldots, m, \tag{43}$$

$$\limsup_{t\to+\infty} \sum_{j=1}^{m} \frac{\widetilde{d}_j |\widetilde{c}_{ij}(t)| \widetilde{\mu}_{ij}}{d_i\beta_i(t)} < 1, \quad i = 1, \ldots, n, \tag{44}$$

$$\limsup_{t\to+\infty} \sum_{i=1}^{n} \frac{d_i |c_{ji}(t)| \mu_{ji}}{\widetilde{d}_j\widetilde{\beta}_j(t)} < 1, \quad j = 1, \ldots, m, \tag{45}$$

then the equilibrium point of (40) *is globally attractive.*

Remark 6 The stability of the autonomous version of system (40) that is, when $b_i(t, u) = b_i(u), \widetilde{b}_j(t, u) = \widetilde{b}_j(u)$, and $\widetilde{c}_{ij}(t) = \widetilde{c}_{ij}, c_{ji}(t) = c_{ji}, \tau_{ij}(t) = \tau_{ij}$ with $\widetilde{c}_{ij}, c_{ij} \in \mathbb{R}$ and $\tau_{ij} \geq 0$, was recently studied in [6, 13]. See also [14] for a detailed description and explanation of bidirectional associative memory neural network models. We note that, for the autonomous situation, the set condition (42), (43), (44), and (45) is equivalent to

$$\begin{cases} d_i\beta_i - \sum_{j=1}^{m} \widetilde{d}_j |\widetilde{c}_{ij}| \widetilde{\mu}_{ij} > 0, \ i = 1, \ldots, n, \\ \widetilde{d}_j\widetilde{\beta}_j - \sum_{i=1}^{n} d_i |c_{ji}| \mu_{ji} > 0 \ \ j = 1, \ldots, m, \end{cases}, \tag{46}$$

for some $d = (d_1, \ldots, d_m) > 0$, $\widetilde{d} = (\widetilde{d}_1, \ldots, \widetilde{d}_n) > 0$. In [6], the existence and global asymptotic stability of an equilibrium point of the autonomous version of (40) was proved assuming (46). Thus, Corollary 4.5 generalizes the stability criterion presented in [6].

Acknowledgments The research was supported by the Research Centre of Mathematics of the University of Minho with the Portuguese Funds from the "Fundação para a Ciência e a Tecnologia", through the Project PEstOE/MAT/UI0013/2014. The author thanks the referee for valuable suggestions and comments.

References

1. Baldi, P., Atiya, A.F.: How delays affect neural dynamics and learning. IEEE Trans. Neural Netw. **5**, 612–621 (1994)
2. Cao, J., Wang, L.: Periodic oscillatory solution of bidirectional associative memory networks with delays. Phys. Rev. E **61**(2), 1825–1828 (2000)
3. Chen, Y.: Global asymptotic stability of delayed Cohen-Grossberg neural networks. IEEE Trans. Circ. Syst. **53**, 351–357 (2006)
4. Cohen, M., Grossberg, S.: Absolute stability of global pattern formation and parallel memory storage by competitive neural networks. IEEE Trans. Syst. Man Cybernet. **13**, 815–826 (1983)
5. Esteves, S., Oliveira, J.J.: Global asymptotic stability of nonautonomous Cohen-Grossberg neural network models with infinite delays. Appl. Math. Comput. **265**, 333–346 (2015)
6. Faria, T., Oliveira, J.J.: General criteria for asymptotic and exponential stabilities of neural network models with unbounded delays. Appl. Math. Comput. **217**, 9646–9658 (2011)
7. Haddock, J., Hornor, W.: Precompactness and convergence in norm of positive orbits in a certain fading memory space. Funkcial. Ekvac. **31**, 349–361 (1988)
8. Hale, J.K., Kato, J.: Phase space for retarded equations with infinite delay. Funkcial. Ekvac. **21**, 11–41 (1978)
9. Hien, L.V., Loan, T.T., Trang, B.T.H., Trinh, H.: Existence and global asymptotic stability of positive periodic solution of delayed Cohen-Grossberg neural networks. Appl. Math. Comput. **240**, 200–212 (2014)
10. Hopfield, J.J.: Neural networks with graded response have collective computational properties like those of two-state neurons. Proc. Natl. Acad. Sci. **81**, 3088–3092 (1984)
11. Huang, T., Chan, A., Huang, Y., Cao, J.: Stability of Cohen-Grossberg neural networks with time-varying delays. Neural Netw. **20**, 868–873 (2007)
12. Huang, T., Li, C., Chen, G.: Stability of Cohen-Grossberg neural networks with unbounded distributed delays. Chaos Solitons Fractals **34**, 992–996 (2007)
13. Jiang, H., Cao, J.: BAM-type Cohen-Grossberg neural networks with time delay. Math. Comput. Model. **47**, 92–103 (2008)
14. Kosko, B.: Bidirectional associative memories. IEEE Trans. Syst. Man Cybern. **18**, 49–60 (1988)
15. Li, J., Yan, J., Jia, X.: Dynamical analysis of Cohen-Grossberg neural networks with finite and infinite delays. Appl. Math. Comput. **213**, 529–537 (2009)
16. Marcus, C.M., Westervelt, R.M.: Stability of analogy neural networks with delay. Phys. Rev. A **39**, 347–359 (1989)
17. Oliveira, J.J.: Global asymptotic stability for neural network models with distributed delays. Math. Comput. Model. **50**, 81–91 (2009)
18. Song, Q., Cao, J.: Stability analysis of Cohen-Grossberg neural network with both time-varying and continuously distributed delays. J. Comput. Appl. Math. **197**, 188–203 (2006)
19. Wu, R.: Exponential convergence of BAM neural networks with time-varying coefficients and distributed delays. Nonlinear Anal. RWA **11**, 562–573 (2010)
20. Xiang, H., Cao, J.: Almost periodic solutions of recurrent neural networks with continuously distributed delays. Nonlinear Anal. **71**, 6097–6108 (2009)
21. Zhou, J., Li, Q., Zhang, F.: Convergence behavior of delayed cellular neural networks without periodic coefficients. Appl. Math. Lett. **21**, 1012–1017 (2008)

Phase Transitions and Coarse-Graining for a System of Particles in the Continuum

Elena Pulvirenti and Dimitrios Tsagkarogiannis

Abstract We revisit the proof of the liquid-vapor phase transition for systems with finite-range interaction by Lebowitz et al. (J. Stat. Phys. **94**(5–6), 955–1025, 1999 [1]) and extend it to the case where we additionally include a hard-core interaction to the Hamiltonian. We establish the phase transition for the mean field limit and then we also prove it when the interaction range is long but finite, by perturbing around the mean-field theory. A key step in this procedure is the construction of a density (coarse-grained) model via cluster expansion. In this note we present the overall result but we mainly focus on this last issue.

Keywords Continuum particle system · Mean field theory · Phase transition · Coarse-graining · Pirogov-Sinai theory · Cluster expansion

1 Introduction

One of the main open problems in equilibrium statistical mechanics is to prove the validity of a liquid-vapour phase transition in a continuum particle system. Although this is well observed in experiments as well as in continuum theories, a rigorous proof for particle systems is still lacking. Intermolecular forces are often described by Lennard-Jones interactions, however the difficulty of handling such (or more realistic) systems has promoted the introduction of several simplified models. A good compromise between realistic models of fluids and mathematically treatable systems may consist of particles interacting via a combination of hard spheres (for repulsion) and an attractive long-range Kac interaction. However, the free energy of

E. Pulvirenti (✉)
Mathematical Institute, Leiden University, P.O. Box 9512,
2300 Leiden, RA, The Netherlands
e-mail: pulvirentie@math.leidenuniv.nl; elena.pulvirenti@gmail.com

D. Tsagkarogiannis
Department of Mathematics, University of Sussex, Brighton BN1 9QH, UK
e-mail: D.Tsagkarogiannis@sussex.ac.uk

© Springer International Publishing Switzerland 2016
P. Gonçalves and A.J. Soares (eds.), *From Particle Systems to Partial Differential Equations III*, Springer Proceedings in Mathematics & Statistics 162,
DOI 10.1007/978-3-319-32144-8_13

hard spheres can be studied for very small values of the density, far from the value at which a transition occurs. Hence, we still need to use a long range 4-body repulsive term as in [1] to determine the phase transition point. Then, the hard-core interaction acts just as a perturbation to the mean-field case. In fact, we show that the liquid-vapour transition persists if the volume of the hard spheres is sufficiently small, but finite. Nevertheless, our model presents a richer behaviour and if one manages to deal with a higher density regime, the hard-core interaction will become relevant and responsible for another transition of the gas-solid type.

Our proof will follow Pirogov-Sinai theory in the version proposed by Zahradník [8]. The analysis requires first of all the notions of *coarse-graining* and *contours* which are introduced in Sect. 3 and subsequently, with an argument á la Peierls, one has to prove that contours are improbable which we do in Sect. 4. In this scenario we are able to compute the effective Hamiltonian for the coarse-grained system with a multi-canonical constraint (given by the fixed density in each cell). This computation involves an integration over the positions of the particles in each cell leading to a new measure on the density at the cells. The computations which lead to the effective Hamiltonian are in general very complicated, nevertheless due to the choice of the interaction they can be carried out. The crucial point here is to show convergence of a cluster expansion in the canonical ensemble with hard-core, Kac interaction and contour weights. This is done in Sect. 5.2 by extending the results in [5].

For more details on the proofs we refer to [7], from which the present paper is a follow-up, to [6] and to the monograph of Presutti [4].

2 Model

We consider a system of identical point particles in $\mathbb{R}^d$, $d \geq 2$, and call *particle configuration* a countable, locally finite collection of points in $\mathbb{R}^d$. The *phase space* $\mathcal{Q}^\Lambda$ is the collection of all particle configurations in a bounded region Λ. We use the notation $\mathcal{Q}$ when $\Lambda \equiv \mathbb{R}^d$. We write $q = (q_1, \dots, q_n)$ to indicate a configuration of n particles positioned at points $q_1, \dots, q_n$ (the order is not important) of $\mathbb{R}^d$, while we write q_Λ when we want to specify that the particles are in $\mathcal{Q}^\Lambda$.

We consider a mean field model with an energy density given by:

$$e_\lambda(\rho) = -\lambda\rho - \frac{\rho^2}{2} + \frac{\rho^4}{4!}, \tag{1}$$

where λ is the chemical potential. Here, the density $\rho = n/|\Lambda|$ is set equal to the total density and it is therefore constant. We further define the LMP model, [1], by relaxing to a local mean field: the Hamiltonian (for configurations with finitely many particles) is given by the following function

$$H^{\mathrm{LMP}}_{\lambda,\gamma}(q) = \int_{\mathbb{R}^d} e_\lambda(\rho_\gamma(r; q))\, dr, \tag{2}$$

where

$$\rho_\gamma(r;q) := \sum_{q_i \in q} J_\gamma(r, q_i) \tag{3}$$

is the local particle density at $r \in \mathbb{R}^d$. The local density is defined through Kac potentials, $J_\gamma(r, r') = \gamma^d J(\gamma r, \gamma r')$, where $J(s, t)$ is a symmetric, translation invariant ($J(s, t) = J(0, t - s)$) smooth function which vanishes for $|t - s| \geq 1$. Thus, the range of the interaction has order γ^{-1} (for both repulsive and attractive potentials) and the "Kac scaling parameter" γ is assumed to be small. This choice of the potentials makes the LMP model a perturbation of the mean-field, in the sense that when taking the thermodynamic limit followed by the limit $\gamma \to 0$ the free energy is equivalent to the free energy in the mean-field description (1).

Note that the LMP interaction is the sum of a repulsive four body potential and an attractive two body potential, which can be written in the following way

$$H^{\text{LMP}}_{\lambda,\gamma}(q) = -\lambda|q| - \frac{1}{2!}\sum_{i \neq j} J^{(2)}_\gamma(q_i, q_j) + \frac{1}{4!}\sum_{i_1 \neq \cdots \neq i_4} J^{(4)}_\gamma(q_{i_1}, \ldots, q_{i_4}), \tag{4}$$

where $|\cdot|$ denotes the cardinality of a set and

$$\begin{aligned} J^{(2)}_\gamma(q_i, q_j) &= \int J_\gamma(r, q_i) J_\gamma(r, q_j)\, dr \\ J^{(4)}_\gamma(q_{i_1}, \ldots, q_{i_4}) &= \int J_\gamma(r, q_{i_1}) \cdots J_\gamma(r, q_{i_4})\, dr. \end{aligned} \tag{5}$$

To this model we add an extra hard-core interaction described by a potential $V^R : \mathbb{R}^d \to \mathbb{R}$ such that

$$V^R(q_i, q_j) = \begin{cases} +\infty & \text{if } |q_i - q_j| \leq R \\ 0 & \text{if } |q_i - q_j| > R \end{cases} \tag{6}$$

where $|q_i - q_j|$ denotes the euclidean distance between the two particles in q_i and q_j. R is the radius of the hard spheres and their volume is $\varepsilon = V_d(R)$, i.e., the volume of the d-dimensional sphere of radius R. Note also that the hard-core potential depends on q_i, q_j only through their distance.

Hence, the Hamiltonian of the model (LMP-hc) we consider is the following

$$H_{\gamma,R,\lambda}(q) = \int e_\lambda(J_\gamma * q(r))\, dr + H^{\text{hc}}_R(q), \tag{7}$$

where

$$H^{\text{hc}}_R(q) := \sum_{i<j} V^R(q_i, q_j). \tag{8}$$

Given two configurations q and $\bar{q}$, we will use the following two notations to represent the energy of the particle configuration q in the field generated by $\bar{q}$ and the interaction energy between the particle configuration q and $\bar{q}$

$$H_{\gamma,R,\lambda}(q|\bar{q}) = H_{\gamma,R,\lambda}(q+\bar{q}) - H_{\gamma,R,\lambda}(\bar{q}) \tag{9}$$

$$U_{\gamma,R,\lambda}(q,\bar{q}) = H_{\gamma,R,\lambda}(q+\bar{q}) - H_{\gamma,R,\lambda}(q) - H_{\gamma,R,\lambda}(\bar{q}). \tag{10}$$

respectively, both for configurations with finitely many particles.

The grand-canonical Gibbs measure in the bounded measurable region Λ in $\mathbb{R}^d$ and boundary conditions $\bar{q} \in \mathcal{Q}^{\Lambda^c}$ is the probability measure on $\mathcal{Q}^{\Lambda}$ defined by

$$\mu^{\Lambda}_{\gamma,\beta,R,\lambda,\bar{q}}(dq) = Z^{-1}_{\gamma,\beta,R,\lambda,\bar{q}}(\Lambda)e^{-\beta H_{\gamma,R,\lambda}(q|\bar{q})}\nu^{\Lambda}(dq), \tag{11}$$

where β is the inverse temperature, $\nu^{\Lambda}(dq)$ is the Poisson point process of intensity 1 and $Z_{\gamma,\beta,R,\lambda,\bar{q}}(\Lambda)$ is the grand canonical partition function (defined as the normalization factor for $\mu^{\Lambda}_{\gamma,\beta,R,\lambda,\bar{q}}(dq)$ to be a probability).

2.1 Mean-Field Model

The model introduced above is a perturbation of a mean-field model, which is defined as follows. We consider the space of configurations with hard-core constraint

$$\mathcal{X}^R_{n,\Lambda} = \{(q_1,\ldots,q_n) \in \Lambda^n : \min_{i\neq j} |q_i - q_j| > R\}. \tag{12}$$

Given a configuration $q \equiv (q_1,\ldots,q_n)$ in $\mathcal{X}^R_{n,\Lambda}$, the mean-field Hamiltonian is

$$H^{\mathrm{mf}}_{\Lambda,R,\lambda}(q) = |\Lambda| e_{\lambda}(\rho) \tag{13}$$

where $\rho = n/|\Lambda|$ and $e_{\lambda}(\cdot)$ is given in (1). The mean-field canonical partition function is

$$Z^{\mathrm{mf}}_{n,\Lambda,R} = \frac{1}{n!}\int_{\mathcal{X}^R_{n,\Lambda}} e^{-\beta H^{\mathrm{mf}}_{\Lambda,R,0}(q)} dq_1\cdots dq_n \tag{14}$$

$$= \exp\Big\{-\beta\Big(-\frac{n^2}{2|\Lambda|}+\frac{n^4}{4!|\Lambda|^3}\Big)\Big\}\frac{1}{n!}\int_{\mathcal{X}^R_{n,\Lambda}} dq_1\cdots dq_n. \tag{15}$$

The existence of its thermodynamic limit follows from general arguments and the canonical mean-field free energy is

$$\phi_{\beta,R}(\rho) = \lim_{|\Lambda|,n\to\infty:\frac{n}{|\Lambda|}\to\infty} -\frac{1}{|\Lambda|\beta}\log Z^{\text{mf}}_{n,\Lambda,R} = e_0(\rho) + f^{\text{hc}}_{\beta,R}(\rho) \tag{16}$$

where

$$f^{\text{hc}}_{\beta,R}(\rho) := \lim_{|\Lambda|,n\to\infty:\frac{n}{|\Lambda|}\to\infty} -\frac{1}{|\Lambda|\beta}\log Z^{\text{hc}}_{n,\Lambda,R}, \qquad Z^{\text{hc}}_{n,\Lambda,R} := \frac{1}{n!}\int_{\mathscr{X}^R_{n,\Lambda}} dq_1\cdots dq_n \tag{17}$$

is a convex function of ρ.

The mean-field model shows a phase transition for β large enough, which is reflected in a loss of convexity of $\phi_{\beta,R}(\rho)$. The critical points of $\phi_{\beta,R,\lambda}(\rho) = \phi_{\beta,R}(\rho) - \lambda\rho$, as a function of ρ, are the solutions of the mean-field equation

$$\frac{d}{d\rho}\Big\{e_\lambda(\rho) + f^{\text{hc}}_{\beta,R}(\rho)\Big\} = 0 \tag{18}$$

and have the form

$$\rho = \exp\Big\{-\beta e'_\lambda(\rho) - \psi'_{\beta,R}(\rho)\Big\} := K_{\beta,\lambda,R}(\rho), \tag{19}$$

where $\psi_{\beta,R}(\rho)$ is the free energy minus the entropy of the free system, i.e.,

$$f^{\text{hc}}_{\beta,R}(\rho) - \frac{1}{\beta}\rho(\log\rho - 1). \tag{20}$$

We have the following properties

- There is a critical inverse temperature $\beta_{c,R}$, such that $\phi_{\beta,R}(\rho)$ is convex for $\beta \le \beta_{c,R}$, while for $\beta > \beta_{c,R}$ it has two inflection points $0 < s_-(\beta) < s_+(\beta)$, being concave for $\rho \in (s_-(\beta), s_+(\beta))$ and convex for $\rho \notin (s_-(\beta), s_+(\beta))$.
- For any $\beta > \beta_{c,R}$, there is $\lambda(\beta, R)$ so that $\phi_{\beta,\lambda(\beta,R),R}(\cdot)$ has two global minimizers, $\rho_{\beta,R,-} < \rho_{\beta,R,+}$ (and a local maximum at $\rho_{\beta,R,0}$). For $\lambda \neq \lambda(\beta, R)$ and for $\beta \le \beta_{c,R}$ the minimizer is unique.
- For any $\beta > \beta_{c,R}$ there is an interval $(\lambda_-(\beta, R), \lambda_+(\beta, R))$ containing $\lambda(\beta, R)$ and for any λ in the interval $\phi_{\beta,\lambda,R}(\cdot)$ it has two local minima $\rho_{\beta,\lambda,R,\pm}$ which are differentiable functions of λ and $\frac{d}{d\lambda}(\phi_{\beta,\lambda,R}(\rho_{\beta,\lambda,R,+}) - \phi_{\beta,\lambda,R}(\rho_{\beta,\lambda,R,-})) = \rho_{\beta,\lambda,R,-} - \rho_{\beta,\lambda,R,+} < 0$. For all $\beta > \beta_{c,R}$,

$$\frac{d}{d\rho}K_{\beta,\lambda(\beta,R),R}(\rho)\Big|_{\rho=\rho_{\beta,R,\pm}} \equiv K'_{\beta,\lambda(\beta,R),R}(\rho_{\beta,R,\pm}) < 1, \tag{21}$$

the condition (21) being equivalent to $\phi''_{\beta,\lambda(\beta),R}(\rho_{\beta,R,\pm}) > 0$. Moreover, there exists $\beta_{0,R} > \beta_{c,R}$ such that

$$K'_{\beta,\lambda(\beta,R),R}(\rho_{\beta,R,\pm}) > -1, \qquad \text{for all } \beta \in (\beta_{c,R}, \beta_{0,R}). \tag{22}$$

- We have an expansion for $\beta_{c,R}$ in powers of $\varepsilon = V_d(R)$

$$\beta_{c,R} = \beta_c^{\text{LMP}} - \varepsilon\,(\beta_c^{\text{LMP}})^{2/3} + O(\varepsilon^2),$$

$\beta_c^{\text{LMP}} = \frac{3}{2}^{\frac{3}{2}}$ being the critical inverse temperature for the LMP mean-field model. Note that while $\beta_{c,R}$ has the meaning of critical inverse temperature, $\beta_{0,R}$ has no physical meaning, but it is introduced for technical reasons. In fact $\beta_{0,R}$ is necessary for (22) to be true and depends on the choice of the mean-field Hamiltonian (13).

3 Contour Model

To prove the phase transition in the LMP-hc model we study perturbations of the homogeneous states with densities $\rho_{\beta,R,\pm}$ which appear in the limit $\gamma \to 0$. We follow an argument à la Peierls, which relies (as for the Ising model) on the possibility to rewrite the partition function of the model as the partition function of an "abstract contour model". To implement this strategy we need to introduce several scaling parameters and phase indicators. Namely, we introduce two scales $\ell_\pm = \gamma^{-(1\pm\alpha)}$ and an accuracy parameter $\zeta = \gamma^a$, with $1 \gg \alpha \gg a > 0$. We define $\mathcal{D}^{(\ell)}$ a partition of $\mathbb{R}^d$ into cubes of side ℓ and we denote $C_r^{(\ell)}$ the cube of $\mathcal{D}^{(\ell)}$ which contains r.

The first phase indicator is defined as

$$\eta^{(\zeta,\ell_-)}(q;r) = \begin{cases} \pm 1 & \text{if } \left|\rho^{(\ell_-)}(q;r) - \rho_{\beta,R,\pm}\right| \le \zeta \\ 0 & \text{otherwise} \end{cases}$$

where $\rho^{(\ell)}(q;r) = |C_r^{(\ell)} \cap q|\ell^{-d}$ is the empirical density in a cube of side ℓ containing r given a configuration q.

Thus $\eta^{(\zeta,\ell_-)}(q;r)$ indicates the phase (or its absence) on the small scale ℓ_-. Because of statistical fluctuations, we must allow for deviations from the ideal plus configurations $\eta^{(\zeta,\ell_-)}(q;r) = 1$. We thus need to define which regions are still in the plus phase and which are those destroyed by the fluctuations. The fact that $\eta^{(\zeta,\ell_-)}(q;r) = 1$ does not qualify r being in the $+$ phase, implies that we need a stronger condition which is defined in terms of two more phase indicators which describe the local phase of the system in increasing degree of accuracy. We have

$$\theta^{(\zeta,\ell_-,\ell_+)}(q;r) = \begin{cases} \pm 1 & \text{if } \eta^{(\zeta,\ell_-)}(q;r') = \pm 1 \quad \forall r' \in C_r^{(\ell_+)} \\ 0 & \text{otherwise} \end{cases}$$

$$\Theta^{(\zeta,\ell_-,\ell_+)}(q;r) = \begin{cases} \pm 1 & \text{if } \eta^{(\zeta,\ell_-)}(q;r') = \pm 1 \quad \forall r' \in C_r^{(\ell_+)} \cup \delta_{\text{out}}^{\ell_+}[C_r^{(\ell_+)}] \\ 0 & \text{otherwise} \end{cases}$$

where $\delta^{\ell}_{\text{out}}[\Lambda]$ of a $\mathscr{D}^{(\ell)}$-measurable region Λ is the union of all the cubes $C \in \mathscr{D}^{(\ell)}$ next to Λ. For simplicity, from now on we drop the superscript from the notation of $\eta^{(\zeta,\ell_-)}$, $\theta^{(\zeta,\ell_-,\ell_+)}$, $\Theta^{(\zeta,\ell_-,\ell_+)}$.

With these definitions, given a configuration q, the "plus phase" is the region $\{r : \Theta(q; r) = 1\}$ while the "minus phase" is the region $\{r : \Theta(q; r) = -1\}$. We call $q^{\pm}$ a $\pm$ boundary conditions relative to a region Λ, if it belongs to the ensemble $\eta(q; r) = \pm 1$ for r on the frame of width $2\gamma^{-1}$ around Λ.

Two sets are connected if their closures have non empty intersection; hence, two cubes with a common vertex are connected. In this way, the plus and the minus regions are separated by zero-phase regions $\{r : \Theta(q; r) = 0\}$.

Definition 1 A contour is a pair $\Gamma = \big(\text{sp}(\Gamma), \eta_\Gamma\big)$, where $\text{sp}(\Gamma)$ is a maximal connected component of the "incorrect set" $\{r \in \mathbb{R}^d : \Theta(q; r) = 0\}$ and η_Γ is the restriction to $\text{sp}(\Gamma)$ of $\eta(q; \cdot)$.

The exterior, $\text{ext}(\Gamma)$, of Γ is the unbounded, maximal connected component of $\text{sp}(\Gamma)^c$. The interior is the set $\text{int}(\Gamma) = \text{sp}(\Gamma)^c \setminus \text{ext}(\Gamma)$; we denote by $\text{int}_i(\Gamma)$ the maximal connected components of $\text{int}(\Gamma)$. Let $c(\Gamma) = \text{sp}(\Gamma) \cup \text{int}(\Gamma)$ and note that $\text{int}_i(\Gamma)$ and $c(\Gamma)$ are both simply connected. The outer boundaries of Γ are the sets

$$A(\Gamma) := \delta^{\ell_+}_{\text{out}}[\text{sp}(\Gamma)] \cap \text{int}(\Gamma), \quad A_{\text{ext}}(\Gamma) := \delta^{\ell_+}_{\text{out}}[c(\Gamma)]. \tag{23}$$

We will also call $A_i(\Gamma) = A(\Gamma) \cap \text{int}_i(\Gamma)$.

Definition 2 Γ is a plus/minus, contour if $\Theta(q; r) = \pm 1$ on $A_{\text{ext}}(\Gamma)$.

We add a superscript $\pm$ to $A_i(\Gamma)$ to indicate the sign of Θ and we write $\text{int}^{\pm}_i(\Gamma)$ if $\text{int}_i(\Gamma)$ contains $A^{\pm}_i(\Gamma)$. Note that Θ is constant on $A_{\text{ext}}(\Gamma)$ and $A_i(\Gamma)$ and its value is determined by η.

Definition 3 Given a plus contour Γ and a plus boundary condition q^+ for $c(\Gamma)$, we define the weight $W^+_{\gamma,R,\lambda}(\Gamma; \bar{q})$ of Γ as equal to

$$\frac{\mu^{c(\Gamma)}_{\gamma,\beta,R,\lambda,q^+}\Big(\eta(q_{c(\Gamma)}; r) = \eta_\Gamma(r), r \in \text{sp}(\Gamma);\ \Theta(q_{c(\Gamma)}; r) = \pm 1, r \in A^{\pm}(\Gamma)\Big)}{\mu^{c(\Gamma)}_{\gamma,\beta,R,\lambda,q^+}\Big(\eta(q_{c(\Gamma)}; r) = 1, r \in \text{sp}(\Gamma);\ \Theta(q_{c(\Gamma)}; r) = 1, r \in A^{\pm}(\Gamma)\Big)} \tag{24}$$

where the measure $\mu^{c(\Gamma)}_{\gamma,\beta,R,\lambda,q^+}$ has been defined in (11). Analogously, we can define the weight of a minus contour.

Thus, the numerator is the probability of the contour Γ conditioned to the outside of sp(Γ) while the denominator is the probability that the contour Γ is absent and replaced by the plus configurations (with the same conditioning to the outside).

The weight $W^-_{\gamma,R,\lambda}(\Gamma; q^-)$ of a minus contour Γ is defined analogously. The weight $W^{\pm}_{\gamma,R,\lambda}(\Gamma; q^{\pm})$ depends only on q^D, i.e., the restriction of $q^{\pm}$ to $D \equiv \{r \in c(\Gamma)^c : \text{dist}(r, c(\Gamma)) \le 2\gamma^{-1}\}$.

Definition 4 The plus diluted Gibbs measure in a bounded $\mathscr{D}^{(\ell_+)}$-measurable region Λ with plus boundary conditions $\bar{q}$ is

$$\mu^{\Lambda,+}_{\gamma,\beta,R,\lambda,\bar{q}}(dq_\Lambda) := \frac{1}{Z^+_{\gamma,\beta,R,\lambda,\bar{q}}(\Lambda)} e^{-\beta H_{\gamma,R,\lambda}(q_\Lambda|\bar{q}_{\Lambda^c})} \mathbf{1}_{\Theta((q_\Lambda+q^+_{\Lambda^c});r)=1 \text{ r } \in \delta^{\ell+}_{\text{out}}[\Lambda^c]} \nu^\Lambda(dq_\Lambda) \tag{25}$$

where $q^+ \in \mathscr{Q}^+ = \{q : \eta(q;r) = 1, r \in \mathbb{R}^d\}$ and $Z^+_{\gamma,\beta,R,\lambda,\bar{q}}(\Lambda)$ is the normalization, also called the plus diluted partition function. A similar definition holds for the minus diluted Gibbs measure.

We end this section by writing the ratio (24) of probabilities in the definition of the weight of a contour as a ratio of two partition functions. By writing explicitly the contributions coming from the support of a contour and those coming from the interior, we have for a plus contour Γ

$$W^+_{\gamma,R,\lambda}(\Gamma; q^+) = \frac{\mathscr{N}^+_{\gamma,R,\lambda}(\Gamma, q^+)}{\mathscr{D}^+_{\gamma,R,\lambda}(\Gamma, q^+)} \tag{26}$$

where:

$$\begin{aligned}\mathscr{N}^+_{\gamma,R,\lambda}(\Gamma, q^+) = \int_{q_{\text{sp}(\Gamma)}:\eta(q_{\text{sp}(\Gamma)};r)=\eta_\Gamma(r), r\in\text{sp}(\Gamma)} & e^{-\beta H_{\gamma,R,\lambda,\text{sp}(\Gamma)}(q_{\text{sp}(\Gamma)}|q^+_{A_{\text{ext}}})} \\ & \times Z^-_{\gamma,\beta,R,\lambda,q_{\text{sp}(\Gamma)}}(\text{int}^-(\Gamma))\, Z^+_{\gamma,\beta,R,\lambda,q_{\text{sp}(\Gamma)}}(\text{int}^+(\Gamma))\end{aligned} \tag{27}$$

$$\begin{aligned}\mathscr{D}^+_{\gamma,R,\lambda}(\Gamma, q^+) = \int_{q_{\text{sp}(\Gamma)}:\eta(q_{\text{sp}(\Gamma)};r)=1, r\in\text{sp}(\Gamma)} & e^{-\beta H_{\gamma,R,\lambda,\text{sp}(\Gamma)}(q_{\text{sp}(\Gamma)}|q^+_{A_{\text{ext}}})} \\ & \times Z^+_{\gamma,\beta,R,\lambda,q_{\text{sp}(\Gamma)}}(\text{int}^-(\Gamma))\, Z^+_{\gamma,\beta,R,\lambda,q_{\text{sp}(\Gamma)}}(\text{int}^+(\Gamma)).\end{aligned} \tag{28}$$

4 The Main Results

Our main theorem states that the system undergoes a first-order phase transition. This means that for β large enough the Gibbs state at the thermodynamic limit, i.e., $\Lambda \to \mathbb{R}^d$, is not unique. It is possible to fix plus/minus boundary conditions such that, if R and γ are small and for some values of β, λ, uniformly in Λ, the typical configurations of the corresponding diluted Gibbs measures are close to the plus/minus phase. This is quantified in the following theorem.

Theorem 1 (Liquid-vapor phase transition) *Consider the LMP-hc model in dimensions $d \geq 2$. For such a model there are R_0, $\beta_{c,R}$, $\beta_{0,R}$ and for any $0 < R \leq R_0$ and $\beta \in (\beta_{c,R}, \beta_{0,R})$ there is $\gamma_{\beta,R} > 0$ so that for any $\gamma \leq \gamma_{\beta,R}$ there is $\lambda_{\beta,\gamma,R}$ such that:*

There are two distinct infinite-volume measures $\mu^{\pm}_{\beta,\gamma,R}$ *with chemical potential* $\lambda_{\beta,\gamma,R}$ *and inverse temperature* β *and two different densities:* $0 < \rho_{\beta,\gamma,R,-} < \rho_{\beta,\gamma,R,+}$.

In the theorem, $\mu^{\pm}_{\beta,\gamma,R}$ are the infinite-volume limits of (25), while $\beta_{c,R}$, $\beta_{0,R}$ are the two inverse temperatures introduced in Sect. 2.1.

We prove the existence of two distinct states, which are interpreted as the two pure phases of the system: $\mu^{+}_{\beta,\gamma,R}$ describes the liquid phase with density $\rho_{\beta,\gamma,R,+}$ while $\mu^{-}_{\beta,\gamma,R}$ describes the vapor phase, with the smaller density $\rho_{\beta,\gamma,R,-}$. Furthermore we have

$$\lim_{\gamma\to 0}\rho_{\beta,\gamma,R,\pm} = \rho_{\beta,R,\pm}, \quad \rho_{\beta,R,-} < \rho_{\beta,R,+}, \qquad \lim_{\gamma\to 0}\lambda_{\beta,\gamma,R} = \lambda(\beta, R)$$

which are the densities and the chemical potential for which there is a phase transition in the mean-field model (see again Sect. 2.1).

The main technical point in the proof of Theorem 1 is to prove that contours are improbable. In particular, they satisfy Peierls estimates which proves that the probability of a contour decays exponentially with its volume.

Theorem 2 *There exists* R_0 *such that for any* $R \le R_0$ *and any* $\beta \in (\beta_{c,R}, \beta_{0,R})$ *there exist* $c > 0$, $\gamma_{\beta,R} > 0$, *so that for any* $\gamma \le \gamma_{\beta,R}$, $\pm$ *contour* Γ *and any* $\pm$ *boundary condition* $q^{\pm}$ *relative to* $c(\Gamma)$,

$$W^{\pm}_{\gamma,R,\lambda}(\Gamma; q^{\pm}) \le \exp\Big\{ -\beta c\,(\zeta^2 \ell^d_-)\, N_\Gamma \Big\} \tag{29}$$

where $\lambda = \lambda_{\beta,\gamma,R}$ *and*

$$N_\Gamma = \frac{|\mathrm{sp}(\Gamma)|}{\ell^d_+} \tag{30}$$

is the number of cubes of the partition $\mathcal{D}^{(\ell_+)}$ *contained in* sp(Γ).

As a corollary of Theorem 2 we have

Corollary 1 *There exists* R_0 *such that for any* $R \le R_0$, *any* $\beta \in (\beta_c, \beta_0)$ *and letting* c, $\gamma_{\beta,R}$, γ *and* $\lambda_{\beta,\gamma,R}$ *as in Theorem 2, we have that for any bounded, simply connected,* $\mathcal{D}^{(\ell_+)}$ *measurable region* Λ, *any* $\pm$ *boundary condition* $q^{\pm}$ *and any* $r \in \Lambda$, *the following holds*

$$\mu^{\Lambda,\pm}_{\gamma,\beta,R,\lambda_{\beta,\gamma,R},q^{\pm}}(\{\Theta(q; r) = \pm 1\}) \ge 1 - \exp\Big\{ -\beta\frac{c}{2}\,(\zeta^2 \ell^d_-)\Big\}. \tag{31}$$

Theorem 1 implies that for any $R \le R_0$ and γ small enough (chosen according to R) the difference between the diluted Gibbs measures $\mu^{\Lambda,+}_{\gamma,\beta,R,\lambda_{\beta,\gamma,R},q^+}(dq)$ and $\mu^{\Lambda,-}_{\gamma,\beta,R,\lambda_{\beta,\gamma,R},q^-}(dq)$ survives in the thermodynamic limit $\Lambda \nearrow \mathbb{R}^d$ and a phase transition occurs.

The main difficulty in proving (29) is that both numerator and denominator in (24) are defined in terms of expressions which involve not only the support of Γ but also its whole interior. They are therefore "bulk quantities" while the desired bound involves only the volume of the support of Γ, which for some contours, at least, is a "surface quantity". The main issue here is to find cancellations of the bulk terms between the numerator and the denominator. This is easy when special symmetries allow to relate the $+$ and $-$ ensembles, as in the ferromagnetic Ising model. Such simplifications are not present here and this is one of the issues which makes continuum models difficult to study. We overcome this difficulty using the Pirogov-Sinai theory [3] which covers cases where this symmetry is broken.

A central point of the Pirogov-Sinai theory is a change of measure. The idea is to introduce a new Gibbs measure (simpler than the original one), but which gives the same properties. The diluted partition function in a region Λ can be written as a partition function in $\mathcal{Q}_+^\Lambda = \{q \in \mathcal{Q}_\Lambda : \eta(q, r) = 1, r \in \Lambda\}$. Namely, for any bounded $\mathcal{D}^{(\ell_+)}$-measurable region Λ and any plus b.c. q^+, we have that

$$Z^+_{\gamma,\beta,R,\lambda,q^+}(\Lambda) = \sum_{\underline{\Gamma} \in \mathcal{B}^+_\Lambda} \int_{q_\Lambda \in \mathcal{Q}_+^\Lambda} W^+_{\gamma,R,\lambda}(\underline{\Gamma}, q_\Lambda)\, e^{-\beta H_{\gamma,R,\lambda}(q_\Lambda | q^+_{\Lambda^c})}, \tag{32}$$

where $q^+_{\Lambda^c}$ is made of all particles of q^+ which are in Λ^c. $\mathcal{B}^+_\Lambda$ is the space of all finite subsets of collection of plus contours made of elements which are mutually disconnected and with spatial support not connected to Λ^c. Furthermore if $\underline{\Gamma} = (\Gamma_1, \ldots, \Gamma_n)$, we use the notation

$$W^\pm_{\gamma,R,\lambda}(\underline{\Gamma}, q) = \prod_{i=1}^n W^\pm_{\gamma,R,\lambda}(\Gamma_i, q). \tag{33}$$

A similar expression holds for the diluted minus partition function.

In order to prove Peierls bounds, we follow the version of the Pirogov-Sinai theory proposed by Zahradnik [8]. In this picture large contours are less likely to be observed and this is implemented by fixing a constraint which literally forbids contours larger than some given value. We introduce therefore a new class of systems, where the contour weights are modified, their values depending on some "cutoff" parameter. In the stable phase the cutoff (if properly chosen) is not reached and the state is not modified by this procedure.

Therefore we choose $\hat{W}^\pm_{\gamma,R,\lambda}(\Gamma; q^\pm)$, positive numbers which depend only on the restriction of $q^\pm$ to $\{r \in c(\Gamma)^c : \text{dist}(r, c(\Gamma)) \le 2\gamma^{-1}\}$ and such that for any $\pm$ contour Γ and any $q^\pm$,

$$\hat{W}^\pm_{\gamma,R,\lambda}(\Gamma; q^\pm) = \min\left\{ \frac{\hat{\mathcal{N}}^\pm_{\gamma,R,\lambda}(\Gamma, q)}{\hat{\mathcal{D}}^\pm_{\gamma,R,\lambda}(\Gamma, q)},\, e^{-\beta \frac{c}{100}(\zeta^2 \ell_-^d) N_\Gamma} \right\}. \tag{34}$$

Here, $\hat{\mathcal{N}}^{\pm}_{\gamma,R,\lambda}(\Gamma, q)$ and $\hat{\mathcal{D}}^{\pm}_{\gamma,R,\lambda}(\Gamma, q)$ are as in (27) and (28) but depend uniquely on the weights $\hat{W}^{\pm}_{\gamma,R,\lambda}(\cdot;\cdot)$. With this new choice of contours weights, if we prove Peierls bounds, i.e., (29) on definition (34), we have Peierls bounds also on the "true" weights defined in (26). We write $\hat{Z}^{+}_{\gamma,\beta,R,\lambda,q^+}(\Lambda)$ to denote the new diluted partition function. For more details one can see in [4].

5 Outline of the Proof

In this section we want to give a sketch of the proof of (29) for the case of the cutoff contours as defined above. For the complete proof of the argument see [4].

The first step is to prove that it is possible to separate in (27) and (28) the estimate in $\text{int}(\Gamma)$ from the one in $\text{sp}(\Gamma)$ with "negligible error". Then one needs to bound a constrained partition function in $\text{sp}(\Gamma)$, which yields the gain factor $e^{-\beta(c\zeta^2 - c'\gamma^{1/2-2\alpha d})\ell_2^d N_\Gamma}$. Hence, we prove that there are $c, c' > 0$ so that given γ small enough, for $R < R_0$,

$$\frac{\hat{\mathcal{N}}^{+}_{\gamma,R,\lambda_{\beta,R,\gamma}}(\Gamma, q^+)}{\hat{\mathcal{D}}^{+}_{\gamma,R,\lambda_{\beta,R,\gamma}}(\Gamma, q^+)} \le e^{-\beta(c\zeta^2 - c'\gamma^{1/2-2\alpha d})\ell_2^d N_\Gamma} \frac{e^{\beta I^{-}_{\gamma,\lambda(\beta,R)}(\text{int}^-(\Gamma))} \hat{Z}^{-}_{\gamma,R,\lambda_{\beta,\gamma,R},\chi^-}(\text{int}^-(\Gamma))}{e^{\beta I^{+}_{\gamma,\lambda(\beta,R)}(\text{int}^-(\Gamma))} \hat{Z}^{+}_{\gamma,R,\lambda_{\beta,\gamma,R},\chi^+}(\text{int}^-(\Gamma))} \tag{35}$$

where we use the shorthand notation

$$\chi^{\pm}_{\Delta}(r) = \rho_{\beta,\pm}\mathbf{1}_{r\in\Delta}, \quad \chi^{\pm} = \chi^{\pm}_{\mathbb{R}^d} \tag{36}$$

and where $I^{\pm}_{\gamma,\lambda(\beta,R)}(\Lambda)$ is a surface term

$$I^{\pm}_{\gamma,\lambda(\beta,R)}(\Lambda) = \int_{\Lambda^c} \{e_{\lambda(\beta,R)}(\rho_{\beta,R,\pm}) - e_{\lambda(\beta,R)}(J_\gamma * \rho_{\beta,R,\pm}\mathbf{1}_{\Lambda^c})\} \tag{37}$$

$$- \int_{\Lambda} e_{\lambda(\beta,R)}(J_\gamma * \rho_{\beta,R,\pm}\mathbf{1}_{\Lambda^c}). \tag{38}$$

The main tool used in this part of the proof is a coarse-graining argument and an analysis à la Lebowitz and Penrose [2]. The error in doing a coarse-graining is bounded by $e^{\beta c\gamma^{1/2}|\text{sp}(\Gamma)|} = e^{\beta c\gamma^{1/2-2\alpha d}\ell_2^d N_\Gamma}$, which is the "negligible factor" mentioned above, as it is a small fraction of the gain term in the Peierls bounds. Thus, in this step we have a reduction, after coarse-graining, to variational problems with the LMP free energy functional. They involve two different regions, one is at the boundary between $\text{int}(\Gamma)$ and $\text{sp}(\Gamma)$, the other is in the bulk of the spatial support. In the former we exploit the definition of contours which implies that the boundary of $\text{int}^{\pm}(\Gamma)$ is in the middle of a "large region" (of size ℓ_+) where $\eta(\cdot;\cdot)$ is identically equal to ± 1, respectively. By the strong stability properties of the LMP free energy functional, the minimizers are then proved to converge exponentially to $\rho_{\beta,R,\pm}$ with the distance from the boundaries. Here we use the assumption that $\beta \in (\beta_{c,R}, \beta_{0,R})$, i.e., where the mean-field operator $K_{\beta,\lambda(\beta,R),R}$ is a contraction, see (21) and (22). We then conclude that with a negligible error we have "thick corridors" where the minimizers are equal to $\rho_{\beta,R,\pm}$ thus separating the regions outside and inside the corridors.

After this step we have plus/minus partition functions in $\text{int}^{\pm}(\Gamma)$ with boundary conditions $\rho_{\beta,R,\pm}$ and still a variational problem in the region $\text{sp}(\Gamma)$ with the constraint that profiles should be compatible with the presence of the contour Γ. The analysis of such a minimization problem leads to the gain factor in the Peierls bounds.

To complete the proof for Peierls bounds we then need to prove the following theorem.

Theorem 3 *There exists R_0 such that for any $R \le R_0$ and any $\beta \in (\beta_{c,R}, \beta_{0,R})$ there are $c > 0$, $\gamma_{\beta,R} > 0$ and $\lambda_{\beta,\gamma,R}$, such that for all $\gamma \le \gamma_\beta$, $|\lambda(\beta, R) - \lambda_{\beta,R,\gamma}| \le c\gamma^{1/2}$, and any bounded $\mathscr{D}^{(\ell_+)}$-measurable region Λ, the following bound holds*

$$\frac{e^{\beta I^-_{\gamma,\lambda(\beta,R)}(\Lambda)}\hat{Z}^-_{\gamma,R,\lambda_{\beta,\gamma,R},\chi^-_{\Lambda^c}}(\Lambda)}{e^{\beta I^+_{\gamma,\lambda(\beta,R)}(\Lambda)}\hat{Z}^+_{\gamma,R,\lambda_{\beta,\gamma,R},\chi^+_{\Lambda^c}}(\Lambda)} \le e^{c\gamma^{1/2}|\delta^{\ell_+}_{\text{out}}[\Lambda]|}. \tag{39}$$

The idea in the proof of (39) is that the leading term in the partition function is

$$\hat{Z}^{\pm}_{\gamma,\beta,R,\lambda,q_{\text{sp}(\Gamma)}}(\text{int}^{\pm}(\Gamma)) \approx e^{\beta P^{\pm}_{\gamma,R,\lambda}|\text{int}^{\pm}(\Gamma)|}, \tag{40}$$

where $P^{\pm}_{\gamma,R,\lambda}$ is the thermodynamic pressure given, for any van Hove sequence of $\mathscr{D}^{(\ell_+)}$-measurable regions Λ_n and any $\pm$ Λ_n-boundary conditions $q_n^{\pm}$, by the following limit

$$\lim_{n\to\infty} \frac{1}{\beta|\Lambda_n|} \log \hat{Z}^{\pm}_{\gamma,R,\lambda,q_n^{\pm}}(\Lambda_n) = P^{\pm}_{\gamma,R,\lambda}. \tag{41}$$

Although (40) is a rough approximation, we need to prove equality of $\pm$ pressures in the bulk terms in $\hat{W}^{\pm}_{\gamma,R,\lambda}(\Gamma; q^{\pm})$ to allow for the numerator and the denominator to cancel. Again for more details we refer the reader to [4].

We now prove that the next term, i.e., the surface corrections to the pressure, are small as $e^{c''\gamma^{1/2}\ell_+^d N_\Gamma}$ at least when the boundary conditions "are perfect", i.e., given by $\chi^{\pm}$. The most difficult step in the proof of Theorem 3 are estimates involving terms which are localized in the bulk of the interior. These rely on a more delicate property of decay of correlations (Theorem 4), whose proof requires a whole new set of ideas.

Theorem 4 (Exponential decay of correlations) *Let Λ be a bounded $\mathscr{D}^{(\ell_+)}$ measurable region. Let x_i be the centers of the cubes $C^{(\ell_-)} \in \mathscr{D}^{(\ell_-)}$; then we define*

$$f_{x_1,\dots,x_n} = \int_{\{r_i \in C^{(\ell_-)}_{x_i}, 1\le i\le n\}} q^{\otimes n}(dr_1 \dots dr_n) J^{(n)}_\gamma(r_1, \dots, r_n) \tag{42}$$

where we use the notation

$$q^{\otimes n}(dr_1 \dots dr_n) = \frac{1}{n!} \sum_{i_1 \neq \dots \neq i_n} \delta_{q_{i_1}}(r_1)\, dr_1 \cdots \delta_{q_{i_n}}(r_n)\, dr_n. \tag{43}$$

There are positive constants δ, c' and c so that for all $f_{x_1,\dots,x_n}$

$$\left| E_{\mu^1}\left(f_{x_1,\dots,x_n}\right) - E_{\mu^2}\left(f_{x_1,\dots,x_n}\right) \right| \le c' e^{-c[\gamma^{-\delta}\ell_+^{-1} dist(C_{x_1}^{(\ell_-)}, \Lambda^c)]} \tag{44}$$

where E_{μ^i}, $i = 1, 2$, are the expectations with respect to the following two measures: μ^1 is the finite-volume Gibbs measure in Λ with b.c. $\bar{q}$ and μ^2 the finite-volume Gibbs measure on a torus $\mathcal{T}$ much larger than Λ.

We compute the expectations in (44) in two steps. We first do a coarse-graining by fixing the number of particles in the cubes $C^{(\ell_-)}$ and integrate over their positions; then, in the second step, we sum over the particle numbers. By its very nature, the Kac assumption makes the first step simple: in fact, to first order the energy is independent of the positions of the particles inside each cube. Neglecting the higher order terms, the energy drops out of the integrals (with fixed particle numbers) which can then be computed explicitly. The result is the phase space volume of the set of configurations with the given particle numbers: this is an entropy factor which, together with the energy, reconstructs the mesoscopic energy functional.

By using cluster expansion techniques, we will show here that it is possible to compute exactly the correction due to the dependence of the energy on the actual positions of the particles in each cube. For the hard-core part of the interaction we can use again a cluster expansion technique, using the result [5] obtained for a system with a single canonical constraint and therefore extending it to the present case of multi-canonical constraints.

Once we are left with an "effective Hamiltonian" we still have to sum over the particle numbers. Since we work in a contour model, the particle densities are close to the mean-field values $\rho_{\beta,R,\pm}$ so that the marginal of the Gibbs measure over the coarse-grained model is Gibbsian and it is a small perturbation of a Hamiltonian given by the mean-field free energy functional restricted to a neighborhood of the mean-field equilibrium density. In such a setup we manage to prove the validity of the Dobrushin uniqueness condition, where we take into account the contribution of the hard-core as a cluster expansion sum.

5.1 Coarse-Graining

To carry out this plan, we need to prove that $\hat{Z}^+_{\gamma,\beta,R,\lambda,\bar{q}}(\Lambda)$ can be written as the partition function of a Hamiltonian which depends on variables ρ_x, $x \in X_\Lambda^{(\ell_-)}$, $X_\Lambda^{(\ell_-)}$ the set of centers of cubes $C^{(\ell_-)}$ in Λ. The new energy of a density configuration $\rho = \{\rho_x\}_{x \in X_\Lambda^{(\ell_-)}}$ is defined as

$$h(\rho|\bar{q}) = -\log \sum_{\underline{\Gamma}\in\mathscr{B}_\Lambda^+} \int_{\mathscr{Q}_+^\Lambda} \nu^\Lambda(dq)\mathbf{1}_{\rho^{(\ell_-)}(q)=\rho}\, e^{-\beta H_{\gamma,R,\lambda}(q|\bar{q})}\,\hat{W}(\underline{\Gamma}|q) \tag{45}$$

so that

$$\hat{Z}^+_{\gamma,\beta,R,\lambda,\bar{q}}(\Lambda) = \sum_\rho e^{-h(\rho|\bar{q})}.$$

Setting $n_x = \ell_-^d \rho_x$, we multiply and divide, inside the argument of the log in (45), by

$$\prod_{x\in X_\Lambda} \frac{\ell_-^{dn_x}}{n_x!}.$$

We denote by $\{q_{x,i},\, i = 1,\dots,n_x,\, x\in X_\Lambda\}$, the particles in $C_x^{(\ell_-)}$. Thus particles are now labelled by the pair (x,i), x specifies the cube $C_x^{(\ell_-)}$ to which the particle "belongs", i distinguishes among the particles in $C_x^{(\ell_-)}$. The corresponding free measure, whose expectation is denoted by E_ρ^0, is the product of the probabilities which give uniform distribution to the positions $q_{x,i}$ in their boxes $C_x^{(\ell_-)}$ divided by $n_x!$ since the particles in each box $C_x^{(\ell_-)}$ are indistinguishable. Note that when we change from labeling of all particles in Λ to labeling separately the particles in each box we have to multiply by $\frac{N!}{\prod_{x\in X_\Lambda^{(\ell_-)}} n_x!}$ for all such possibilities.

We define a new a priori measure for the particles in a given box $C_x^{(\ell_-)}$, $x\in X_\Lambda$, as

$$\frac{dq_{x,1}\cdots dq_{x,n_x} e^{-\beta U^{\text{hc}}(q^{(C_x)},\bar{q})}}{\int dq_{x,1}\cdots dq_{x,n_x} e^{-\beta U^{\text{hc}}(q^{(C_x)},\bar{q})}} Z_{x,\bar{q}}(\rho_x) \tag{46}$$

where $q^{(C_x)}$ denotes the configuration of the particles in $C_x^{(\ell_-)}$, each integral in the denominator is over $C_x^{(\ell_-)}$ with the constraint $\mathscr{Q}_+^\Lambda$ and where

$$Z_{x,\bar{q}}(\rho_x) = \int_{\mathscr{Q}_+^\Lambda} \frac{dq_{x,1}}{\ell_-^d}\cdots\frac{dq_{x,n_x}}{\ell_-^d}\, e^{-\beta U^{\text{hc}}(q^{(C_x)},\bar{q})} \tag{47}$$

is the extra factor coming from the change of measure and contributing for each cube with

$$U^{\text{hc}}(q^{(C_x)},\bar{q}) := \sum_{i=1}^{n_x}\sum_{j=1}^{|\bar{q}|} V^R(q_{x,i}-\bar{q}_j). \tag{48}$$

The corresponding expectation will be denoted by $E^0_{\rho,\bar{q}}$. We then have

$$h(\rho|\bar{q}) = -\sum_x \log\frac{\ell_-^{dn_x}}{n_x!} - \sum_{x:\,C_x^{(\ell_-)}\in\partial\Lambda^{\text{int}}} \log Z_{x,\bar{q}}(\rho_x) \tag{49}$$

$$- \log E^0_{\rho,\bar{q}}\Big(e^{-\beta H_\gamma(q|\bar{q})} e^{-\beta H_R^{\rm hc}(q)} \sum_{{\rm sp}(\Gamma)\subseteq\Lambda^0} W(\Gamma|q)\Big)$$

where $\partial\Lambda^{\rm int}$ is the set of the $\mathscr{D}^{(\ell_-)}$ boxes adjacent to Λ^c (i.e., the interior boxes of Λ). Note that the total normalization is a product of the normalizations in each cube and that, because of the hard-core interaction, $Z_{x,\bar{q}}(\rho_x)$ for a given box C_x gives the following contribution:

$$Z_{x,\bar{q}}(\rho_x) = \Big(\int_{C_x} \frac{dq}{\ell_-^d} \mathbf{1}_{q\in C_x^{\bar{q}}}\Big)^{n_x} = \frac{|C_x^{\bar{q}}|^{n_x}}{\ell_-^{dn_x}} \tag{50}$$

where: $C_x^{\bar{q}} = \{r \in C_x : {\rm dist}(r, \bar{q}_i) > R, \forall i\}$. This means that, because of the presence of the hard-core, the new measure “reduces” the admissible volume for the particles in each box.

Let $H^{(\ell_-)}(q|\bar{q})$ be the coarse-grained Hamiltonian on scale ℓ_-. It is obtained by replacing $J_\gamma^{(n)}$ by $\tilde{J}_\gamma^{(n)}$, where

$$\tilde{J}_\gamma^{(n)}(r_1,\ldots,r_n) = \frac{1}{|C^{(\ell_-)}|^n} \int_{C_{r_1}^{(\ell_-)}} dq_1 \cdots \int_{C_{r_n}^{(\ell_-)}} dq_n\, J_\gamma^{(n)}(q_1,\ldots,q_n) \tag{51}$$

are the coarse-grained potentials.

It depends only on the particle numbers n_x (or the densities ρ_x) and we can thus write

$$h^0(\rho|\bar{\rho}) = H^{(\ell_-)}(q|\bar{q}), \qquad \rho_x = \rho_x^{(\ell_-)}(q),\ \bar{\rho}_x = \rho_x^{(\ell_-)}(\bar{q}). \tag{52}$$

Setting

$$\Delta H(q|\bar{q}) = H_\gamma(q|\bar{q}) - H^{(\ell_-)}(q|\bar{q}) \tag{53}$$

we have

$$h(\rho|\bar{q}) = -\sum_x \log \frac{\ell_-^{dn_x}}{n_x!} - \sum_{x:\, C_x^{(\ell_-)}\in\partial\Lambda^{\rm int}} \log Z_{x,\bar{q}} + \beta h^0(\rho|\bar{\rho}) + \delta h(\rho|\bar{q}) \tag{54}$$

where

$$\delta h(\rho|\bar{q}) = -\log E^0_{\rho,\bar{q}}\Big(e^{-\beta\Delta H(q|\bar{q})} e^{-\beta H_R^{\rm hc}(q)} \sum_{{\rm sp}(\Gamma)\subseteq\Lambda^0} \hat{W}(\Gamma|q)\Big). \tag{55}$$

It is convenient to split $\delta h(\rho|\bar{q})$ in three parts

$$\delta h(\rho|\bar{q}) = h^p(\rho|\bar{q}) + h^c(\rho|\bar{q}) \tag{56}$$

where

$$h^p(\rho|\bar{q}) = -\log E^0_{\rho,\bar{q}}\big(e^{-\beta\Delta H(q|\bar{q})} e^{-\beta H_R^{\rm hc}(q)}\big) \tag{57}$$

$$h^c(\rho|\bar{q}) = -\log E_{\rho,\bar{q}}\Big(\sum_{\mathrm{sp}(\Gamma)\subseteq\Lambda^0} \hat{W}(\Gamma|q)\Big) \tag{58}$$

$$E_{\rho,\bar{q}}(f) = \frac{E^0_{\rho,\bar{q}}\big(e^{-\beta\Delta H(q|\bar{q})}e^{-\beta H^{\mathrm{hc}}_R(q)} f\big)}{E^0_\rho\big(e^{-\beta\Delta H(q|\bar{q})}e^{-\beta H^{\mathrm{hc}}_R(q)}\big)}. \tag{59}$$

In words, $h^p(\rho|\bar{q})$ is the contribution to the effective Hamiltonian coming from the average over the measure (46) of the hard-core interaction $H^{\mathrm{hc}}_R(q)$ and the coarse-grained correction $\Delta H(q|\bar{q})$ (defined in (53)). It will have an expansion in polymers as we will show in Sect. 5.2. $h^c(\rho|\bar{q})$ is the same average to which is also added the contribution of the contours and it can also be expressed in terms of another class of polymers.

5.2 Cluster Expansion

In order to find an expression for h^p and h^c we perform a cluster expansion which involves both hard-core, Kac interaction and contours. Let us start from h^p, which is easier since there are no contours. We define diagrams which will be the *polymers* of the cluster expansion. Let $L^{(2)} = (i_1, i_2)$ and $L^{(4)} = (i_1, i_2, i_3, i_4)$ denote a pair (resp. a quadruple) of mutually distinct particle labels. They will be called 2-links and 4-links. We will refer to the two types of 2-links by calling them respectively γ-links and R-links.

Definition 5 A diagram θ is a collection of 2- and 4-links, i.e., an ordered triple $\theta \equiv \big(\mathscr{L}^{(2)}_R(\theta), \mathscr{L}^{(2)}_\gamma(\theta), \mathscr{L}^{(4)}(\theta)\big)$, where we denote by $\mathscr{L}^{(2)}_R(\theta)$, $\mathscr{L}^{(2)}_\gamma(\theta)$ and $\mathscr{L}^{(4)}(\theta)$ the set of 2-links (of type R and γ) and of 4-links in θ. Note that one can have a repetition of links, i.e., the same link $L^{(2)}$ can belong to both sets $\mathscr{L}^{(2)}_\gamma(\theta)$ and $\mathscr{L}^{(2)}_R(\theta)$. We use $\mathscr{L}^{(2)}(\theta)$ for the set of 2-links (which eventually contains twice a link when it is both a γ-link and a R-link) and Θ for the set of all such diagrams.

We construct the set of polymers starting from the diagrams defined above, but eliminating some of their links. Indeed, to work with cluster expansion an "a priori" estimate of some links is needed in order to reduce the complexity of the diagrams that we consider. This is an essential step to assure convergence of the cluster expansion. To this scope, we are going to define a new set of diagrams. The procedure is the following: We first get rid of all the R-links which appear over γ-links and we extract a subdiagram $\hat{\theta}$. Let $\hat{\Theta} \subset \Theta$ be the set of all the diagrams which do not have double 2-links, i.e., $\hat{\Theta} := \{\hat{\theta} : \mathscr{L}^{(2)}_\gamma(\hat{\theta}) \cap \mathscr{L}^{(2)}_R(\hat{\theta}) = \emptyset\}$.

The next step is to obtain a diagram which is at most a tree in R.

Definition 6 (*Partial ordering relation $\prec$ on a diagram θ*). For $L^{(2)}_1, L^{(2)}_2 \in \mathscr{L}^{(2)}_R(\theta)$ we have that $L^{(2)}_1 \prec L^{(2)}_2$ according to lexicographic ordering (i.e., we start by comparing the first index and if the same we compare the next etc.). We say that a diagram is ordered if the set of its R-links is ordered according to this definition. We can endow an ordered diagram with the usual notion of distance. We will write $d(v)$ to indicate the distance of a vertex v to the first vertex in the previous order relation.

Definition 7 (*Redundant link*). Given an ordered diagram θ, we say that a link $L^{(2)} \in \mathscr{L}_R^{(2)}(\theta)$ is redundant in the following two cases:

- If $L^{(2)} = \{i, j\}$ with $d(i) = d(j)$;
- If $L_1^{(2)} = \{i_1, j\}$ with $d(i_1) = d(j) - 1$ and it exists $L_2^{(2)} = \{i_2, j\} \in \mathscr{L}_R^{(2)}(\theta)$, with $d(i_2) = d(j) - 1$, such that: $L_2^{(2)} \prec L_1^{(2)}$ (i.e., $i_2 < i_1$).

We denote the set of the redundant links of a diagram θ by: $\mathscr{R}_R^{(2)}(\theta)$.

We call $\bar{\Theta} \subset \hat{\Theta}$ the set of diagrams with no double 2-links and with no redundant links. In formulas: $\bar{\Theta} := \{\bar{\theta} : \bar{\theta} \in \hat{\Theta}, \mathscr{R}_R^{(2)}(\bar{\theta}) = \emptyset\}$.

Two diagrams θ and θ' are *compatible* ($\theta \sim \theta'$) if the set of their common labels is empty.

Theorem 5 *For all γ and R small enough, there exist functions $z_{\gamma,R}^T(\pi; \rho; \bar{q})$ such that*

$$h^p(\rho|\bar{q}) = -\sum_{\pi} z_{\gamma,R}^T(\pi; \rho; \bar{q}), \tag{60}$$

where π is a collection of non-compatible diagrams in the space $\bar{\Theta}$.

Let us now find an expansion for h^c defined in (58). Let us fix a collection $\underline{\Gamma} = \{\Gamma_i\}_{i=1}^n$, where $\Gamma_i \equiv (\text{sp}(\Gamma_i), \eta_{\Gamma_i})$. As said after (24), the weights $W_{\gamma,R,\lambda}^{\pm}(\Gamma_i; q^{\pm})$ depend only on q^{D_i}, i.e., the restriction of $q^{\pm}$ to $D_i = \{r \in c(\Gamma)^c : \text{dist}(r, c(\Gamma)) \leq 2\gamma^{-1}\}$. We also let $D := \cup_{i=1}^n D_i$. We then have, for the numerator of (58),

$$E^0_{\rho^D,\bar{q}}\Big(W(\Gamma|q)e^{-h^p(\rho^{\Lambda\setminus D}|\bar{q}\cup q^D)}\Big). \tag{61}$$

We write h^p as a sum of clusters using (60). Due to the dependence of the a priori measure on $\bar{q}$ (now on both $\bar{q}$ and q^D), the clusters involving a particle in a neighboring ℓ_--cell to D will also depend on q^D. We denote the union of the set D with the frame consisting of the neighboring ℓ_--cells by $D^* \in \mathscr{D}^{(\ell_2)}$.

To distinguish between clusters we introduce $\bar{D}_i := D_i \cup \{r : \text{dist}(r, D_i) \leq \ell_+/4\} \in \mathscr{D}^{(\ell_-)}$ and we call $\mathscr{B}_i$ the set of all clusters π whose points are all in $\bar{D}_i$. As the distance between contours is $\geq \ell_+$, the sets $\mathscr{B}_i$ are mutually disjoint; we call $\mathscr{B}$ their union. Note that they depend on $\underline{\Gamma}$ through the domain where they are constructed. By $\mathscr{R}_i$ we denote the set of π which have points both in D_i^* (so that they depend on q^D) and in the complement of $\bar{D}_i$ (such π are therefore not in $\mathscr{B}_i$). There may be points of $\pi \in \mathscr{R}_i$ which are in D_j^*, $j \neq i$, hence also $\pi \in \mathscr{R}_j$, so that the sets $\mathscr{R}_i$ are not disjoint. We call $\mathscr{R}$ their union.

For any given $\underline{\Gamma}$ we do analogous splitting on the polymers appearing when developing the denominator of (59) thus defining the sets $\mathscr{B}'_i, \mathscr{B}', \mathscr{R}'_i, \mathscr{R}'$. The clusters that appear in the numerator and denominator of (59) are different, however those not in $\mathscr{B} \cup \mathscr{R}$ (i.e., those that do not involve q^D) are common to the corresponding ones in the denominator of (59) (i.e., those not in $\mathscr{B}' \cup \mathscr{R}'$) and have same statistical weights, hence they cancel.

The clusters $\pi \in \mathscr{B}$ can be grouped together and absorbed by a renormalization of the measure in $E^0_{\rho^D,\bar{q}}$, since they do not involve interactions between different contours. Thus, they will be part of the *activities* in the expansion, while the *polymers* will be defined in terms of elements of $\mathscr{R}$ and $\mathscr{R}'$.

Hence, to formulate the problem into the general context of the abstract polymer model we define as *connected* polymer P a set of contours with "connections" consisting of elements of $\mathscr{R} \cup \mathscr{R}'$ which necessarily "connect" all contours in the given set and "decorations" consisting of clusters in $\mathscr{R} \cup \mathscr{R}'$ not necessarily connecting contours. We denote by $\mathscr{P}$ the space of all such elements

$$\mathscr{P} := \big\{ P \equiv (\underline{\Gamma}(P), R(P)), \forall \Gamma_i, \Gamma_j \in \underline{\Gamma}(P), \exists \pi \in R(P) \subset \mathscr{R} \cup \mathscr{R}' \\ \text{connecting } D^*_i, D^*_j \in D^*(\underline{\Gamma}) \big\}. \tag{62}$$

We use $D(P)$, $D^*(P)$ to denote the set of frames corresponding to the contours in P and $R(P)$ to denote the set of clusters. We also introduce $A(\pi)$ to denote the union of the $C^{(\ell_2)}$ cells which correspond to the labels of π. Similarly, let $A(P) := \cup_{\Gamma \in \underline{\Gamma}(P)} D^*(\Gamma) \cup_{\pi \in R(P)} A(\pi)$. A compatible collection of polymers consists of mutually compatible polymers.

Theorem 6 *For all γ and R small enough, there exist functions $\zeta^T_{\gamma,R}(C;\rho)$ such that*

$$h^c(\rho|\bar{q}) = -\sum_C \zeta^T_{\gamma,R}(C;\rho), \tag{63}$$

where C is a collection of non-compatible polymers P in the space $\mathscr{P}$.

With these two theorems we can define a new measure on the space of the density configurations $\rho = \{\rho_x\}_{x \in X^{(\ell_-)}_\Lambda}$, where the new Hamiltonian is

$$h(\rho|\bar{q}) = -\sum_x \log \frac{\ell_2^{dn_x}}{n_x!} - \sum_{x:\, C_x^{(\ell_2)} \in \partial\Lambda^{\text{int}}} \log Z_{x,\bar{q}} + \beta h^0(\rho|\bar{\rho}) + h^p(\rho|\bar{q}) + h^c(\rho|\bar{q}). \tag{64}$$

We then use the notation $\mathbb{E}$ for the expectation w.r.t. this new coarse-grained measure ν.

To estimate the difference $E_{\mu^1}\left(f_{x_1,\dots,x_n}\right) - E_{\mu^2}\left(f_{x_1,\dots,x_n}\right)$ in Theorem 4, we split $E_{\mu^i}\left(f_{x_1,\dots,x_n}\right)$ into two parts, one which is of order one and one which is exponentially small. However the order one parts will be small when we consider their difference. The main idea is the following.

Given $x_1, \dots, x_n$, for $n = 1, 2, 4$, and such that each x_i, x_j are not more distant than γ^{-1}, we choose a box of side $2\ell_+$ that contains all of them and is far enough from the boundary. The contribution of clusters attached to any subset of $x_1, \dots, x_n$ inside the box will be denoted by g and the ones attached to any subset of $x_1, \dots, x_n$ inside the box and going out of it will be denoted by R. This latter contribution is exponentially small, as a corollary of the above theorems.

We first prove this splitting in the following Lemma:

Lemma 1 *Let $f_{x_1,\dots,x_n}$ be as in (42), then*

$$E_{\mu^i}\big(f_{x_1,\dots,x_n}\big) = \mathbb{E}_{\nu^i}(g) + R_i, \qquad i = 1, 2 \tag{65}$$

where g is a function of $\{\rho_x\}$ with $x \in X_\Lambda$ contained in the cube of side $2\ell_+$ and R_i are remainder terms. Moreover, there are $\delta > 0$ and constants c_1, c_2, c so that

$$\|g\|_\infty \le c_1, \qquad \|R_i\|_\infty \le c_2 e^{-c\gamma^{-\delta}}. \tag{66}$$

To conclude the proof of Theorem 4 we need to estimate the difference $\mathbb{E}_1(g) - \mathbb{E}_2(g)$. In order to do this, we work in the coarse-grained model, i.e., in the space

$$\mathscr{X}^\Lambda = \Big\{\underline{n} = (n_x)_{x\in X_\Lambda} \in \mathbb{N}^{X_\Lambda} : |\ell_-^{-d} n_x - \rho_{\beta,+}| \le \zeta, \text{ for all } x \in X_\Lambda\Big\}, \quad n_x = \ell_- \rho_x. \tag{67}$$

The goal is to prove that there exists a joint representation $\mathscr{P}(\underline{n}^1, \underline{n}^2 | \bar{q}^1, \bar{q}^2)$ of the measures ν^1 and ν^2 on $\mathscr{X}^\Lambda$ such that, for any $x \in X_\Lambda$, and denoting by $\mathscr{E}$ the expectation w.r.t. to $\mathscr{P}$, we can bound the difference $\mathbb{E}_{\nu^1}(g) - \mathbb{E}_{\nu^2}(g)$ with $\mathscr{E}\big[d(n_x^1, n_x^2)\big]$, where $d(n_x^1, n_x^2)$ is an appropriate distance that we have to define and where $\mathscr{E}\big[d(n_x^1, n_x^2)\big]$ has the desired exponential decay property.

To complete the proof we then need to find a bound for $\mathscr{E}\big[d(n_x^1, n_x^2)\big]$. This comes from a Dobrushin uniqueness condition. We want to bound the Vaserstein distance between two Gibbs measures with the same Hamiltonian (64) but with different b.c. $\bar{q}^i, i = 1, 2$. It is convenient to define the Vaserstein distance in terms of the following cost functions

$$d(\underline{n}^1, \underline{n}^2) = \sum_{x\in X_\Lambda} d(n_x^1, n_x^2), \qquad d(n_x^1, n_x^2) = |n_x^1 - \underline{n}_x^2| \tag{68}$$

and if we suppose $\bar{q}^1 = (\bar{q}_1^1, \dots, \bar{q}_n^1)$ and $\bar{q}^2 = (\bar{q}_1^2, \dots, \bar{q}_{n+p}^2)$,

$$D_z(\bar{q}^1, \bar{q}^2) := p + \min_{\{j_\ell\}} \sum_{\ell=1}^{n} \mathbf{1}_{\bar{q}_\ell^1 \ne \bar{q}_{j_\ell}^2}, \tag{69}$$

the min being over all the subsets $\{j_\ell\}$ of $\{1, \dots, n+p\}$ which have cardinality n. Following Dobrushin, we need to estimate the Vaserstein distance between conditional probabilities at a single site. We thus fix arbitrarily $x \in \Lambda$, $\underline{n}^i$, $i = 1, 2$, in

$\mathcal{X}^{\Lambda\setminus x}$, call $\rho^i := \ell_2^{-d}\underline{n}^i$; $\bar{q}^i$ are the b.c. outside Λ. The energy in x plus the interaction with the outside is, as usual,

$$h(\rho_x|\rho^i, \bar{q}^i) = h\big(\{\rho_x, \rho^i\}|\bar{q}^i\big) - h(\rho^i|\bar{q}^i), \tag{70}$$

where the first term on the r.h.s. is the energy of the configuration $\{\rho_x, \rho^i\}$ (with $\bar{q}^i$ outside Λ). The second term is the energy in $\Lambda \setminus x$ of ρ^i with nothing in x and $\bar{q}^i$ outside Λ. The conditional Gibbs measures are then the following probabilities on $\mathcal{X}^x$ (for $i = 1, 2$)

$$p(n_x|\rho^i, \bar{q}^i) = \frac{1}{Z_x(\rho^i, \bar{q}^i)} \exp\big\{ - h(\rho_x|\rho^i, \bar{q}^i)\big\}, \tag{71}$$

and their Vaserstein distance is

$$R\Big(p(\cdot|\rho^1, \bar{q}^1), p(\cdot|\rho^2, \bar{q}^2)\Big) := \inf_Q \sum_{n_x^1, n_x^2} Q(n_x^1, n_x^2) d(n_x^1, n_x^2), \tag{72}$$

where the inf is over all the joint representations Q of $p(\rho_x|\rho^i, \bar{q}^i)$, $i = 1, 2$. The key bound for the Dobrushin scheme to work is the following theorem.

Theorem 7 *There are $u, c_1, c_2 > 0$ s.t. $\forall x \in \Lambda$*

$$R\Big(p^1(\cdot), p^2(\cdot)\Big) \le \sum_{z \in X_\Lambda, z \neq x} r_{\gamma,R}(x,z) d(n_z^1, n_z^2) + \sum_{z \in X_{\Lambda^c}} r_{\gamma,R}(x,z) D_z(\bar{q}^1, \bar{q}^2)$$

with

$$\sum_z r_{\gamma,R}(x,z) \le u < 1$$

$$r_{\gamma,R}(x,z) \le c_1 e^{-c_2\gamma|z-x|}, \quad |z-x| \ge \ell_+.$$

Remark 1 The reduction to an abstract contour model allows us to deal with a coarse-grained system in which the configurations we look at are those chosen in the restricted ensembles, roughly speaking those which should be seen under the effects of a double-well potential once we restrict to its minima. In this scenario, after we compute the effective Hamiltonian for the coarse-grained system, we have a new Gibbs measure which depends only on the cells variables. But now these variables are close to the mean-field value and in such a setup it is possible to prove the validity of the Dobrushin uniqueness theory.

Acknowledgments We would like to thank Errico Presutti for introducing us to this subject. EP is supported by ERC Advanced Grant 267356-VARIS.

References

1. Lebowitz, J.L., Mazel, A., Presutti, E.: Liquid-vapor phase transitions for systems with finite-range interactions. J. Stat. Phys. **94**(5–6), 955–1025 (1999)
2. Lebowitz, J.L., Penrose, O.: Rigorous treatment of the Van der Waals Maxwell theory of the liquid-vapor transition. J. Math. Phys **7**, 98–113 (1966)
3. Pirogov, S.A., Sinai, Ya.G.: Phase diagrams of classical lattice systems. Theor. Math. Phys. **25**, 358–369, 1185–1192 (1975)
4. Presutti, E.: Scaling Limits in Statistical Mechanics and Microstructures in Continuum Mechanics. Springer (2009)
5. Pulvirenti, E., Tsagkarogiannis, D.: Cluster expansion in the canonical ensemble. Commun. Math. Phys. **316**(2), 289–306 (2012)
6. Pulvirenti, E., Tsagkarogiannis, D.: A unified model for phase transition in the continuum (manuscript in preparation)
7. Pulvirenti, E.: Ph.D. thesis (2013)
8. Zahradnik, M.: An alternate version of Pirogov-Sinai theory. Commun. Math. Phys. **93**, 559–581 (1984)

Modelling of Systems with a Dispersed Phase: "Measuring" Small Sets in the Presence of Elliptic Operators

Valeria Ricci

Abstract When modelling systems with a dispersed phase involving elliptic operators, as is the case of the Stokes or Navier-Stokes problem or the heat equation in a bounded domain, the geometrical structure of the space occupied by the dispersed phase enters in the homogenization process through its capacity, a quantity which can be used to define the equivalence classes in H^1. We shall review the relationship between capacity and homogenization terms in the limit when the number of inclusions becomes large, focusing in particular on the situation where the distribution of inclusions is not necessarily too regular (i.e. it is not periodic).

Keywords Two-component systems · Large number limit · Capacity

1 Introduction

In some previous paper ([4, 5, 12]), we dealt with systems where the interplay between two phases is modelled by a boundary value problem. In such a kind of problems, one of the phases (fluid) is described by hydrodynamic equations in a perforated domain and the other one (dispersed phase) is represented by solid inclusions occupying the holes in the perforated domain; the field variable takes, on the boundary of the inclusions, different values which are constant on each inclusion with respect to the space variable.

In these systems, so as in the more commonly studied systems where the boundary condition on the inclusion are homogeneous, the homogenization process when the number of inclusions grows to infinity and their size shrinks to zero gives different results according to a specific characteristic of the set occupied by the inclusions. This characteristic is not the volume density (or the volume fraction) of the dispersed phase: for these systems these quantities, if different or not, vanish asymptotically,

V. Ricci (✉)
Dipartimento di Matematica e Informatica, Università di Palermo,
Via Archirafi 34, 90123 Palermo, Italy
e-mail: valeria.ricci@unipa.it

© Springer International Publishing Switzerland 2016
P. Gonçalves and A.J. Soares (eds.), *From Particle Systems to Partial Differential Equations III*, Springer Proceedings in Mathematics & Statistics 162,
DOI 10.1007/978-3-319-32144-8_14

because they are assumed to be negligible in the final model. Indeed, the choice of an asymptotics such that the Lebesgue measure of the set occupied by the inclusions vanishes in the limit is not sufficient to guarantee that the influence of the dispersed phase will be negligible in the limit evolution of the fluid phase, i.e. in the final homogenised equations.

The entity measuring (in some sense) the effective size of the set A occupied by the inclusions in such a kind of systems is the so called *capacity* $\mathrm{Cap}(A)$ of the set A.

The notion of capacity goes back to the potential theory, a discipline developed in the 19th century as the theory of potentials satisfying the Laplace equation ([7, 8]), and has been later generalised, with slightly different meanings, both in analysis and in probability theory (see e.g. [1, 2, 6, 9]). Although the capacity itself is not a measure, this quantity (in its many acceptations) can be used to characterise and sometimes measure (at least indirectly) the size of a variety of sets, in particular singular sets (e.g. [11] Chap. IV sec. 3 p. 72).

The potential theory has a so long history, that it would be hard to cite all relevant articles and books written about this subject: we just mentioned a minimal bibliography, since we shall not go deep into this subject.

Keeping in mind that the word "capacity" is used in different contexts for different (from a mathematical point of view) entities, we shall refer in this paper to one of the classical definitions of capacity, the one which is physically related to condensers, i.e. the *Green capacity* (see, e.g. [9, 10]). We shall use the notion of capacity to describe in a general way the asymptotics studied in [4, 5] not because this adds anything new to the theory, which has already been analysed in books like [10], but because it allows the reader to understand how the procedure adopted in those papers can be used in different dimensions and for different shapes of the inclusions representing the solid phase.

2 Capacity and Correctors

We summarize here the general frame of the papers [4, 5], generalising the space dimension and the shape of the holes in the perforated domain. All constants that do not need to be specified shall be denoted by the same letter, C.

We consider an open domain $\Omega \subset \mathbb{R}^d$, $d \geq 2$ and a distribution of centres $X_N = (x_i, \ldots, x_N)$, for $x_i \in \Omega$, $i = 1, \ldots N$, such that the empirical measure associated to the centres converges weakly (in the sense of measure) to some regular limit measure:

$$\rho_N(x) = \frac{1}{N} \sum_{i=1}^{N} \delta_{x_i} \rightharpoonup \rho(x) \in C_b(\bar{\Omega}).$$

The open domain Ω contains compact, non intersecting, simply connected subsets T_i^ε, $i = 1, \ldots, N$ of size ε,

$$\text{diam}\, T_i^\varepsilon = \varepsilon \qquad \text{and} \qquad \frac{\int_{\mathbb{R}^d} x I_{T_i^\varepsilon}}{|T_i^\varepsilon|} = x_i \in \Omega,$$

which we call *inclusions*. The internal radius of the inclusions, i.e.

$$\varepsilon_i' = \sup\{a : B_a(x_i) \subset T_i^\varepsilon\}, i = 1, \ldots, N$$

is such that, for some positive $b \in \mathbb{R}$,

$$\frac{\varepsilon_i'}{\varepsilon} \geq b > 0 \quad \forall i = 1, \ldots, N.$$

Denoting $T_\varepsilon^{X_N} = \bigcup_{i=1}^{N} T_i^\varepsilon$, the subset of Ω which is free from inclusions is

$$\Omega_\varepsilon = \Omega \setminus T_\varepsilon^{X_N}.$$

We assume that $\partial\Omega$ and $\partial T_\varepsilon^{X_N}$ are submanifolds of dimension $d-1$ of R^d sufficiently regular and that $T_\varepsilon^{X_N} \bigcap \Omega = \emptyset$, so that all mathematical quantities involved in the equations can be properly defined. The unit normal field n on the boundary of $T_\varepsilon^{X_N}$ is oriented towards Ω_ε (i.e. n is oriented externally to $T_\varepsilon^{X_N}$).

In general, the homogenization problem analysed in [4, 5] consists in studying the asymptotic behaviour, when $\varepsilon \to 0$, $N \to \infty$ and $\varepsilon N^{d-2} = C$, of the field $F_\varepsilon : \mathbb{R}^d \to \mathbb{R}^{d'}$ (in those paper d' is either 1 or 3, but of course we can consider the general case), belonging to an assigned subspace $\mathfrak{h}_F(\Omega_\varepsilon)$ of $(H^1(\Omega))^{d'}$ and satisfying a boundary (or initial-boundary) value problem, with non homogeneous boundary conditions, where the differential operator acting on the space variable is a second order operator of divergence form, more specifically in those papers the Laplacian operator (but it could be a more general one). The problem may possibly include suitable constraints on F_ε (e.g. div $F_\varepsilon = 0$), which in any case do not modify too much the general procedure we are going to illustrate.

We explained in [4, 5] how the chosen scaling can be heuristically justified: we shall now sketch how it can be derived from the mathematical properties of the problem, and we shall illustrate the connection between the "size" of the set occupied by the inclusions and the capacity.

We consider the simple problem

$$\begin{cases} -\text{div}\,(\nabla F_\varepsilon) = g\,, & x \in \Omega_\varepsilon\,, \\ F_\varepsilon(x) = 0\,, & x \in \partial\Omega\,, \\ F_\varepsilon(x) = F_{i,\varepsilon}\,, & x \in \partial T_i^\varepsilon(x_i)\,,\ 1 \leq i \leq N\,. \end{cases} \tag{1}$$

Simple nonstationary problems, as the one considered in [5], can be considered in a similar way.

The weak formulation of the problem is then as follows:

$$\int_{\Omega_\varepsilon} \nabla F_\varepsilon \cdot \nabla W_\varepsilon = \int_{\Omega_\varepsilon} g \cdot W_\varepsilon \tag{2}$$

for $F_\varepsilon \in \mathfrak{H}_F(\Omega_\varepsilon) = \{f \in (H^1(\Omega_\varepsilon))^{d'}$: the boundary conditions are satisfied$\}$ and for all test functions $W_\varepsilon \in (H_0^1(\Omega_\varepsilon))^{d'}$.

The space $\mathfrak{H}_F(\Omega_\varepsilon)$ in (1) can be $(H_0^1(\Omega_\varepsilon))^{d'}$ or a subspace of $(H^1(\Omega_\varepsilon))^{d'}$, depending on the selected boundary conditions, homogeneous in the first case and non homogeneous in the second.

To the boundary problem (1) we associate the empirical measure on $\Omega \times \mathbb{R}^{d'}$,

$$\mathfrak{F}_N = \frac{1}{N} \sum_{i=1}^{N} \delta_{x_i, F_i}(x, f),$$

which describes the distribution of inclusions with respect to their position and the field value on their boundary.

Of course, the density of centres is given by $\rho_N = \int df \mathfrak{F}_N$ and, in order to be able to perform the limit in the large number/small size asymptotics of the system, we require this measure to be such that:

$$\int_{\mathbb{R}^{d'}} f \mathfrak{F}_N df \rightharpoonup \rho \mathscr{F} \qquad \text{and} \qquad \sup_N \int dx df \mathfrak{F}_N f^2 < C.$$

In order to perform the limit, all functions involved in the weak formulation have to be extended to the whole Ω.

We then associate to the solution F_ε to (1) its natural extension on Ω:

$$\bar{F}_\varepsilon(x) = \begin{cases} F_\varepsilon(x) & x \in \Omega_\varepsilon \\ F_\varepsilon(x)|_{\partial T_i^\varepsilon} = F_{i,\varepsilon} & x \in T_i^\varepsilon, \end{cases}$$

and we choose classes of test functions in such a way to cover in the limit the whole target space $(H_0^1(\Omega))^{d'}$.

So, for each $w \in (\mathscr{D}(\Omega))^{d'}$, we define a test function $\bar{W}_\varepsilon \overset{(H_0^1(\Omega))^{d'}}{\rightharpoonup} w \in (\mathscr{D}(\Omega))^{d'}$ weakly by writing:

$$\bar{W}_\varepsilon = w - \mathscr{B}_\varepsilon,$$

where $\mathscr{B}_\varepsilon \in (H_0^1(\Omega))^{d'}$ is a corrector whose role is to assure that W_ε verifies the required boundary conditions on T_ε^N at each value of ε.

The possibility to define good correctors, in the sense of correctors which allow to pass to the limit in the right space, depends on the structure of the holes. Since we chose holes with a regular boundary, this should be possible, provided we choose a good asymptotics for N and ε (the weak convergence of the empirical measure associated to the set of centroids of the holes is also a regularity assumption).

To summarize, at this stage, we require the corrector to be such that:

(a) $\mathscr{B}_\varepsilon|_{T_i^\varepsilon} = w|_{T_i^\varepsilon}$ (the function $\bar{W}_\varepsilon$ is then s.t. $\bar{W}_\varepsilon|_{T_i^\varepsilon} = 0$)

(b) $\mathscr{B}_\varepsilon \overset{(H_0^1(\Omega))^{d'}}{\rightharpoonup} 0$ weakly (the corrector has no influence on the limit function)

The function associated on the whole space at the end of the limit process to W_ε will be then its desired limit w.

There are more requirements we ask to the correctors then.

First, we want to deal with the correctors in an additive manner, i.e. we want that, in adding a new hole to the system, the corrector is modified only locally around the added hole, leaving all contributions due to the holes already present unmodified. This means that the correctors have to be the sum over the set of centroids of single contributions of holes.

Second, we want W_ε and w to be at the minimal possible distance in $H_0^1(\Omega)$, at each step and for a given r_ε. We choose therefore correctors minimizing the energy among the functions which satisfy property (a).

These two last requirements are summarized in the following list:

(c) For each ε, we may find $r_\varepsilon > \varepsilon$ such that supp $B_\varepsilon \subset \bigcup_{i=1}^{N} B_{r_\varepsilon}(x_i) = B_{r_\varepsilon}^{X_N}$, with $B_{r_\varepsilon}(x_i) \cap B_{r_\varepsilon}(x_j) = \emptyset$ for $i \neq j$ (for each ε the holes are well separated).
When this happens, the corrector $\mathscr{B}_\varepsilon$ is a sum of terms each one supported on the sphere of radius r_ε centred in a centroid, as showed in Fig. 1 (the picture is given in dimension $d = 2$).
Therefore, when passing from N to $N + 1$ and adding a new hole with centroid in x_{N+1}, it is sufficient to add a new term supportend in the sphere $B_{r_\varepsilon}(x_{N+1})$ without modifying the other terms in the corrector.
This condition adds of course a requirement on the distribution of holes, i.e. that $\forall i, j = 1, \ldots, N$ such that $i \neq j$ we have $|x_i - x_j| > 2r_\varepsilon$.

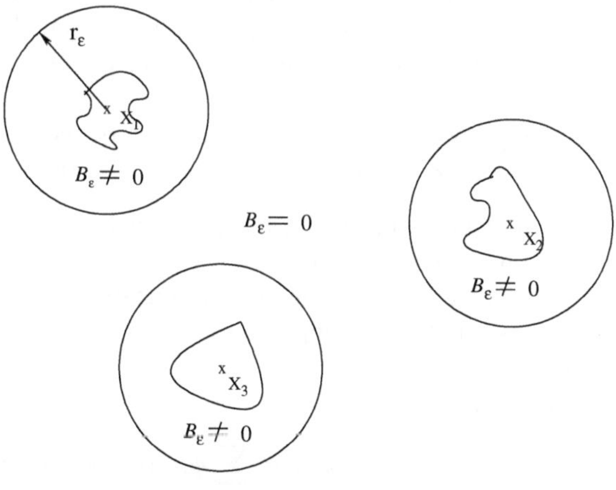

Fig. 1 Schematic representation of the support of the corrector B_ε in d=2

(d) $\int_{B_{r_\varepsilon}^{X_N}} |\nabla \mathscr{B}_\varepsilon|^2 = \min_{f \in \mathscr{I}_A} \left\{ \int_{B_{r_\varepsilon}^{X_N}} |\nabla f|^2 \right\}$, where

$$\mathscr{I}_A = \left\{ \begin{array}{l} f \in (H_0^1(B_{r_\varepsilon}^{X_N}))^{d'} : \exists v_n = (v_{n^1}, \ldots . v_{n^{d'}}) \overset{(H_0^1(B_{r_\varepsilon}^{X_N}))^{d'}}{\rightarrow} f \text{ s.t. } |v_{n^i}| \geq \sum_{i=1}^{N} |w^i| I_{T_i^\varepsilon} \text{ and } \operatorname{sign} v_{n^i}|_{T_i^\varepsilon} = \operatorname{sign} w^i|_{T_i^\varepsilon} \\ \quad \text{for } i = 1, \ldots N \text{ a.e. in a neighbourhood of } B_\varepsilon^{X_N} \end{array} \right\}.$$

The solution to this variational problem is known (see e.g. [9]), so that this last condition fix the form of the corrector as

$$\mathscr{B}_\varepsilon(x) = \sum_{i=1}^{N} \psi_{\varepsilon, r_\varepsilon}[w(\cdot + x_i)](x - x_i) \tag{3}$$

where, for a given sufficiently regular function $W = W(x)$, the function $\psi_{\varepsilon, r_\varepsilon}[W]$ is the solution to

$$\left\{ \begin{array}{ll} \Delta \psi_{\varepsilon, r_\varepsilon}[W] = 0\,, & x \in B_{r_\varepsilon}(0) \setminus T_i^\varepsilon\,, \\ \psi_{\varepsilon, r_\varepsilon}[W] = 0\,, & x \in B_{r_\varepsilon}^C(0)\,, \\ \psi_{\varepsilon, r_\varepsilon}[W] = W\,, & x \in T_i^\varepsilon, \end{array} \right. \tag{4}$$

When we require $W_\varepsilon|_{T_i^\varepsilon} = 0$, then $W = w$.

Remark 1 When the formulation of the boundary problem requires test functions which assume constant values (in general different from 0) on the boundary of the inclusions $\partial \Omega_\varepsilon$, as in [5], this same procedure can be adopted to build correctors; in this case, the choice of the function W is given by $W = w - w(x_i)$, so that $W_\varepsilon|_{T_i^\varepsilon} = w(x_i)$.

The radius r_ε has two constraints: because the balls do not have to superimpose (condition (c)), $Nr_\varepsilon^d \leq C < \infty$ and, since we do not want to have any contribution due to discontinuities of the correctors on the exterior spheres $\partial B_{r_\varepsilon}$, we require $\frac{\varepsilon}{r_\varepsilon} \to 0$ when $\varepsilon \to 0$. We may moreover add a further condition, if we want to minimize the corrector with respect to r_ε too. Indeed, the solution to (4) is such that $\|\nabla \psi_{\varepsilon, r_\varepsilon}\|_{L^2(B_{r_\varepsilon}(0) \setminus T_i^\varepsilon}$ is decreasing in r_ε, so choosing the maximal possible value of r_ε will select the corrector giving the minimal possible distance in $H_0^1(\Omega)$ between W_ε and w.

Once proposed the form of the correctors, we can analyse the behaviour of the test function W_ε in the asymptotic when $\varepsilon \to 0$: at this stage, we have still to find the asymptotics which assures that (3) is indeed a corrector (i.e. that $\mathscr{B}_\varepsilon \overset{(H_0^1(\Omega))^{d'}}{\rightharpoonup} 0$), and this will be done by connecting the H^1-norm of the solution to (4) to the notion of capacity we mentioned in the introduction.

We first single out a simple convergence property of a part of the corrector defined previously.

Denoting

$$\mathscr{R}_{1,\varepsilon} = \sum_{i=1}^{N} \psi_{\varepsilon, r_\varepsilon}[w(\cdot + x_i) - w(x_i)],$$

and recalling that $\varepsilon = \operatorname{diam} T^\varepsilon$, by the maximum principle we get

$$\|\mathscr{R}_{1,\varepsilon}\|_{L^2(\mathbb{R}^d)} \leq \sum_{i=1}^{N} \|\psi_{\varepsilon,r_\varepsilon}[w(\cdot + x_i) - w(x_i)]\|_{L^2(\mathbb{R}^d)} \leq \|\nabla w\|_{L^\infty} C N \varepsilon^2 r_\varepsilon^d.$$

Moreover, because

$$\left\| \int_{\partial T^\varepsilon} \partial_n \psi_{\varepsilon,r_\varepsilon}[w] \right\|_{L^\infty} \leq \varepsilon^{d-2} \left\| \int_{\partial T^1} \partial_n \psi_{1,\frac{r_\varepsilon}{\varepsilon}}[w] \right\|_{L^\infty} < C\varepsilon^{d-2},$$

we have

$$\|\nabla \mathscr{R}_{1,\varepsilon}\|_{L^2(\mathbb{R}^d)} \leq \sum_{i=1}^{N} \|\nabla \psi_{\varepsilon,r_\varepsilon}[w(\cdot + x_i) - w(x_i)]\|_{L^2(\mathbb{R}^d)} \leq \|\nabla w\|_{L^\infty} C N \varepsilon^{d-1} :$$

Therefore, we can rewrite the corrector as

$$\mathscr{B}_\varepsilon(x) = \sum_{i=1}^{N} \psi_{\varepsilon,r_\varepsilon}[w(x_i)](x - x_i) + \mathscr{R}_{1,\varepsilon} = \mathscr{R}_{2,\varepsilon} + \mathscr{R}_{1,\varepsilon}$$

where $\mathscr{R}_{1,\varepsilon} \to 0$ in $(H_0^1(\Omega))^{d'}$ strong whenever $N\varepsilon^{d-1} \to 0$ and $\mathscr{R}_{1,\varepsilon} \rightharpoonup 0$ in $(H_0^1(\Omega))^{d'}$ weak if $N\varepsilon^{d-1} = C$.

So, a first sufficient condition for (3) to define a corrector is that $N\varepsilon^{d-1} = C$.

Remark 2 When we have non homogeneous constant boundary conditions on the holes, as in ([5]), $\mathscr{B}_\varepsilon(x) = \mathscr{R}_{1,\varepsilon}$.

Recalling the definition of *Green capacity*, i.e. the relative capacity with respect to a set $S \subset \mathbb{R}^d$ of a subset $\Sigma \subset S$ (see [10], p. 43 formula 2.28 for the definition in $d = 2$ or [9] p. 182 for a definition with a normalised equilibrium measure),

$$\operatorname*{Cap}_S(\Sigma) = \inf\{\|\nabla f\|^2_{L^2(S)} : H_0^1(S) \ni f \geq 1 \text{ a.e. in a neighbourhood of } \Sigma\},$$

we notice that:

$$\|\nabla \psi_{\varepsilon,r_\varepsilon}[w(x_i)]\|^2_{L^2(B_{r_\varepsilon}(0))} = |w(x_i)|^2 \operatorname*{Cap}_{B_{r_\varepsilon}(x_i)}(T_i^\varepsilon).$$

Indeed, in this case $\psi_{s,r}[1]$ is the so called *capacitary potential* relative to B_r associated to the set T^s of diameter s.

Denoting as

$$\operatorname*{Cap}_{B_{r_\varepsilon}}(T^\varepsilon)(x_i) = \operatorname*{Cap}_{B_{r_\varepsilon}(x_i)}(T_i^\varepsilon)$$

we may rewrite:

$$\|\nabla \mathscr{B}_\varepsilon\|^2_{(\Omega)} = \sum_{i=1}^{N} |w(x_i)|^2 \operatorname*{Cap}_{B_{r_\varepsilon}}(T^\varepsilon)(x_i) + \|\nabla \mathscr{R}_{1,\varepsilon}\|^2_{L^2(\mathbb{R}^d)} + 2(\nabla \mathscr{R}_{2,\varepsilon}, \nabla \mathscr{R}_{1,\varepsilon})_{L^2(R^d)} \tag{5}$$

(here $(\cdot,\cdot)_{L^2(R^d)}$ denotes the L^2 scalar product) and, noticing that, since by construction $T_i^\varepsilon \subseteqq \bar{B}_\varepsilon(x_i)$,

$$\operatorname*{Cap}_{B_{r_\varepsilon}}(T^\varepsilon)(x_i) \le \operatorname*{Cap}_{B_{r_\varepsilon}(0)}(\bar{B}_\varepsilon(0)) = \begin{cases} \frac{d(d-1)}{|B_1(0)|} \frac{\varepsilon^{d-2} r_\varepsilon^{d-2}}{r_\varepsilon^{d-2} - \varepsilon^{d-2}} & d \ge 3 \\ \frac{2\pi}{\log \frac{r_\varepsilon}{\varepsilon}} & d = 2 \end{cases}$$

(the sphere has maximal capacity among simply connected compacts with diameter ε), we may use the Poincaré's inequality to estimate the first and the third term in (5), getting

$$\|\nabla \mathscr{R}_{2,\varepsilon}\|_{L^2(\mathbb{R}^d)} = \sum_{i=1}^{N} |w(x_i)|^2 \operatorname*{Cap}_{B_{r_\varepsilon}}(T^\varepsilon)(x_i) \le CN \begin{cases} \varepsilon^{d-2} & d \ge 3 \\ \frac{1}{\log \frac{r_\varepsilon}{\varepsilon}} & d = 2 \end{cases}$$

$$(\nabla \mathscr{R}_{2,\varepsilon}, \nabla \mathscr{R}_{1,\varepsilon})_{L^2(\mathbb{R}^d)} \le CN\varepsilon^{d-1}.$$

We have also for $\mathscr{R}_{2,\varepsilon}$:

$$\|\mathscr{R}_{2,\varepsilon}\|_{L^2(\mathbb{R}^d)} \le CN \begin{cases} \varepsilon^d \max(1, (\frac{\varepsilon}{r_\varepsilon})^{d-4}) & d \ge 3 \\ \varepsilon^2 & d = 2. \end{cases}$$

Moreover, $\|\nabla \mathscr{R}_{2,\varepsilon}\|_{L^2(\Omega)}$ (and therefore $\|\nabla \mathscr{B}_\varepsilon\|_{L^2(\mathbb{R}^d)}$) can be bound from below, since for $i = 1, \ldots, N$ we have $T_i^\varepsilon \supseteq B_{\varepsilon_i^{'}}(x_i)$, where $\varepsilon_i^{'} = \sup\{r : B_r(x_i) \subset T_i^\varepsilon\}$. We have then

$$\operatorname*{Cap}_{B_{r_\varepsilon}(x_i)}(T_i^\varepsilon) \ge \operatorname*{Cap}_{B_{r_\varepsilon(0)}}(\bar{B}_{\varepsilon'}(0)).$$

and, since $0 < a \le \frac{\varepsilon_i^{'}}{\varepsilon} \le 1$, when ε vanishes (recalling that $\frac{\varepsilon}{r_\varepsilon}$ vanishes also), we get

$$\|\nabla \mathscr{R}_{2,\varepsilon}\|_{L^2(\mathbb{R}^d)} \sim N \begin{cases} \varepsilon^{d-2} & d \ge 3 \\ \frac{1}{|\log \varepsilon|} & d = 2 \end{cases} \tag{6}$$

Given the previous estimates we observe that when $d \ge 3$ (resp. $d = 2$), whenever $N\varepsilon^{d-2} \to 0$ (resp. $\frac{N}{|\log \varepsilon|} \to 0$), $\mathscr{R}_{2,\varepsilon} \to 0$ strongly in $(H_0^1(\Omega))^{d'}$, while when $N\varepsilon^{d-2} = C$ (resp. $\frac{N}{|\log \varepsilon|} = C$) the convergence is weak, $\mathscr{R}_{2,\varepsilon} \rightharpoonup 0$ in $(H_0^1(\Omega))^{d'}$, and, because the behaviour of $\|\nabla \mathscr{R}_{2,\varepsilon}\|_{L^2(\mathbb{R}^d)}$ is given by (6), all the remaining options imply divergence in $(H_0^1(\Omega))^{d'}$.

Finally, we conclude (with suitable assumptions on the involved limit functions) that when $N\varepsilon^{d-1} \to 0$ we get:

$$\|\nabla \mathscr{B}_\varepsilon\|_{L^2(\Omega)} \to$$

$$\lim_{\varepsilon\to 0} \frac{1}{N} \sum_{i=1}^{N} |w(x_i)|^2 (N \operatorname*{Cap}_{B_{r_\varepsilon}})(T^\varepsilon)(x_i) = \int |w(x)|^2 \rho(x) \lim_{\varepsilon\to 0} (N \operatorname*{Cap}_{B_{r_\varepsilon}})(T^\varepsilon)(x)$$

$$\leq [\lim_{\varepsilon\to 0} N \operatorname*{Cap}_{B_{r_\varepsilon}(0)} (\bar{B}_\varepsilon(0))] \int |w(x)|^2 \rho(x) = \begin{cases} \frac{d(d-1)}{|B_1(0)|} \lim_{\varepsilon\to 0} N\varepsilon^{d-2} \int |w(x)|^2 \rho(x) & d \geq 3 \\ 2\pi \lim_{\varepsilon\to 0} \frac{N}{\log \frac{r_\varepsilon}{\varepsilon}} \int |w(x)|^2 \rho(x) & d = 2 \end{cases}. \tag{7}$$

Let now discuss the relationship with the homogenization problem.

We reformulate the weak formulation (2) on the whole Ω as:

$$\int_\Omega \nabla \bar{F}_\varepsilon \cdot \nabla \bar{W}_\varepsilon = \int_\Omega g \cdot \bar{W}_\varepsilon,$$

i.e. as

$$\int_\Omega \nabla \bar{F}_\varepsilon \cdot \nabla w - \int_\Omega \nabla \bar{F}_\varepsilon \cdot \nabla \mathscr{B}_\varepsilon = \int_\Omega g \cdot (w - \mathscr{B}_\varepsilon),$$

where we substituted to the solution F_ε its natural extension $\bar{F}_\varepsilon$ and to the test function W_ε the test function $\bar{W}_\varepsilon$. Assuming that we can easily prove the weak convergence of $\bar{F}_\varepsilon$ to a limit function, $\bar{F}_\varepsilon \rightharpoonup F$ in $(H_0^1(\Omega))^{d'}$, (which is true for (2) with homogeneous boundary conditions) and that we have at least the weak convergence $\mathscr{B}_\varepsilon \rightharpoonup 0$ in $(H_0^1(\Omega))^{d'}$, we deduce immediately

$$\int_\Omega \nabla \bar{\mathscr{F}}_\varepsilon \cdot \nabla w \to \int_\Omega \nabla F \cdot \nabla w$$

and

$$\int_\Omega g \cdot (w - \mathscr{B}_\varepsilon) \to \int_\Omega g \cdot w.$$

The remaining term is the one which should originate, in a suitable asymptotics, the homogenized term representing the inclusions.

Of course, if the scaling is such that the corrector converges strongly to 0 in $(H_0^1(\Omega))^{d'}$ the homogenised term vanishes and no trace of the inclusions is left in the limit: in this case, the homogenised equation does not have additional terms with respect to the original problem. As we can see from (7), this happens when the total capacity of the set occupied by the inclusions vanishes in the limit.

Since we also saw from the previous discussion that, when the total capacity diverges, $\mathscr{B}_\varepsilon$ as defined by (3) cannot be considered a corrector, we see that the effect of the inclusions permains in the limit, through a nonvanishing homogenization term, if the capacity of the set occupied by them has a finite, nonzero limit.

We have then that, if N and ε are such that $N\varepsilon^{d-2} \to 0$, when $d \geq 3$, or $\frac{N}{|\log \varepsilon|} \to 0$, when $d = 2$, $\|\mathscr{B}_\varepsilon\|_{H_0^1(\Omega)} \to 0$ strongly and the homogenized solution solves on Ω the same equation as (2), i.e. the equation $\int_\Omega \nabla F \cdot \nabla W = \int_\Omega g \cdot W$, with test functions $W \in (H_0^1(\Omega))^{d'}$.

We summarize the possible results we can get when performing different asymptotic limits:

As1 $\lim_{\varepsilon\to 0} N\varepsilon^{d-2} = 0$, $d \geq 3$, or $\lim_{\varepsilon\to 0} \frac{N}{|\log \varepsilon|} = 0$, $d = 2$
The homogenization process does not add terms to the original equation.
As2 $\lim_{\varepsilon\to 0} N\varepsilon^{d-2} = C \neq 0$, $d \geq 3$, or $\lim_{\varepsilon\to 0} \frac{N}{|\log \varepsilon|} = C \neq 0$, $d = 2$
The homogenization process adds terms to the original equation in the limit.
As3 $\lim_{\varepsilon\to 0} N\varepsilon^{d-2} = \infty$, $d \geq 3$, or $\lim_{\varepsilon\to 0} \frac{N}{|\log \varepsilon|} = \infty$, $d = 2$
The definition (3) does not define correctors.

We may now describe the homogenization process by using the previous formalism. In the next subsections we shall select the critical asymptotics given in (As2): this choice imposes the bound $r_\varepsilon \leq \varepsilon^{\frac{d-2}{d}}$.

In order to be able to pass to the limit in the equations, we shall assume

$$\frac{\mathrm{Cap}_{B_{r_\varepsilon}}(T^\varepsilon)}{|B_{r_\varepsilon}(0)|}\rho_N(x) \rightharpoonup K(x)\rho(x)$$

in the sense of measure, where we define:

$$\mathop{\mathrm{Cap}}_{B_{r_\varepsilon}}(T^\varepsilon)(x) = \sum_{i=1}^{N} \mathop{\mathrm{Cap}}_{B_{r_\varepsilon}}(T^\varepsilon)(x_i) I_{B_{r_\varepsilon}(x_i)}(x).$$

Since in the given hypothesis for the inclusions

$$\frac{\mathrm{Cap}_{B_{r_\varepsilon}}(T^\varepsilon)(x_i)}{|B_{r_\varepsilon}(0)|} \sim \frac{\varepsilon^{d-2}}{r_\varepsilon^d},$$

this requirement fixes $r_\varepsilon^d = \varepsilon^{d-2}$.

Since there are some small difference in the procedure, depending on the selected boundary conditions, we shall discuss separately the case with homogeneous boundary conditions and the case with nonhomogeneous boundary conditions.

3 Homogeneous Boundary Conditions

We consider the homogeneous case ($\mathfrak{H}_F(\Omega_\varepsilon) = (H_0^1(\Omega_\varepsilon))^{d'}$): in this case the space of test functions and the space of solutions is the same, so, from the Eq. (2), we can get the estimate

$$\|\nabla \bar{F}_\varepsilon\|_{L_2(\Omega)} \leq \|g\|_{L^2(\Omega)}$$

and we can deduce $\bar{F}_\varepsilon \rightharpoonup F \in (H_0^1(\Omega))^{d'}$ (by subsequences).

As already seen, Eq. (2) can be reformulated in the whole Ω as:

$$\int_\Omega \nabla \bar{F}_\varepsilon \cdot \nabla w - \int_\Omega \nabla \bar{F}_\varepsilon \cdot \nabla \mathscr{B}_\varepsilon = \int_\Omega g \cdot (w - \mathscr{B}_\varepsilon)$$

where $\mathscr{B}_\varepsilon = \mathscr{R}_{2,\varepsilon} + \mathscr{R}_{1,\varepsilon}$, and the first term on the left-hand side and the term on the right-hand side converge respect. to $\int_\Omega \nabla F \cdot \nabla w$ and to $\int_\Omega g \cdot w$.

As for the second term on the right-hand side, we have

$$\int_\Omega \nabla \bar{F}_\varepsilon \cdot \nabla \mathscr{B}_\varepsilon = \int_\Omega \nabla \bar{F}_\varepsilon \cdot \nabla \mathscr{R}_{2,\varepsilon} + \int_\Omega \nabla \bar{F}_\varepsilon \cdot \nabla \mathscr{R}_{1,\varepsilon},$$

and the strong convergence in $(H_0^1(\Omega))^{d'}$ of $\mathscr{R}_{1,\varepsilon} \to 0$ and the weak convergence in $(H_0^1(\Omega))^{d'}$ of $\bar{F}_\varepsilon \rightharpoonup F$ imply

$$\int_\Omega \nabla \bar{F}_\varepsilon \cdot \nabla \mathscr{R}_{1,\varepsilon} \to 0.$$

To evaluate the term $\int_\Omega \nabla \bar{F}_\varepsilon \cdot \nabla \mathscr{R}_{2,\varepsilon}$ we proceed as follows: we recall [9] that we can associate to the set $T \subset A$ a measure μ_T (*equilibrium measure*) s.t.

$$\mu_T(A) = \operatorname*{Cap}_A(T).$$

We have moreover, for T compact:

$$\mu_T = -\partial_n \psi_T|_{\partial T} \delta_{\partial T}$$

where ψ_T is the already mentioned capacitary potential of T and n is the exterior unit vector on ∂T.

By construction, we have (on the whole Ω):

$$\Delta \psi_{\varepsilon, r_\varepsilon}[1](x - x_i) = -\partial_n \psi_{\varepsilon, r_\varepsilon}[1]|_{\partial B_{r_\varepsilon}(x_i)} \delta_{\partial B_{r_\varepsilon}(x_i)} - \mu_{T_i^\varepsilon}$$

where

$$-\int_\Omega \partial_n \psi_{\varepsilon, r_\varepsilon}[1]|_{\partial B_{r_\varepsilon}(x_i)} \delta_{\partial B_{r_\varepsilon}(x_i)} = \mu_{T_i^\varepsilon}(\Omega) = \operatorname*{Cap}_{B_{r_\varepsilon}}(T^\varepsilon)(x_i)$$

and, since $\bar{F}|_{\partial B_{r_\varepsilon}(x_i)} = 0$, we may write:

$$\int_\Omega \nabla \bar{F}_\varepsilon \cdot \nabla \mathscr{R}_{2,\varepsilon} = -\sum_{i=1}^{N} \int_\Omega \bar{F}_\varepsilon \cdot \Delta \psi_{\varepsilon,r_\varepsilon}[w(x_i)](x - x_i) =$$

$$= \sum_{i=1}^{N} \int_\Omega \bar{F}_\varepsilon \cdot (w(x_i)\partial_n \psi_{\varepsilon,r_\varepsilon}[1]|_{\partial B_{r_\varepsilon}(x_i)})\delta_{\partial B_{r_\varepsilon}(x_i)}$$

where, because of (As2), we have the bound

$$-\sum_{i=1}^{N} \int_\Omega \partial_n \psi_{\varepsilon,r_\varepsilon}[1]|_{\partial B_{r_\varepsilon}(x_i)})\delta_{\partial B_{r_\varepsilon}(x_i)} = \sum_{i=1}^{N} \underset{B_{r_\varepsilon}}{\mathrm{Cap}}(T^\varepsilon)(x_i) \leq C.$$

Since $\psi_{\varepsilon,r_\varepsilon}$ is the solution to the Laplace equation, we have the asymptotic equality

$$-\partial_n \psi_{\varepsilon,r_\varepsilon}[1]|_{\partial B_{r_\varepsilon}(x_i)} = \frac{\mathrm{Cap}_{B_{r_\varepsilon}}(T^\varepsilon)(x_i)}{|\partial B_{r_\varepsilon}(0)|}\left(1 + O\left(\frac{\varepsilon}{r_\varepsilon}\right)\right)$$

and therefore

$$-\sum_{i=1}^{N} \int_\Omega w(x_i)\partial_n \psi_{\varepsilon,r_\varepsilon}[1]|_{\partial B_{r_\varepsilon}(x_i)}\delta_{\partial B_{r_\varepsilon}(x_i)} \approx$$

$$\sum_{i=1}^{N} \int_\Omega w(x_i)\frac{\mathrm{Cap}_{B_{r_\varepsilon}}(T^\varepsilon)(x_i)}{|\partial B_{r_\varepsilon}(0)|}\delta_{\partial B_{r_\varepsilon}(x_i)}.$$

So, the last step is to prove the strong convergence in H^{-1} of

$$\sum_{i=1}^{N} \int_\Omega w(x_i)\frac{\mathrm{Cap}_{B_{r_\varepsilon}}(T^\varepsilon)(x_i)}{|\partial B_{r_\varepsilon}(0)|}\delta_{\partial B_{r_\varepsilon}(x_i)} \to K(x)\rho(x)w(x).$$

This can be done, when $\frac{\mathrm{Cap}_{B_{r_\varepsilon}}(T^\varepsilon)}{|B_{r_\varepsilon}(0)|}\rho_N(x) \rightharpoonup K(x)\rho(x)$ in the sense of measure, in the same way as in the first paragraph of the Appendix of [4], following a procedure already used in [3].

More generally, given a function $G \in (C_b(\Omega))^{d'}$, using the auxiliary problem in $\cup_{i=1}^{N} B_{r_\varepsilon}(x_i)$ (extended by 0 in $\Omega \setminus \cup_{i=1}^{N} B_{r_\varepsilon}(x_i)$):

$$\begin{cases} -\Delta \xi = -G_\varepsilon(x_i)\dfrac{\mathrm{Cap}_{B_{r_\varepsilon}}(T^\varepsilon)(x_i)}{|B_{r_\varepsilon}(0)|} \\ \partial_n \xi\,|_{\partial B_{r_\varepsilon}(x_i)} = G_\varepsilon(x_i)\dfrac{\mathrm{Cap}_{B_{r_\varepsilon}}(T^\varepsilon)(x_i)}{|\partial B_{r_\varepsilon}(0)|} \end{cases}$$

it is possible to prove that

$$\sum_{i=1}^{N} G_\varepsilon(x_i) \frac{\mathrm{Cap}_{B_{r_\varepsilon}}(T^\varepsilon)(x_i)}{|\partial B_{r_\varepsilon}(0)|} \delta_{\partial B_{r_\varepsilon}} \to G(x)K(x)\rho(x)$$

in H^{-1} strong.

Therefore, we have

$$\int_\Omega \nabla \bar{F}_\varepsilon \cdot \nabla \mathscr{R}_{2,\varepsilon} \to \int_\Omega K\rho(x)F \cdot w.$$

and we obtain, as a limit equation,

$$\int_\Omega \nabla F \cdot \nabla w + \int_\Omega K\rho(x)F \cdot w = \int_\Omega g \cdot w$$

4 Non-homogeneous Boundary Conditions

In the non-homogeneous case,

$$\mathfrak{H}_F(\Omega_\varepsilon) = \{f \in (H^1(\Omega_\varepsilon))^{d'} : f|_{\partial T_i^\varepsilon}, i = 1, \ldots, N \quad \text{are} \quad \text{constant}\},$$

so that test functions and solutions do not belong to the same space and, because of the non zero boundary terms, it is not possible to get easily an a priori bound on $\|\nabla \bar{F}_\varepsilon\|_{L^2(\Omega_\varepsilon)}$.

We need then to introduce correctors also into the natural extension, rewriting it as $\bar{F}_\varepsilon = \mathscr{F}_\varepsilon + \mathscr{A}_\varepsilon$, the sum of a function $\mathscr{F}_\varepsilon \in \mathfrak{H}_W = (H_0^1(\Omega_\varepsilon))^{d'}$ plus a corrector $\mathscr{A}_\varepsilon$, having the same role as the corrector $\mathscr{B}_\varepsilon$, i.e. such that at each step $\mathscr{F}_\varepsilon|_{\partial\Omega_\varepsilon} = 0$.

$\mathscr{A}_\varepsilon \in (H_0^1(\Omega))^{d'}$ enjoys properties which are similar to the properties of $\mathscr{B}_\varepsilon$, i.e.:

(a') $\mathscr{A}_\varepsilon|_{B_\varepsilon(x_i)} = F_i$
(b') $\mathscr{A}_\varepsilon| \rightharpoonup 0$ in $(H_0^1(\Omega))^{d'}$ weakly
(c') supp $\mathscr{A}_\varepsilon \subset B_{r_\varepsilon}^{X_N}$ with $B_{r_\varepsilon}(x_i) \cap B_{r_\varepsilon}(x_i) = \emptyset$ for $i \neq j$
(d') $\mathscr{A}_\varepsilon$ has minimal energy among the functions satisfying condition (a').

With this definition:

$$\mathscr{A}_\varepsilon = \sum_{i=1}^{N} \psi_{\varepsilon,r_\varepsilon}[F_i](x - x_i) = \mathscr{R}_{2,\varepsilon}[\mathbf{F}_N],$$

and we can get estimates on $\mathscr{F}_\varepsilon$ using Eq. (2). Indeed :

$$\int_{\Omega_\varepsilon} \nabla \bar{F}_\varepsilon \cdot \nabla \mathscr{F}_\varepsilon = \int_\Omega |\nabla \mathscr{F}_\varepsilon|^2 + \int_\Omega \mathscr{A}_\varepsilon \cdot \nabla \mathscr{F}_\varepsilon$$

so that we have:

$$\|\nabla \mathscr{F}_\varepsilon\|_{L^2(\Omega)} \leq C\|g\|_{L^2(\Omega)} + \|\nabla \mathscr{A}_\varepsilon\|_{L^2(\Omega)} < C$$

and we obtain $\mathscr{F}_\varepsilon \overset{(H_0^1(\Omega))^{d'}}{\rightharpoonup} F \in (H^1(\Omega))^{d'}$ (by subsequences).

We rewrite now the equation in the whole Ω:

$$\int_\Omega \nabla w \cdot \nabla \mathscr{F}_\varepsilon + \int_\Omega \nabla \mathscr{A}_\varepsilon \cdot \nabla w - \int_\Omega \nabla \mathscr{B}_\varepsilon \cdot \nabla \mathscr{F}_\varepsilon - \int_\Omega \nabla \mathscr{B}_\varepsilon \cdot \nabla \mathscr{A}_\varepsilon = \int_\Omega g \cdot (w - \mathscr{B}_\varepsilon).$$

The first three terms on the left-hand side and the right-hand side term have the same behaviour as the corresponding terms in the homogeneous case ($\mathscr{A}_\varepsilon$ has the same compactness properties as $\mathscr{B}_\varepsilon$); we need then to analyse only the last term on the left-hand side, $\int_\Omega \nabla \mathscr{B}_\varepsilon \cdot \nabla \mathscr{A}_\varepsilon$.

Using the same notation as in the previous section,

$$\int_\Omega \nabla \mathscr{B}_\varepsilon \cdot \nabla \mathscr{A}_\varepsilon = \int_\Omega \nabla \mathscr{R}_{1,\varepsilon} \cdot \nabla \mathscr{A}_\varepsilon + \int_\Omega \nabla \mathscr{R}_{2,\varepsilon} \cdot \nabla \mathscr{R}_{2,\varepsilon}[\mathbf{F}_N].$$

Thanks to the strong convergence $\mathscr{R}_{1,\varepsilon} \to 0$, the first term vanishes in the chosen asymptotics.

As for the second term. observe that:

$$\int_\Omega \nabla \mathscr{R}_{2,\varepsilon} \cdot \nabla \mathscr{R}_{2,\varepsilon}[\mathbf{F}_N] = \sum_{i=1}^N w(x_i) \cdot F_i \int_{B_{r_\varepsilon} \setminus T_i^\varepsilon} |\nabla \psi_{\varepsilon, r_\varepsilon}[1]|^2 (x - x_i)$$

$$= \sum_{i=1}^N w(x_i) \cdot F_i \operatorname*{Cap}_{B_{r_\varepsilon}}(T_i^\varepsilon) \to \int_\Omega w(x) \cdot (\rho \mathscr{F})(x) K(x)$$

so that the limit equation is

$$\int_\Omega \nabla w \cdot \nabla F + \int_\Omega w(x) \cdot \rho(x) F K(x) - \int_\Omega w(x) \cdot (\rho \mathscr{F})(x) K(x) = \int_\Omega g \cdot w.$$

5 Adding Constraints on the Solutions

Let us add a last remark, related to the paper [4]: when the original problem (1) imposes suitable (linear) constraints to the solution ($Op(F_\varepsilon) = 0$ in Ω_ε), like e.g. Div $F_\varepsilon = 0$ for $x \in \Omega_\varepsilon$ in the Stokes or Navier-Stokes equations, we may perform a similar procedure by adding the constraint on the space where we minimise the energy to build the correctors.

This means that we can build correctors verifying the same constraints as the solution, by redefining condition (d) (and (d')) as:

D) $\int_{B_{r_\varepsilon}^{X_N}} |\nabla \mathscr{B}_\varepsilon|^2 = \min_{f \in \mathscr{I}'} \left\{ \int_{B_{r_\varepsilon}^{X_N}} |\nabla f|^2 \right\}$, where

$$\mathscr{I}' = \mathscr{I}_A \cap \left\{ f \in (H_0^1(B_{r_\varepsilon}^{X_N}))^{d'} : Op(f) = 0 \right\}.$$

The steps which follow are then the same as the ones described in the previous sections, provided we substitute to the "free" capacity $\mathrm{Cap}_B(S)$ a capacity with constraints $\mathrm{Cap}_B^{vin}(S)$, which will be of course bigger than $\mathrm{Cap}_B(S)$, $\mathrm{Cap}_B^{vin}(S) \geq \mathrm{Cap}_B(S)$.

Because the test functions verify the same constraints as the solutions, in the paper [4] it will not be required an explicit estimate of the pressure in the Stokes or Navier-Stokes equations (i.e. of the "Lagrange multiplier" associated to the constraint Div $F = 0$) for the solution in the weak formulation of the equation. The constraint will nevertheless be present when computing explicitly the capacitary potential.

6 Conclusion

We gave a general picture of the procedure adopted in [4, 5] for deriving the homogenised limit of some particular equations describing two-component mixtures containing a dispersed phase. In those mixtures, the dispersed phase is modelled through solid particles which are not periodically distributed; the fluid phase is modelled through a field, satisfying equations where the space differential operator involved in the equations is a second order operator of divergence form, which assumes constant values on the boundary of the solid particles. In order to give a simplified description of the steps adopted to get the homogenised limit and of the relevant size of the set occupied by the dispersed phase leading to an effective term which permains in that limit, we connected the procedure to the notion of capacity. In generalising the procedure, we extended it to a general dimension $d \geq 2$ and to general shapes of the inclusions. Finally we gave a short comment on the case, also met in [4], in which simple constraints are present.

References

1. Adams, D.R., Hedberg, L.H.: Function Space and Potential Theory. Springer (1999)
2. Brelot, M.: Lectures on Potential Theory, 2nd edn. Tata Institute (1967)
3. Cioranescu, D., Murat, F.: Un terme étrange venu d'ailleurs. In: Nonlinear Partial Differential Equations and their Applications, Collège de France Seminar, vol. 2 (Paris 1979–1980). Research Notes in Mathematics, vol. 60, pp. 98–138. Pitman, Boston, London (1982)
4. Desvillettes, L., Golse, F., Ricci, V.: The mean-field limit for solid particles in a Navier-stokes flow. J. Stat. Phys. **131**, 941–967 (2008)
5. Desvillettes, L., Golse, F., Ricci, V.: Derivation of a homogenized two-temperature model from the heat equation. M2AN **48**, 1583–1613 (2014)
6. Doob, J.L.: Classical Potential Theory and Its Probabilistic Counterpart. Springer (1984)
7. Günther, N.M.: La théorie du potentiel et ses applications aux problémes fondamentaux de la physique mathématique. In: Collections de monographies sur la théorie des fonctions, 303 pp. Gauthier-Villars, Paris (1934)
8. Kellogg, O.D.: Foundations of Potential Theory. Springer, Berlin (1929)
9. Landkof, N.S.: Foundations of Modern Potential Theory. Spinger, Berlin (1972)
10. Marchenko, V.A., Khruslov, E.Y.: Homogenization of Partial Differential Equations. Birkhäuser, Boston (2006)
11. Falconer, K.: Fractal Geometry: Mathematical Foundations and Applications. 2nd edn. Wiley (2003)
12. Ricci, V.: Large number asymptotics for two–component systems with self–consistent coupling. In: From Particle Systems to Partial Differential Equations. Springer Proceedings in Mathematics & Statistics, vol. 75, pp. 285–300. Springer (2014)

Derivation of the Boltzmann Equation: Hard Spheres, Short-Range Potentials and Beyond

Chiara Saffirio

Abstract We review some results concerning the derivation of the Boltzmann equation starting from the many-body classical Hamiltonian dynamics. In particular, the celebrated paper by Lanford III [21] and the more recent papers [13, 23] are discussed.

Keywords Boltzmann equation · Many-body particle systems

1 Introduction

A central question in non-equilibrium statistical mechanics is to investigate the rigorous derivation of effective macroscopic equations starting from the fundamental laws of classical mechanics. Though we are still very far from a complete understanding, considerable progress has been made in the last years in developing new mathematical methods. In particular, several interesting questions regarding classical systems in the mean-field and low-density limits are now approachable by a rigorous mathematical analysis. The aim of this paper is to give an overview on the derivation of the classical Boltzmann equation, in light of recent developments.

The Boltzmann equation. At the end of the XIXth century Maxwell [22] and Boltzmann [5] addressed independently the problem of the mathematical description of classical dilute gases, in an attempt to produce a reduced kinetic picture emerging from the microscopic fundamental laws of classical mechanics. A kinetic description holds at a mesoscopic level, that is on quantities which averages are susceptible of measurement. The equation for the evolution of a rarefied gas, that nowadays bears the name of Boltzmann, reads

$$(\partial_t + v \cdot \nabla_x) f = Q(f, f). \tag{1}$$

C. Saffirio (✉)
Institute of Mathematics, University of Zürich, Winterthurerstrasse 190, 8057 Zürich, Switzerland
e-mail: chiara.saffirio@math.uzh.ch

© Springer International Publishing Switzerland 2016

P. Gonçalves and A.J. Soares (eds.), *From Particle Systems to Partial Differential Equations III*, Springer Proceedings in Mathematics & Statistics 162,
DOI 10.1007/978-3-319-32144-8_15

The unknown $f : \mathbb{R}_+ \times \mathbb{R}^3 \times \mathbb{R}^3 \to \mathbb{R}_+$ is the probability density of finding a particle with position x and velocity v at time t.

The l.h.s. in the Boltzmann Eq. (1) is the free transport operator, representing the free flow of particles in absence of external forces. The r.h.s. is a non-linear (quadratic) operator which describes the elastic binary collisions among particles:

$$Q(f,f)(t,x,v) = \int_{S^2} d\omega \int_{\mathbb{R}^3} dv_* B(v - v_*, \omega)\{f(t,x,v')f(t,x,v'_*) - f(t,x,v)f(t,x,v_*)\}, \tag{2}$$

where S^2 is the unit sphere in $\mathbb{R}^3$, $\omega \in S^2$ is the scattering vector, v' and v'_* are obtained as functions of v and v_* by the following scattering laws:

$$\begin{cases} v' = v - [(v - v_*) \cdot \omega]\omega, \\ v'_* = v_* + [(v - v_*) \cdot \omega]\omega. \end{cases} \tag{3}$$

The integral kernel $B(\cdot, \cdot)$ is proportional to the differential cross-section.

In particular, in [5] Boltzmann established Eq. (1) by taking into account the interactions among particles which occur through elastic binary collisions, which are localised in space and time. Precisely, when the particles interact as hard spheres (in other words as billiard balls), the kernel assumes the simple and explicit form $B(v - v_*, \omega) = \omega \cdot (v - v_*)$. In this case, the scattering vector ω is equal to $\boldsymbol{\nu} \in S^2_+ := \{\theta \in S^2 \mid \theta \cdot (v - v_*) \geq 0\}$, the unit vector pointing from the particle with velocity v to the particle with velocity v_*.

Then the Boltzmann collision operator for hard spheres reads

$$Q(f,f)(t,x,v) = \int_{S^2_+} d\boldsymbol{\nu} \int_{\mathbb{R}^3} dv_* \, \boldsymbol{\nu} \cdot (v - v_*) \{f(t,x,v'_*)f(t,x,v') - f(t,x,v_*)f(t,x,v)\}. \tag{4}$$

The peculiarity of Eq. (1) is the following: on the one hand, it purports to describe the evolution of the density of a rarefied gas, whose dynamics is time-reversible at a microscopic level; on the other hand, the equation itself has an irreversible behaviour, with an increasing entropy (the celebrated H-Theorem[1]) and trend to equilibrium.

The derivation problem. The issue of derivation consists in determining whether the theory of Boltzmann is only a phenomenological observation or a rigorous consequence of the laws of mechanics. The question is: is it possible to derive mathematically an irreversible dynamics, such as the Boltzmann dynamics, starting from

[1]The H-Theorem asserts that the kinetic entropy associated to the solution $f(t)$ of the Boltzmann Eq. (1) decreases in time. More precisely, let $H(f)$ be the H-functional defined as the information entropy with a negative sign:

$$H(f(t)) = \int_{\mathbb{R}^3 \times \mathbb{R}^3} dx\, dv\, f(t,x,v)[\log f(t,x,v) - 1].$$

A straightforward computation shows that $H(f(t)) \leq H(f(0))$.

the microscopic reversible classical dynamics? A positive answer to this question would show rigorously that there is no contradiction between the reversibility of the molecular dynamics and the irreversibility implied by the H-Theorem.

The mathematical formulation of the derivation problem was given by Grad (more than fifty years after Boltzmann). Indeed, in [14] Grad formulated for the first time the question of the validity of the Boltzmann equation as a limit—involving the number of particles—in which it is expected to hold. To state Grad's idea, we introduce the key ingredients for the description of a microscopic classical dynamics of a system of particles.

The time evolution of a configuration of N particles in the phase space

$$\mathcal{M}_N = \{(q_1, \ldots, q_N, v_1, \ldots, v_N) \in (\mathbb{R}^3 \times \mathbb{R}^3)^N \ : \ |q_i - q_k| > 0, \ i, k = 1, \ldots, N, \ k \neq i\}, \tag{5}$$

is given by the Newton equations:

$$\begin{cases} \dot{q}_i(\tau) = v_i(\tau), \\ \dot{v}_i(\tau) = -\sum_{j\neq i}^N \nabla\Phi(q_i(\tau) - q_j(\tau)), \end{cases} \qquad i = 1, \ldots, N \tag{6}$$

where τ is the time variable, $(q_1, \ldots, q_N) \in \mathbb{R}^{3N}$ and $(v_1, \ldots, v_N) \in \mathbb{R}^{3N}$ are respectively the position and velocity variables of the N particles, and Φ is a smooth two–body interaction potential.[2] We introduce the Hamiltonian associated to (6):

$$H(\tau, q_1, \ldots, q_N, v_1, \ldots, v_N) = \frac{1}{2}\sum_{i=1}^N |v_i(\tau)|^2 + \sum_{\substack{i,j=1 \\ j\neq i}}^N \Phi(q_i(\tau) - q_j(\tau)). \tag{7}$$

which is constant in time.

We stress that, in order to get a kinetic picture, we are not interested in the detailed analysis of the motion of each particle, but in the collective behaviour of the system. For this reason it is useful to adopt a statistical viewpoint: consider a probability density W_0^N on the phase space $\mathbb{R}^{3N} \times \mathbb{R}^{3N}$ and denote by $\Psi(\tau)$ the Newtonian flow. Then, $W^N(\tau) := W_0^N \circ \Psi(-\tau)$ solves the Liouville equation

$$\partial_\tau W^N(\tau) + \sum_{i=1}^N v_i \cdot \nabla_{q_i} W^N(\tau) - \sum_{j\neq i}^N \nabla_{q_i}\Phi(q_i - q_j) \cdot \nabla_{v_i} W^N(\tau) = 0 \tag{8}$$

for a probability density W^N on the phase space $\mathcal{M}_N$, with $W^N \in \mathscr{C}^1(\mathbb{R}_+ \times \mathcal{M}_N)$, $v_i \cdot \nabla_{q_i} W^N$, $\nabla\Phi \cdot \nabla_{v_i} W^N \in L^1(\mathcal{M}_N)$. Note moreover that since the particles are identical, W^N is symmetric w.r.t. permutation of particles.

[2] For simplicity, the potential is assumed to be smooth, to ensure the existence and uniqueness of the solution to the Newton equations (6).

In [14], Grad remarked that the solution of (8) can be approximated by the Boltzmann equation in the following regime, and in the following sense. Let $\varepsilon > 0$ be a scale parameter, which represents the ratio between macroscopic and microscopic units. Let us scale time and space according to

$$t = \varepsilon\tau, \quad x_i = \varepsilon q_i, \ \forall i = 1, \ldots, N. \tag{9}$$

Sometimes we will use the shortened notation $z_i := (x_i, v_i) \in \mathbb{R}^3 \times \mathbb{R}^3$, for $i = 1, \ldots, N$. In the limit of N large, with $N\varepsilon^2 = 1$, the solution of (8) with approximately factorised initial data remains approximately factorised. Note that approximate factorisation has to be understood in the sense of the marginal distributions

$$f_j^N(t, z_1, \ldots, z_j) := \int_{\mathbb{R}^{6(N-j)}} W^N(t, z_1, \ldots, z_N)\, dz_{j+1} \ldots dz_N. \tag{10}$$

More precisely, if $f_j^N(0, x_1, \ldots, x_j, v_1 \ldots, v_j) \simeq f_0^{\otimes j}(x_i, v_i)$, then

$$f_j^N(t, x_1, \ldots, x_j, v_1 \ldots, v_j) \simeq f(t)^{\otimes j}(x_i, v_i), \tag{11}$$

where f solves the Boltzmann Eq. (1) with initial datum f_0. This approximation, called *propagation of chaos*, is specified in Theorem 1.

The regime

$$N \to \infty, \qquad N\varepsilon^2 = 1$$

is called the low-density limit (or *Boltzmann-Grad limit*, BG-limit from now on). The underlying idea is that, on the one hand, we want to describe the physical situation in which the gas is rarefied. Hence a tagged particle undergoes a finite number of collisions in a macroscopic time, implying that the density $N\varepsilon^3$ vanishes in the limit of large N. On the other hand, we want the collisional structure of the microscopic system to survive in the limit, that is

$$\frac{\text{number of interactions}}{\text{time unit}} = O(1).$$

Therefore, it is important to keep in mind that the number of particles N is linked to the scale parameter ε by the relation $N\varepsilon^2 = O(1)$ (for simplicity we have chosen $N\varepsilon^2 = 1$), so that the limit $\varepsilon \to 0$ is equivalent to $N \to \infty$. Observe that, if Φ has compact support, the parameter ε represents the range of the interaction at a macroscopic scale thanks to the scaling (9).

The crucial assumption to be verified is that Eq. (11) holds for positive times once it is assumed to be true at time zero. The lack of correlation between two particles (i.e. factorisation (11)) up to the moment in which they collide is the way to conciliate a microscopic time-reversible dynamics with an evolution equation exhibiting increase of entropy and trend to equilibrium.

The evolution of the j-particle marginal distribution (10) is given by the following set of equations, called BBGKY hierarchy (after Bogoliubov [4], Born and Green [6], Kirkwood [19], Yvon [32]). It is obtained by integrating the Liouville equation (8) with respect to the variables $dz_{j+1}\dots dz_N$:

$$\begin{aligned}\partial_t f_j^N(t,z_1,\dots,z_j)&+\sum_{i=1}^{j} v_i\cdot\nabla_{x_i} f_j^N(t,z_1,\dots,z_j)\\&=(\mathscr{L}_j^{\varepsilon} f_j^N)(t,z_1,\dots,z_j)+(\mathscr{C}_{j+1}^{\varepsilon} f_{j+1}^N)(t,z_1,\dots,z_j),\end{aligned} \tag{12}$$

where

$$(\mathscr{L}_j^{\varepsilon} f_j^N)(t,z_1,\dots,z_j):=\frac{1}{\varepsilon}\sum_{i=1}^{j}\sum_{\substack{k=1\\k\neq i}}^{j}\nabla\Phi\left(\frac{x_i-x_k}{\varepsilon}\right)\cdot\nabla_{v_i} f_j^N(t,z_1,\dots,z_j), \tag{13}$$

$$\begin{aligned}&(\mathscr{C}_{j+1}^{\varepsilon} f_{j+1}^N)(t,z_1,\dots,z_j)\\&\quad:=\frac{(N-j)}{\varepsilon}\sum_{i=1}^{j}\int\nabla\Phi\left(\frac{x_i-x_{j+1}}{\varepsilon}\right)\cdot\nabla_{v_i} f_{j+1}^N(t,z_1,\dots,z_{j+1})\,dz_{j+1},\end{aligned} \tag{14}$$

where (14) is called *collision operator*.

Lanford's theorem. The first rigorous derivation of the Boltzmann equation in the low-density limit was given by Lanford [21], for hard-sphere potentials. To prove his result, Lanford studied the BBGKY hierarchy, describing the evolution of the marginals f_j^N, $j=1,\dots N$, and expressed $f_j^N(t)$ as a sum of operators acting on the sequence of initial data $f_j^N(0)$. To state in a precise way the result, we need to introduce some functional normed spaces, on which these operators act.

Definition 1 Let $X_{j,\beta}$ be the space of Borel functions f_j on $\mathscr{M}_j$ such that

$$\|f_j\|_{j,\beta}=\sup_{(z_1,\dots,z_j)\in\mathscr{M}_j}|f_j(z_1,\dots,z_j)|(\beta/2\pi)^{-\frac{3}{2}j}\exp(\beta H(z_1,\dots,z_j))<\infty,$$

where $H(z_1,\dots,z_j)$ is the Hamiltonian of the j particle system.

Definition 2 For all $b>0$, we define the space $X_{b,\beta}$ of sequences of functions $f=\{f_j\}_{j\geq 1}$ such that the following norm is finite

$$\|f\|_{b,\beta}=\sup_j b^{-j}\|f_j\|_{j,\beta}.$$

Remark 1 Observe that β and b can be interpreted respectively as the inverse of the temperature and the activity of the j-particle system, see [25].

Now we have all the ingredients to state Lanford's theorem:

Theorem 1 (Lanford 1975) *Given a system of N identical hard spheres of diameter* ε *and the set* $f^N = \{f_j^N\}_{1\leq j\leq N}$ *of associated j-particle marginals. Assume that:*

(i) *there exist positive constants b and* β *such that*

$$\|f^N(0)\|_{b,\beta} \leq C,$$

where C is an absolute constant, independent of N;

(ii) f_j *is continuous on the phase space* $\mathcal{M}_j$ *and*

$$\lim_{\varepsilon\to 0} f_j^N(0, x_1, \ldots, x_j, v_1, \ldots, v_j) = f(0)^{\otimes j}(x, v)$$

uniformly on compact sets in $\mathcal{M}_j$.

Then, there exists a strictly positive time $t_0 := [K\,\pi\, N\varepsilon^2 b\,\beta^{-1/2}]^{-1}$, *with K a positive constant, such that for* $0 < t < t_0$,

$$f_j^N(t, x_1, \ldots, x_j, v_1, \ldots, v_j) \to f(t)^{\otimes j}(x, v)$$

a.e. in the BG-limit, with $f(t, \cdot, \cdot)$ *the solution of the Boltzmann equation with initial datum* $f(0, \cdot, \cdot)$.

Remark 2 Notice that in the hard-sphere case, the system (6) is defined for a singular potential on the phase space

$$\mathcal{M}_N(\varepsilon) := \{(x_1, \ldots, x_N, v_1, \ldots, v_N) \in \mathbb{R}^{3N} \times \mathbb{R}^{3N} \ :\ |x_i - x_j| \geq \varepsilon,\ \text{for } i \neq j\} \tag{15}$$

and the j-particle marginals are defined accordingly to this constraint.

Remark 3 Observe that the Boltzmann equation only describes likely configurations, i.e. there are configurations which are out of the picture painted by Boltzmann. This justifies the almost everywhere convergence.

Although all the ideas of the proof were present in [21], some details were missing and they have been analysed in [8, 9, 13, 23, 27–30]. We will give a sketch of the proof of Theorem 1 in Sect. 2. A slightly different but detailed argument can be found in [13], Part II.

We stress that Lanford's result holds only for short time intervals, which are of the same order as the mean free time. This is a severe limitation, since in the applications of the Boltzmann equation a long-time behaviour of the solution is involved. One of the difficulties in extending the proof for long times is to prove that, once the j-particle marginals $f_j^N(0)$ are smooth, their evolutions $f_j^N(t)$ do not develop singularities. To our knowledge, the only situation in which the validity result for the non linear Boltzmann equation has been proved globally in time is the one analysed in [16, 17], where a rare cloud of gas expanding in the vacuum is considered. Nevertheless, the positive time t_0 in Lanford's theorem is large enough to observe a decrease of entropy in the Boltzmann H-functional. It is worth mentioning that recently a new quantitative point of view to study the correlations has been introduced in [24], where the authors consider a system of hard spheres in the BG-limit and introduce a set of functions measuring the correlation error. Although these objects seem to be more appropriate to identify and isolate the dynamical events responsible for the breakdown of propagation of chaos, the extension to long times of the validity of the Boltzmann equation is still far from being achieved. Recently, the validity for long times has been achieved in [2, 3], in the context of the linear Boltzmann equation in any dimension $d \geq 2$ and the linearised Boltzmann equation in dimension $d = 2$.

The second limitation of Lanford's theorem is the restriction to the hard-sphere interaction. In 1975 King presented a PhD thesis (unpublished, [18]) on the derivation of Eq. (1) for smooth short-range potentials. This problem has been considered a simple extension of Lanford proof, until it was recently reconsidered in [13]: there the authors proved rigorously that Eq. (1) can be obtained from a system of particles interacting via smooth positive short-range potentials. This is done through a sophisticated analysis and some further restrictions on the potential are needed. Hence, in [13] the authors have shown that the extension from hard-spheres to short-range potentials is a delicate and non trivial task.

The paper is organised as follows: in Sect. 2 we give a sketch of the proof of Theorem 1, where the derivation of the Boltzmann equation for the hard-sphere dynamics is presented; in Sect. 3 we give an idea of the main difficulties in extending Theorem 1 to the case of short-range potentials and we review the recent results obtained in [13, 23]; Sect. 4 is devoted to the open problem of the derivation of Eq. (1) in the case of long-range interactions.

2 Hard-Sphere Interaction

The aim of this section is to give an overview on the steps of Lanford's proof [21].

We consider a system of N particles, interacting as hard spheres of diameter ε on the phase space (15); we define the j-particle marginals associated to it and we compute their evolution in time, according to (12). An important observation is that, as already pointed out in [7], there is a formal similarity between the BBGKY hierarchy for hard spheres and the Boltzmann equation. Indeed, in the case of a hard-sphere interaction, the BBGKY reads as (12), with $\mathscr{L}_j^\varepsilon$ replaced by appropriate boundary conditions $\tilde{\mathscr{L}}_j^\varepsilon$ and C_{j+1}^ε is replaced by

$$(\tilde{\mathscr{C}}_{j+1}^\varepsilon f_{j+1}^N)(t, z_1, \dots, z_j) = (N-j)\varepsilon^2 \sum_{i=1}^{j} \int d\omega \int dv_{j+1}\, \omega \cdot (v_{j+1} - v_i) \times f_{j+1}^N(t, x_1, \dots, x_j, x_i + \varepsilon\omega, v_1, \dots, v_{j+1}). \tag{16}$$

To deal with the full differential hierarchy is a hard task. Indeed one has to deal with a family of N integro-differential equations, in the limit of N large. Hence, the idea of Lanford is to proceed in a perturbative way, by considering the temporal series solution of the BBGKY hierarchy, i. e. the Duhamel series

$$f_j^N(t, z_1, \dots, z_j) = \sum_{n=0}^{N-j} \alpha_n^\varepsilon(j) \int_0^t dt_1 \int_0^{t_1} dt_2 \dots \int_0^{t_{n-1}} dt_n \mathscr{S}_j^\varepsilon(t-t_1)\tilde{\mathscr{C}}_{j+1}^\varepsilon \mathscr{S}_{j+1}^\varepsilon(t_1-t_2) \dots \tilde{\mathscr{C}}_{j+n}^\varepsilon \mathscr{S}_{j+n}^\varepsilon(t_n) f_{j+n}^N(0, z_1, \dots, z_{j+n}), \tag{17}$$

where $\alpha_n^\varepsilon(j) := \varepsilon^{2n}(N-j)(N-j-1)\dots(N-j-n+1)$ is $O(1)$ in the BG-limit and $\mathscr{S}_j^\varepsilon(t)$ is the flow operator of the j-body hard-sphere dynamics. Roughly speaking, it behaves as the free flow up to the first impact time, then a collision occurs according to the scattering and the dynamics restarts as a free flow with the new outgoing configuration as initial condition up to the next impact time. Notice that, by conservation of energy, the operator $\mathscr{S}_j^\varepsilon(t)$ acts as a one-parameter group of isometries on the functional space $X_{j,\beta}$, i. e. $\|\mathscr{S}_j^\varepsilon(t) f_j\|_{j,\beta} = \|f_j\|_{j,\beta}$ for any β.

We want to compare (17) with $f_j(t, z_1, \dots, z_j)$, obtained as follows: let $f(t, z)$ be a solution to the Boltzmann equation, then

$$f_j(t, z_1, \dots, z_j) := \prod_{i=1}^j f(t, z_i)$$

is a solution to the following hierarchy of equations

$$\partial_t f_j + \sum_{i=1}^j v_i \cdot \nabla_{x_i} f_j = \mathscr{C}_{j+1} f_{j+1} \tag{18}$$

where

$$\mathscr{C}_{j+1} = \sum_{k=1}^j \mathscr{C}_{k,j+1}, \quad \mathscr{C}_{k,j+1} = \mathscr{C}_{k,j+1}^+ - \mathscr{C}_{k,j+1}^- \tag{19}$$

$$\mathscr{C}_{k,j+1}^+ f_{j+1}(t, z_1, \dots, z_j) = \int_{S_+^2} d\omega \int_{\mathbb{R}^3} dv_{j+1} \omega \cdot (v_k - v_{j+1}) f_{j+1}(t, z_1, \cdots, x_k, v_k', \cdots, z_j, x_k, v_{j+1}'),$$

$$\mathscr{C}_{k,j+1}^- f_{j+1}(t, z_1, \dots, z_j) = \int_{S_+^2} d\omega \int_{\mathbb{R}^3} dv_{j+1} \omega \cdot (v_k - v_{j+1}) f_{j+1}(t, z_1, \cdots, x_k, v_k, \cdots, z_j, x_k, v_{j+1}).$$

Again, we can apply iteratively Duhamel formula to get the series expansion

$$f_j(t, z_1, \ldots, z_j) = \sum_{n\geq 0} \int_0^t dt_1 \int_0^{t_1} dt_2 \ldots \int_0^{t_{n-1}} dt_n \mathscr{S}_j(t - t_1)\mathscr{C}_{j+1}\mathscr{S}_{j+1}(t_1 - t_2) \\ \ldots \mathscr{C}_{j+n}\mathscr{S}_{j+n}(t_n) f_{j+n}(0, z_1, \ldots, z_{j+n}), \tag{20}$$

where $\mathscr{S}_j(t)$ is the free-flow of j particles.

Lanford's proof is made of two parts:

(*a*) a proof of the absolute convergence of series expansions (17) and (20), uniformly in ε;
(*b*) a proof of the term by term convergence of one series to the other in the BG-limit.

The limitation to small times arises from point (*a*). Indeed, we prove the absolute convergence of the series by bounding the series by the geometric series $\sum_n (Ct)^n$, for which the convergence is achieved only when $|t| < 1/C$.

Step (*a*): absolute convergence of the series. In the first step we show that the series solutions exist, at least in a small time interval, by proving the absolute convergence of (17) and (20). We focus on Eq. (17), the procedure for Eq. (20) is analogous.

We observe that Eq. (17) expresses $f_j^N(t)$ as a sum of operators acting on the sequence of initial data $f_j^N(0)$, hence it is useful to set the problem on the functional spaces introduced in Definitions 1 and 2, on which the operators $\mathscr{S}_j^\varepsilon$ and $\tilde{\mathscr{C}}_{j+1}^\varepsilon$ act.

To prove the absolute convergence of the series we first prove the following

Proposition 1 *Let* $\beta > \beta' > 0$ *and* $b' > (\beta'/\beta)^{3/2} b$. *Then, for all* $f^N = \{f_j^N\}_{j\geq 1} \in X_{b,\beta}$, *there exists a constant* $K = K(\beta'/\beta, b'/b)$ *such that*

$$\sup_j b'^{-j} \| \mathscr{S}_j^\varepsilon(t - t_1)\tilde{\mathscr{C}}_{j+1}^\varepsilon \mathscr{S}_{j+1}^\varepsilon(t_1 - t_2) \ldots \tilde{\mathscr{C}}_{j+n}^\varepsilon \mathscr{S}_{j+n}^\varepsilon(t_n) f_{j+n}^N(0) \|_{j,\beta'} \leq n!\, t_0^{-n} \| f^N \|_{b,\beta}, \tag{21}$$

where $t_0 = [K\pi N\varepsilon^2 b\beta^{-1/2}]^{-1}$.

Remark 4 The time of validity t_0 is of the order of the mean-free time, defined as the ratio between the mean-free path and the mean square velocity.

Remark 5 Observe that we started from a functional space $X_{b,\beta}$ and we obtained $X_{b',\beta'}$. The loss is quantised by $\beta - \beta'$ and $b - b'$ and it will be compensated by integration in time (see Corollary 1). This is typical of Cauchy-Kovaleskaya proofs (see also [30]).

Proof We first estimate the term $\| \tilde{\mathscr{C}}_{j+1}^\varepsilon f_{j+1}^N \|_{b',\beta'}$. We have

$$|f_{j+1}^N(z_1, \ldots, z_{j+1})| \leq \| f_{j+1}^N \|_{j+1,\beta} (\beta/2\pi)^{\frac{3}{2}j} e^{-\beta \sum_{i=1}^{j+1} \frac{v_i^2}{2}}.$$

Hence,

$$|\tilde{\mathscr{C}}^{\varepsilon}_{j+1} f^N_{j+1}(z_1,\dots,z_j)| \\ \leq \pi N \varepsilon^2 \|f^N_{j+1}\|_{j+1,\beta} \int dv_{j+1} \sum_{i=1}^{j} (|v_i| + |v_{j+1}|) \left(\frac{\beta}{2\pi}\right)^{\frac{3}{2}j} e^{-\beta \sum_{i=1}^{j+1} \frac{v_i^2}{2}}.$$

Therefore, simple computations show that

$$\|\tilde{\mathscr{C}}^{\varepsilon}_{j+1} f^N_{j+1}\|_{j,\beta'} \leq \pi N \varepsilon^2 \left(\frac{\beta}{\beta'}\right)^{\frac{3}{2}j} \left[(4\pi/\beta)^{3/2} \frac{\sqrt{j}}{\sqrt{\beta-\beta'}} + j\frac{8\pi}{\beta^2}\right] \|f^N_{j+1}\|_{j+1,\beta} \tag{22}$$

so that $\tilde{\mathscr{C}}^{\varepsilon}_{j+1}$ is a bounded operator from $X_{j+1,\beta}$ to $X_{j,\beta'}$ for any $\beta > \beta'$. The bound we got depends on j as $j(\beta/\beta')^{\frac{3}{2}j}$, hence, for $b' > (\beta/\beta')^{3/2} b$, the sequence of operators $\{\tilde{\mathscr{C}}^{\varepsilon}_{j+1}\}_{j\geq 1}$ is a bounded operator from $X_{b,\beta}$ to $X_{b',\beta'}$.

Since $\mathscr{S}^{\varepsilon}_j(t)$ is an isometry on $X_{j,\beta}$, by iterating the argument above, we obtain the bound (21).

Corollary 1 *Let $b, b', \beta, \beta', t_0$ given as in Proposition 1. Then the nth term in* (17)

$$T^{\varepsilon}(j,n) = \mathscr{S}^{\varepsilon}_j(t-t_1)\tilde{\mathscr{C}}^{\varepsilon}_{j+1}\mathscr{S}^{\varepsilon}_{j+1}(t_1-t_2)\dots\tilde{\mathscr{C}}^{\varepsilon}_{j+n}\mathscr{S}^{\varepsilon}_{j+n}(t_n)$$

is an operator $T^{\varepsilon}(j,n) : X_{b,\beta} \to X_{b',\beta'}$, such that the following bound holds

$$\|T^{\varepsilon}(j,n) f^N\|_{b,\beta} \leq C\left(\frac{|t|}{t_0}\right)^n. \tag{23}$$

Therefore the series (17) *converges uniformly in ε, N and j, for $|t| < t_0$.*

Remark 6 Observe that we started from a functional space $X_{b,\beta}$ and we obtained $X_{b',\beta'}$. The loss is quantised by $\beta - \beta'$ and $b - b'$ and it will be compensated by integration in time (see Corollary 1). This is typical of Cauchy-Kovaleskaya proofs (see also [30]). Due to the singularity of the interaction, in the hard-sphere case this is a delicate argument that has been made rigours in [26] and in the erratum of [13].

Proof The time integrals can be easily bounded as follows

$$\int_0^t dt_1 \int_0^{t_1} dt_2 \dots \int_0^{t_{n-1}} dt_n \leq \frac{t^n}{n!}.$$

Observe that in the BG-limit $\alpha^{\varepsilon}_n(j) = O(1)$ uniformly in j. By virtue of Proposition 1, estimate (23) follows.

Step (b)**: term by term convergence**. To prove the convergence of each term of (17) to the corresponding term in (20), we need to look at the structure of each term of the series expansion. In order to have a clearer picture, it is useful to rewrite first (20) in a handier way, which expresses the terms of the series through binary trees. The analysis which follows is called *tree expansion* and it relays strongly on an interpretational effort while considering (20) (or (17)). Basically, we look at (20) and we consider the j particles $(z_1, \ldots, z_j)$ to have known positions and velocities. In the r.h.s., the integrand describes a collision process, in which a particle $j+1$ is added to the j fixed particles via the definition (19) of collision operator. The tree expansion is based on this interpretation of the iteration of transport flow and collision process. Precisely, for each j and n, we denote by $\Gamma(j,n)$ the binary tree with j roots and n nodes. For fixed j and n, each tree $\Gamma(j,n)$ represents a class of backwards trajectories

$$\zeta(s) = (\xi(s), \eta(s)), \quad s \in (0,t),$$

called the Boltzmann backwards flow (BBF), and specified by the collection of variables in the r.h.s. of (17). the j-particle configuration at time t is denoted by $(z_1, \ldots, z_j)$; n is the number of added particles; $t_1, \ldots, t_n$ are the times of creation of the added particles; $\boldsymbol{\nu}_1, \ldots, \boldsymbol{\nu}_n$ are the impact vectors of the added particles; $v_{j+1}, \ldots, v_{j+n}$ are the velocities of the added particles; $\sigma_1, \ldots, \sigma_n$ indicate the type of creation, i. e. outgoing when $\sigma_i = +$ or incoming when $\sigma_i = -$.

Then the Boltzmann series (20) can be rewritten as

$$f_j(t, z_1, \ldots, z_j) = \sum_{n \geq 0} \mathscr{T}(j,n), \tag{24}$$

with

$$\mathscr{T}(j,n) := \sum_{\Gamma(j,n)} \prod_{i=1}^{n} \sigma_i \int d\Lambda \left(\prod_{i=1}^{n} B_i \right) f_{j+n}(0, \zeta(0)),$$

where

$$d\Lambda = \mathbf{1}_{\{t_1 > t_2 > \cdots > t_n\}} dt_1 \ldots dt_n \, d\boldsymbol{\nu}_1 \ldots d\boldsymbol{\nu}_n \, dv_{j+1} \ldots dv_{j+n}$$

and for $i = 1, \ldots, n$

$$B_i = |\boldsymbol{\nu}_i \cdot (v_{j+i} - \eta_{k_i}(t_i^+))| \mathbf{1}_{\{\sigma_i \boldsymbol{\nu}_i \cdot (v_{j+i} - \eta_{k_i}(t_i^+)) \geq 0\}},$$

with k_i the index of the progenitor of particle $j+i$ in the binary tree (see [23]).

Analogously, Eq. (17) can be rewritten as

$$f_j^N(t, z_1, \ldots, z_j) = \sum_{n=0}^{N-j} \alpha_n^{\varepsilon}(j) \mathscr{T}^{\varepsilon}(j,n), \tag{25}$$

with

$$\mathcal{T}^{\varepsilon}(j,n) := \sum_{\Gamma(j,n)} \prod_{i=1}^{n} \sigma_i \int d\Lambda \left(\prod_{i=1}^{n} B_i^{\varepsilon} \right) f_{j+n}^{N}(0, \zeta^{\varepsilon}(0)),$$

where $\zeta^{\varepsilon}(s) = (\xi^{\varepsilon}(s), \eta^{\varepsilon}(s))$ is a backward in time flow associated to the particle dynamics, called the interacting backwards flow (IBF); the integral kernel is given by

$$B_i^{\varepsilon} = |\boldsymbol{v}_i \cdot (v_{j+1} - \eta_{k_i}(t_i^{+}))| \mathbf{1}_{\{\sigma_i \, \boldsymbol{v}_i \cdot (v_{j+1} - \eta_{k_i}(t_i^{+})) \geq 0\}} \mathbf{1}_{\{|\xi_{j+i}^{\varepsilon}(t_i) - \xi_k^{\varepsilon}(t_i)| > \varepsilon, \, \forall k \neq k_i\}},$$

with k_i the index of the progenitor of particle i in the binary tree. That is, the IBF ζ^{ε} is constructed analogously to the BBF ζ with the difference that, between two creations, the trajectories evolve according to the interaction operator $\mathcal{S}^{\varepsilon}$. Moreover, the created particles are added at distance ε from their progenitors in the tree.

By means of this expansion, the proof of step (b) reduces, via dominated convergence arguments, to the proof of a.e. convergence of the IBF to the BBF:

$$\zeta^{\varepsilon}(s) \to \zeta(s), \quad \text{a.e. with respect to } d\Lambda, \text{ for every } s \in (0,t) \tag{26}$$

where $t < t_0$ is given, with t_0 the limiting time obtained in step (a).

We on the generic terms of the two series (25) and (24), for fixed j and n, we consider the difference

$$|\alpha_n^{\varepsilon}(j)\mathcal{T}^{\varepsilon}(j,n) - \mathcal{T}(j,n)| \leq |(\alpha_n^{\varepsilon}(j) - 1)\mathcal{T}^{\varepsilon}(j,n)| + |\mathcal{T}^{\varepsilon}(j,n) - \mathcal{T}(j,n)|. \tag{27}$$

Since $\alpha_n^{\varepsilon}(j) \to 1$ in the BG-limit, the first term in the r.h.s. of Eq. (27) vanishes as $N \to \infty$ and $N\varepsilon^2 = O(1)$. As for the second term in the r.h.s. of Eq. (27), we split it as follows:

$$\begin{aligned}
&|\mathcal{T}^{\varepsilon}(j,n) - \mathcal{T}(j,n)| \\
&\quad \leq \Big| \sum_{\Gamma(j,n)} \sum_{\sigma_1,\ldots,\sigma_n} (-1)^{|\sigma|} \int d\Lambda \Big[\prod_{i=1}^{n} B_i^{\varepsilon} - \prod_{i=1}^{n} B_i\Big] f_{j+n}^{N}(0, \zeta^{\varepsilon}(0))\Big| \\
&\quad + \Big| \sum_{\Gamma(j,n)} \sum_{\sigma_1,\ldots,\sigma_n} (-1)^{|\sigma|} \int d\Lambda \prod_{i=1}^{n} B_i \,[f_{j+n}^{N}(0, \zeta^{\varepsilon}(0)) - f_{j+n}(0, \zeta^{\varepsilon}(0))]\Big| \\
&\quad + \Big| \sum_{\Gamma(j,n)} \sum_{\sigma_1,\ldots,\sigma_n} (-1)^{|\sigma|} \int d\Lambda \prod_{i=1}^{n} B_i \,[f_{j+n}(0, \zeta^{\varepsilon}(0)) - f_{j+n}(0, \zeta(0))]\Big|.
\end{aligned} \tag{28}$$

The second term on the r.h.s. of Eq. (28) vanishes thanks to hypothesis (ii) in Theorem 1; the first and the third term vanish by continuity and dominated convergence once we assume (26).

Hence, it remains to prove Eq. (26). We observe that, by construction, the IBF ζ^ε differs from the BBF ζ because from the one hand the particle flow is sensitive to small perturbations, so that a small variation of velocities may prevent a collision, producing a drastically different flow; on the other hand in the IBF two particles may undergo a recollision, that is a collision which is not a creation (i.e. a node of the binary tree), while in the BBF all the scattering events are creations of a new particle. The final argument in the proof consists in the verification that the set of integrated variables such that one of the two situations above occurs, has measure zero with respect to $d\Lambda$. The control of the recollision set is a delicate task and one has to do an accurate analysis of the recollision set. Observe that the recollision set is a non-countable union of zero-measure sets for $\varepsilon > 0$ ([8]). The proof is concluded by showing that the set of velocities and impact vectors of the particles added in the binary tree which lead to a recollision has a vanishing measure in the BG-limit.

3 Short-Range Interactions

In this section we present the recent results obtained in [13, 23]. These papers rely strongly on the ideas presented in [18, 21] and make use of the reduced marginals introduced by Grad. For this propose, we revert to general interaction potentials Φ, with the property of being compactly supported.

The new difficulties one has to face are essentially three: the long time scattering, the multiple collisions and the recollisions, which require a more careful analysis with respect to the hard-sphere case addressed by Lanford. Moreover, the appropriate objects to study are not the particle marginals, but the so called *reduced particle marginals*, as already noted in [18]. The notion of reduced marginal was introduced by Grad in [14] and it is asymptotically (in the BG-limit) equivalent to the one of marginal. The j-particle reduced marginal is defined as follows:

$$\tilde{f}_j^N(z_1, \ldots, z_j) := \int_{S(x_1,\ldots,x_j)^{N-j}} W^N(z_1, \ldots, z_N)\, dz_{j+1} \ldots dz_N, \tag{29}$$

where $S(x_1, \ldots, x_j) = \{(x, v) \in \mathbb{R}^3 \times \mathbb{R}^3 : |x - x_i| > \varepsilon, \text{ for all } i = 1, \ldots, j\}$. The evolution equations for the reduced marginals $\tilde{f}_j^N$ are obtained by integrating the Liouville Eq. (8) on the domain $S(x_1, \ldots, x_j)^{N-j}$ with respect to $dz_{j+1} \ldots dz_N$. This procedure leads to the following hierarchy:

$$(\partial_t + \sum_{i=1}^{j} v_i \cdot \nabla_{x_i})\tilde{f}_j^N(t) = \mathscr{L}_j^\varepsilon \tilde{f}_j^N + \sum_{m=0}^{N-j-1} \mathscr{A}_{j+1+m}^\varepsilon \tilde{f}_{j+1+m}^N, \tag{30}$$

with

$$\mathscr{A}^{\varepsilon}_{j+1+m}\tilde{f}^{N}_{j+1+m}(t)$$
$$= \alpha^{\varepsilon}_{m+1}(j)\sum_{i=1}^{j}\varepsilon^{2}\int_{S^{2}} d\boldsymbol{v}\mathbf{1}_{\{\min_{l=1,\dots,j;l\neq i}|x_i+\boldsymbol{v}\varepsilon-x_l|>\varepsilon\}}(\boldsymbol{v})\int_{\mathbb{R}^3} dv_{j+1}\,\boldsymbol{v}\cdot(v_{j+1}-v_i)$$
$$\times\int_{\Delta_m(x_{j+1})}\varepsilon^{-2m}\frac{dz_{j+2}\dots dz_m}{m!}\tilde{f}^{N}_{j+1+m}(t,z_1,\dots,z_j,x_i+\varepsilon\boldsymbol{v},v_{j+1},z_{j+2},\dots,z_m)$$

where $\alpha^{\varepsilon}_{m}(j)=\varepsilon^{2m}(N-j)(N-j-1)(\dots)(N-j-m+1)$ and

$$\Delta_m(x_1,\dots,x_j,x_i+\boldsymbol{v}\varepsilon):=$$
$$\{(z_{j+2},\dots,z_m)\in S(x_1,\dots,x_j)^m\ :\ \forall\, l=j+2,\dots,j+1+m,$$
$$\exists\, h_1,\dots,h_r\in\{j+2,\dots,j+1+m\}\text{such that }|x_l-x_{h_1}|\le\varepsilon,$$
$$|x_{h_{k-1}}-x_{h_k}|\le\varepsilon,\ \text{ for } k=2,\dots,r \text{ and } \min_{i\in\{l,h_1,\dots,h_r\}}|x_i-x_{j+1}|\le\varepsilon\}.$$

In particular, for $m=0$

$$\mathscr{A}^{\varepsilon}_{j+1}\tilde{f}^{N}_{j+1}=\varepsilon^{2}(N-j)\sum_{i=1}^{j}\int_{S^2}d\boldsymbol{v}\int_{\mathbb{R}^3}dv_{j+1}\,\mathbf{1}_{\{\min_{l=1,\dots,j;\ l\neq i}|x_i+\varepsilon\boldsymbol{v}-x_l|>\varepsilon\}}(\boldsymbol{v})$$
$$\times\,\boldsymbol{v}\cdot(v_{j+1}-v_i)\tilde{f}^{N}_{j+1}(t,z_1,\dots,z_j,x_i+\varepsilon\boldsymbol{v},v_{j+1})$$
$$=\varepsilon^{2}(N-j)C^{\varepsilon}_{j+1}\tilde{f}^{N}_{j+1}(t,z_1,\dots,z_j).$$

It is not difficult to prove that the contributions given by $m\ge 1$ (corresponding to multiple collisions) are negligible in the BG-limit (indeed, clearly $|\Delta_m|\sim O(\varepsilon^{3m})$ and $\mathscr{A}^{\varepsilon}_{j+1+m}\sim N^{m+1}\varepsilon^{2}\varepsilon^{3m}\sim\varepsilon^{m}$). For details, see Chap. 10 Part III in [13] or Sect. 3.1 in [23].

First, we report the main result achieved in [13]:

Theorem 2 (Theorem 5 in [13]) *Assume the repulsive potential Φ satisfies the following assumptions:*

(i) $\Phi:\mathbb{R}^3\to\mathbb{R}$ *is a radial, nonnegative, non increasing function supported in the unit ball of* $\mathbb{R}^3$*, of class* $\mathscr{C}^2$ *in* $\{x\in\mathbb{R}^3,\ 0<|x|<1\}$*, unbounded near zero, approaches zero as* $|x|\to 1^-$ *with bounded derivatives, and* $\nabla\Phi$ *vanishes only on* $|x|=1$*;*
(ii) *for* $|x|\in(0,1)$*,*

$$|x|\Phi''(|x|)+2\Phi'(|x|)\ge 0. \tag{31}$$

Let $f(0):\mathbb{R}^3\times\mathbb{R}^3\to\mathbb{R}_+$ *be a continuous density of probability such that for* $\beta>0$

$$\|f(0)\exp[\frac{\beta}{2}|v|^2]\|_{L^\infty}<\infty.$$

Consider the system of N particles, initially distributed according to f_0 and asymptotically independent, governed by Eq. (6). *Then, in the BG-limit, its distribution function converges to the solution of the Boltzmann equation* (1) *with a bounded cross-section, depending on Φ implicitly, and with initial data f_0, in the sense of observables, for short times.*

Remark 7 The convergence established by Theorem 2 is "in the sense of observables", that means convergence uniformly in t and x, after testing against a compactly supported function of v. Precisely, we say that f_j^N converges to f_j in the sense of the observables if, for any $\varphi \in \mathscr{C}_c^0(\mathbb{R}^{3j})$,

$$\int \varphi(v_1, \dots, v_j) f_j^N(z_1, \dots, z_N)\, dv_1 \dots dv_j \to \int \varphi(v_1, \dots, v_j) f_j(z_1, \dots, z_N)\, dv_1 \dots dv_j.$$

Remark 8 Item (ii) in Theorem 2 is a technical assumption due to the strategy adopted in the proof. Indeed, the authors need the scattering angle to be invertible in the impact parameter variable. Condition (31) ensures that the scattering angle is a monotone function of the impact parameter, and hence invertible (see also Appendix in [23] for a detailed explanation).

From a physical point of view, the assumptions on the class of potentials for which the Boltzmann equation has been proved to hold is not satisfying, since it is heuristically expected to be valid independently of the details of the scattering.

We are now ready to state the following

Theorem 3 (Theorem 1 in [23]) *Consider a two-body radial potential $\Phi : \mathbb{R}^3 \to \mathbb{R}$ supported in $|q| < 1$ and non increasing in $|q|$. We assume*

(i) either $\Phi \in \mathscr{C}^2(\mathbb{R}^3)$, or $\Phi \in \mathscr{C}^2(\mathbb{R}^3 \setminus \{0\})$ and $\Phi(|x|) \to \infty$, as $|x| \to 0$;

(ii) the initial data of the Boltzmann equation $f(0) : \mathbb{R}^3 \times \mathbb{R}^3 \to \mathbb{R}_+$ is a probability density, continuous and such that, for $\beta > 0$,

$$\| f(0) \exp[\frac{\beta}{2}|v|^2] \|_{L^\infty(\mathbb{R}^3 \times \mathbb{R}^3)} < \infty.$$

(iii) for any N, $W^N(0)$ is a probability density on the phase space $\mathscr{M}_N$, symmetric in the exchange of particles, with reduced marginals $\{\tilde{f}_j^N(0)\}_{\{j=1\}}^N$ such that $\|\tilde{f}_j^N(0)\|_{j,\beta} < e^{bj}$, for $b, \beta > 0$ and given $\tilde{f}_j^N(0)$ and $f_j(0) = f^{\otimes j}(0)$, we assume

$$\lim_{\varepsilon \to 0} \tilde{f}_j^N(0) = f_j(0),$$

in the BG-limit, uniformly on compact sets in $\mathscr{M}_j$.

Then, there exists $t_0 > 0$ such that, $\forall t < t_0$ and $\forall j \in \mathbb{N}$, $\tilde{f}_j^N(t)$ and $f_j(t) = f(t)^{\otimes j}$ exist and

$$\lim_{\varepsilon \to 0} \tilde{f}_j^N(t) = f_j(t),$$

in the BG-limit, uniformly on compact sets in $\Omega_j = \{(z_1, \ldots, z_j) \in \mathcal{M}_j : (x_i - x_k) \wedge (v_i - v_k) \neq 0\}$, *with* $f(t)$ *solution to* (1) *with initial datum* $f(0)$.

The key ingredient here is to consider the formulation (4) for the collision operator, which does not require the inverse of the scattering angle to exist as a single-valued function. Roughly speaking, the problem in considering the formulation (2) is to invert the map $\boldsymbol{\nu} \to \omega$. This is a big technical and conceptual difference with respect to the hard-sphere case, in which $\omega = \boldsymbol{\nu}$.

Sketch of the proof of Theorem 3. Following Lanford's proof, we want to compare $\tilde{f}_j^N$ and f_j. Step (a) is achieved exactly as in Sect. 2, according to the new definition (29) and the formulation (30). Obviously, the time restriction in Theorem 3 is a consequence of step (a). As for step (b), we have to compare the IBF and the BBF. In particular, we want to show that, even for smooth short-range potentials, the sets which lead to a dynamics which is not close to the one of the Boltzmann flow are negligible in the BG-limit.

In the case of short-range potentials, the IBF differs from the BBF because:

- collisions occur at distance ε;
- recollisions may occur;
- the scattering is not instantaneous;
- multiple collisions may occur.

As consequence of the third point, we have to carefully analyse the low energy collisions, the high energy collisions, the central collisions and the recollisions. In particular, a dramatic difference may occur if: a particle created in the IBF interacts for long time with its progenitor; a couple of particles in the IBF undergoes a recollision. We study each event separately:

(a) *A particle in the IBF interacts for long time with its progenitor*. This issue was not present in the hard-sphere case. It is overcome by cutting-off the impact vectors and the velocities $(\boldsymbol{\nu}_i, v_{j+1})$ leading to the singular scattering and by showing that the contribution they give to the integrals is small in the BG-limit. To prove that, we need to estimate the scattering time t_*. As it is shown in Lemma 1 in [23], it can be bounded as follows:

$$t_* \leq \frac{A}{\rho V}\varepsilon, \tag{32}$$

where A is an absolute constant, ρ is the impact parameter and V is the relative velocity before the scattering takes place. Hence the scattering time may be too long if the relative velocity involved in the bound (32) is small or if the impact parameter ρ is close to zero (i. e. a central collision occurs). To avoid these pathologies, we cut off small relative velocities and the parameters (ν_i, v_{j+1}) leading to a central collision. The contribution given to the integrals by the set of cut off variables is negligible in the BG-limit.

(b) *A particle has a very large velocity*. This occurence is present in the hard-sphere case too and it is controlled in the same way, by cutting-off the large values of $|(v_{j+1}, \ldots, v_{j+n})|$. The integral over the cut-off region is small because $f_{j+n}^N \in X_{j+n,\beta}$.

(*c*) *A couple of particles in the IBF undergoes a recollision.* This is the most delicate task, because concentrations of measure in the differential cross-section may occur so that the integral over negligible sets can give a contribution of positive measure. Here we just give an idea of the main issues and we refer to [23] Sect. 7.2 for a detailed description of the technical part. We need to demonstrate that the contribution of recolliding trajectories is negligible in the limit $\varepsilon \to 0$. To do that, the strategy adopted in [23] is based on three main ideas: (i) to work on the BBF instead of looking at the IBF and to exploit its simpler structure; (ii) to perform the integrals on the time variables; (iii) to keep using $\boldsymbol{\nu}$ instead of switching to ω. Because of this latter point, the Boltzmann collision operator emerges in the form (4) rather than in the usual formulation given by (2).

First, we define by words the set

$$\mathscr{N}(\delta) := \{\text{a couple of particles in the BBF is getting closer than } \delta > \varepsilon\},$$

where δ is chosen as a function of ε, vanishing as $\varepsilon \to 0$. We observe that

$$\lim_{\delta \to 0} \mathbf{1}_{\mathscr{N}(\delta)} = \mathbf{1}_{\mathscr{N}},$$

with $\mathscr{N} :=$ {couples of particles in the BBF which recollide pointwise} and $\mathscr{N}$ is a zero-measure set with respect to the measure $d\Lambda$. This is shown by making use of the time integrals $\int_0^t dt_1 \int_0^{t_1} dt_2 \ldots \int_0^{t_{n-1}} dt_n$. Hence, we are left with the control of the contribution given by the complement of the set $\mathscr{N}(\delta)$, defined as $\mathscr{N}(\delta)^c$. We notice that in $\mathscr{N}(\delta)^c$ the BBF is close to the IBF when the scattering times are small and the velocities are not large (we already cut off the long scattering times in (a) and large velocities in (b)). □

Following the sketch of proof above, it is possible to extend Theorem 3 to stable short-range potentials:

Theorem 4 (Theorem 1' in [23]) *Let $\Phi(q)$ be a stable radial potential, with support $|q| < 1$. Under the Hypotheses $(i) - -(iii)$ in Theorem 3, there exists $t_0 > 0$ such that, for any positive $t < t_0$ and $j \in \mathbb{N}$, the series expansions are absolutely convergent (uniformly in ε), and*

$$\lim_{\substack{\varepsilon \to 0 \\ N\varepsilon^2 = 1}} f_j^N(t) = f_j(t) \tag{33}$$

uniformly on compact sets in Ω_j.

Remark 9 The lack of explicit estimates in the proof of the above Theorem is due to the difficulty in reproducing a bound of type (32) in the case of stable potentials, due to the possible presence of trapping orbits in the attractive region.

Under the assumptions of Theorem 3 on the potential and further assumptions on the initial data, it is possible to compute explicitly the rate of convergence:

Theorem 5 (Theorem 2 in [23]) *Assume the hypotheses of Theorem 3 to hold. Moreover, assume that the potential Φ is non-increasing and that:*

$$\sup_{|x_i - x_k| > \varepsilon} e^{\beta \sum_{i=1}^{j} \frac{v_i^2}{2}} |f_j^N(0) - f(0)^{\otimes j}| \leq C^j \varepsilon,$$

$$e^{\beta \frac{v^2}{2}} |f(0, x, v) - f(0, x', v)| \leq L|x - x'|, \text{ for some } L \geq 0.$$

Then, for $t \in [0, t_0)$, $(z_1, \ldots, z_j) \in \Omega_j$, there exist constants $C, \gamma > 0$, such that for any $j \geq 1$ and ε small enough, the following estimates hold

$$|f_j^N(t, z_1, \ldots, z_j) - f(t)^{\otimes j}(z_1, \ldots, z_j)| \leq C^j \varepsilon^\gamma, \quad \gamma < \frac{1}{6}.$$

4 Beyond the Short Range

Apart from the long time validity, the other interesting and natural open question concerning the derivation of the Boltzmann equation is whether the results [13, 21, 23] can be extended to the case of long-range interactions. From a phenomenological point of view, it should be possible to show that the Boltzmann equation emerges from the microscopic classical dynamics, at least for potentials of the form $\Phi(|x|) = \frac{1}{|x|^\alpha}$, for an appropriate choice of α. Heuristically, this was justified by Maxwell in his paper [22], where he proposed Eq. (1) (tested against a smooth function of the velocity variable) to be a good approximation of the dynamics of a rarefied gas with intermolecular force an inverse power law potential. The question here is to make rigorous Maxwell's argument, for a reasonable class of long-range potentials. It has been investigated in [11] in the simpler linear case.

In this section we want to give an overlook on this open problem, underlining the difficulties one has to face.

The first obstacle one has to cope with is to define the scaling limit. Indeed, in the case of hard-core and short-range potentials, the scale parameter ε, which goes to zero, represents the diameter of particles or the range of the interaction, respectively. Because of the long tail of the potential, ε cannot represent the range of interaction anymore. Therefore, a revised version of the BG-limit seems to be necessary to state the problem in a rigorous mathematical way, taking into account the mean-field effects appearing at large distances in the long-range interaction.

The second difficulty is to show the well-posedness for the BBGKY hierarchy. In fact, consider the particular case of an interaction given by an hard-sphere dynamics plus a long tail Φ for $|x| > \varepsilon$. Under the usual hyperbolic scaling of space and time (9), a new term related to the long tail of the potential appears in the hierarchy:

$$(\partial_t + \sum_{i=1}^{j} v_i \cdot \nabla_{x_i}) f_j^N = \tilde{\mathscr{L}}_j^\varepsilon f_j^N + \tilde{\mathscr{C}}_{j+1}^\varepsilon f_{j+1}^N + \mathscr{L}_j^\varepsilon f_j^N$$
$$+ \frac{N-j}{\varepsilon} \sum_{i=1}^{j} \int dx_{j+1} \int dv_{j+1} \nabla_{x_i} \Phi \left(\frac{x_i - x_{j+1}}{\varepsilon} \right) \cdot \nabla_{v_i} f_{j+1}^N, \tag{34}$$

where we used the notations introduced in Sect. 2. The difficulty here is to get a priori estimates on the derivative with respect to v_i of the reduced marginal f_{j+1}^N. One could use the ideas proposed by Maxwell in his heuristic presentation of the Boltzmann equation, based on the convergence in the sense of observables (as defined in Remark 7), i.e. the weak formulation may help to give sense to the third term on the r.h.s. of Eq. (34).

The situation becomes even more problematic when looking at the Coulomb potential $\Phi(|x|) = \frac{1}{|x|}$ ($\alpha = 1$). In this case the collision integral in the r.h.s. of the Boltzmann equation makes no sense whatever choice of f. This suggests to replace the Boltzmann equation by a different model. Indeed, the slow decay at infinity of the potential makes the so-called grazing collisions to be of leading importance in the macroscopic behaviour of the gas. This problem was pointed out in 1936 by Landau [20], who proposed a modified equation to describe the effect of grazing collisions. The Landau equation reads

$$(\partial_t + v \cdot \nabla_x) f(t,x,v)$$
$$= \nabla_v \cdot \int_{\mathbb{R}^3} dv_* \frac{P_{(v-v_*)^\perp}}{|v - v_*|} \{ f(t,x,v_*) \nabla_v f(t,x,v) - f(t,x,v) \nabla_{v_*} f(t,x,v_*) \}. \tag{35}$$

where $P_{(w)^\perp}$ is the orthogonal projection on the subspace orthogonal to $w \in \mathbb{R}^3$.

Up to now, there are only very few mathematical results about the Landau model (35). The validity problem and the well-posedness of the equation are open questions of primary interest and importance, especially so because of the several applications involving the Landau equation. The Cauchy problem associated to the homogeneous Landau equation has been studied in [10, 31], where weak solutions are proven to exist. Uniqueness is proved in [12] once the solution is known to belong to L^∞. In the non-homogeneous case, the only available result is due to Guo [15], who proved that there exists a global unique classical solution of (35) for small perturbations of the equilibrium.

As for the validity problem, the only attempt to derive the Landau equation from the Hamiltonian dynamics is contained in [1], where a consistency result is achieved, in the weak-coupling limit starting from a system of N particles interacting via a rescaled smooth short-range potential. In this regime, a given particle undergoes a huge number of collisions in the kinetic time, but the two-body potential is weakened, and hence the variance of the total momentum variation remains finite. The key idea in [1] is based on the fact that f_j^N cannot be smooth. If it were, we would have a trivial free dynamics. Hence we make the ansatz

$$f_j^N = g_j^N + \gamma_j^N$$

where g_j^N is smooth and γ_j^N is strongly oscillating. This allows to find a system of coupled equations for g_j^N and γ_j^N. In particular, the equation for γ_j^N can be solved in terms of g_j^N. This leads to the following hierarchy:

$$\begin{aligned} g_j^N(t) &= \mathscr{S}(t) f_j(0) + \frac{N-j}{\sqrt{\varepsilon}} \int_0^t d\tau\, \mathscr{S}(t-\tau)\, \mathscr{C}_{j+1}^\varepsilon g_{j+1}^N(\tau) \\ &+ \frac{N-j}{\varepsilon} \int_0^t d\tau \int_0^\tau d\sigma\, \mathscr{S}(t-\tau)\, \mathscr{C}_{j+1}^\varepsilon\, \mathscr{U}_{j+1}^\varepsilon(\tau-\sigma)\, \mathscr{T}_{j+1}^\varepsilon\, g_{j+1}^N(\sigma), \end{aligned} \tag{36}$$

where $\mathscr{S}$ is the generator of the free-flow, $\mathscr{U}_{j+1}^\varepsilon$ is the generator of the evolution of γ_j^N, $\mathscr{C}_{j+1}^\varepsilon$ is a collision operator and $\mathscr{T}_{j+1}^\varepsilon$ is the Liouville operator restricted to $j+1$ particles. By perturbing (36) up to the second order in time, we obtain $g_j^N(t) \to f(t)^{\otimes j}$, where $f(t)$ is a solution to the Landau equation, and $\gamma_j^N \to 0$, where the convergence has to be understood in distributional sense.

This result is not fully satisfactory because of the lack of control of higher order terms. Therefore, the rigorous mathematical validity of the Landau equation is an open problem, even for small time intervals.

Acknowledgments The author is supported by the fund "Forschungskredit UZH FK-15-108".

References

1. Bobylev, A.V., Pulvirenti, M., Saffirio, C.: From particle systems to the Landau equation: a consistency result. Comm. Math. Phys. **319**(3), 683–702 (2013)
2. Bodineau, T., Gallagher, I., Saint-Raymond, L.: The Brownian motion as the limit of a deterministic system of hard-spheres. arXiv:1305.3397
3. Bodineau, T., Gallagher, I., Saint-Raymond, L. : From hard sphere dynamics to the Stokes-Fourier equations: an L^2 analysis of the Boltzmann-Grad limit. arXiv:1511.03057
4. Bogoliubov, N.: Problems of dynamical theory in statistical physics. In: de Boer, J., Uhlenbeck, G.E. (eds.) Studies in Statistical Mechanics. Interscience, New York (1962)
5. Boltzmann, L.: Lectures on Gas Theory. English edition annotated by Brush, S. University of California Press, Berkeley (1964, reprint)
6. Born, M., Green, H.S.: A general kinetic theory of liquids. I. The molecular distribution functions. Proc. Roy. Soc. Lond. A **188**, 10–18 (1946)
7. Cercignani, C.: On the Boltzmann equation for rigid spheres. Transp. Theor. Stat. Phys. **2**, 211–225 (1972)
8. Cercignani, C., Gerasimenko, V.I., Petrina, D.I.: Many-Particle Dynamics and Kinetic Equations. Kluwer Academic Publishers, Netherlands (1997)
9. Cercignani, C., Illner, R., Pulvirenti, M.: The Mathematical Theory of Dilute Gases. Applied Mathematical Sciences, vol. 106. Springer, New York (1994)
10. Desvillettes, L.: Entropy dissipation estimates for the Landau equation in the Coulomb case and applications. arXiv:1408.6025
11. Desvillettes, L., Pulvirenti, M.: The linear Boltzmann equation for long-range forces: a derivation from particle systems. Model. Methods Appl. Sci. **9**, 1123–1145 (1999)

12. Fournier, N.: Uniqueness of bounded solutions for the homogeneous Landau equation with a Coulomb potential. Commun. Math. Phys. **299**(3), 765–782 (2010)
13. Gallagher, I., Saint Raymond, L., Texier, B.: From Newton to Boltzmann: Hard Spheres and Short-Range Potentials. Zürich Lectures in Advanced Mathematics. European Mathematical Society (EMS), Zürich (2013)
14. Grad, H.: On the kinetic theory of rarefied gases. Commun. Pure Appl. Math. **2**, 331–407 (1949)
15. Guo, Y.: The Landau equation in a periodic box. Commun. Math. Phys. **231**(3), 391–434 (2002)
16. Illner, R., Pulvirenti, M.: Global validity of the Boltzmann equation for a two-dimensional rare gas in the vacuum. Commun. Math. Phys. **105**, 189–203 (1986)
17. Illner, R., Pulvirenti, M.: Global validity of the Boltzmann equation for a two- and three-dimensional rare gas in vacuum: erratum and improved result. Commun. Math. Phys. **121**, 143–146 (1989)
18. King, F.: BBGKY Hierarchy for Positive Potentials, Ph.D. thesis, Department of Mathematics, University of California, Berkeley (1975)
19. Kirkwood, J.G.: The statistical mechanical theory of transport process I. General theory. J. Chem. Phys **14**, 180–202 (1946)
20. Landau, L.D.: Kinetic equation in the case of Coulomb interaction. Phys. Zs. Sow. Union **10**, 154 (1936). (in German)
21. Lanford, O.E.: Time evolution of large classical systems. In: Moser, J. (ed.) Dynamical Systems, Theory and Applications. Lecture Notes in Physics, vol. 38, pp. 1–111. Springer, Berlin (1975)
22. Maxwell, J.C.: On the dynamical theory of gases. Philos. Trans. Roy. Soc. Lond. Ser. A **157**, 49–88 (1867)
23. Pulvirenti, M., Saffirio, C., Simonella, S.: On the validity of the Boltzmann equation for short range potentials. Rev. Math. Phys. **26**(2), 64 (2014)
24. Pulvirenti, M., Simonella, S.: The Boltzmann-Grad limit of a hard sphere system: analysis of the correlation error. arXiv:1405.4676
25. Ruelle, D.: Statistical Mechanics: Rigorous Results. Imperial College Press, London (1999). Reprint of the 1989 edition. World Scientific Publishing Co. Inc., River Edge, NJ
26. Simonella, S.: Evolution of correlation functions in the hard sphere dynamics. J. Stat. Phys. **155**(6), 1191–1221 (2004)
27. Spohn, H.: Boltzmann equation and Boltzmann hierarchy. In: Cercignani, C. (ed.) Kinetic Theories and the Boltzmann Equation. Lecture Notes in Mathematics, vol. 1048, pp. 207–220. Springer, Berlin (1984)
28. Spohn, H.: On the Integrated Form of the BBGKY Hierarchy for Hard Spheres. arXiv: 0605068v1
29. Uchiyama, K.: Derivation of the Boltzmann equation from particle dynamics. Hiroshima Math. J. **18**, 245–297 (1988)
30. Ukai, S.: The Boltzmann-Grad limit and Cauchy-Kovalevskaya theorem. Jpn. J. Ind. Appl. Math. **18**, 383–392 (2001)
31. Villani, C.: On a new class of weak solutions to the spatially homogeneous Boltzmann and Landau equations. Arch. Rational Mech. Anal. **143**(3), 273–307 (1998)
32. Yvon, J.: La théorie statistique des fluides et l'équation d'état. Actualités scientifiques et industrielles. Hermann, Paris (1935)

Duality Relations for the Periodic ASEP Conditioned on a Low Current

G.M. Schütz

Abstract We consider the asymmetric simple exclusion process (ASEP) on a finite lattice with periodic boundary conditions, conditioned to carry an atypically low current. For an infinite discrete set of currents, parametrized by the driving strength s_K, $K \geq 1$, we prove duality relations which arise from the quantum algebra $U_q[\mathfrak{gl}(2)]$ symmetry of the generator of the process with reflecting boundary conditions. Using these duality relations we prove on microscopic level a travelling-wave property of the conditioned process for a family of shock-antishock measures for $N > K$ particles: If the initial measure is a member of this family with K microscopic shocks at positions $(x_1, \ldots, x_K)$, then the measure at any time $t > 0$ of the process with driving strength s_K is a convex combination of such measures with shocks at positions $(y_1, \ldots, y_K)$, which can be expressed in terms of K-particle transition probabilities of the conditioned ASEP with driving strength s_N.

Keywords Asymmetric simple exclusion process · Duality · Quantum algebra

1 Introduction

In the asymmetric simple exclusion process (ASEP) [26, 27, 34, 36] each lattice site k on a lattice $\Lambda = (1, \ldots, L)$ is occupied by at most one particle, indicated by occupation numbers $\eta(k) \in \mathbb{S} = \{0, 1\}$. We denote by $\eta = (\eta(1), \ldots, \eta(L)) \in \Omega = \mathbb{S}^L$ a configuration η of the particle system. Informally speaking, in one dimension particles try to jump to the right with rate $r = wq$ and to the left with rate $\ell = wq^{-1}$. The jump attempt is successful if the target site is empty, otherwise the jump attempt is rejected. The invariant measures of the ASEP with periodic boundary conditions are well-known: For fixed particle number N these are the uniform measures. From these one can construct the grand canonical Bernoulli product measures with fugacity $z = \rho/(1-\rho)$ where $\rho = N/L$ is the particle density on the torus. For these measures,

G.M. Schütz (✉)
Institute of Complex Systems II, Forschungszentrum Jülich, 52425 Jülich, Germany
e-mail: g.schuetz@fz-juelich.de

© Springer International Publishing Switzerland 2016

P. Gonçalves and A.J. Soares (eds.), *From Particle Systems to Partial Differential Equations III*, Springer Proceedings in Mathematics & Statistics 162,
DOI 10.1007/978-3-319-32144-8_16

where each lattice site k is occupied with probability ρ independently of all other sites, one has a stationary particle current $j^* = (r - \ell)\rho(1 - \rho)$, corresponding to an expected mean time-integrated current $\langle J(t) \rangle / t = j^*$.

In the context of macroscopic fluctuation theory [8] one is interested in conditioning the process on fluctuations around some atypical mean time-integrated current $j \neq j^*$. A question of fundamental interest is then which macroscopic density profile is most likely to realize such a large deviation of the current inside a very large (more precisely: infinite) time interval of conditioning. This large-deviation problem thus concerns an untypical ensemble of trajectories of the process. This ensemble is usually not defined by j but via Legendre transformation in terms of the canonically conjugate driving strength $s(j)$ with $s(j^*) = 0$. Interestingly, for conditioning on a lower-than-typical current (i.e., for $s < 0$), it was found by Bodineau and Derrida [9] for the weakly asymmetric simple exclusion process that there is a dynamical phase transition: For currents slightly below the typical value j^* the optimal macroscopic profile is constant as it is for j^*. However, below a critical threshold $j_c < j^*$ (corresponding to some $s_c < 0$) the optimal macroscopic profile is a travelling wave with a shape resembling a smoothened shock/antishock pair.

More recently, in a similar setting, but for finite duration t of conditioning, the microscopic structure of a travelling wave in the ASEP (not weakly!) was elucidated in detail for a specific choice of negative driving strength [5]: One considers a certain family of inhomogeneous product measures μ_k, indexed by a lattice site k, where the microscopic density profile as function of the position on the lattice has a density-jump at position k on the torus, analogous to a shock on macroscopic scale. At time $t = 0$ N particles (N arbitrary) are distributed according to the restricted measure $\mu_k^N \propto \mu_k \delta_{\sum_k \eta(k), N}$. Then at any future time $t > 0$ of the conditioned dynamics the measure is a convex combination $\mu_k^N(t) = \sum_l c(l, t|k, 0)\mu_l^N$ of such measures. The weights $c(l, t|k, 0)$ are the transition probabilities of a single biased random walk, thus suggesting that a shock in a macroscopic travelling wave performs a biased random walk on microscopic level.

In this work we trace back the mathematical origin of this rigorous result to certain algebraic properties of the generator of the process. Then, using these properties, we go beyond [5] to derive a family of duality relations that allow us to construct more complex microscopic structures corresponding to more general macroscopic optimal profiles. The starting point is the well-known fact that for reflecting boundaries, where the process is reversible, the generator of the process commutes with the generators of the quantum algebra $U_q[\mathfrak{gl}(2)]$ [1]. This fact has been used in [31] to construct the canonical reversible measures and in [33] to derive self-duality relations for the unconditioned ASEP.[1]

[1] We mention that the deep link between duality of Markov processes and symmetries of its generator, first noted in [32], that we exploit here was given a systematic abstract treatment in [21]. More recently many concrete symmetry-based dualities for interacting particle systems were derived using this approach [7, 10, 12–15, 17, 25, 29].

On the torus, however, the symmetry of the generator under $U_q[\mathfrak{gl}(2)]$ breaks down. Nevertheless, some time ago Pasquier and Saleur [30] found intertwining relations involving the generators of $U_q[\mathfrak{gl}(2)]$ and the Heisenberg quantum Hamiltonian with a boundary twist. This quantum Hamiltonian operator became later to be known to be closely related to the generator of the conditioned ASEP [18]. Here we present a new proof for the results of [33] (and correct some typos there) for reflecting boundaries and make use of intertwining relations of [30] (correcting some typos also in that paper) to derive an infinite discrete family of duality relations for the ASEP with *periodic* boundary conditions. These new duality relations apply to the process conditioned on fluctuations around some untypically low mean time-integrated current.

The simplest of these duality relations proves that the homogeneous Bernoulli measure is the invariant measure for the unconditioned ASEP with periodic boundary conditions. The derivation of this well-known fact from the $U_q[\mathfrak{gl}(2)]$-symmetry of the process with reflecting boundary questions is remarkable in so far as it raises the interesting question whether one can construct the matrix product measures [19, 20] of the periodic multi-species ASEP from the $U_q[\mathfrak{gl}(n)]$-symmetry of that process with reflecting boundaries [1, 6].

From the non-trivial higher order duality relations we obtain an infinite discrete family of new microscopic "travelling waves" for the conditioned process.

2 Definitions and Notation

It is convenient to work with the quantum Hamiltonian formalism [28, 34] where the generator of the process is represented by a matrix which in a judiciously chosen basis turns out to be closely related to the Hamiltonian operator of a physical quantum system. We first introduce some notation and then describe in some detail the tools required for the quantum Hamiltonian formalism for the benefit of readers not familiar with this approach.

2.1 State Space and Configurations

We say that a site $k \in \Lambda$ is occupied by a particle if $\eta(k) = 1$ or that it is empty if $\eta(k) = 0$. The fact that a site can be occupied by at most one particle is the exclusion principle. Occasionally we denote configurations with a fixed number of N particles by η_N. The set of all configurations with N particles is denoted Ω_N. We also define

$$\upsilon(k) := 1 - \eta(k) \tag{1}$$

and the particle numbers

$$N(\eta) = \sum_{k=1}^{L} \eta(k), \quad V(\eta) = \sum_{k=1}^{L} \upsilon(k) = L - N(\eta). \tag{2}$$

A useful alternative way of presenting uniquely of configuration η_N is obtained by labelling the particles consecutively from left to right (clockwise) by 1 to N and their positions on Λ by $x_i \bmod L$. A configuration η is then represented by the set $\mathbf{x} := \{x : \eta(x) = 1\}$. We call this notation the position representation. We shall use interchangeably the arguments η, $\mathbf{x}$, for functions of the configurations. When the argument is clear from context it may be omitted. We note the trivial, but frequently used identities $N(\eta) \equiv N(\mathbf{x}) = |\mathbf{x}|$ and

$$\eta(k) = \sum_{i=1}^{N(\mathbf{x})} \delta_{x_i,k}. \tag{3}$$

For a configuration $\eta \equiv \mathbf{x}$ we also define the number $N_k(\eta)$ of particles to the left of a particle at site k

$$N_k(\eta) := \sum_{i=1}^{k-1} \eta(i) = \sum_{i=1}^{N(\eta)} \sum_{l=1}^{k-1} \delta_{x_i,l}. \tag{4}$$

Furthermore, for $1 \leq k \leq L-1$ we define the local permutation

$$\pi^{kk+1}(\eta) = \{\eta(1), \ldots \eta(k-1), \eta(k+1), \eta(k), \eta(k+2), \ldots, \eta(L)\} =: \eta^{kk+1}, \tag{5}$$

and for $k = L$ we define

$$\pi^{L1}(\eta) = \{\eta(L), \ldots, \eta(k), \ldots, \eta(1)\} =: \eta^{L1}. \tag{6}$$

The space reflection is defined by

$$R(\eta) = \{\eta(L), \eta(L-1), \ldots, \eta(1)\} \tag{7}$$

corresponding to $R(\eta(k)) = \eta(L+1-k)$ for the occupation numbers.

2.2 Definition of the ASEP

For functions $f : \mathbb{S}^L \to \mathbb{C}$ the ASEP η_t with periodic boundary conditions and hopping asymmetry q is defined by the generator

$$\mathscr{L} f(\eta) := {\sum}'_{\eta' \in \mathbb{S}^L} w(\eta \to \eta')[f(\eta') - f(\eta)] \tag{8}$$

where the transition rates between configurations

$$w(\eta \to \eta') = \sum_{k=1}^{L} w^{kk+1}(\eta)\delta_{\eta',\eta^{kk+1}} \tag{9}$$

are defined in terms of the local hopping rates

$$w^{kk+1}(\eta) = w\left[q\eta(k)\upsilon(k+1) + q^{-1}\upsilon(k)\eta(k+1)\right]. \tag{10}$$

The prime at the summation symbol (8) indicates the absence of the term $\eta' = \eta$ which is omitted since $w(\eta \to \eta)$ is not defined.[2] The transition rates are non-zero only for a transition from a configuration η to a configuration $\eta' = \eta^{kk+1}$ defined by (5).

We shall assume partially asymmetric hopping $q \neq 0, 1, \infty$. The constant $w \neq 0$ sets the time scale of the process. On the torus we identify increasing order of the lattice index with the clockwise direction. In the case of reflecting boundary conditions no jumps from site 1 to the left and no jumps from site L to the right are allowed. Increasing order of the lattice index is identified with the direction left to right. The upper summation limit L in (9) has to be replaced by $L-1$, giving rise to a generator that we denote by $\tilde{\mathscr{L}}$.

In order to study fluctuations around some untypical integrated current, parameterized in terms of the driving strength s, we define the weighted transition rates

$$w_s^{kk+1}(\eta) = w\left[q\mathrm{e}^{s}\eta(k)\upsilon(k+1) + q^{-1}\mathrm{e}^{-s}\upsilon(k)\eta(k+1)\right], \quad 1 \leq k \leq L-1 \tag{11}$$

$$w_{s,\bar{s}}^{L1}(\eta) = w\left[q\mathrm{e}^{s+\bar{s}}\eta(L)\upsilon(1) + q^{-1}\mathrm{e}^{-s-\bar{s}}\upsilon(L)\eta(1)\right], \quad k = L. \tag{12}$$

This leads us to define the weighted generators

$$\tilde{\mathscr{L}}_s f(\eta) := \sum_{k=1}^{L-1} w_s^{kk+1}(\eta) f(\eta^{kk+1}) - w^{kk+1}(\eta) f(\eta) \tag{13}$$

$$\mathscr{L}_{s,\bar{s}} f(\eta) := \tilde{\mathscr{L}}_s f(\eta) + w_{s,\bar{s}}^{L1}(\eta) f(\eta^{L1}) - w^{L1}(\eta) f(\eta). \tag{14}$$

The weighted generators give a weight e^s (e^{-s}) to each particle jump to the right (left) anywhere on the lattice and for the process with periodic boundary conditions an extra weight $\mathrm{e}^{\bar{s}}$ ($\mathrm{e}^{-\bar{s}}$) to each particle jump to the right (left) across bond $(L, 1)$. Thus each random trajectory of the process is given a weight $\mathrm{e}^{sJ(t)+\bar{s}J_L(t)}$ where $J(t)$ is the time-integrated total current, i.e., the total number of all particle jumps to the right up to time t minus the total number of all particle jumps to the left up to time t and $J_L(t)$ is the time-integrated current across bond $(L, 1)$. Studying the evolution

[2] When the summation is over $\Omega = \mathbb{S}^L$ we shall usually omit the set $\mathbb{S}^L$ under the summation symbol and simply write $\sum_\eta$.

under the weighted generator provides a description of the process under conditioning on some non-typical current parametrized by the parameters s or $\bar{s}$. Choosing $s < 0$ corresponds to conditioning on a time-integrated total current that is smaller than the typical total current. Similarly, $\bar{s} < 0$ yields a conditioning on a smaller than typical time-integrated local current across bond $(L, 1)$. For details on this construction see e.g. [16, 22, 23] and specifically for the present context see [35].

We fix more notation and summarize some well-known basic facts from the theory of Markov processes. For a probability distribution $P(\eta)$ we denote the expectation of a continuous function $f(\eta)$ by $\langle f \rangle_P := \sum_\eta f(\eta)P(\eta)$. The transposed generator is defined by $\mathscr{L}^T f(\eta) := \sum'_{\eta' \in \mathbb{S}^L} f(\eta')\mathscr{L}\mathbf{1}_{\eta'}(\eta)$ where $\mathbf{1}_{\eta'}(\eta) = \delta_{\eta,\eta'}$. With this definition (8) yields for a probability distribution $P(\eta)$ the *master equation*

$$\mathscr{L}^T P(\eta) = \sum\nolimits'_{\eta'} [w(\eta' \to \eta)P(\eta') - w(\eta \to \eta')P(\eta)]. \tag{15}$$

An invariant measure is denoted $\pi^*(\eta)$ and defined by

$$\mathscr{L}^T \pi^*(\eta) = 0 \tag{16}$$

and the normalization $\sum_\eta \pi^*(\eta) = 1$. An unnormalized measure with the property (16) is denoted $\pi(\eta)$. The time-reversed process is defined by

$$\mathscr{L}^{rev} f(\eta) := \sum_{\eta'}{}' w^{rev}(\eta \to \eta')[f(\eta') - f(\eta)] \tag{17}$$

with $w^{rev}(\eta \to \eta') = w(\eta' \to \eta)\pi(\eta')/\pi(\eta)$. The process is reversible if $\mathscr{L}^{rev} = \mathscr{L}$ which means that the rates satisfy the detailed balance condition $\pi(\eta)w(\eta \to \eta') = w(\eta' \to \eta)\pi(\eta')$. A probability distribution satisfying the detailed balance condition is a reversible measure. It is easily verified that the ASEP with reflecting boundary conditions is reversible with reversible measure

$$\pi(\eta) = q^{\sum_{k=1}^{L}(2k+\mu)\eta(k)} = e^{\mu N(\eta)} q^{2\sum_{i=1}^{N(\eta)} x_i}. \tag{18}$$

for any $\mu \in \mathbb{R}$.

We define the transition matrix H of the process by the matrix elements

$$H_{\eta'\eta} = \begin{cases} -w(\eta \to \eta') & \eta \neq \eta' \\ \sum'_{\eta'} w(\eta \to \eta') & \eta = \eta'. \end{cases} \tag{19}$$

with $w(\eta \to \eta')$ given by (9). One has

$$\mathscr{L} f(\eta) = -\sum_{\eta'} f(\eta')H_{\eta'\eta}, \quad \mathscr{L}^T P(\eta) = -\sum_{\eta'} H_{\eta\eta'} P(\eta'). \tag{20}$$

Notice that here the sum includes the term $\eta' = \eta$. In slight abuse of language we shall also call H the generator of the process. Analogously we also define the weighted transition matrix where the off-diagonal elements are replaced by the weighted rates (11), (12).

For an unnormalized stationary distribution we define the diagonal matrix $\hat{\pi}$ with the stationary weights $\pi(\eta)$ on the diagonal. For ergodic processes with finite state space one has $0 < \pi(\eta) < \infty$ for all η. In terms of this diagonal matrix we can write the generator of the reversed dynamics as $H^{rev} = \hat{\pi} H^T \hat{\pi}^{-1}$. The reversibility condition $H^{rev} = H$ then reads

$$\hat{\pi}^{-1} H \hat{\pi} = H^T. \tag{21}$$

Therefore, if one finds a diagonal matrix with the property (21) then this matrix defines a reversible measure.

2.3 *Representation of the Generator in the Natural Tensor Basis*

In order to write the matrix H explicitly we assign to each configuration η a canonical basis vector $|\,\eta\,\rangle$. We choose the binary ordering $\iota(\eta) = 1 + \sum_{k=1}^{L} \eta(k) 2^{k-1}$ of the basis. Defining single-site basis vectors of dimension 2

$$|0) := \begin{pmatrix} 1 \\ 0 \end{pmatrix}, \quad |1) := \begin{pmatrix} 0 \\ 1 \end{pmatrix} \tag{22}$$

one then has $|\,\eta\,\rangle = |\eta(1)) \otimes \cdots \otimes |\eta(L))$ where $\otimes$ denotes the tensor product. These basis vectors span the complex vector space $(\mathbb{C}^2)^{\otimes L}$ of dimension $d = 2^L$. We also define transposed basis vectors $\langle\,\eta\,| := |\,\eta\,\rangle^T$ and the inner product $\langle\, v \,|\, w \,\rangle := \sum_\eta v(\eta) w(\eta)$.

Furthermore we define the two-by-two Pauli matrices

$$\sigma^x := \begin{pmatrix} 0 & 1 \\ 1 & 0 \end{pmatrix}, \quad \sigma^y := \begin{pmatrix} 0 & -i \\ i & 0 \end{pmatrix}, \quad \sigma^z := \begin{pmatrix} 1 & 0 \\ 0 & -1 \end{pmatrix} \tag{23}$$

and the two-dimensional unit matrix $\mathbb{1}$. From these we construct

$$\sigma^\pm = \frac{1}{2}(\sigma^x \pm i\sigma^y), \quad \hat{n} = \frac{1}{2}(\mathbb{1} - \sigma^z), \quad \hat{\upsilon} = \frac{1}{2}(\mathbb{1} + \sigma^z). \tag{24}$$

These matrices satisfy the following relations:

$$\begin{aligned} &\sigma^+\sigma^- = \hat{\upsilon},\ \sigma^-\sigma^+ = \hat{n},\ \hat{n}\hat{\upsilon} = 0, \quad \hat{\upsilon}\hat{n} = 0 \\ &\sigma^+\hat{n} = \sigma^+,\ \hat{n}\sigma^+ = 0, \quad \sigma^+\hat{\upsilon} = 0, \quad \hat{\upsilon}\sigma^+ = \sigma^+, \\ &\sigma^-\hat{n} = 0, \quad \hat{n}\sigma^- = \sigma^-,\ \sigma^-\hat{\upsilon} = \sigma^-,\ \hat{\upsilon}\sigma^- = 0. \end{aligned} \tag{25}$$

With the occupation variables (1) for a single site we have the projector property

$$\hat{n}|\eta) = \eta|\eta), \quad \hat{\upsilon}|\eta) = \upsilon|\eta). \tag{26}$$

Having in mind the action of these operators to the right on a column vector, we call σ^- a creation operator, and σ^+ annihilation operators. When acting to the left on a bra-vector the roles are interchanged: σ^+ acts as creation operator and σ^- as annihilation operator.

For $L > 1$ and any linear combination u of these matrices we define the tensor operators $u_k := \mathbb{1}^{\otimes k-1} \otimes u \otimes \mathbb{1}^{\otimes L-k}$. By convention the zero'th tensor power of any matrix is the real number 1 and $u^{\otimes 1} = u$. We note that also the tensor occupation operators $\hat{n}_k$ act as projectors

$$\hat{n}_k| \,\eta\,\rangle = \eta(k)| \,\eta\,\rangle = \sum_{i=1}^{N(\eta)} \delta_{x_i,k}| \,\eta\,\rangle, \tag{27}$$

with the occupation variables $\eta(k)$ or particle coordinates x_i respectively understood as functions of η. The proof is trivial: The first equality is inherited from (26) by multilinearity of the tensor product, the second equality follows from (3). Multilinearity of the tensor product also yields $u_k v_{k+1} = \mathbb{1}^{\otimes(k-1)} \otimes [(u \otimes \mathbb{1})(\mathbb{1} \otimes v)] \otimes \mathbb{1}^{\otimes(L-k-1)} = \mathbb{1}^{\otimes(k-1)} \otimes (u \otimes v) \otimes \mathbb{1}^{\otimes(L-k-1)}$ and the commutator property $u_k v_l = v_l u_k$ for $k \neq l$. For $k = l$ one has relations analogous to (25).

It turns out to be convenient to introduce parameters $\alpha = q\mathrm{e}^s$ and $\beta = \mathrm{e}^{\bar{s}}$ and express for periodic boundary conditions the weighted generator as $H(q, \alpha, \beta)$ with the convention $H(q, q, 1) = H$ for the unweighted generator. Similarly one writes $\tilde{H}(q, \alpha)$ for the weighted generator with reflecting boundary conditions with the convention $\tilde{H}(q, q) = \tilde{H}$. With these definitions the weighted generators $\tilde{H}(q, \alpha)$ and $H(q, \alpha, \beta)$ defined by (13) and (14) resp. become

$$\tilde{H}(q, \alpha) = \sum_{k=1}^{L-1} h_{k,k+1}(q, \alpha) \tag{28}$$

$$H(q, \alpha, \beta) = \tilde{H}(q, \alpha) + h_{L,1}(q, \alpha, \beta) \tag{29}$$

with the hopping matrices

$$h_{k,k+1}(q, \alpha) = -w \left[\alpha \sigma_k^+ \sigma_{k+1}^- - q \hat{n}_k \hat{\upsilon}_{k+1} + \alpha^{-1} \sigma_k^- \sigma_{k+1}^+ - q^{-1} \hat{\upsilon}_k \hat{n}_{k+1}\right] \tag{30}$$

and

$$h_{L,1}(q, \alpha, \beta) = -w \left[\alpha\beta \sigma_k^+ \sigma_{k+1}^- - q \hat{n}_k \hat{\upsilon}_{k+1} + (\alpha\beta)^{-1} \sigma_k^- \sigma_{k+1}^+ - q^{-1} \hat{\upsilon}_k \hat{n}_{k+1})\right]. \tag{31}$$

It is useful to introduce the space-reflection operator $\hat{R}$ defined by

$$\hat{R} u_k \hat{R}^{-1} = u_{L+1-k} \tag{32}$$

for local one-site operators u_k and the diagonal transformations

$$V(\gamma) = \gamma^{\frac{1}{4}\sum_{k=1}^{L}(2k-L-1)\sigma_k^z} = \gamma^{-\frac{1}{2}\sum_{k=1}^{L}(2k-L-1)\hat{n}_k} \tag{33}$$

$$W(z) = z^{\hat{N}} \tag{34}$$

with the number operator $\hat{N} = \sum_{K=1}^{L} \hat{n}_k$. We note the properties

$$\hat{R}V(\gamma)\hat{R}^{-1} = V^{-1}(\gamma) = V(\gamma^{-1}), \tag{35}$$

$$\hat{R}W(z)\hat{R}^{-1} = W(z) = W^{-1}(z^{-1}), \tag{36}$$

$$\hat{R}H(q,\alpha,\beta)\hat{R}^{-1} = H(q,\alpha^{-1},\beta^{-1}) \tag{37}$$

$$H^T(q,\alpha,\beta) = H(q,\alpha^{-1},\beta^{-1}) \tag{38}$$

$$WH(q,\alpha,\beta)W^{-1} = H(q,\alpha,\beta). \tag{39}$$

Moreover, the transformation property

$$V(\gamma)\sigma_k^{\pm}V^{-1}(\gamma) = \gamma^{\pm\frac{1}{2}(2k-L-1)}\sigma_k^{\pm} \tag{40}$$

yields

$$V(\gamma)H(q,\alpha,\beta)V^{-1}(\gamma) = H\left(q,\alpha\gamma^{-1},\beta\gamma^{L}\right) \tag{41}$$

$$V(\gamma)\tilde{H}(q,\alpha)V^{-1}(\gamma) = \tilde{H}\left(q,\alpha\gamma^{-1}\right). \tag{42}$$

Thus for periodic boundary conditions global conditioning and local conditioning are related by a similarity transformation, while for reflecting boundary conditions the conditioning can be completely absorbed into a similarity transformation. One also finds with $\gamma = q^2, \alpha = q$ the reversibility relation $V(q^2)\tilde{H}(q)V^{-1}(q^2) = \tilde{H}^T$. By (21) this shows that $\hat{\pi} = V^{-1}(q^2)$ is the matrix form of the reversible measure (18) with $\mu = -L-1$ [33].[3]

For driving strength $s_0 := -\ln(q)$ corresponding to $\alpha = 1$ one finds for $1 \le k \le L-1$

$$h_{k,k+1}(q,1) = -\frac{w}{2}\left[\sigma_k^x\sigma_{k+1}^x + \sigma_k^y\sigma_{k+1}^y + \Delta(\sigma_k^z\sigma_{k+1}^z - \mathbf{1}) + h(\sigma_k^z - \sigma_{k+1}^z)\right] \tag{43}$$

with

$$\Delta = \frac{1}{2}(q+q^{-1}), \quad h = \frac{1}{2}(q-q^{-1}) \tag{44}$$

[3]This is equivalent to Eq. (2.14) in [33], which, however, has a sign error and should read $H^T = V^{-2}HV^2$.

and the unit-matrix $\mathbf{1}$ of dimension 2^L. Notice that for periodic boundary conditions the local divergence terms $\sigma_k^z - \sigma_{k+1}^z$ cancel. For reflecting boundaries the local divergence term contributes opposite boundary fields $h(\sigma_L^z - \sigma_1^z)$. With the further choice $\bar{s}_0 = 0$ corresponding to $\beta = 1$ the weighted generator (29) becomes the Hamiltonian operator $H(q, 1, 1)$ of the ferromagnetic Heisenberg spin-1/2 quantum chain, while for $\beta \neq 1$ one has the Heisenberg chain $H(q, 1, \beta)$ with twisted boundary conditions [30].

Since particle number is conserved the process is trivially reducible. For each particle number N one has an irreducible process $\eta_{N,t}$ on the state space Ω_N. We define the projector

$$\hat{\mathbf{1}}_N := \sum_{\eta \in \Omega_N} |\, \eta \,\rangle\langle\, \eta \,| \tag{45}$$

where we have used the quantum mechanical ket-bra convention $|\, \eta \,\rangle\langle\, \eta \,| \equiv |\, \eta \,\rangle \otimes \langle\, \eta \,|$ for the tensor product of two vectors. Thus one obtains the generator

$$H_N(q, \alpha, \beta) := \hat{\mathbf{1}}_N \; H(q, \alpha, \beta) \; \hat{\mathbf{1}}_N \tag{46}$$

for the N-particle weighted ASEP.

Notice that $\hat{\mathbf{1}}_N$ acts as unit matrix on the irreducible subspace corresponding to particle number N. The unit matrix $\mathbf{1}$ in the full space has the useful representation

$$\mathbf{1} = \sum_{\eta \in \Omega} |\, \eta \,\rangle\langle\, \eta \,|. \tag{47}$$

2.4 The Quantum Algebra $U_q[\mathfrak{gl}(2)]$

The quantum algebra $U_q[\mathfrak{gl}(2)]$ is the q-deformed universal enveloping algebra of the Lie algebra $\mathfrak{gl}(2)$. This associative algebra over $\mathbb{C}$ is generated by $\mathbf{L}_i^{\pm 1}$, $i = 1, 2$ and $\mathbf{S}^{\pm}$ with the relations [11, 24]

$$[\, \mathbf{L}_i \,,\, \mathbf{L}_j \,] = 0 \tag{48}$$

$$\mathbf{L}_i \mathbf{S}^{\pm} = q^{\pm(\delta_{i,2} - \delta_{i,1})} \mathbf{S}^{\pm} \mathbf{L}_i \tag{49}$$

$$[\, \mathbf{S}^+ \,,\, \mathbf{S}^- \,] = \frac{(\mathbf{L}_2 \mathbf{L}_1^{-1})^2 - (\mathbf{L}_2 \mathbf{L}_1^{-1})^{-2}}{q - q^{-1}} \tag{50}$$

Notice the replacement $q^2 \to q$ that we made in the definitions of [11].

It is convenient to work also with the subalgebra $U_q[\mathfrak{sl}(2)]$. We introduce the generators $\mathbf{N}$ and $\mathbf{V}$ via $q^{-\mathbf{N}/2} = \mathbf{L}_1$, $q^{-\mathbf{V}/2} = \mathbf{L}_2$ and define

$$\mathbf{S}^z = \frac{1}{2}(\mathbf{N} - \mathbf{V}) \tag{51}$$

and the identity I. Then the quantum algebra $U_q[\mathfrak{sl}(2)]$ is the subalgebra generated by $q^{\pm \mathbf{S}^z}$ and $\mathbf{S}^{\pm}$ with relations

$$q^{\mathbf{S}^z} q^{-\mathbf{S}^z} = q^{-\mathbf{S}^z} q^{\mathbf{S}^z} = I \tag{52}$$

$$q^{\mathbf{S}^z} \mathbf{S}^{\pm} q^{-\mathbf{S}^z} = q^{\pm 1} \mathbf{S}^{\pm} \tag{53}$$

$$[\,\mathbf{S}^+ \,,\, \mathbf{S}^- \,] = \frac{q^{2\mathbf{S}^z} - q^{-2\mathbf{S}^z}}{q - q^{-1}} \tag{54}$$

Observing that $\mathbf{N} + \mathbf{V}$ belongs to the center of $U_q[\mathfrak{gl}(2)]$ [24] one sees that $U_q[\mathfrak{sl}(2)]$ is a subalgebra of $U_q[\mathfrak{gl}(2)]$.

It is trivial to verify that $U_q[\mathfrak{gl}(2)]$ has the two-dimensional fundamental representation $\mathbf{S}^{\pm} \to \sigma^{\pm}$, $\mathbf{N} \to \hat{n}$, $\mathbf{V} \to \hat{\upsilon}$ given by the matrices (24). Then $\sigma^{\pm}$ and $\sigma^z/2$ form the two-dimensional fundamental representation of $U_q[\mathfrak{sl}(2)]$. Reducible higher-dimensional representations can be constructed using the coproduct [24]

$$\Delta(\mathbf{S}^{\pm}) = \mathbf{S}^{\pm} \otimes q^{-\mathbf{S}^z} + q^{\mathbf{S}^z} \otimes \mathbf{S}^{\pm} \tag{55}$$

$$\Delta(\mathbf{S}^z) = \mathbf{S}^z \otimes \mathbb{1} + \mathbb{1} \otimes \mathbf{S}^z. \tag{56}$$

By repeatedly applying the coproduct to the fundamental representation we obtain

$$S^{\pm}(k) = q^{\frac{1}{2}\sum_{j=1}^{k-1}\sigma_j^z - \frac{1}{2}\sum_{j=k+1}^{L}\sigma_j^z} \sigma_k^{\pm} \tag{57}$$

$$S^z(k) = \frac{1}{2}\sigma_k^z. \tag{58}$$

One has

$$S^{\pm}(k)S^{\pm}(l) = \begin{cases} q^{\pm 2} S^{\pm}(l) S^{\pm}(k) & k > l \\ 0 & k = l \\ q^{\mp 2} S^{\pm}(l) S^{\pm}(k) & k < l \end{cases} \tag{59}$$

Thus the spatial order in which particles are created (or annihilated) by applying the operators $S^{\pm}(k)$ gives rise to combinatorial issues when building many-particle configurations from the reference state corresponding to the empty lattice.

From the coproduct one obtains the tensor representations of $U_q[\mathfrak{sl}(2)]$, denoted by capital letters,

$$S^{\pm} = \sum_{k=1}^{L} S^{\pm}(k), \quad S^z = \sum_{k=1}^{L} S^z(k). \tag{60}$$

For the full quantum algebra $U_q[\mathfrak{gl}(2)]$ the tensor generators are $S^{\pm}$ and $\hat{N} = \sum_{k=1}^{L} \hat{n}_k$, $\hat{V} = \sum_{k=1}^{L} \hat{\upsilon}_k$. The unit I is represented by the 2^L-dimensional unit matrix $\mathbf{1} := \mathbb{1}^{\otimes L}$.

For reflecting boundary conditions the Heisenberg Hamiltonian $\tilde{H}(q, 1)$ is symmetric under the action of $U_q[\mathfrak{gl}(2)]$ [1, 30]. This symmetry property is the origin of the duality relations derived in [33] and will also be used extensively below. In fact, for $1 \le k \le L-1$ one has $[\, h_{k,k+1}(q,1)\,,\ S^{\pm}\,] = [\, h_{k,k+1}(q,\alpha)\,,\ S^{z}\,] = 0$, which imply

$$[\, \tilde{H}(q,1)\,,\ S^{\pm}\,] = 0 \tag{61}$$

and, equivalently to (39), the diagonal symmetries $[\, \tilde{H}(q,\alpha)\,,\ \hat{N}\,] = [\, \tilde{H}(q,\alpha)\,,\ \hat{V}\,] = 0$, thus giving rise to the $U_q[\mathfrak{gl}(2)]$ symmetry of $\tilde{H}(q,1)$.

We stress that $[\, h_{L,1}(q,1)\,,\ S^{\pm}\,] \neq 0$. One the other hand $[\, h_{L,1}(q,\alpha,\beta)\,,\ S^{z}\,] = 0$. Hence for periodic boundary conditions the symmetry breaks down to only a residual $U(1)$ symmetry $[\, H(q,\alpha,\beta)\,,\ S^{z}\,] = 0$ generated by S^z, which corresponds to particle number conservation since the z-component of the total spin S^z is related to the particle number operator $\hat{N}$ through $S^z = L/2 - \hat{N}$.

We also define

$$S^{\pm}(q,\alpha) = \sum_{k=1}^{L} S_k^{\pm}(q,\alpha) \tag{62}$$

where

$$S_k^{\pm}(q,\alpha) = \alpha^{\pm\frac{1}{2}(L+1-2k)} q^{\frac{1}{2}\sum_{i=1}^{k-1}\sigma_i^z - \frac{1}{2}\sum_{i=k+1}^{L}\sigma_i^z} \sigma_k^{\pm}. \tag{63}$$

The diagonal transformation (40) and the defining relation (53) yield

$$V(\gamma)S^{\pm}(q,\alpha)V^{-1}(\gamma) = S^{\pm}\left(q,\alpha\gamma^{-1}\right) \tag{64}$$

$$W(z)S^{\pm}(q,\alpha)W^{-1}(z) = z^{\mp 1}S^{\pm}(q,\alpha). \tag{65}$$

Notice that $S^{\pm}(q,1) = S^{\pm}$ as defined in (60). Hence $S^{\pm}(q,\alpha)$ and S^z also form a representation of $U_q[\mathfrak{sl}(2)]$. Since according to (61) $\tilde{H}(q,1)$ commutes with the generators $S^{\pm} = S^{\pm}(q,1)$ we conclude from (42) that $\tilde{H}(q,\alpha)$ commutes with $S^{\pm}(q,\alpha)$, which together with S^z form an equivalent representation of $U_q[\mathfrak{sl}(2)]$. In particular, the generator of the ASEP with reflecting boundary conditions $\tilde{H} = \tilde{H}(q,q)$ commutes with $\tilde{S}^{\pm} := S^{\pm}(q,q)$.

We note that

$$(S^{\pm}(q,\alpha))^T = S^{\mp}(q,\alpha^{-1}) \tag{66}$$

$$\hat{R}S^{\pm}(q,\alpha)\hat{R}^{-1} = S^{\pm}(q^{-1},\alpha^{-1}). \tag{67}$$

To prove the second equality one uses $\hat{R}S_k^{\pm}(q,\alpha)\hat{R}^{-1} = S_{L+1-k}^{\pm}(q^{-1},\alpha^{-1})$ which comes from (32).

Finally we introduce the symmetric q-number

$$[x]_q := \frac{q^x - q^{-x}}{q - q^{-1}} \tag{68}$$

for q, $q^{-1} \neq 0$ and $x \in \mathbb{C}$. This definition can be applied straightforwardly to finite-dimensional matrices through the Taylor expansion of the exponential. For integers we also define the q-factorial

$$[n]_q! := \begin{cases} 1 & n = 0 \\ \prod_{k=1}^{n} [k]_q & n \geq 1. \end{cases} \tag{69}$$

2.5 Duality in the Quantum Hamiltonian Formalism

For self-containedness we briefly review how to express expectation values using the matrix representation of the generator which allows to state the notion of duality in a neat matrix form [21, 37].

A probability measure $P(\eta)$ is represented by the column vector

$$|\, P \,\rangle = \sum_{\eta} P(\eta) |\, \eta \,\rangle. \tag{70}$$

Next we define the *summation vector*

$$\langle\, s \,| := \sum_{\eta} \langle\, \eta \,| \tag{71}$$

which is the row vector where all components are equal to 1. The expectation $\langle\, f \,\rangle_P$ of a function $f(\eta)$ with respect to a probability distribution $P(\eta)$ is the inner product

$$\langle\, f \,\rangle_P = \langle\, f \,|\, P \,\rangle = \langle\, s \,|\hat{f}|\, P \,\rangle \tag{72}$$

where

$$\hat{f} := \sum_{\eta} f(\eta) |\, \eta \,\rangle\langle\, \eta \,| \tag{73}$$

is a diagonal matrix with diagonal elements $f(\eta)$. Notice that

$$f(\eta) = \langle\, \eta \,|\hat{f}|\, \eta \,\rangle = \langle\, s \,|\hat{f}|\, \eta \,\rangle. \tag{74}$$

One obtains the diagonal matrix $\hat{f}$ (73) corresponding to a function $f(\eta)$ by substituting in $f(\eta)$ the variable $\eta(k)$ by the diagonal matrix $\hat{n}_k$.

For a Markov process η_t the master equation (15) for a probability measure $P(\eta_t) := \text{Prob}\,[\, \eta_t = \eta)\,]$ reads

$$\frac{\mathrm{d}}{\mathrm{d}t} |\, P(t) \,\rangle = -H |\, P(t) \,\rangle \tag{75}$$

which implies

$$| P(t) \rangle = \mathrm{e}^{-Ht} | P_0 \rangle \tag{76}$$

for an initial probability measure $P_0(\eta) \equiv P(\eta_0)$ at time $t = 0$. We write the expectation of a function $f(\eta_t)$ as

$$\langle f(t) \rangle := \sum_{\eta} f(\eta) P(\eta_t) = \sum_{\eta} f(\eta) \langle \eta | \mathrm{e}^{-Ht} | P_0 \rangle = \langle s | \hat{f} \mathrm{e}^{-Ht} | P_0 \rangle . \tag{77}$$

If the initial distribution needs to be specified we use an upper index $\langle f(t) \rangle^{P_0}$. Normalization implies $\langle s | P(t) \rangle = 1$ for all $t \geq 0$ and therefore $\langle s | H = 0$. A stationary distribution, denoted by $| \pi^* \rangle$, is a right eigenvector of H with eigenvalue 0, i.e., $H | \pi^* \rangle = 0$ and normalization $\langle s | \pi^* \rangle = 1$. For the ergodic subspaces with fixed particle number N it is unique.

In order to introduce duality we consider a process ξ_t with generator H and a process x_t with generator G which may have different countable state spaces Ω_A and Ω_B. Consider also a family of functions $f^x : \Omega_A \mapsto \mathbb{C}$ indexed by $x \in \Omega_B$ and a family of functions $g^\xi : \Omega_B \mapsto \mathbb{C}$ indexed by $\xi \in \Omega_A$ such that $f^x(\xi) = g^\xi(x) =: D(x, \xi)$. Let the process ξ_t start at some fixed $\xi \in \Omega_A$ and let x_t start at some fixed $x \in \Omega_B$. Then the two processes are said to be dual with respect to the duality function $D(x, \xi)$ if [26]

$$\langle f^x(t) \rangle^\xi = \langle g^\xi(t) \rangle^x . \tag{78}$$

As pointed out in [21] this property can be stated neatly in terms of the generators as

$$DH = G^T D \tag{79}$$

where the duality matrix D is defined by

$$D = \sum_{\xi \in \Omega_A} \sum_{x \in \Omega_B} D(x, \xi) | x \rangle \langle \xi | \tag{80}$$

By construction one has $D(x, \xi) = \langle x | D | \xi \rangle$.

2.6 Shock/Antishock Measures

In vector notation a product measure with marginals ρ_k is a tensor vector $| \{\rho_i\} \rangle = |\rho_1) \otimes \cdots \otimes |\rho_L)$ with the single-site column vectors $|\rho_k) = (1 - \rho_k, \rho_k)^T$. It is convenient to introduce the local fugacity

$$z_k = \frac{\rho_k}{1 - \rho_k} \tag{81}$$

and write the product measure in the form $|\{z_i\}\rangle = |z_1) \otimes \cdots \otimes |z_L)/Z_L$ with $|z_k) = (1, z_k)^T$ and normalization $Z_L = \prod_{k=1}^{L}[z_k/(1+z_k)]$.

Specifically, we define for a set $\mathbf{x}$ of lattice sites with cardinality $K = |\mathbf{x}|$ the following family $|\nu_{\mathbf{x}}\rangle = |z_1) \otimes \cdots \otimes |z_L)/Z_L$ of shock/antishock measures, or SAM's for short, in terms of the fugacities

$$z_k = \begin{cases} zq^{2l} & \text{for } x_l < k < x_{l+1}, \quad l \in \{0, 1, \dots, K+1\} \\ \infty & \text{for } k \in \mathbf{x} \end{cases} \tag{82}$$

with $x_0 = 0$ and $x_{K+1} = L + 1$. On coarse-grained scale with $\xi = k/L$ and $\xi_i^s = x_i/L$ the macroscopic density profile $\rho(\xi)$ corresponding to the fugacities z_k has discontinuities at $\xi = \xi_i^s$ with constant fugacity ratios $z_i^+/z_i^- = q^2$ where $z_i^{\pm} = z_{\xi_i^s \pm 1}$. In forward (clockwise) direction and for $q > 1$ this is an upward step, corresponding to a shock profile for the ASEP (with positive bias $q > 1$). Between site L and site 1 there is downward jump with fugacity ratio q^{-2K}. On macroscopic scale this constitutes an antishock at position $\xi^a = \xi^s - (1+\kappa)/2 \mod 1$, hence the term SAM. These shock measures are closely related to the shock measures defined in [4] and also to the infinite-volume shock measures studied in [3] where the shock positions x_i are occupied by second-class particles.

With a different normalization factor the general SAM (82) with constant fugacity jumps q^2 can be written as

$$|\bar{\mu}_{\mathbf{x}}\rangle := \prod_{j=1}^{K} z^{-1} q^{-\sum_{i=1}^{x_j-1}\hat{n}_i + \sum_{i=x_j+1}^{L}\hat{n}_i} \hat{n}_{x_j} |z\rangle \propto |\nu_{\mathbf{x}}\rangle. \tag{83}$$

Here

$$|z\rangle := |z)^{\otimes L} \tag{84}$$

for $K = 0$ is the *unnormalized* homogeneous product measure corresponding to the Bernoulli product measure $|\rho\rangle = |z\rangle/(1+z)^L$ where $\rho = z/(1+z)$.

From the SAM defined by (83) we construct a second type of SAM'S using the transformations (33) and (34)

$$|\mu_{\mathbf{x}}\rangle := z^{-K} V(q^{\frac{2K}{L}}) \prod_{j=1}^{K} W(q^{\frac{2x_j-L-1}{L}}) |\bar{\mu}_{\mathbf{x}}\rangle \tag{85}$$

$$= \prod_{j=1}^{K} \left[z^{-1} q^{\frac{2}{L}\sum_{l=1}^{L}(x_j-l)\hat{n}_l - \sum_{i=1}^{x_j-1}\hat{n}_i + \sum_{i=x_j+1}^{L}\hat{n}_i} \hat{n}_{x_j} \right] |z\rangle. \tag{86}$$

These are product measures that have a shock discontinuity at the same positions as the shock measures of the first type, but the density between shocks is not constant. We illustrate the definition for $K = 1$ and $K = 2$.

For $K = 1$ the SAM (85) is closely related to the type-II shock measures defined in [5]. In explicit form it reads

$$|\,\mu_x\,\rangle := z^{-1} q^{\frac{2}{L}\sum_{l=1}^{L}(x-l)\hat{n}_l - \sum_{i=1}^{x-1}\hat{n}_i + \sum_{i=x+1}^{L}\hat{n}_i}\,\hat{n}_x\,|\,z\,\rangle \tag{87}$$

with $1 \le x \le L$. This is a product measure with space-dependent fugacities

$$z_k = \begin{cases} zq^{-\frac{2(k-x)+L}{L}} & \text{for } 1 \le k < x \\ \infty & \text{for } k = x \\ zq^{-\frac{2(k-x)-L}{L}} & \text{for } x < k \le L \end{cases} \tag{88}$$

corresponding to space-dependent densities

$$\rho_k = \begin{cases} \frac{1}{2}\left[1 - \tanh\left(\frac{E}{L}(k - x + L\frac{\kappa+1}{2})\right)\right] & \text{for } k < x \\ 1 & \text{for } k = x \\ \frac{1}{2}\left[1 - \tanh\left(\frac{E}{L}(k - x + L\frac{\kappa-1}{2})\right)\right] & \text{for } k > x \end{cases} \tag{89}$$

where $E = \ln q$ and $\kappa = \ln z/E$ (corresponding to $z = q^\kappa$). The point x where the density is 1 marks the microscopic shock position. On coarse-grained scale with $\xi = k/L$ and $\xi^s = x/L$ the macroscopic density profile $\rho(\xi)$ has a discontinuity at $\xi = \xi^s$ with amplitude $A^s := \rho^+ - \rho^- = \tanh(E(\kappa+1)/2) - \tanh(E(\kappa-1)/2)$, where $\rho^\pm = \lim_{\epsilon\to 0}\rho(\xi^s \pm \epsilon)$. In forward (clockwise) direction and for $E > 0$ this is an upward step with fugacity ratio $z^+/z^- = q^2$, corresponding to a shock profile for the ASEP (with positive bias $E > 0$). For strong asymmetry $E = \epsilon L$ one has near $k = x - L(1+\kappa)/2 \bmod L$ a smoothened downward "step" with an intrinsic width $\propto 1/\epsilon$ on lattice scale. On macroscopic scale this constitutes an antishock at position $\xi^a = \xi^s - (1+\kappa)/2 \bmod 1$.

For $K = 2$ the SAM $|\,\mu_{x,y}\,\rangle$ (85) with $1 \le x < y \le L$ has local fugacities

$$z_k = \begin{cases} zq^{-\frac{2(2k-x-y+L)}{L}} & \text{for } 1 \le k < x \\ \infty & \text{for } k = x \\ zq^{-\frac{2(2k-x-y)}{L}} & \text{for } x < k < y \\ \infty & \text{for } k = y \\ zq^{-\frac{2(2k-x-y-L)}{L}} & \text{for } y < k \le L \end{cases} \tag{90}$$

corresponding to densities

$$\rho_k = \begin{cases} \frac{1}{2}\left[1 - \tanh\left(\frac{E}{L}(2k - x - y + L\frac{\kappa+2}{2})\right)\right] & \text{for } 1 \le k < x \\ 1 & \text{for } k = x \\ \frac{1}{2}\left[1 - \tanh\left(\frac{E}{L}(2k - x - y + L\frac{\kappa}{2})\right)\right] & \text{for } x < k < y \\ 1 & \text{for } k = y \\ \frac{1}{2}\left[1 - \tanh\left(\frac{E}{L}(2k - x - y + L\frac{\kappa-2}{2})\right)\right] & \text{for } y < k \le L \end{cases} \tag{91}$$

On macroscopic scale this density profile has two shock discontinuities at $\xi_1^s = x/L \bmod 1$ and $\xi_2^s = y/L \bmod 1$. Both fugacity ratios are of magnitude q^2. For strong asymmetry $E = \epsilon L$ there are two antishocks at $\xi_1^a = (\xi_1^s + \xi_2^s)/2 - (\kappa + 2)/4 \bmod 1$ and $\xi_2^a = (\xi_1^s + \xi_2^s)/2 - (\kappa + 2)/4 \bmod 1$.

3 Results

Before stating the new results we recall the duality relation for the ASEP with reflecting boundary conditions derived in [33], Eq. (3.12). We reformulate this duality relation slightly and correct a sign error in Eq. (3.12) of [33]. In the next section we also present a new proof, parts of which are then used to prove the new results given below.

Theorem 1 (Schütz, [33]) *The ASEP with reflecting boundary conditions and asymmetry parameter q is self-dual w.r.t. the duality function*

$$D(\mathbf{x}, \eta) = \prod_{j=1}^{|\mathbf{x}|} q^{-2x_j} Q_{x_j}(\eta) \tag{92}$$

where

$$Q_{x_j}(\eta) = q^{\sum_{i=1}^{x_j-1} \eta(i) - \sum_{i=x_j+1}^{L} \eta(i)} \eta(x_j). \tag{93}$$

Remark 1 Because of particle number conservation also

$$\tilde{D}(\mathbf{x}, \eta) = q^{|\mathbf{x}|(N(\eta)-1)} D(\mathbf{x}, \eta) = \prod_{j=1}^{|\mathbf{x}|} q^{2N_{x_j}(\eta) - 2x_j} \eta(x_j) \tag{94}$$

is a duality function with the particle numbers $N(\eta)$ (2) and $N_x(\eta)$ (4). This is the duality function (3.12) of [33].[4]

Now we present the main results of this work.

Theorem 2 *Let* $H_N(\cdot, \cdot, \cdot)$ *be the conditioned generator* (46) *of the ASEP with N particles and periodic boundary conditions. Then the matrix*

$$D_K^{K\pm n}(q, \alpha) = \mathbf{1}_{K\mp n} (S^{\pm}(q, \alpha))^n \mathbf{1}_K \tag{95}$$

yields the duality relation

$$D_K^{K\mp n}(q, \alpha) H_K(q, \alpha, q^{2n} \beta^{\pm}) = \left(H_{K\mp n}(q, \alpha^{-1}, \beta^{\mp})\right)^T D_K^{K\mp n}(q, \alpha). \tag{96}$$

[4]Notice a sign error in front of the term $2k_i$ in Eq. (3.12) of [33] and pay attention to the different convention $q \leftrightarrow q^{-1}$.

As an application of this duality we focus now on global conditioning ($\alpha \neq q,\ \beta = 1$) and local conditioning $\alpha = q,\ \beta \neq 1$.

Theorem 3 *Let* $H_N^K := \mathbf{1}_N H(q, q^{1-\frac{2K}{L}}, 1)\mathbf{1}_N$ *be the generator* (46) *of the globally conditioned ASEP with N particles and periodic boundary conditions and driving strength* $s = -2K/L \ln q$. *Furthermore, let* $|\, \mu_{\mathbf{x}}^N \,\rangle = \mathbf{1}_N |\, \mu_{\mathbf{x}} \,\rangle$ *be the unnormalized shock-antishock measure* (85) *restricted to N particles and*

$$|\, \mu_{\mathbf{x}}^N(t) \,\rangle := \mathrm{e}^{-H_N^K t} |\, \mu_{\mathbf{x}}^N \,\rangle \tag{97}$$

with $K = |\mathbf{x}|$. *Then*

$$|\, \mu_{\mathbf{x}}^N(t) \,\rangle = \sum_{\mathbf{y} \in \Omega_K} P^N(\mathbf{y}, t|\mathbf{x}, 0) |\, \mu_{\mathbf{y}}^N \,\rangle \tag{98}$$

where $P^N(\mathbf{y}, t|\mathbf{x}, 0) := \langle\, \mathbf{y} \,| \mathrm{e}^{-H_K^N t} |\, \mathbf{x} \,\rangle$ *is the conditioned K-particle transition probability from* $\mathbf{x}$ *to* $\mathbf{y}$ *at time t with driving strength* $s' = -2N/L \ln q$.

Remark 2 The significance of this result lies in the fact that the conditioned evolution of an N-particle SAM is fully determined by the conditioned transition probability of only K particles, in analogy to the evolution of shocks in the infinite lattice explored in [3, 4].

Remark 3 For $K = 1$ a related result was obtained in [5] for a normalized and slightly different definition of the shock measures. The proof of [5] is by explicit computations relying on the presence of a single shock. The present proof for the generalized case $K \geq 1$ shows that the mathematical origin of the conditioned shock motion is the duality relation (114).

Theorem 4 *Let* $\bar{H}_N^K := \mathbf{1}_N H(q, q, q^{-2K})\mathbf{1}_N$ *be the generator* (46) *of the locally conditioned ASEP with N particles and periodic boundary conditions and boundary driving strength* $\bar{s} = -2K \ln q$. *Furthermore, let* $|\, \bar{\mu}_{\mathbf{x}}^N \,\rangle = \mathbf{1}_N |\, \bar{\mu}_{\mathbf{x}} \,\rangle$ *be the unnormalized shock-antishock measure* (82) *restricted to N particles and*

$$|\, \bar{\mu}_{\mathbf{x}}^N(t) \,\rangle := \mathrm{e}^{-\bar{H}_N^K t} |\, \bar{\mu}_{\mathbf{x}}^N \,\rangle \tag{99}$$

with $K = |\mathbf{x}|$. *Then*

$$|\, \bar{\mu}_{\mathbf{x}}^N(t) \,\rangle = \sum_{\mathbf{y} \in \Omega_K} \bar{P}^N(\mathbf{y}, t|\mathbf{x}, 0) |\, \bar{\mu}_{\mathbf{y}}^N \,\rangle \tag{100}$$

where $\bar{P}^N(\mathbf{y}, t|\mathbf{x}, 0) := \langle\, \mathbf{y} \,| \mathrm{e}^{-\bar{H}_K^N t} |\, \mathbf{x} \,\rangle$ *is the boundary-conditioned K-particle transition probability from* $\mathbf{x}$ *to* $\mathbf{y}$ *at time t with driving strength* $\bar{s}' = -2N \ln q$.

Remark 4 In both theorems the conditioning parameters s, s' and $\bar{s}$, $\bar{s}'$ respectively are negative, corresponding to conditioning on a time-integrated current that is lower than typical.

4 Proofs

4.1 Proof of Theorem 1

Proof We first note

Lemma 1 *Let*

$$\tilde{S} = \sum_{n=0}^{L} \frac{\tilde{S}^{+}}{[n]_q!}, \quad \hat{Q}_x = q^{\sum_{i=1}^{x-1} \hat{n}_i - \sum_{i=x+1}^{L} \hat{n}_i} \hat{n}_x. \tag{101}$$

Then for a configuration $\mathbf{x} \in \Omega_N$ *with* $N = |\mathbf{x}|$ *particles one has*

$$\langle\, \mathbf{x} \,| \tilde{S} = \langle\, s \,| \prod_{i=1}^{|\mathbf{x}|} \hat{Q}_{x_i}. \tag{102}$$

The proof is completely analogous to the proof in [7] of (164) with $\mathbf{y} = \emptyset$. Now we observe that with the reversible measure (18) and with (74) we can write

$$D(\mathbf{x}, \eta) = \pi^{-1}(\mathbf{x}) \langle\, s \,| \prod_{i=1}^{|\mathbf{x}|} \hat{Q}_{x_i} |\, \eta \,\rangle = f^{\mathbf{x}}(\eta) = g^{\eta}(\mathbf{x}). \tag{103}$$

Then the following chain of equalities holds and proves the theorem:

$$\begin{aligned}
\langle\, f^{\mathbf{x}}(t) \,\rangle^{\eta} &:= \sum_{\xi} f^{\mathbf{x}}(\xi) \langle\, \xi \,| \mathrm{e}^{-\tilde{H}t} |\, \eta \,\rangle && (104)\\
&= \sum_{\xi} \pi^{-1}(\mathbf{x}) \langle\, s \,| \prod_{i=1}^{|\mathbf{x}|} \hat{Q}_{x_i} |\, \xi \,\rangle \langle\, \xi \,| \mathrm{e}^{-\tilde{H}t} |\, \eta \,\rangle && (105)\\
&= \pi^{-1}(\mathbf{x}) \langle\, s \,| \prod_{i=1}^{|\mathbf{x}|} \hat{Q}_{x_i} \mathrm{e}^{-\tilde{H}t} |\, \eta \,\rangle && (106)\\
&= \pi^{-1}(\mathbf{x}) \langle\, \mathbf{x} \,| \tilde{S} \mathrm{e}^{-\tilde{H}t} |\, \eta \,\rangle && (107)\\
&= \pi^{-1}(\mathbf{x}) \langle\, \mathbf{x} \,| \mathrm{e}^{-\tilde{H}t} \tilde{S} |\, \eta \,\rangle && (108)\\
&= \pi^{-1}(\mathbf{x}) \sum_{\mathbf{y}} \langle\, \mathbf{x} \,| \mathrm{e}^{-\tilde{H}t} |\, \mathbf{y} \,\rangle \langle\, \mathbf{y} \,| \tilde{S} |\, \eta \,\rangle && (109)\\
&= \sum_{\mathbf{y} \in \Omega_N} \langle\, \mathbf{x} \,| \hat{\pi}^{-1} \mathrm{e}^{-\tilde{H}t} \hat{\pi} |\, \mathbf{y} \,\rangle \pi^{-1}(\mathbf{y}) \langle\, s \,| \prod_{i=1}^{|\mathbf{y}|} \hat{Q}_{y_i} |\, \eta \,\rangle && (110)\\
&= \sum_{\mathbf{y} \in \Omega_N} \langle\, \mathbf{y} \,| \mathrm{e}^{-\tilde{H}t} |\, \mathbf{x} \,\rangle \pi^{-1}(\mathbf{y}) \langle\, s \,| \prod_{i=1}^{|\mathbf{y}|} \hat{Q}_{y_i} |\, \eta \,\rangle && (111)
\end{aligned}$$

$$= \sum_{\mathbf{y}\in\Omega_N} g^{\eta}(\mathbf{y})\langle\, \mathbf{y}\,|\mathrm{e}^{-\tilde{H}t}|\,\mathbf{x}\,\rangle \tag{112}$$

$$=: \langle\, g^{\eta}(t)\,\rangle^{\mathbf{x}} \tag{113}$$

The following ingredients were used: Eqs. (104) and (113): The expressions (77) for expectations; Eqs. (105) and (112): The expression (103) for the duality function; Eq. (106): The expressions (77) and the representation (47) of the unit matrix of dimension 2^L; Eq. (107): The expression (102) of part of the duality function in terms of the symmetry operator $\tilde{S}$; Eq. (108): The $U_q[\mathfrak{sl}(2)]$symmetry (61); Eq. (109): Particle number conservation and the representation (45) of the unit matrix in the subspace of N particles; Eq. (110): The diagonal matrix representation of the reversible measure (18); Eq. (111): Reversibility (21). □

4.2 Proof of Theorem 2

We first present a generalized and slightly reformulated version of the intertwiner relation Eq. (2.62b) of [30] for periodic boundary conditions.

Proposition 1 *Let $H(\cdot,\cdot,\cdot)$ be the conditioned generator* (29) *of the ASEP with periodic boundary conditions and let $\eta_K \in \Omega_K$ be any configuration with K particles. Then for $0 \leq n \leq L-K$ one has the intertwiner relation*

$$\left[(S^{\pm}(q,\alpha))^n H(q,\alpha,q^{2n}\beta^{\pm}) - H(q,\alpha,\beta^{\pm})(S^{\pm}(q,\alpha))^n\right]|\,\eta_K\,\rangle = 0 \tag{114}$$

with

$$\beta^{\pm} = q^{\pm(L-2K)}\alpha^{-L} \tag{115}$$

and the generators $S^{\pm}(q,\alpha)$ (62) of $U_q[\mathfrak{gl}(2)]$.

Remark 5 For $\alpha = 1$ this is the result (2.62b) of [30].[5] The proof of Proposition (1) is entirely analogous to the derivation of (2.62b) given in [30] since the generalized form (114) follows trivially from the result of [30] for $\alpha = 1$ through the similarity transformation (33). Some ingredients of the proof, with sign errors in [30] corrected, are presented in the appendix.

Defining the duality matrix $D_K^{K\pm n} = \mathbf{1}_{K\mp n}(S^{\pm}(q,\alpha))^n\mathbf{1}_K$ with the projector (45) and using (38) the intertwiner relation (114) then yields the duality relation (96) by projection on the sectors with fixed particle number.

[5] Eqs. (2.62a) and (2.62b) of [30] have some sign errors which are corrected in Proposition (1).

4.3 Proofs of Theorems 3 and 4

Before we set out to prove Theorems 3 and 4 we show that the SAM (85) can be generated by the action of the particle creation operator $U_q[\mathfrak{sl}(2)]$.

Lemma 2 *Let* **x** *be a configuration of* $K = |\mathbf{x}|$ *particles and let*

$$\bar{\mu}_{\mathbf{x}}^{N} := \bar{\mu}_{\mathbf{x}} \delta_{\sum_{k=1}^{L} \eta(k),N}, \quad \mu_{\mathbf{x}}^{N} := \mu_{\mathbf{x}} \delta_{\sum_{k=1}^{L} \eta(k),N} \tag{116}$$

be the SAM's defined by (82), (85) *restricted to* $N \geq K$ *particles. Then the vector representations* $|\,\bar{\mu}_{\mathbf{x}}^{N}\,\rangle := \mathbf{1}_N |\,\bar{\mu}_{\mathbf{x}}\,\rangle$ *and* $|\,\mu_{\mathbf{x}}^{N}\,\rangle := \mathbf{1}_N |\,\mu_{\mathbf{x}}\,\rangle$ *can be written as*

$$|\,\bar{\mu}_{\mathbf{x}}^{N}\,\rangle = z^{N-K} \frac{\left(S^{-}(q^{-1}, q)\right)^{N-K}}{[N-K]_q!} |\,\mathbf{x}\,\rangle \tag{117}$$

$$|\,\mu_{\mathbf{x}}^{N}\,\rangle = z^{N-K} V(q^{\frac{2(K-N)}{L}}) \frac{\left(S^{-}(q^{-1}, q^{1-\frac{N}{L}})\right)^{N-K}}{[N-K]_q!} |\,\mathbf{x}\,\rangle \tag{118}$$

in terms of the generators (62) *of* $U_q[\mathfrak{sl}(2)]$ *and the transformation* (33).

Proof From Lemma 1 and (66) one finds

$$\sum_{n=0}^{L-K} \frac{\left(S^{-}(q^{-1}, q)\right)^{n}}{[n]_q!} |\,\mathbf{x}\,\rangle = \prod_{j=1}^{K} q^{-\sum_{i=1}^{x_j-1} \hat{n}_i + \sum_{i=x_j+1}^{L} \hat{n}_i} \hat{n}_{x_j} |\,s\,\rangle. \tag{119}$$

Notice that for $z = 1$ one has $|\,z = 1\,\rangle = |\,s\,\rangle$.

The transformation (33) yields $V(\gamma)|\,\mathbf{x}\,\rangle = \gamma^{-\frac{1}{2}\sum_{j=1}^{K}(2x_j-L-1)} |\,\mathbf{x}\,\rangle$ and (64) gives $S^{-}(q^{-1}, q) = V^{-1}(\lambda) S^{-}(q^{-1}, q\lambda^{-1}) V(\lambda)$. Putting this together and using (65) turns (119) into

$$\begin{aligned} &V(\gamma) \sum_{n=0}^{L-K} \frac{z^n \left(S^{-}(q^{-1}, q\lambda^{-1})\right)^{n}}{[n]_q!} |\,\mathbf{x}\,\rangle \\ &= V(\gamma\lambda) \prod_{j=1}^{K} z^{-1} \lambda^{\frac{1}{2}(2x_j-L-1)} q^{-\sum_{i=1}^{x_j-1} \hat{n}_i + \sum_{i=x_j+1}^{L} \hat{n}_i} \hat{n}_{x_j} |\,z\,\rangle. \end{aligned} \tag{120}$$

Now we choose $\lambda = q^{\frac{2N}{L}}$ and $\gamma = q^{\frac{2(K-N)}{L}}$ to obtain

$$\begin{aligned} &V(q^{\frac{2(K-N)}{L}}) \sum_{n=0}^{L-K} \frac{z^n \left(S^{-}(q^{-1}, q^{1-\frac{2N}{L}})\right)^{n}}{[n]_q!} |\,\mathbf{x}\,\rangle \\ &= \prod_{j=1}^{K} z^{-1} q^{\frac{N}{L}(2x_j-L-1) - \frac{1}{L}\sum_{l=1}^{L}(2l-L-1)\hat{n}_l} q^{-\sum_{i=1}^{x_j-1} \hat{n}_i + \sum_{i=x_j+1}^{L} \hat{n}_i} \hat{n}_{x_j} |\,z\,\rangle. \end{aligned} \tag{121}$$

Finally one applies the projector $\mathbf{1}_N$ on both sides of the equation. On the l.h.s. this projects out the term with $n = N - K$, corresponding to the r.h.s. of (118). On the r.h.s. the projection allows us to substitute the number N in the first power of q by the number operator $\hat{N}$ (34). Thus the terms proportional to $L + 1$ cancel and the expression (86) remains under the projection operator. Therefore the r.h.s. is equal to $|\,\mu_{\mathbf{x}}^N\,\rangle$. Similarly one chooses $\gamma = \lambda = 1$ to obtain (117). □

4.3.1 Proof of Theorem 3

Now we are in a position to prove (98).

Consider a K-particle configuration $\mathbf{x}$ and the duality relation (114) with $n = N - K$ and $\beta = 1$:

$$\begin{aligned}
&\frac{(S^-(q, q^{2\frac{K}{L}-1}))^{N-K}}{[N-K]_q!} H(q, q^{2\frac{K}{L}-1}, q^{2N-2K})|\,\mathbf{x}\,\rangle \\
&\quad = H(q, q^{2\frac{K}{L}-1}, 1)\frac{(S^-(q, q^{2\frac{K}{L}-1}))^{N-K}}{[N-K]_q!}|\,\mathbf{x}\,\rangle. \qquad (122)
\end{aligned}$$

With the transformation (33) with $\gamma^L = q^{2K-2N}$ one uses (41) to cast this in the form

$$\begin{aligned}
&H(q, q^{2\frac{K}{L}-1}, 1)\frac{(S^-(q, q^{2\frac{K}{L}-1}))^{N-K}}{[N-K]_q!}|\,\mathbf{x}\,\rangle \\
&= \frac{(S^-(q, q^{2\frac{K}{L}-1}))^{N-K}}{[N-K]_q!} V^{-1}(\gamma) H(q, q^{2\frac{N}{L}-1}, 1) V(\gamma)|\,\mathbf{x}\,\rangle \qquad (123)
\end{aligned}$$

or, alternatively,

$$\begin{aligned}
&H(q, q^{2\frac{K}{L}-1}, 1) V^{-1}(\gamma)\frac{(S^-(q, q^{2\frac{N}{L}-1}))^{N-K}}{[N-K]_q!}|\,\mathbf{x}\,\rangle \\
&\quad = V^{-1}(\gamma)\frac{(S^-(q, q^{2\frac{N}{L}-1}))^{N-K}}{[N-K]_q!} H(q, q^{2\frac{N}{L}-1}, 1)|\,\mathbf{x}\,\rangle. \qquad (124)
\end{aligned}$$

Applying (37), (67), (35) this turns into

$$\begin{aligned}
&H(q, q^{1-2\frac{K}{L}}, 1) V(\gamma)\frac{(S^-(q^{-1}, q^{1-2\frac{N}{L}}))^{N-K}}{[N-K]_q!}|\,\tilde{\mathbf{x}}\,\rangle \\
&\quad = V(\gamma)\frac{(S^-(q^{-1}, q^{1-2\frac{N}{L}}))^{N-K}}{[N-K]_q!} H(q, q^{1-2\frac{N}{L}}, 1)|\,\tilde{\mathbf{x}}\,\rangle. \qquad (125)
\end{aligned}$$

where $|\,\tilde{\mathbf{x}}\,\rangle = V^{-1}(\gamma)|\,\mathbf{x}\,\rangle$ is an arbitrary K-particle configuration.

Since H conserves particle number, this relation remains valid for any power of H. Thus we find

$$\begin{aligned}
&\mathrm{e}^{-H_N^K t} V(\gamma) \frac{(S^-(q^{-1}, q^{1-2\frac{N}{L}}))^{N-K}}{[N-K]_q!} |\, \tilde{\mathbf{x}}\,\rangle \\
&= V(\gamma) \frac{(S^-(q^{-1}, q^{1-2\frac{N}{L}}))^{N-K}}{[N-K]_q!} \mathrm{e}^{-H_K^N t} |\, \tilde{\mathbf{x}}\,\rangle \qquad (126) \\
&= \sum_{\eta'_K} V(\gamma) \frac{(S^-(q^{-1}, q^{1-2\frac{N}{L}}))^{N-K}}{[N-K]_q!} |\, \eta'_K\,\rangle\langle\, \eta'_K\,| \mathrm{e}^{-H_K^N t} |\, \tilde{\mathbf{x}}\,\rangle \qquad (127)
\end{aligned}$$

where in the last equality we have inserted the unit operator restricted to K-particle states. Using (118) of Lemma 2 then proves (98). □

4.3.2 Proof of Theorem 4

The proof of Theorem 4 is similar. For $\alpha = q^{-1}$ where $\beta = q^{2K}$ one has

$$\frac{(S^-(q, q^{-1}))^{N-K}}{[N-K]_q!} H(q, q^{-1}, q^{2N}) |\, \mathbf{x}\,\rangle = H(q, q^{-1}, q^{2K}) \frac{(S^-(q, q^{-1}))^{N-K}}{[N-K]_q!} |\, \mathbf{x}\,\rangle. \qquad (128)$$

Applying space reflection (37), (67) this becomes

$$\frac{(S^-(q^{-1}, q))^{N-K}}{[N-K]_q!} H(q, q, q^{-2N}) |\, \mathbf{x}\,\rangle = H(q, q, q^{-2K}) \frac{(S^-(q^{-1}, q))^{N-K}}{[N-K]_q!} |\, \mathbf{x}\,\rangle. \qquad (129)$$

Here we dropped the tilde over the configuration $\mathbf{x}$ since it is arbitrary.

Projecting on N particles and iterating this duality over powers of $\bar{H}_N^K$ yields

$$\mathrm{e}^{-\bar{H}_N^K t} \frac{(S^-(q^{-1}, q))^{N-K}}{[N-K]_q!} |\, \mathbf{x}\,\rangle = \frac{(S^-(q^{-1}, q))^{N-K}}{[N-K]_q!} \mathrm{e}^{-\bar{H}_K^N t} |\, \mathbf{x}\,\rangle. \qquad (130)$$

Inserting the unit operator restricted to K-particle states and Using (117) of Lemma 2 then proves (100). □

Acknowledgments GMS thanks DFG and FAPESP for financial support.

Appendix

We present some details of the proof of Proposition (1) which are not shown in [30] and from which Proposition (1) follows by the similarity transformation (33).

We define $e_L(\cdot,\cdot,\cdot) := h_{L,1}(\cdot,\cdot,\cdot)$, see (31). By explicit matrix multiplications one finds from the relations (25) for the bulk operators

$$S_k^{\pm}(q,\alpha)e_L(\alpha',q',\beta) = e_L(\alpha',q',\beta q^{-2})S_k^{\pm}(q,\alpha) \quad 2 \le k \le L-1 \tag{131}$$
$$e_L(\alpha',q',\beta)S_k^{\pm}(q,\alpha) = S_k^{\pm}(q,\alpha)e_L(\alpha',q',\beta q^{2}) \quad 2 \le k \le L-1 \tag{132}$$

and for the boundary operators

$$S_1^{+}(q,\alpha)e_L(\alpha',q',\beta) = q^{-1/2}\alpha^{\frac{1}{2}(L-1)}\left[(q')^{-1}\sigma_1^{+}\hat{v}_L - \alpha'\beta\hat{v}_1\sigma_L^{+}\right]q^{-S^z} \tag{133}$$
$$S_1^{-}(q,\alpha)e_L(\alpha',q',\beta) = q^{1/2}\alpha^{-\frac{1}{2}(L-1)}\left[q'\sigma_1^{-}\hat{n}_L - (\alpha'\beta)^{-1}\hat{n}_1\sigma_L^{-}\right]q^{-S^z} \tag{134}$$
$$S_L^{+}(q,\alpha)e_L(\alpha',q',\beta) = q^{1/2}\alpha^{-\frac{1}{2}(L-1)}\left[q'\hat{v}_1\sigma_L^{+} - (\alpha'\beta)^{-1}\sigma_1^{+}\hat{v}_L\right]q^{S^z} \tag{135}$$
$$S_L^{-}(q,\alpha)e_L(\alpha',q',\beta) = q^{-1/2}\alpha^{\frac{1}{2}(L-1)}\left[(q')^{-1}\hat{n}_1\sigma_L^{-} - \alpha'\beta\sigma_1^{-}\hat{n}_L\right]q^{S^z} \tag{136}$$

and

$$e_L(\alpha',q',\beta)S_1^{+}(q,\alpha) = q^{-1/2}\alpha^{\frac{1}{2}(L-1)}\left[q'\sigma_1^{+}\hat{n}_L - \alpha'\beta\hat{n}_1\sigma_L^{+}\right]q^{-S^z} \tag{137}$$
$$e_L(\alpha',q',\beta)S_1^{-}(q,\alpha) = q^{1/2}\alpha^{-\frac{1}{2}(L-1)}\left[(q')^{-1}\sigma_1^{-}\hat{v}_L - (\alpha'\beta)^{-1}\hat{v}_1\sigma_L^{-}\right]q^{-S^z} \tag{138}$$
$$e_L(\alpha',q',\beta)S_L^{+}(q,\alpha) = q^{1/2}\alpha^{-\frac{1}{2}(L-1)}\left[(q')^{-1}\hat{n}_1\sigma_L^{+} - (\alpha'\beta)^{-1}\sigma_1^{+}\hat{n}_L\right]q^{S^z} \tag{139}$$
$$e_L(\alpha',q',\beta)S_L^{-}(q,\alpha) = q^{-1/2}\alpha^{\frac{1}{2}(L-1)}\left[q'\hat{v}_1\sigma_L^{-} - \alpha'\beta\sigma_1^{-}\hat{v}_L\right]q^{S^z}. \tag{140}$$

Consider now $q = q'$ and $\alpha = \alpha'$. From the quantum algebra symmetry and from the previous relations one obtains (omitting the q,α-dependence)

$$S^{\pm}H(\beta) - H(\beta')S^{\pm} = S^{\pm}e_L(\beta) - e_L(\beta')S^{\pm} \tag{141}$$
$$= \left[e_L(q^{-2}\beta) - e_L(\beta')\right]\sum_{k=2}^{L-1} S_k^{\pm}$$
$$+ \left(S_1^{\pm} + S_L^{\pm}\right)e_L(\beta) - e_L(\beta')\left(S_1^{\pm} + S_L^{\pm}\right). \tag{142}$$

Observe that

$$S_1^{+} = q^{-1/2}\alpha^{\frac{1}{2}(L-1)}\sigma_1^{+}q^{-S^z}, \quad S_1^{-} = q^{1/2}\alpha^{-\frac{1}{2}(L-1)}\sigma_1^{-}q^{-S^z}, \tag{143}$$
$$S_L^{+} = q^{1/2}\alpha^{-\frac{1}{2}(L-1)}\sigma_L^{+}q^{S^z}, \quad S_L^{-} = q^{-1/2}\alpha^{\frac{1}{2}(L-1)}\sigma_L^{-}q^{S^z} \tag{144}$$

and the auxiliary relations

$$\sigma_1^{+}e_L(\beta) = q^{-1}\sigma_1^{+}\hat{v}_L - \alpha\beta\hat{v}_1\sigma_L^{+}, \tag{145}$$
$$e_L(\beta)\sigma_1^{+} = q\sigma_1^{+}\hat{n}_L - \alpha\beta\hat{n}_1\sigma_L^{+} \tag{146}$$
$$\sigma_L^{+}e_L(\beta) = q\hat{v}_1\sigma_L^{+} - (\alpha\beta)^{-1}\sigma_1^{+}\hat{v}_L, \tag{147}$$
$$e_L(\beta)\sigma_L^{+} = q^{-1}\hat{n}_1\sigma_L^{+} - (\alpha\beta)^{-1}\sigma_1^{+}\hat{n}_L, \tag{148}$$

and

$$\sigma_1^- e_L(\beta) = q\sigma_1^- \hat{n}_L - (\alpha\beta)^{-1}\hat{n}_1\sigma_L^-, \tag{149}$$
$$e_L(\beta)\sigma_1^- = q^{-1}\sigma_1^- \hat{v}_L - (\alpha\beta)^{-1}\hat{v}_1\sigma_L^- \tag{150}$$
$$\sigma_L^- e_L(\beta) = q^{-1}\hat{n}_1\sigma_L^- - \alpha\beta\sigma_1^- \hat{n}_L, \tag{151}$$
$$e_L(\beta)\sigma_L^- = q\hat{v}_1\sigma_L^- - \alpha\beta\sigma_1^- \hat{v}_L. \tag{152}$$

Thus one obtains

$$\begin{aligned}&\left(S_1^+ + S_L^+\right) e_L(\beta) - e_L(\beta') \left(S_1^+ + S_L^+\right) \\ &= A^+(\beta,\beta')\beta^{1/2}\alpha^{L/2}q^{-S^z-1} + B^+(\beta,\beta')\beta^{-1/2}\alpha^{-L/2}q^{S^z+1}\end{aligned} \tag{153}$$

with

$$A^+(\beta,\beta') = \frac{q^{1/2}}{(\alpha\beta)^{1/2}}\sigma_1^+\left[q^{-1}\hat{v}_L - q\hat{n}_L\right] - \frac{(\alpha\beta)^{1/2}}{q^{1/2}}\sigma_L^+\left[q\hat{v}_1 - q\frac{\beta'}{\beta}\hat{n}_1\right] \tag{154}$$
$$B^+(\beta,\beta') = \frac{q^{1/2}}{(\alpha\beta)^{1/2}}\sigma_1^+\left[q^{-1}\frac{\beta}{\beta'}\hat{n}_L - q^{-1}\hat{v}_L\right] - \frac{(\alpha\beta)^{1/2}}{q^{1/2}}\sigma_L^+\left[q^{-1}\hat{n}_1 - q\hat{v}_1\right] \tag{155}$$

and

$$\begin{aligned}&\left(S_1^- + S_L^-\right) e_L(\beta) - e_L(\beta') \left(S_1^- + S_L^-\right) \\ &= A^-(\beta,\beta')\beta^{-1/2}\alpha^{-L/2}q^{-S^z+1} + B^-(\beta,\beta')\beta^{1/2}\alpha^{L/2}q^{S^z-1}\end{aligned} \tag{156}$$

with

$$A^-(\beta,\beta') = \frac{(\alpha\beta)^{1/2}}{q^{1/2}}\sigma_1^-\left[q\hat{n}_L - q^{-1}\hat{v}_L\right] - \frac{q^{1/2}}{(\alpha\beta)^{1/2}}\sigma_L^-\left[q^{-1}\hat{n}_1 - q^{-1}\frac{\beta}{\beta'}\hat{v}_1\right] \tag{157}$$
$$B^-(\beta,\beta') = \frac{(\alpha\beta)^{1/2}}{q^{1/2}}\sigma_1^-\left[q\frac{\beta'}{\beta}\hat{v}_L - q\hat{n}_L\right] - \frac{q^{1/2}}{(\alpha\beta)^{1/2}}\sigma_L^-\left[q\hat{v}_1 - q^{-1}\hat{n}_1\right]. \tag{158}$$

With the choice $\beta' = q^{-2}\beta$ (142) reduces to

$$S^\pm H(\beta) - H(q^{-2}\beta)S^\pm = \left(S_1^\pm + S_L^\pm\right) e_L(\beta) - e_L(q^{-2}\beta)\left(S_1^\pm + S_L^\pm\right). \tag{159}$$

For S^+ the r.h.s. reduces to

$$\begin{aligned}&\left\{\frac{q^{1/2}}{(\alpha\beta)^{1/2}}\sigma_1^+\left[q^{-1}\hat{v}_L - q\hat{n}_L\right] - \frac{(\alpha\beta)^{1/2}}{q^{1/2}}\sigma_L^+\left[q\hat{v}_1 - q^{-1}\hat{n}_1\right]\right\} \\ &\times\left[\beta^{1/2}\alpha^{L/2}q^{-S^z-1} - \beta^{-1/2}\alpha^{-L/2}q^{S^z+1}\right]\end{aligned}$$

With (143), (144) one thus arrives at

$$\begin{aligned} S^+H(\beta) - H(q^{-2}\beta)S^+ &= \left[q^{-1}\hat{v}_L - q\hat{n}_L\right] S_1^+ \left[1 - \beta^{-1}\alpha^{-L}q^{2S^z+2}\right] \\ &+ \left[q\hat{v}_1 - q^{-1}\hat{n}_1\right] S_L^+ \left[1 - \beta\alpha^L q^{-2S^z-2}\right]. \end{aligned} \tag{160}$$

Notice that the action of the pseudo commutator on states with particle number satisfying

$$q^{L-2N+2} = \beta\alpha^L \tag{161}$$

vanishes.

Similarly one obtains for S^- the r.h.s. of (159)

$$\begin{aligned} &\left\{ \frac{(\alpha\beta)^{1/2}}{q^{1/2}} \sigma_1^- \left[q^{-1}\hat{v}_L - q\hat{n}_L\right] - \frac{q^{1/2}}{(\alpha\beta)^{1/2}} \sigma_L^- \left[q\hat{v}_1 - q^{-1}\hat{n}_1\right] \right\} \\ &\times \left[\beta^{1/2}\alpha^{L/2}q^{S^z-1} - \beta^{-1/2}\alpha^{-L/2}q^{-S^z+1}\right] \end{aligned}$$

which yields

$$\begin{aligned} S^-H(\beta) - H(q^{-2}\beta)S^- &= -\left[q^{-1}\hat{v}_L - q\hat{n}_L\right] S_1^- \left[1 - \beta\alpha^L q^{2S^z-2}\right] \\ &- \left[q\hat{v}_1 - q^{-1}\hat{n}_1\right] S_L^- \left[1 - \beta^{-1}\alpha^{-L}q^{-2S^z+2}\right]. \end{aligned} \tag{162}$$

Notice that the action of the pseudo commutator on states with particle number satisfying

$$q^{-L+2N+2} = \beta\alpha^L \tag{163}$$

vanishes.

In compact form (159) can thus be written

$$\begin{aligned} S^\pm H(\beta) - H(q^{-2}\beta)S^\pm &= \pm\left[q^{-1}\hat{v}_L - q\hat{n}_L\right] S_1^\pm \left[1 - \beta^{\mp 1}\alpha^{\mp L}q^{2S^z\pm 2}\right] \\ &\pm \left[q\hat{v}_1 - q^{-1}\hat{n}_1\right] S_L^\pm \left[1 - \beta^{\pm 1}\alpha^{\pm L}q^{-2S^z\mp 2}\right]. \end{aligned} \tag{164}$$

One can iterate. E.g. for $(S^-)^2$ one obtains

$$\begin{aligned} &(S^-)^2H(\beta) - H(q^{-4}\beta)(S^-)^2 \\ &= (1+q^{-2})\left[q\hat{n}_L - q^{-1}\hat{v}_L\right] S_1^- \left(\sum_{k=2}^{L-1} S_k^-\right)\left[1 - \beta\alpha^L q^{2S^z-4}\right] \\ &+ (1+q^{-2})\left[q^{-1}\hat{n}_1 - q\hat{v}_1\right]\left(\sum_{k=2}^{L-1} S_k^-\right) S_L^- \left[1 - \beta^{-1}\alpha^{-L}q^{-2S^z+4}\right]. \end{aligned} \tag{165}$$

Iterating further as in [30] one arrives at Proposition 1.

References

1. Alcaraz, F.C., Rittenberg, V.: Reaction-diffusion processes as physical realizations of Hecke algebras. Phys. Lett. B **314**, 377–380 (1993)
2. Andrews, G.E.: The Theory of Partitions. Encyclopedia of Mathematics and its Applications, vol. 2. Addison Wesley, London (1976)
3. Balázs, M., Farkas, G., Kovács, P., Rákos, A.: Random walk of second class particles in product shock measures. J. Stat. Phys. **139**(2), 252–279 (2010)
4. Belitsky, V., Schütz, G.M.: Diffusion and scattering of shocks in the partially asymmetric simple exclusion process. Electron. J. Probab. **7**, paper 11, 1–21 (2002)
5. Belitsky, V., Schütz, G.M.: Microscopic structure of shocks and antishocks in the ASEP conditioned on low current. J. Stat. Phys. **152**, 93–111 (2013)
6. Belitsky, V., Schütz, G.M.: Quantum algebra symmetry of the ASEP with second-class particles. J. Stat. Phys. **161**, 821–842 (2015)
7. Belitsky, V., Schütz, G.M.: Self-duality for the two-component asymmetric simple exclusion process. J. Math. Phys. **56**, 083302 (2015)
8. Bertini, L., De Sole, A., Gabrielli, D., Jona Lasinio, G., Landim, C.: Macroscopic fluctuation theory. Rev. Mod. Phys. **87**, 593–636 (2015)
9. Bodineau, T., Derrida, B.: Distribution of current in non-equilibrium diffusive systems and phase transitions. Phys. Rev. E **72**, 066110 (2005)
10. Borodin, A., Corwin, I., Sasamoto, T.: From duality to determinants for Q-TASEP and ASEP. Ann. Probab. **42**, 2314–2382 (2014)
11. Burdík, Č., Havlíček, M., Vančura, A.: Irreducible highest weight representations of quantum groups $U(gl(n, \mathbb{C}))$. Commun. Math. Phys. **148**(2), 417–423 (1992)
12. Carinci, G., Giardinà, C., Giberti, C., Redig, F.: Duality for stochastic models of transport. J. Stat. Phys. **152**(4), 657–697 (2013)
13. Carinci, G., Giardinà, Redig, F., and Sasamoto, T.: A generalized Asymmetric Exclusion Process with $U_q(\mathfrak{sl}_2)$ stochastic duality. Probab. Theory Relat. Fields (2015). doi:10.1007/s00440-015-0674-0
14. Carinci, G., Giardinà, C., Giberti, C., Redig, F.: Dualities in population genetics: a fresh look with new dualities. Stoch. Process. Appl. **125**(3), 941–969 (2015)
15. Carinci, G., Giardinà, Redig, F., and Sasamoto, T.: Asymmetric stochastic transport models with $\mathcal{U}_q(\mathfrak{su}(1,1))$ symmetry (2015). arXiv:1507.01478
16. Chetrite, R., Touchette, H.: Nonequilibrium Markov processes conditioned on large deviations. Annales Henri Poincaré, **16**(9), 2005–2057 (2015)
17. Corwin, I., and Petrov, L.: Stochastic higher spin vertex models on the line (2015). arXiv:1502.07374
18. Derrida, B., Lebowitz, J.L.: Exact large deviation function in the asymmetric exclusion process. Phys. Rev. Lett. **80**(2), 209–213 (1998)
19. Evans, M.R., Ferrari, P.A., Mallick, K.: Matrix representation of the stationary measure for the multispecies TASEP. J. Stat. Phys. **135**(2), 217–239 (2009)
20. Ferrari, P.A., Martin, J.B.: Stationary distributions of multi-type totally asymmetric exclusion processes. Ann. Probab. **35**(3), 807–832 (2007)
21. Giardinà, C., Kurchan, J., Redig, F., Vafayi, K.: Duality and Hidden Symmetries in Interacting Particle Systems. J. Stat. Phys. **135**, 25–55 (2009)
22. Harris, R.J., and Schütz, G.M.: Fluctuation theorems for stochastic dynamics. J. Stat. Mech., P07020 (2007)
23. Jack, R.L., Sollich, P.: Large deviations and ensembles of trajectories in stochastic models. Prog. Theor. Phys. Supp. **184**, 304–317 (2010)
24. Jimbo, M.: A q-analogue of $U(\mathfrak{gl}(N \mid 1))$, Hecke algebra, and the Yang-Baxter equation. Lett. Math. Phys. **11**, 247–252 (1986)
25. Kuan, J.: Stochastic duality of ASEP with two particle types via symmetry of quantum groups of rank two, J. Phys. A: Math. Theor. **49**(11), 115002 (2016)

26. Liggett, T.M.: Interacting Particle Systems. Springer, Berlin (1985)
27. Liggett, T.M.: Stochastic Interacting Systems: Contact Voter and Exclusion Processes. Springer, Berlin (1999)
28. Lloyd P., Sudbury A., Donnelly P.: Quantum operators in classical probability theory: I. "Quantum spin" techniques and the exclusion model of diffusion. Stoch. Process. Appl. **61**(2), 205–221 (1996)
29. Ohkubo, J.: Duality in interacting particle systems and boson representation. J. Stat. Phys. **139**, 454–465 (2010)
30. Pasquier, V., Saleur, H.: Common structures between finite systems and conformal field theories through quantum groups. Nucl. Phys. B **330**, 523–556 (1990)
31. Sandow, S., Schütz, G.: On $U_q[SU(2)]$-symmetric driven diffusion. Europhys. Lett. **26**, 7–12 (1994)
32. Schütz, G., Sandow, S.: Non-abelian symmetries of stochastic processes: derivation of correlation functions for random vertex models and disordered interacting many-particle systems. Phys. Rev. E **49**, 2726–2744 (1994)
33. Schütz, G.M.: Duality relations for the asymmetric exclusion process. J. Stat. Phys. **86**(5/6), 1265–1287 (1997)
34. Schütz, G.M.: Exactly solvable models for many-body systems far from equilibrium. In: Domb, C., Lebowitz, J. (eds.) Phase Transitions and Critical Phenomena, vol. 19, pp. 1–251. Academic Press, London (2001)
35. Schütz, G.M.: Conditioned stochastic particle systems and integrable quantum spin systems. In: Gonçalves, P., Soares, A.J. (eds.) From Particle Systems to Partial Differential Equations II. Springer Proceedings in Mathematics & Statistics, vol. 129. Springer, Switzerland (2015)
36. Spitzer, F.: Interaction of Markov processes. Adv. Math. **5**, 246–290 (1970)
37. Sudbury, A., Lloyd, P.: Quantum operators in classical probability theory. II: the concept of duality in interacting particle systems. Ann. Probab. **23**(4), 1816–1830 (1995)

MIX
Papier aus verantwortungsvollen Quellen
Paper from responsible sources
FSC® C105338

FSC
www.fsc.org

If you have any concerns about our products,
you can contact us on
ProductSafety@springernature.com

In case Publisher is established outside the EU,
the EU authorized representative is:
Springer Nature Customer Service Center GmbH
Europaplatz 3, 69115 Heidelberg, Germany

Printed by Libri Plureos GmbH
in Hamburg, Germany